Innovations in Next-Generation Energy Storage Solutions

Mohsen Mhadhbi
National Institute of Research and Physicochemical Analysis, Tunisia

IGI Global
Scientific Publishing
Publishing Tomorrow's Research Today

Vice President of Editorial	Melissa Wagner
Director of Acquisitions	Mikaela Felty
Director of Book Development	Jocelynn Hessler
Production Manager	Mike Brehm
Cover Design	Phillip Shickler

Published in the United States of America by
IGI Global Scientific Publishing
701 East Chocolate Avenue
Hershey, PA, 17033, USA
Tel: 717-533-8845
Fax: 717-533-7115
Website: https://www.igi-global.com E-mail: cust@igi-global.com

Library of Congress Cataloging-in-Publication Data

LCCN: 2024055142 (CIP Data Pending)
ISBN13: 979-8-3693-9316-1
Isbn13Softcover: 979-8-3693-9317-8
EISBN13: 979-8-3693-9318-5
British Cataloguing in Publication Data
A Cataloguing in Publication record for this book is available from the British Library.

Table of Contents

Detailed Table of Contents

renewable energy sources. Like fossil fuel energy, hydrogen energy, helps to generate electricity but without any negative impact on environment, Hence, all over the world, researchers are seeing hydrogen energy as the main sustainable source of energy. Still, there are some challenges creating barriers to adoption of hydrogen energy in residential, commercial and industrial sectors. In this chapter, the challenges of hydrogen energy adoption are found in WASPAS (weighted aggregated sum and product assessment method) is used to find the most important challenges such that the stakeholders can frame new policies and develop new methodologies with advanced technology to mitigate these challenges and familiarize hydrogen use as sustainable energy source.

Advanced materials are key to battery and energy storage technology improvements, which are a cornerstone of sustainable energy for the future and are the topic of this chapter. It explores advances in solid-state electrolytes, lithium-sulfuric and sodium-ion batteries, nanomaterials and organic compounds, which all have the potential to enhance energy density, cycle life and environmental sustainability. These materials hold great promise, as they may overcome current limitations in battery performance, safety, and cost, the authors say. The chapter also explores the economic and environmental implications of these innovations, spotlighting their role in the global transition to renewable energy. Given ongoing research efforts and favourable policies, next-generation energy storage systems will play an essential role in advancing clean energy technologies in areas from electric vehicles to electric grid storage.

This chapter examines hydrogen storage technologies for next-generation energy systems, addressing challenges in creating efficient, secure, and affordable solutions. With hydrogen's energy density of 142 MJ/kg, it offers significant potential as a clean energy carrier. The research evaluates four storage methods: compressed hydrogen, liquid hydrogen, metal hydrides, and chemical storage, analyzing efficiency, cost, scalability, and safety. Recent breakthroughs include cryogenic systems achieving densities of 72.53 kg/m^3, metal hydride architectures enhancing capacity by 66.22%, and Metal-Organic Framework materials demonstrating storage up to 10.1 wt%. Integration with renewables shows intelligent systems can reduce cost variability by 42%. Economic analysis reveals storage costs of approximately €0.80/kg, with projections for reductions. The chapter concludes with recommendations to accelerate hydrogen's adoption as a sustainable energy solution.

AI plays a significant role in enhancing solar cell performance and has implications for sustainability and renewable energy integration. Additionally, it investigates how ML & AI may be applied to optimize energy production, boost solar cell efficiency, and facilitate intelligent grid integration. This chapter focuses on using AI-driven algorithms for solar cell material design and development. It shows how AI could accelerate the search for and development of materials with increased stability and effectiveness. It also explores the application of AI in developing solar cell technologies. The chapter discusses AI-powered control systems and prediction models for tracking and preserving solar cell performance. To improve the overall dependability of solar energy systems, it emphasizes the function of AI algorithms in identifying anomalies, predicting energy usage, and maximizing energy production. The chapter talks about the challenges of manufacturing and implementing solar cells. It draws attention to the ways AI could boost the sustainability of solar energy capacity.

This study explores the transformative impact of Artificial Intelligence (AI) in renewable energy, particularly in optimizing energy storage and harvesting solutions. The primary objective is to analyze AI-driven predictive maintenance techniques, with a focus on battery Remaining Useful Life (RUL) prediction. Targeted towards researchers, industry professionals, and policymakers, this work employs a literature

G. Boopathy, Vel Tech Rangarajan Dr. Sagunthala R&D Institute of Science and Technology, India
V. Srinivasan, Vel Tech Rangarajan Dr. Sagunthala R&D Institute of Science and Technology, India
Balaji Ganesan, Hindustan Institute of Technology and Science, India
Sivaprakasam Palani, Addis Ababa Science and Technology University, Ethiopia

Our pressing need to build better energy storage and conversion systems grows as our world develops faster. Energy storage systems perform better through nanomaterials which have unique physicochemical traits and redefine how these technologies work. This chapter evaluates how nanomaterials strengthen energy technology through product types and develops fabrication processes for particular solutions. Materials at nanoscale show better energy storage because of their huge surface area and flexible design. Nanomaterials provide multiple solutions for storing hydrogen as well as performing electro catalysis and photovoltaics. Our chapter shows that energy technology requires exact engineering along with modeling and environmental-friendly nanomaterial manufacturing to help turn scientific discoveries into real-world applications and next-level energy solutions.

Surjit Singha, Kristu Jayanti College, India
Ranjit Singha, Christ University, India
Shruti Jose, Christ University, India
V. Muthu Ruben, Christ University, India
R. Sreethi Rebeka, Mount Carmel College, India
Alphonsa Diana Haokip, Don Bosco College, Maram, India

In this chapter, we reviewed works related to renewable energy storage technologies with a focus on innovative solutions that can increase the efficiency, scalability and sustainability of energy systems. This includes advanced materials such as solid-state and lithium-air batteries, hybrid storage systems combining multiple technologies, and decentralised storage supporting grid resilience and energy security. By storing energy generated from renewable sources such as the sun and wind, these innovations seek to reduce carbon footprints and enable a transition to a more sustainable energy future by solving the issue of the intermittency of renewables. The chapter examines the emerging trends and technological developments, highlighting the critical role that next-generation energy storage systems can play in driving the global shift towards renewables and transforming energy infrastructure into a cleaner, more resilient world.

The transition to renewable energy systems demands advanced materials capable of enhancing the efficiency and performance of solar cells, photocatalysis, and hydrogen storage technologies. This chapter explores innovative materials that are shaping the future of sustainable energy. In solar cells, new generations of materials such as perovskites and organic photovoltaics are achieving higher efficiencies with lower production costs. These materials exhibit superior light absorption and charge carrier mobility, making them promising for next-generation solar energy harvesting. In photocatalysis, nanostructured semiconductors, including titanium dioxide (TiO_2) and graphene-based composites, are emerging as catalysts for environmental purification and hydrogen production through water splitting.By focusing on material innovations in these areas, this chapter highlights the pivotal role of cutting-edge research in advancing renewable energy technologies and paving the way for sustainable, high-performance energy systems

Metal oxide composites have newly become an attractive topic of particular interest for efficient and sustainable energy storage systems. These composites are essential for next-generation energy technologies as they provide considerable progress in electrical conductivity, reversible capacity, and cycling efficiency. In this chapter, the essential properties of metal oxide composites, such as their electrical conductivity and the enhancement by using conductive materials such as graphene or carbon nanotubes are discussed. Overcoming the current limitations necessitates innovative synthesis approaches, advanced nanostructuring and multi-disciplinary collaborative efforts. Ultimately, this chapter highlights the transformative potential of metal oxide composites in energy storage technologies, promoting sustained research efforts to fully exploit their capabilities in advancing a future powered by renewable energy.

This chapter delves into the critical role of renewable energy in addressing the sustainable global energy demands. It emphasizes the importance of renewable sources such as light, wind, and mechanical energy in reducing the carbon emissions and mitigating climate change. The chapter thoroughly explains the various renewable energy technologies and their transduction mechanisms, exploring for both renewable energy and hybrid energy harvesters. The hybrid system's ability to harness energy from multiple sources and convert it into electricity is highlighted as a promising solution to meet the power demands. Furthermore, the chapter discusses about the factors affecting energy conservation and the applications of these technologies in domestic and industrial settings. Through a detailed comparison, the efficacy of hybrid mechanisms in enhancing energy conversion efficiency and their potential benefits for future generations are highlighted. The conclusion includes a comparative analysis of hybrid mechanisms, potential applications, and the market availability of these energy devices.

The transition to renewable energy sources is critical for achieving sustainability and combating climate change. However, the intermittent nature of solar and wind energy presents significant challenges for reliable energy supply. "Harnessing Tomorrow: Pioneering Advances in Energy Storage for Renewable Futures" explores innovative energy storage solutions that are essential for integrating renewable energy into the grid effectively. This chapter highlights the latest advancements in energy storage technologies, including lithium-ion batteries, flow batteries, supercapacitors, and emerging technologies such as solid-state batteries and hydrogen storage systems. By examining case studies and pilot projects, the research emphasizes the role of energy storage in enhancing grid stability, enabling peak shaving, and providing backup power. Furthermore, the chapter discusses the economic implications of these technologies and their potential to reduce costs, improve efficiency, and support the widespread adoption of renewable energy.

Graphene-TiO2 composites are promising due to their synergistic properties, combining graphene's exceptional electronic and optical characteristics with TiO2's high photocatalytic efficiency. This review covers synthesis methods like chemical vapor deposition (CVD), sol-gel, and hydrothermal techniques, highlighting their impact on composite structure and properties. Key factors such as graphene loading, morphology control, and interface engineering are discussed for their role in enhancing photocatalytic activity and charge transfer efficiency. Environmental applications include water purification through pollutant degradation and air filtration for VOCs removal. In energy, these composites improve solar cell efficiency and lithium-ion battery electrode performance. Challenges like scalability, stability, and cost are addressed, with future research directions focusing on innovative synthesis and interface engineering for commercialization.

This chapter describes how energy storage technologies provide critical services for the integration of renewable energy sources, as those sources are variable and intermittent (e.g. solar and wind). It reviews different types of storage solutions — electrochemical, mechanical and thermal — including their relative advantages and drawbacks, as well as their appropriateness for various applications. It highlights the importance of adopting different energy storage types that are tailored to different energy needs. Global Sustainability and Energy Storage: Discussing the broader implications of energy storage systems for global sustainability, such as reducing reliance on fossil fuels, lowering carbon emissions, and improving access to energy in underserved regions. Energy storage technologies are crucial to a reliable, sustainable, low-carbon energy future.

The quest for higher, more sustained, and affordable solar cells has opened the floodgates to an incredible period of innovation in energy storage technologies. This chapter, "Revolutionizing Solar Cells: Novel Materials in Next-Generation Energy Storage Solutions," reviews frontier materials driving this evolution of solar energy systems. It focuses on the new material science developments that achieve higher efficiency and durability in solar cells by developing materials that have huge challenges of efficiency, reliability, and scalability. The chapter reviews in detail a good number of materials that have been used, such as perovskites, organic photovoltaics, and quantum dots, and their respective evolutions of advanced nanomaterials, and discusses their impact on energy storage. It provides insights from cutting-edge research and practical applications in the quest for sustainable and futuristic solar energy solutions.

While the study does a good job of highlighting the most important advantages of carbon-based supercapacitors, it would be beneficial to conduct a more in-depth analysis of their limits, such as the energy density of carbon-based supercapacitors in comparison to typical lithium-ion batteries. To have a more nuanced knowledge of the trade-offs that are involved in the process of incorporating these systems into electric vehicle architectures, a balanced perspective would be beneficial. A solid foundation for understanding the potential of this technology is provided by the carbon-based supercapacitors that are used in electric vehicles for regenerative braking systems. It opens how to prospective research and innovation in this area of energy storage by concentrating on their benefits while also admitting the difficulties that they face along the process.

Preface

The escalating global energy demand—driven by rapid population growth and accelerating economic development—has intensified the need for innovative, efficient, and sustainable energy storage and harvesting solutions. In response to these growing challenges, *Innovations in Next-Generation Energy Storage Solutions* presents a comprehensive exploration of the latest advancements in materials and technologies that are shaping the future of energy systems.

This book delves into significant new findings in the synthesis, fabrication, structure, and performance of advanced materials, with a particular focus on their technological applications in energy harvesting and storage. It also addresses the strategic and policy dimensions that support the integration of these technologies into practical use.

Covering both fundamental and applied research, the chapters span a wide array of topics including metal oxide materials, 2D materials, nanomaterials, materials for solar cells, electrocatalytic and photocatalytic systems, energy storage devices such as batteries and supercapacitors, membranes for energy conversion, hydrogen storage, advanced and solid-state electrolytes, and battery modeling and simulation.

Here's a chapter overview based on the abstracts you provided, highlighting the main focus and contributions of each chapter in a cohesive format suitable for a book preface or introduction:

Chapter 1

This chapter introduces graphene and other 2D materials, showcasing their groundbreaking electrical, thermal, and mechanical properties. Graphene's flexibility and conductivity position as a key material for next-gen electronics, sensors, and wearable tech. The chapter also touches on other materials like tungsten disulfide for applications in field-effect transistors and optoelectronics.

Chapter 2

Hydrogen's potential to revolutionize electricity generation and storage is explored, with emphasis on green hydrogen produced via water electrolysis. The chapter applies the WASPAS method to identify key adoption challenges in various sectors, providing actionable insights for policymakers and technologists.

Chapter 3

This chapter investigates novel materials like lithium-sulfur, sodium-ion, and solid-state electrolytes, which aim to improve energy density, safety, and sustainability in batteries. It emphasizes the vital role of material innovation in enabling the clean energy transition.

Chapter 4

Focusing on storage methods like compressed, liquid, and chemically bonded hydrogen, this chapter analyzes technical breakthroughs, such as cryogenic systems and metal-organic frameworks, while also assessing economic feasibility and integration with renewables.

Chapter 5

AI and machine learning are applied to optimize solar cell design, performance tracking, and intelligent grid integration. This chapter outlines how data-driven models enhance energy yield and system reliability, while also addressing challenges in solar cell manufacturing.

Chapter 6

This chapter discusses AI's role in predictive maintenance and energy efficiency, especially in battery management and smart grid systems. It also examines regulatory challenges and cybersecurity risks, providing recommendations for widespread AI adoption in energy sectors.

Chapter 7

Introducing MXenes as next-generation nanomaterials, this chapter reviews their electrochemical performance in energy devices. It details synthesis methods and structural advantages that make MXenes suitable for high-power energy storage systems.

Chapter 8

Nanomaterials' large surface area and customizable structures make them ideal for hydrogen storage, photocatalysis, and electrochemical applications. This chapter discusses fabrication techniques and environmental benefits for scalable, next-level energy technologies.

Chapter 9

Covering innovations like hybrid and lithium-air batteries, this chapter emphasizes the need for decentralized, scalable, and environmentally friendly energy storage solutions that support global renewable energy adoption and grid resilience.

Chapter 10

This chapter showcases breakthroughs in solar materials like perovskites and organic photovoltaics, and explores advanced photocatalytic and hydrogen storage materials, highlighting their role in pushing the boundaries of renewable energy technologies.

Chapter 11

Metal oxide composites enhance conductivity and cycle stability in energy devices. This chapter discusses their synthesis, integration with materials like graphene, and the collaborative efforts needed to overcome current technological barriers.

Chapter 12

Focusing on systems that combine multiple energy sources (solar, wind, mechanical), this chapter discusses energy transduction mechanisms and real-world applications. It provides a comparative analysis of energy efficiency and market readiness.

Chapter 13

From flow batteries to hydrogen storage, this chapter reviews advanced storage technologies essential for mitigating renewable energy intermittency. It emphasizes cost reduction, grid stability, and scalability through case studies and economic analysis.

Chapter 14

The synergy between graphene and TiO_2 improves photocatalytic activity for water purification and energy storage. This chapter discusses composite synthesis methods and commercial viability, highlighting environmental and energy applications.

Chapter 15

This chapter evaluates a range of storage technologies—electrochemical, thermal, and mechanical—needed for managing solar and wind intermittency. It advocates for a diversified, global strategy to ensure sustainability and energy equity.

Chapter 16

Focusing on materials like perovskites and quantum dots, this chapter highlights how innovations in material science are addressing the scalability and efficiency challenges of solar energy storage systems.

Chapter 17

Supercapacitors offer fast charge-discharge cycles and long life spans. This chapter examines their use in electric vehicles, especially for regenerative braking, and suggests a balanced analysis of their limitations versus lithium-ion batteries.

Chapter 18

A broad overview of global energy demands, and the environmental damage caused by fossil fuels. The chapter calls for accelerated investment in renewable energy to mitigate climate change, improve air quality, and enhance energy security.

Chapter 19

Detailing how supercapacitors store energy via electric double layers, this chapter compares aqueous and organic electrolytes and analyzes performance characteristics like cycle life and charge-discharge rates.

This book is designed for a diverse audience—from students and early-career researchers to seasoned academics and professionals in the fields of physics, materials science, engineering, and chemistry. It is also relevant to policy-makers, non-governmental organizations, and activists engaged in energy research and sustainability initiatives.

By bringing together recent scientific and technological breakthroughs, this volume aims to inspire new directions in research and contribute to the global effort toward cleaner, more reliable, and more efficient energy systems.

Mohsen Mhadhbi

National Institute of Research and Physicochemical Analysis, Tunisia

Acknowledgment

I would like to express my special thank to the authors for their excellent contributions their expertise, their perseverance, and time.

I would like to acknowledge the help of all the people involved in this project and, more specifically, to the authors and reviewers that took part in the review process. Without their support, this book would not have become a reality.

I also would like to acknowledge the valuable contributions of the reviewers concerning the improvement of quality, coherence, and content of chapters. Further, I want to express my gratitude to all staffs of IGI Global for their huge efforts and their availability.

I would like to thank IGI Global Scientific Publishing for providing the opportunity to write this book.

I would like to thank my family for their patience and understanding during the work on this book project.

Finally, a big thanks to everyone who said they'd like to read a book like mine. Every reader is a gift.

Chapter 1
2D Materials for Energy Storage and Conversion

Nisha Malik
https://orcid.org/0009-0007-1063-9739
Chandigarh Group of Colleges, Jhanjeri, India

E. Nirupama
https://orcid.org/0000-0002-0810-1805
Saveetha School of Management, Saveetha Institute of Medical and Technical Sciences, India

P. Selvakumar
https://orcid.org/0000-0002-3650-4548
Department of Science and Humanities, Nehru Institute of Technology, Coimbatore, India

Umesh Sankpal
R.P. Gogate College of Arts and Science, India & R.V. Jogalekar College of Commerce, India

Santosh Kumar Nathsharma
https://orcid.org/0000-0003-3935-2938
Stewart Science College, India

Mohit Sharma
https://orcid.org/0009-0007-2280-8077
Maharshi Dayanand University, Rohtak, India

T. C. Manjunath
https://orcid.org/0000-0003-2545-9160
Rajarajeswari College of Engineering, India

ABSTRACT

2D materials, characterized by their single-layer thickness, have revolutionized wide-ranging applications. hexagonal lattice, which has set the benchmark for 2D materials. Graphene's exceptional electrical, thermal, strength—have spurred extensive research into its potential applications and inspired the exploration of other 2D materials.Graphene's electrical properties are particularly striking. It r next-generation devices. Its ability to conduct electricity with minimal resistance opens up possibilities for ultrafast transistors, flexible electronics, and advanced sensors. Additionally, graphene's high thermal conductivity enhances its suitability

DOI: 10.4018/979-8-3693-9316-1.ch001

s. Its mechanical strength, being 100 times stronger than steel, combined with its flexibility, enables, from to wearable technology.Beyond tungsten disulfide (WS). (FETs). Their strong interaction with light enables high-performance photonic and optoelectronic devices, including applications in solar cells and imaging technologies.

INTRODUCTION TO 2D MATERIALS: PROPERTIES AND APPLICATIONS

2D materials, characterized by their single-layer thickness, have revolutionized wide-ranging applications. hexagonal lattice, which has set the benchmark for 2D materials. Graphene's exceptional electrical, thermal, strength—have spurred extensive research into its potential applications and inspired the exploration of other 2D materials.Graphene's electrical properties are particularly striking. It r next-generation devices. Its ability to conduct electricity with minimal resistance opens up possibilities for ultrafast transistors, flexible electronics, and advanced sensors. Additionally, graphene's high thermal conductivity enhances its suitability s. Its mechanical strength, being 100 times stronger than steel, combined with its flexibility, enables, from to wearable technology. Beyond tungsten disulfide (WS). (FETs). Their strong interaction with light enables high-performance photonic and optoelectronic devices, including applications in solar cells and imaging technologies.Another important class of 2D materials is hexagonal boron nitride, often to as "white graphene" h-BN is an excellent electrical insulator with a wide bandgap, making it ideal for use as a dielectric material in electronic devices. Its high thermal conductivity and chemical stability also lend it to applications in thermal management and as a protective layer in various technologies. Furthermore, the layered o be used as a substrate or encapsulation performance and stability, such as graphene and TMDs.2D materials also extend to other categories, such as the 2D polymeric materials and metal-organic frameworks (MOFs). These materials offer diverse functionalities due to their tunable structural and chemical properties. 2D polymers, like poly(phenylene vinylene) (PPV), have been explored for their optical and electronic properties in organic electronics and flexible displays. MOFs, which can be synthesized into 2D structures, provide and tunable porosity, in gas storage, separation, and catalysis.In addition to their electronic, optical, and mechanical applications, 2D materials have promising uses. The conductivity of graphene-based materials improve charge storage capacity and energy density, leading to more efficient energy storage solutions. Similarly, TMDs and other 2D .The integration of 2D materials into composite materials further expands their application potential. Combining 2D materials with other substances can result in composite materials with enhanced improved mechanical components, and multifunctional devices.

Despite their remarkable properties, the practical application of 2D materials faces challenges related to production, scalability, and integration. Synthesizing high-quality, large-area 2D materials remains a significant challenge, and developing cost-effective methods for their Additionally, integrating 2D materials into existing technologies and devices requires overcoming technical hurdles related to material compatibility, stability, and performance.

GRAPHENE: THE PIONEER 2D MATERIAL - STRUCTURE, SYNTHESIS, AND PROPERTIES

Graphene, the pioneering 2D material, has garnered immense serves as a foundational material for the broader category of 2D materials, demonstrating unparalleled potential across various fields.Structurally, planar structure, while the fourth valence electron forms a delocalized pi bond above and below the plane. This conjugated system (Vijayakumar G, et al., 2024) of pi bonds imparts graphene with its unique electronic properties. The lattice structure gives graphene its extraordinary mechanical strength and flexibility, with carbon-carbon bonds exhibiting a bond length of approximately 1.42 Å, contributing to its overall stability and robustness. The honeycomb arrangement of carbon atoms results in a material that is both lightweight and incredibly strong, boasting a tensile strength of over 130 GigaPascals (GPa) and an intrinsic stiffness.The synthesis of graphene has evolved through several key methods, each with its advantages and challenges. Mechanical exfoliation, or the "Scotch tape" method, was the original CVD and substrate interactions. Despite their remarkable properties, the practical application of 2D materials faces challenges related to production, scalability, and integration. Synthesizing high-quality, large-area 2D materials remains a significant challenge, and developing cost-effective methods for their Additionally, integrating 2D materials into existing technologies and devices requires overcoming technical hurdles related to material compatibility, stability, and performance.

Chemical reduction of graphene oxide is another method, wherein graphene oxide is chemically reduced to obtain graphene. This method allows easily accessible graphite oxide, although the resulting material may contain defects and oxygen functional groups that can alter its properties.The exceptional and bonding. Electrically, graphene exhibits high carrier mobility, with electrons traveling through the material with near-relativistic speeds.

This high mobility results in graphene's remarkable electrical conductivity, making it a promising candidate for advanced electronic high-speed transistors and flexible electronics. Graphene's c band structure is, which results in its zero bandgap and makes it a zero-gap semiconductor(Ansari et al. 2021). This property is beneficial for certain applications but poses challenges for others, such as traditional semiconductor devices that rely on a well-defined bandgap.Thermally, graphene is an excellent conductor, exhibiting a thermal conductivity of about 5000 W/mK, which is superior to that of most other materials.(Nejad & Gladwin, 2016) This property is attributed Mechanically, graphene's strength is exceptional. It is often described as being approximately 130 GPa and a Young's modulus of around 1 TeraPascal (TPa). This strength, combined with its lightweight and flexible nature, enables and composite materials. Additionally, graphene's flexibility allows it to be bent, stretched, and deformed without losing its integrity, which is advantageous for applications. This optical absorption property is utilized in various optoelectronic applications, including photodetectors and transparent conductive films. The interaction of graphene with light can be tuned by altering its electronic structure or by stacking multiple layers, thereby modifying its optical properties for specific applications. In summary, graphene stands as a pioneering 2D material with a distinctive structure that endows it with a suite of unique optical characteristics make it a versatile material with broad applications across electronics, energy, materials science, and nanotechnology(Aruna et al. 2019). As research progresses, ongoing efforts to refine synthesis techniques and integrate graphene into practical applications will likely unlock new possibilities and drive innovations across multiple domains.

TRANSITION METAL DICHALCOGENIDES: ELECTRONIC AND OPTICAL PROPERTIES

Transition metal dichalcogenides (TMDs) layered gained considerable element tungsten, or niobium) sandwiched between two chalcogenide layers (like sulfur, selenium, or tellurium), range of fascinating characteristics diverse in electronics, optoelectronics, and beyond.One of the most compelling features of TMDs is their ability to transition between different electronic phases depending on the specific composition For example, in bulk form, many TMDs are indirect bandgap semiconductors, points in momentum space. However, when thinned down to monolayer or few-layer forms, TMDs often exhibit a direct bandgap, which is a crucial property for various(Ashok et al. 2001) . This electrons and holes interact with light. Monolayer MoS_2 (molybdenum disulfide) and WS_2 (tungsten disulfide) are prime examples of TMDs that exhibit this direct bandgap behavior, enabling high-efficiency photodetectors, light-emitting devices, and photovoltaic cells.The electronic properties of TMDs are further characterized by their unique band structures, which often feature notable spin-orbit coupling effects. its orbital motion, leads to spin-split bands in TMDs, giving rise to phenomena such as valley polarization and spin-valley coupling. of both the electron's spin and its valley index, opening up possibilities for advanced spintronic and valleytronic applications(Balachander et al. 2021).Optically, TMDs exhibit remarkable properties due to their strong light-matter interactions. The direct bandgap nature of monolayer TMDs results in strong excitonic effects, where electron-hole pairs, or excitons, form bound states with significant binding energies.

This strong excitonic binding is a direct consequence of the reduced dielectric screening in monolayers compared to bulk materials, making excitons more stable and increasing their binding energy.(Lai et al., 2015) Consequently, TMDs exhibit distinct photoluminescence and absorption spectra, with sharp peaks corresponding to excitonic transitions. This high exciton binding energy also facilitates the observation of various excitonic states, such as charged excitons (trions) and biexcitons, which have implications for optoelectronic devices.TMDs are also known for their nonlinear optical properties, which can be utilized in applications such as frequency doubling and optical switching.(Kanipriya et al.,2024) The strong light-matter interaction in these materials enables the generation of higher harmonics and the modulation of optical signals, which are valuable for developing advanced optical communication systems and photonic devices.Another notable aspect of TMDs is their potential for integration with other 2D materials and heterogeneous systems. The ability to stack TMDs tailored electronic and optical properties.(from TMDs), which are useful for applications in high-speed electronics and optoelectronics.In summary,that exhibit a range of unique electronic and optical properties, including tunable bandgaps, strong excitonic effects, spin-valley coupling, and nonlinear optical responses. Their ability to transition between different electronic phases, coupled with their strong light-matter interactions, makes TMDs highly versatile and suitable optoelectronics, and beyond. As research into TMDs continues to advance, their integration into novel device architectures and their combination with other 2D materials will likely unlock new possibilities and drive innovation across various technological domains.

HEXAGONAL BORON NITRIDE: A 2D INSULATOR WITH UNIQUE PROPERTIES

The structure of hexagonal boron nitride (h-BN) features planar, two-dimensional sheet that, while structurally similar to graphene, differs significantly in its electronic properties due to the different atomic constituents. The bonding in h-BN results in a wide bandgap, typically around 5.9 eV, which classifies it as an excellent electrical insulator. This large bandgap is a direct consequence of the strong ionic character of the B-N bonds and the absence of free charge carriers in the material, making h-BN particularly useful for applications where electrical isolation is crucial.One of the most significant applications of candidate for use as a gate dielectric in field-effect transistors (FETs) and as an insulating layer in various electronic components. In these applications, h-BN provides excellent electrical insulation, preventing leakage currents and improving device performance. Additionally, its l management of electronic devices.h-BN also serves as an effective substrate and encapsulation(Basu et al. 2016) The smooth and flat surface of h-BN provides an ideal environment for the growth and transfer of these materials, enhancing their When used as a substrate, h-BN can improve the devices by reducing charge scattering and increasing carrier mobility. As an encapsulation layer, h-BN protects sensitive 2D materials from environmental degrade The optical properties of h-BN are also noteworthy. While h-BN itself is a wide-bandgap insulator, it exhibits interesting optical characteristics due to its lattice structure. It can in the deep ultraviolet (UV) range, which is useful for applications in UV optoelectronics and sensing. Additionally, h-BN can be employed as a host material for quantum emitters, such as point defects, which have potential applications in quantum computing and quantum communication. Another intriguing property of h-BN is its potential for use in friction and lubrication. high-quality, large-area h-BN films on metal substrates, while mechanical exfoliation can produce thinner flakes of h-BN for research purposes(Bartlett et al. 2015). Both methods have their advantages, with CVD providing scalability and uniformity, and mechanical exfoliation offering high purity and ease of use for small-scale applications.Despite its advantageous properties, there are challenges associated with the integration and manipulation of h-BN in practical applications (Mahadev Madgule, et al.,2023).

The production of large-area, high-quality h-BN films requires precise control over growth conditions and substrate interactions. Additionally, the handling and transfer of h-BN layers need to be carefully managed to avoid contamination and preserve the material's properties. Despite their remarkable properties, the practical application of 2D materials faces challenges related to production, scalability, and integration. Synthesizing high-quality, large-area 2D materials remains a significant challenge, and developing cost-effective methods for their Additionally, integrating 2D materials into existing technologies and devices requires overcoming technical hurdles related to material compatibility, stability, and performance.In summary, . Its wide bandgap and high dielectric strength enable its use as an electrical insulator and dielectric material, while its role as a substrate and encapsulation layer enhances the performance of other 2D materials. The optical and frictional properties of h-BN further expand its application potential in UV optoelectronics and lubrication(Karuppannan et al.,2024). As research and development continue, overcoming the challenges associated with h-BN synthesis and integration will unlock new possibilities and drive innovation in electronics, materials science, and beyond.

2D METAL-ORGANIC FRAMEWORKS: DESIGN, SYNTHESIS, AND APPLICATIONS

2D metal-organic frameworks (MOFs) represent an innovative subclass of materials characterized by their extended, two-dimensional structures ligands. Combine area and tunable of traditional MOFs with unique advantages of 2D layered structures, offering significant design of 2D MOFs and organic linkers. Metal nodes, often comprising transition metals or lanthanides, serve as the primary centers for coordinating the organic ligands. The choice of metal affects the stability, functionality, and electronic properties of the framework. Organic ligands, which are typically multifunctional organic molecules, connect the metal nodes and dictate the topology and dimensionality of the framework. In 2D MOFs, these ligands are arranged in a planar configuration, resulting in sheets or layers that are connected through metal-ligand interactions. and organic ligands in a solvent, followed by heating the mixture under high pressure to promote crystal growth. This method is versatile and widely used for synthesizing high-quality 2D MOFs. Hydrothermal synthesis is similar but uses water as the solvent, often yielding MOFs with different properties due to the use of aqueous conditions. Electrochemical synthesis, a more recent method, involves applying an electric current to induce the formation of MOFs from metal salts and organic ligands, offering control over the growth process and the potential for fabricating MOFs on conductive substrates.The resulting 2D MOFs exhibit a range of intriguing properties due to porosity. The layered structure of 2D MOFs provides abundant surface area for adsorption, . For example, 2D MOFs have shown promise in storing gases like hydrogen and methane, where their high surface area facilitates efficient gas adsorption and release. Their tunable pore size also enables selective separation of different gas molecules(Corcau et al. 2020), which is valuable in industrial processes and environmental applications.In catalysis, 2D MOFs offer several advantages. Their large surface area and the availability of metal centers for catalytic reactions make them effective as heterogeneous catalysts. The metal nodes in 2D MOFs can various catalytic processes, such as oxidation, reduction, and coupling reactions. Moreover, the tunability of the organic ligands allows for the introduction of functional groups that can enhance catalytic activity or provide specific fine chemical synthesis(Ramakrishnan T.,et al.,2022).The sensing capabilities of 2D MOFs are also notable. Their the organic ligands allow for the detection of various analytes through changes in their optical, electrical, or chemical properties. make them suitable for use in supercapacitors and batteries, where they can enhance charge storage and cycling stability. In energy conversion, 2D MOFs can act as catalysts for or carbon dioxide reduction, contributing to sustainable energy technologies. Despite their promising applications, the development and commercialization of 2D MOFs face several challenges. The synthesis of large-area, high-quality 2D MOFs can be complex and costly, requiring (Dai et al. 2021)precise control over reaction conditions and precursor quality. Additionally, integrating 2D MOFs into practical devices and systems involves addressing issues related to stability, scalability, and reproducibility. Researchers are actively working on overcoming these challenges by developing more efficient synthesis methods, exploring novel materials, and improving the understanding of structure-property relationships.

PHOSPHORENE AND OTHER 2D PHOSPHORUS MATERIALS: SYNTHESIS AND PROPERTIES

Phosphorene, a single layer of material exhibits intriguing characteristics that distinguish it from other materials like graphene and transition metal dichalcogenides (TMDs), offering new opportunities in electronics, optoelectronics, and beyond. Phosphorene, which is derived from bulk black phosphorus, layeredconsisting of arranged in a puckered honeycomb lattice. This structure introduces a distinct anisotropy to phosphorene, resulting in different properties along its armchair and zigzag directions. . Mechanical exfoliation, akin to the method used for graphene, involves peeling off thin layers of black phosphorus using adhesive tape or other techniques, resulting in monolayer or few-layer phosphorene. growth of high-quality phosphorene films by depositing phosphorus precursors onto substrates, while black phosphorus in a solvent and using ultrasonic agitation to produce thin phosphorene flakes.The electronic properties of phosphorene are notable for its direct bandgap, which varies with the number of layers. Monolayer phosphorene exhibits a bandgap of around 1.5 eV, which is in the visible range of the electromagnetic changing the number of layers or applying external strain, offering flexibility in designing devices for specific applications. Additionally, phosphorene displays high carrier mobility(Hassan et al. 2022), though it is anisotropic—significantly higher along the armchair direction compared to the zigzag direction. This anisotropy is a result of the material's puckered structure, which influences electron and hole transport properties.Phosphorene's optical properties are also of interest. The material exhibits strong light-matter interactions, with absorption spanning a broad range of wavelengths. The direct bandgap allows for significant optical absorption, making phosphorene useful in photodetectors, phototransistors, and other optoelectronic devices. Furthermore, phosphorene can show a significant photoluminescence effect, which can be harnessed for applications in imaging and sensing. Its optical properties can be engineered by tuning the number of layers or through chemical modifications, broadening its applicability in various optical technologies.Mechanically, phosphorene is relatively flexible and robust compared to other 2D materials. It has a high Young's modulus and can withstand significant strain before breaking, making phosphorene-based In addition to phosphorene, phosphorus studied, phosphorene oxides derivatives. Phosphorene oxides, for instance, result from the oxidation of phosphorene and exhibit different compared to the pristine material(Lai et al., 2015). These oxides in conversion, where their chemical reactivity and high surface area are advantageous. Phosphorene derivatives, modified by incorporating various functional groups, offer enhanced properties for specific applications, including sensing and catalysis.The integration of phosphorene and other 2D phosphorus materials into practical devices involves addressing several challenges(Dai et al. 2015). phosphorene and other 2D phosphorus materials represent a promising area of research electronics, optoelectronics, and flexible devices. The study of phosphorene derivatives and oxides further expands the potential applications of phosphorus-based 2D materials. As research continues to advance, synthesis, stability, and integration will be key to unlocking the phosphorene and other 2D phosphorus materials in practical applications.

SILICENE AND GERMANENE: 2D SILICON AND GERMANIUM MATERIALS

Silicene and germanene are two-dimensional (2D) materials composed of silicon and germanium, respectively, and are akin to graphene in their structure but distinct in their elemental composition and properties. These materials are part of a broader category of 2D materials that exhibit unique physical and electronic characteristics due to their low-dimensional nature.Silicene, a shares structural similarities with graphene, but with key differences in electronic behavior and stability. Unlike graphene, which is perfectly planar, silicene has a slightly buckled structure due to the sp^2 hybridization of silicon atoms, which introduces an additional layer of complexity to its properties. This buckling results in a small, but significant, bandgap in silicene, making it a semiconductor rather than a zero-gap material. The electronic properties of silicene are influenced by the presence of spin-orbit coupling, which introduces a gap that can be modulated by external factors such as strain or electric fields. This feature makes silicene a candidate for applications in advanced device operation. on appropriate substrates like silver or platinum, which can stabilize the silicene layer and promote its growth. Another method involves the exfoliation of bulk silicon in vacuum or inert atmospheres. Despite these methods, maintaining silicene's stability and integrating it into practical devices remain significant hurdles. applying external fields, which makes it versatile for use Synthesis of germanene can be achieved using methods such as CVD, molecular beam epitaxy (MBE), or chemical exfoliation of bulk germanium. CVD and MBE are particularly effective in producing high-quality, uniform germanene layers, but similar to silicene, maintaining stability under ambient conditions remains a challenge. Research is ongoing to explore various substrate interactions and protective coatings that can enhance the stability of germanene in practical applications(Gong et al. 2015).Both silicene and germanene exhibit intriguing optical properties. Their direct bandgap allows for significant optical absorption, which is advantageous devices. Additionally, the buckling of their structures leads to unique phonon modes and optical responses that can be exploited in advanced photonics. mechanical properties of these materials are influenced by their lattice structure and buckling, which affect their response to external forces and strains.The integration of silicene and germanene into practical devices involves addressing several key challenges. These include maintaining material stability, achieving high-quality synthesis, and ensuring compatibility with existing fabrication technologies. Researchers are also exploring ways to combine these 2D materials with other materials to create heterostructures that leverage their unique properties for enhanced performance in electronic and optoelectronic applications(-Jha et al. 2019).In summary, silicene and germanene represent exciting additions to the family of 2D materials, offering unique properties derived from their elemental composition and structural characteristics. Their buckled lattice structures introduce small bandgaps and significant spin-orbit coupling, providing opportunities for advanced electronic and spintronic applications. While challenges remain in their synthesis and stability, likely to unlock new possibilities and applications for these materials in various technological domains.

2D MATERIALS FOR ELECTRONICS: TRANSISTORS, SENSORS, AND ENERGY STORAGE

Two-dimensional materials solutions across transistors, sensors, and energy storage. These materials, characterized by their single or few-layer thickness, exhibit unique electrical, mechanical open up new possibilities in electronic technology. are fundamental components in electronic devices, and into transistor has led to significant advancements. Graphene, with its exceptionally electrical conductivity, has been a prominent material in this domain. However, its zero bandgap limits its use in traditional digital electronics where a finite bandgap is essential for switching applications. including tunable bandgaps and strong spin-orbit coupling, which can be exploited in advanced transistor designs electronics. molecules, which can lead to detectable changes in electrical resistance or optical properties. Similarly, TMDs and other 2D materials can be engineered to respond to specific chemical or biological targets, enhancing their applicability in various sensing technologies. The flexibility and scalability of 2D materials also facilitate the development of wearable and flexible sensors, which can be integrated into diverse environments for real-time monitoring. is another critical area where 2D materials are making a significant impact. Supercapacitors and batteries, essential for portable electronics and renewable energy storage,, which improve charge storage and power density in supercapacitors. Similarly, TMDs, with their layered structure, provide additional pathways for charge storage and ion intercalation, enhancing the performance of batteries. For example, MoS_2 has been investigated as an anode material in lithium-ion batteries due to its high capacity and stability (Janabi & Wang, 2019). The incorporation of 2D materials into energy storage devices also leads to improvements in charge/discharge rates, cycle stability, and overall energy density. Heterostructures and composite materials** formed by stacking different further enhance device performance. For example, van der Waals heterostructures, created by layering different 2D materials such as graphene and h-BN, offer novel electronic properties and improved device functionalities. These heterostructures can be engineered to create devices with tailored properties for specific applications, such as high-efficiency transistors or sensitive sensors.Challenges associated with the integration of 2D materials into electronic devices include issues related to scalability, uniformity, and material stability. Producing high-quality, large-area 2D materials with consistent properties is crucial for commercial applications. Additionally, the stability of 2D materials under operational conditions must be addressed to ensure long-term reliability and performance. Researchers are actively working on overcoming these challenges by developing improved synthesis methods, protective coatings, and integration techniques to enhance the practicality and functionality of 2D materials in electronic applications(Karuppannan et al.,2024).In summary, 2D materials have transformed the landscape of electronics by offering enhanced performance and new functionalities in transistors, sensors, and energy storage devices. The unique properties of these materials, such as high carrier mobility, large surface area, and tunable electronic characteristics, make them ideal candidates for next-generation electronic technologies(Hazim et al. 2023). As research continues to advance, addressing the challenges of material synthesis and stability will be key to unlocking the full potential of 2D materials and driving innovation in electronics.

OPTOELECTRONIC APPLICATIONS OF 2D MATERIALS: PHOTODETECTORS, LEDS, AND LASERS

Two-dimensional (2D) materials have become prominent optoelectronics their unique, tunable bandgaps, and strong light-matter interactions. These attributes enable innovative applications in photodetectors, light-emitting diodes (LEDs), and lasers, which are critical components in modern electronics and communication systems. into optoelectronic devices offers potential for enhanced performance, flexibility, and novel functionalities. optical properties and high surface-to-volume ratios. Graphene, with its broad optical absorption range and high carrier mobility, has been extensively studied for photodetection applications. However, its zero bandgap limits its efficiency in detecting visible and near-infrared ligh(Jiang et al. 2016)t. . This property allows TMD-based photodetectors to operate efficiently in the visible to near-infrared spectrum, making them suitable for applications such as imaging and environmental monitoring. Additionally, TMDs can exhibit high responsivity and fast response times, crucial for high-speed photodetection. These materials are also amenable to integration with other components in flexible and wearable optoelectronic devices, expanding their applicability in emerging technologies.utilizing 2D materials benefit from their direct bandgaps and strong photoluminescence. Monolayers of TMDs such as MoS_2 also enable the design of flexible and transparent LEDs, which can be integrated into new types of displays and lighting systems.Lasers based on an area for producing coherent light with high efficiency. Monolayer and few-layer TMDs have been shown to exhibit strong optical gain, which is essential for laser operation. For instance, WS_2 and $MoSe_2$ have been demonstrated to emit laser light at room temperature, with tunable wavelengths depending on the material and its number of layers. The ability to precisely control the emission wavelength and optical properties of these materials enables the development of on-chip lasers for telecommunications and sensing applications. Furthermore(Kanthimathi et al. 2015), 2D materials can be integrated into various device architectures, including vertical-cavity surface-emitting lasers (VCSELs) and microcavity lasers, enhancing their versatility and functionality.Challenges in integrating 2D materials into optoelectronic devices include issues related to material stability, uniformity, and scalability. For photodetectors and LEDs, ensuring consistent performance across large-area substrates and integrating these materials into existing fabrication processes is critical. Additionally, the stability of 2D materials under operational conditions, such as exposure to air and moisture, must be addressed to ensure long-term reliability. with tailored optical properties, the integration of 2D materials into novel device architectures, and the exploration of hybrid systems combining 2D materials with other technologies. Advancements crucial role in scaling up the production and integration of 2D materials for commercial applications.In summary, 2D materials offer significant, including high surface area, tunable bandgaps, and strong light-matter interactions. Their use in photodetectors, LEDs, and lasers has the potential to revolutionize various technologies, providing enhanced performance, flexibility, and new functionalities. in optoelectronics will likely be realized, leading to innovative solutions and advancements in lighting, communication, and sensing technologies.

flexibility, and tunable characteristics make them well-suited for applications in batteries, supercapacitors, and solar cells, where they can significantly enhance performance and efficiency.Batteries benefit from the incorporation of 2D materials in several ways. In lithium-ion batteries (LIBs), For instance, graphene-based anodes provide excellent conductivity and mechanical stability, which enhance the cycle life and rate capability of LIBs. Additionally, TMDs like MoS_2 and WS_2These materials can also be combined with other components to create composite electrodes that further boost battery performance. Beyond LIBs, 2D materials are also explored in next-generation batteries such as sodium-ion batteries and lithium-sulfur batteries, where their unique properties can address specific challenges related to energy density, stability, and efficiency. To further enhance supercapacitor performance, researchers have developed graphene-based composites and hybrid materials that incorporate other 2D materials,(Wang et al. 2011) such as transition metal oxides or conductive polymers. These composites can improve capacitance and energy density while maintaining high power density. Similarly, other 2D materials, including MXenes (a family of transition metal carbides and nitrides), have shown potential in supercapacitors due to their high conductivity and layered structure, which facilitate fast ion and electron transport.Solar cells benefit from 2D materials through their unique optical and electronic properties, which can be exploited to enhance light absorption, charge separation, and overall efficiency. Monolayer $_2$ exhibit direct bandgaps that are beneficial for absorbing visible light, making . When integrated into heterostructures with other 2D materials or traditional semiconductors, TMDs can improve the light absorption and charge carrier dynamics of solar cells. Graphene, with its high electrical conductivity and transparency. The combination of 2D materials with conventional photovoltaic materials, such as silicon or perovskites, leads to the development of hybrid solar cells with enhanced performance. These hybrid devices can achieve better light absorption(Wang et al. 2011), charge transport, and overall energy conversion efficiency. enhancing device performance. Developing scalable synthesis and fabrication techniques that ensure uniformity and quality of 2D materials is essential for commercial applications.Future Directions in the use their fundamental properties and interactions, developing new materials with tailored characteristics, and optimizing device architectures. Research is focused on exploring new 2D materials with enhanced electrochemical properties, improving the scalability of fabrication processes, and integrating 2D materials into multifunctional devices(Mansouri et al. 2021). Additionally, efforts are underway to combine 2D materials with other emerging technologies, such as flexible electronics and advanced composites, to create innovative solutions for energy storage and conversion.In summary, 2D materials offer significant advantagesurface tunable properties, and unique electrochemical behavior contribute to enhanced performance and efficiency in these devices. materials and innovative device architectures will drive the progress of energy storage and conversion technologies, leading to more efficient and sustainable solutions.

MECHANICAL PROPERTIES OF 2D MATERIALS: STRENGTH, TOUGHNESS, AND FLEXIBILITY

Understanding these mechanical properties is essential for leveraging 2D materials in practical applications, from flexible electronics to advanced composites. strength is mechanical materials., perhaps the most studied 2D material, exhibits extraordinary tensile strength, largely due to its sp^2 hybridized carbon-carbon bonds arranged in a hexagonal lattice. This structure imparts a tensile strength of around 130 gigapascals (GPa), making graphene one of the strongest known materials. The strength of graphene is a result of the robust covalent bonds like MoS_2, exhibit high mechanical strength, although typically lower than graphene. The strength of TMDs material and the number of layers, but they generally maintain substantial mechanical robustness.Toughness, which varies significantly among 2D materials. Graphene, despite its high strength, has limited toughness because it lacks inherent ductility; it tends to fracture rather than deform plastically. In contrast, materials such as silicene and germanene exhibit greater toughness compared to graphene. The puckered structure of silicene and the layered structure of germanene allow for some degree of out-of-plane deformation, which can enhance their toughness. However, their overall toughness still depends on factors like layer number and the presence of defects or impurities.Flexibility is another critical mechanical property that affects the practical use of 2D materials in flexible and wearable electronics. This flexibility arises from the planar structure of graphene and its ability to maintain electronic properties even when deformed. Similarly, other 2D materials like molybdenum disulfide (MoS_2) and tungsten disulfide (WS_2) are flexible to a certain extent, though their flexibility can be influenced by the number of layers and the presence of structural defects. For example, monolayer TMDs generally show greater flexibility than their multilayer counterparts due to the reduced van der Waals forces between layers(Zhang et al. 2019). where different 2D materials are stacked or combined, can leverage the strengths of individual materials to create devices with superior mechanical properties and performance. related to the mechanical properties of 2D materials include issues related to leading to reduced strength and toughness. Addressing these challenges involves developing improved synthesis methods and fabrication techniques to produce high-quality, defect-free materials.

Additionally, scaling up the production of 2D materials while maintaining their mechanical properties remains a critical challenge for industrial applications. in the study of 2D materials' mechanical(Yuan et al. 2021) properties include exploring new materials with enhanced performance characteristics, understanding the fundamental mechanisms behind their mechanical behavior, and developing innovative applications that exploit their unique properties. Research is focused on improving the mechanical robustness of 2D materials through chemical modifications, be tailored for specific applications. Understanding and optimizing these properties are essential for the development of advanced materials and devices that leverage the unique advantages of 2D materials. As research progresses, addressing challenges related to defects and scalability will be key to fully realizing the potential of 2D materials in practical applications.

THERMAL AND PHONONIC PROPERTIES OF 2D MATERIALS: HEAT MANAGEMENT AND THERMOELECTRICITY

The thermal and phononic to their performance in various applications, including heat management and thermoelectric energy conversion. These properties are intricately linked to the unique atomic structures of 2D materials, which affect how heat is conducted and how phonons, or lattice vibrations, propagate through these materials. Understanding and optimizing these properties is essential for harnessing 2D materials in advanced electronic, optoelectronic, and thermoelectric devices(Nejad et al. 2016).Thermal conductivity is a crucial parameter for heat management in electronic devices. Graphene, known for its exceptional thermal conductivity, demonstrates the ability to conduct heat efficiently due to its strong sp^2-bonded carbon lattice and high carrier mobility. Its thermal conductivity, reported to be around 3000 to 5000 W/m·K, surpasses that of many other materials, making it an excellent candidate for thermal management applications. This high advantageous for dissipating components and preventing overheating. However, this property can be affected by factors such as material defects, grain boundaries, and the presence of interfaces, which can scatter phonons conductivity.In contrast,. These materials possess weaker in-plane bonding and higher interlayer interactions, which can limit their thermal transport properties.The efficient phonon transport in graphene contributes to its high thermal conductivity and is advantageous for applications requiring effective heat dissipation. Conversely, in materials like silicene and germanene, phonon transport can be influenced by their buckled structures and weaker bonding, leading to different thermal behaviors. These variations in phonon transport can be harnessed for specific applications where tuning thermal conductivity is required.Thermoelectric properties are another important aspect of 2D materials, particularly for energy conversion applications. Thermoelectric materials can convert heat gradients into electrical energy and are characterized by For example, the relatively high Seebeck coefficient and electrical conductivity observed in certain TMDs, combined with their relatively low thermal conductivity, can result in favorable thermoelectric performance.(Tao et al. 2022) Materials such as MoS_2 and WSe_2 have shown promising thermoelectric properties, and ongoing research aims to optimize their performance by controlling their structure and composition.Heat management in electronic and optoelectronic devices is another critical area where 2D materials can make a significant impact. The ability to efficiently manage and, the integration of 2D materials with other materials or into composite structures can be used to create thermal management solutions that leverage the strengths of different materials.Challenges related to the thermal and phononic properties of 2D materials include the influence of defects, interfaces, and environmental conditions. Defects and impurities can scatter phonons and reduce thermal conductivity, while interfaces between different materials or layers can introduce additional thermal resistance. methods challenges, such as developing high-quality, defect-free 2D materials, optimizing material interfaces, and employing advanced fabrication techniques(Pahlevani et al. 2020).Future Directions in the study of thermal and phononic properties of 2D materials include developing new materials with tailored thermal and thermoelectric properties, optimizing device architectures for better heat management, and exploring novel applications. Advances in material synthesis, characterization techniques, and computational modeling will further enhance the understanding of these properties and their implications for practical applications. Additionally, combining 2D materials with other advanced materials and technologies could lead to innovative solutions for heat management and energy conversion..

SCALABLE SYNTHESIS AND INTEGRATION OF 2D MATERIALS: CHALLENGES AND SOLUTIONS

Techniques such as chemical vapor deposition (CVD), liquid-phase exfoliation, and chemical reduction have been employed to synthesize these materials. However, scaling these methods up while maintaining the desired quality remains a significant hurdle. For instance, CVD, which is often used to grow graphene or TMDs, requires precise control over temperature, pressure, and gas composition. While small-scale CVD systems can produce high-quality single-layer films, scaling this process to industrial levels introduces variability in these parameters, leading to inconsistencies in material properties and performance.Another significant challenge is where a 2D material is lifted from its growth substrate and placed onto a target substrate, are commonly used. can further complicate the integration.To address these challenges, researchers have been exploring several solutions.(Tan et al. 2011) For scalable synthesis, advancements in CVD technology, such as the development of roll-to-roll systems and large-area reactors, are being investigated to improve the uniformity and quality of 2D materials(Rahman et al. 2016). Additionally, new synthesis methods, including liquid-phase methods and mechanical exfoliation techniques, are being optimized for large-scale production. In terms of integration, new fabrication techniques are being developed to enhance the reliability and reproducibility of advances in cleanroom technologies and contamination control are being applied to minimize defects and ensure high-performance devices.Overall, while the synthesis and integration of 2D materials at a scalable level face significant challenges, ongoing research and technological advancements offer promising solutions. By addressing these issues, it is possible

FUTURE DIRECTIONS AND EMERGING TRENDS IN 2D MATERIALS RESEARCH

The field of 2D materials research is rapidly evolving, with future directions and emerging trends pointing towards transformative advancements across various domains. areas are emerging as pivotal for the next generation of technological innovations. These trends are shaping the trajectory of research and development in 2D materials and promise to unlock new possibilities in electronics, energy, and beyond.One of the foremost directions in 2D materials research is the development of heterostructures and van der Waals (vdW) heterojunctions. By stacking different 2D materials on top of one another, researchers can create complex layered structures that exhibit novel properties not present in individual materials. This approach leverages the unique electronic, optical, and mechanical characteristics of each material to engineer devices with tailored functionalities. For example, combining graphene with transition metal dichalcogenides (TMDs) can yield devices with enhanced electronic properties and improved performance Another significant trend is the exploration of new 2D materials beyond the well-known graphene, h-BN, and TMDs. Researchers are increasingly focusing on discovering and characterizing novel 2D materials with unique properties. For instance, materials such as black phosphorus, which exhibits an anisotropic band structure, and 2D materials with topological properties, like topological insulators, are gaining attention for their potential applications in next-generation electronic and spintronic devices.(Sakhdari et al. 2015). The mechanical flexibility and lightweight nature of 2D materials properties of 2D materials to ensure their performance and durability in flexible device Finally, the intersection of 2D materials with quantum technologies is an exciting frontier. Th. Research in this area aims to harness these properties to develop novel quantum devices and systems with enhanced performance and functionality.In summary, the future directions and emerging trends in 2D materials research are characterized by a focus on heterostructures, new material discovery, advanced synthesis techniques, energy applications, flexible technologies, and quantum innovations. These areas hold the promise of significant technological advancements and the potential to revolutionize various industries.

CONCLUSION

The groundbreaking isolation of graphene, the exploration of various 2D materials has unveiled a diverse array of properties and potential applications. Each material, from graphene to transition metal dalcogenides (TMDs) and beyond, brings unique characteristics that can be harnessed for specific technological advancements. Graphene's exceptional electrical and thermal conductivity, combined with its mechanical strength, has already revolutionized areas such as electronics, energy storage, and materials engineering. Similarly.The interdisciplinary nature of 2D materials research, encompassing physics, chemistry, materials science, and engineering, underscores the collaborative efforts required to drive innovation forward. As researchers continue to explore the fundamental properties and potential applications of 2D materials, the development of new theoretical models and experimental techniques will be essential for unlocking their full potential. In conclusion, the exploration and utilization material offering exciting opportunities for innovation across various fields. While challenges remain, the rapid progress in this area promises a future where the unique properties of 2D materials can be harnessed to create advanced technologies that could reshape our world. The continued pursuit of knowledge and technological development in the realm of 2D materials will undoubtedly yield significant advancements, highlighting the importance of ongoing research and collaboration in this dynamic and evolving field. Graphene and its derivatives, for example, have been incorporated into supercapacitors and batteries to enhance their performance. The conductivity of graphene-based materials improve charge storage capacity and energy density, leading to more efficient energy storage solutions. Similarly, TMDs and other 2D .The integration of 2D materials into composite materials further expands their application potential. Combining 2D materials with other substances can result in composite materials with enhanced improved mechanical components, and multifunctional devices. Despite their remarkable properties, the practical application of 2D materials faces challenges related to production, scalability, and integration. Synthesizing high-quality, large-area 2D materials remains a significant challenge, and developing cost-effective methods for their Additionally, integrating 2D materials into existing technologies and devices requires overcoming technical hurdles related to material compatibility, stability, and performance.

REFERENCES

Ansari, S., Ayob, A., Hossain Lipu, M. S., Hussain, A., & Saad, M. H. M. (2021). Data-driven remaining useful life prediction for lithium-ion batteries using multi-charging profile framework: A recurrent neural network approach. *Sustainability (Basel), 13*(23), 13333. DOI: 10.3390/su132313333

Aruna, P., & Vasan, P. V. (2019, August). Review on energy management system of electric vehicles. In *2019 2nd International Conference on Power and Embedded Drive Control (ICPEDC)* (pp. 371-374). IEEE. DOI: 10.1109/ICPEDC47771.2019.9036689

Ashok, S., & Banerjee, R. (2001). An optimization mode for industrial load management. *IEEE Transactions on Power Systems, 16*(4), 879–884. DOI: 10.1109/59.962440

Balachander, K., Amudha, A., Ramkumar, M. S., Emayavaramban, G., Divyapriy, S., & Nagaveni, P. (2021). Design and analysis of modified CUK converter for electric hybrid vehicle. *Materials Today: Proceedings, 45*, 1691–1695. DOI: 10.1016/j.matpr.2020.08.566

Bartlett, A., Marcicki, J., Onori, S., Rizzoni, G., Yang, X. G., & Miller, T. (2015). Electrochemical model-based state of charge and capacity estimation for a composite electrode lithium-ion battery. *IEEE Transactions on Control Systems Technology, 24*(2), 384–399. DOI: 10.1109/TCST.2015.2446947

Basu, S., Hariharan, K. S., Kolake, S. M., Song, T., Sohn, D. K., & Yeo, T. (2016). Coupled electrochemical thermal modelling of a novel Li-ion battery pack thermal management system. *Applied Energy, 181*, 1–13. DOI: 10.1016/j.apenergy.2016.08.049

Corcau, J. I., & Dinca, L. (2020). Fuzzy energy management scheme for a hybrid power sources of high-altitude pseudosatellite. *Modelling and Simulation in Engineering, 2020*(1), 5459098. DOI: 10.1155/2020/5459098

Dai, H., Jiang, B., Hu, X., Lin, X., Wei, X., & Pecht, M. (2021). Advanced battery management strategies for a sustainable energy future: Multilayer design concepts and research trends. *Renewable & Sustainable Energy Reviews, 138*, 110480. DOI: 10.1016/j.rser.2020.110480

Dai, H., Zhu, L., Zhu, J., Wei, X., & Sun, Z. (2015). Adaptive Kalman filtering based internal temperature estimation with an equivalent electrical network thermal model for hard-cased batteries. *Journal of Power Sources, 293*, 351–365. DOI: 10.1016/j.jpowsour.2015.05.087

Gan, L., Yang, F., Shi, Y. F., & He, H. L. (2017, November). Lithium-ion battery state of function estimation based on fuzzy logic algorithm with associated variables. []. IOP Publishing.]. *IOP Conference Series. Earth and Environmental Science, 94*(1), 012133. DOI: 10.1088/1755-1315/94/1/012133

Gong, X., Xiong, R., & Mi, C. C. (2015). A data-driven bias-correction-method-based lithium-ion battery modeling approach for electric vehicle applications. *IEEE Transactions on Industry Applications, 52*(2), 1759–1765.

Hassan, R. F., Ajel, A. R., Abbas, S. J., & Humaidi, A. J. (2022). FPGA based HIL co-simulation of 2DOF-PID controller tuned by PSO optimization algorithm. *ICIC Express Letters, 16*(12), 1269–1278.

Hazim, H. I., Baharin, K. A., Gan, C. K., Sabry, A. H., & Humaidi, A. J. (2023). Review on optimization techniques of PV/inverter ratio for grid-Tie PV systems. *Applied Sciences (Basel, Switzerland), 13*(5), 3155. DOI: 10.3390/app13053155

Janabi, A., & Wang, B. (2019). Switched-capacitor voltage boost converter for electric and hybrid electric vehicle drives. *IEEE Transactions on Power Electronics, 35*(6), 5615–5624. DOI: 10.1109/TPEL.2019.2949574

Jha, M., Blaabjerg, F., Khan, M. A., Bharath Kurukuru, V. S., & Haque, A. (2019). Intelligent control of converter for electric vehicles charging station. *Energies, 12*(12), 2334. DOI: 10.3390/en12122334

Jiang, J., Ruan, H., Sun, B., Zhang, W., Gao, W., Wang, L. Y., & Zhang, L. (2016). A reduced low-temperature electro-thermal coupled model for lithium-ion batteries. *Applied Energy, 177*, 804–816. DOI: 10.1016/j.apenergy.2016.05.153

Kanipriya, M., Sugumaran, D., Pandiyarajan, V., Yogananda, B. S., Selvakumar, P., & Mohan, S. (2024, January). Analysis of Electromyography Signals for Control of Mechanical Prosthesis using Machine Learning Techniques. In *2024 2nd International Conference on Intelligent Data Communication Technologies and Internet of Things (IDCIoT)* (pp. 1059-1067). IEEE.

Kanthimathi, R., & Kamala, J. (2015, May). Analysis of different flyback Converter topologies. In *2015 international conference on industrial instrumentation and control (ICIC)* (pp. 1248-1252). IEEE. DOI: 10.1109/IIC.2015.7150939

Karuppannan, P. G., Paramasivam, S., Mariappan, P., Shunmugam, S., & Richard Paul, S. J. (2024). Removal of chemicals from effluent using photobioreactor technology to improve environmental and health impacts. *Zeitschrift für Physikalische Chemie, 238*(7), 1295–1302. DOI: 10.1515/zpch-2023-0450

Lai, C. M., Lin, Y. C., & Lee, D. (2015). Study and implementation of a two-phase interleaved bidirectional DC/DC converter for vehicle and dc-microgrid systems. *Energies*, *8*(9), 9969–9991. DOI: 10.3390/en8099969

Lai, X., Yuan, M., Tang, X., Yao, Y., Weng, J., Gao, F., Ma, W., & Zheng, Y. (2022). Co-estimation of state-of-charge and state-of-health for lithium-ion batteries considering temperature and ageing. *Energies*, *15*(19), 7416. DOI: 10.3390/en15197416

Lei, Z., Qin, D., Hou, L., Peng, J., Liu, Y., & Chen, Z. (2020). An adaptive equivalent consumption minimization strategy for plug-in hybrid electric vehicles based on traffic information. *Energy*, *190*, 116409. DOI: 10.1016/j.energy.2019.116409

Madgule, M., Vinayaka, N., Sonkhaskar, Y. M., Patel, D. R., Karthikeyan, R., & Selvakumar, P. (2023). Mechanical properties and microstructure of activated TIG welded similar joints of Inconel alloys by desirability approaches. *Materials Today: Proceedings*, *77*, 528–533. DOI: 10.1016/j.matpr.2022.12.250

Magnor, D., Gerschler, J. B., Ecker, M., Merk, P., & Sauer, D. U. (2009, September). Concept of a battery aging model for lithium-ion batteries considering the lifetime dependency on the operation strategy. In *Proceedings of the European Photovoltaic Solar Energy Conference,* Hamburg, Germany (pp. 21-25).

Mansouri, S. A., Nematbakhsh, E., Javadi, M. S., Jordehi, A. R., Shafie-khah, M., & Catalão, J. P. (2021, September). Resilience enhancement via automatic switching considering direct load control program and energy storage systems. In *2021 IEEE International Conference on Environment and Electrical Engineering and 2021 IEEE Industrial and Commercial Power Systems Europe (EEEIC/I&CPS Europe)* (pp. 1-6). IEEE. DOI: 10.1109/EEEIC/ICPSEurope51590.2021.9584609

Nejad, S., Gladwin, D. T., & Stone, D. A. (2016). A systematic review of lumped-parameter equivalent circuit models for real-time estimation of lithium-ion battery states. *Journal of Power Sources*, *316*, 183–196. DOI: 10.1016/j.jpowsour.2016.03.042

Pahlevani, M., & Jain, P. K. (2020). Soft-switching power electronics technology for electric vehicles: A technology review. *IEEE Journal of Emerging and Selected Topics in Industrial Electronics*, *1*(1), 80–90. DOI: 10.1109/JESTIE.2020.2999590

Rahman, M. A., Anwar, S., & Izadian, A. (2016). Electrochemical model parameter identification of a lithium-ion battery using particle swarm optimization method. *Journal of Power Sources*, *307*, 86–97. DOI: 10.1016/j.jpowsour.2015.12.083

Ramakrishnan, T., Mohan Gift, M. D., Chitradevi, S., Jegan, R., Hency Jose, P. S., Nagaraja, H. N., & Hailegiorgis, S. M. (2022). Study of numerous resins used in polymer matrix composite materials. *Advances in Materials Science and Engineering*, *2022*(1), 1088926. DOI: 10.1155/2022/1088926

Sakhdari, B., & Azad, N. L. (2015). An optimal energy management system for battery electric vehicles. *IFAC-PapersOnLine*, *48*(15), 86–92. DOI: 10.1016/j.ifacol.2015.10.013

Selvakumar, P. (2023). Biomedical applications of polymers in biosensors, cancer vaccines and drug delivery systems. *Biopolimery i Kletka*, *39*(4), 245–256. DOI: 10.7124/bc.000AA0

Tan, N. M. L., Abe, T., & Akagi, H. (2011). Design and performance of a bidirectional isolated DC–DC converter for a battery energy storage system. *IEEE Transactions on Power Electronics*, *27*(3), 1237–1248. DOI: 10.1109/TPEL.2011.2108317

Tao, S., & Yang, X.. (2022). Adaptive fuzzy energy management for hybrid electric vehicles. *IEEE Transactions on Transportation Electrification*, *8*(2), 1860–1870.

Wang, W., Blunier, B., & Miraoui, A. (2011). *Equivalent circuit battery models for EV and HEV applications. In: 2011 IEEE Vehicle Power and Propulsion Conference*.

Yuan, M., Tang, X., & Song, H.. (2021). A novel battery state-of-health estimation method based on incremental capacity analysis using Gaussian process regression. *Applied Sciences (Basel, Switzerland)*, *11*(4), 1756.

Zhang, X., Li, Z., Luo, L., Fan, Y., & Du, Z. (2022). A review on thermal management of lithium-ion batteries for electric vehicles. *Energy*, *238*, 121652. DOI: 10.1016/j.energy.2021.121652

Chapter 2
Adoption Challenges of Hydrogen Energy

Suchismita Satapathy
https://orcid.org/0000-0002-4805-1793
KIIT University, India

ABSTRACT

Hydrogen can bring a revolution to vehicles and to the grid system. Hydrogen can be stored for long periods of time and converted back to electricity when needed, Electricity can be produced by using hydrogen in gas turbines or fuel cells. Green hydrogen can be generated by water electrolysis method and supercharged by renewable energy sources. Like fossil fuel energy, hydrogen energy, helps to generate electricity but without any negative impact on environment, Hence, all over the world, researchers are seeing hydrogen energy as the main sustainable source of energy. Still, there are some challenges creating barriers to adoption of hydrogen energy in residential, commercial and industrial sectors. In this chapter, the challenges of hydrogen energy adoption are found in WASPAS (weighted aggregated sum and product assessment method) is used to find the most important challenges such that the stakeholders can frame new policies and develop new methodologies with advanced technology to mitigate these challenges and familiarize hydrogen use as sustainable energy source.

INTRODUCTION

Day by day, energy demand is rising in every industry. For the improvement of productivity and for operating machines, industries are dependent on electricity energy. For generation of electricity, energy sector is dependent on fossil fuels. On the one hand, fossil fuels help to generate electricity and meet the demand of

DOI: 10.4018/979-8-3693-9316-1.ch002

industries. On the other hand, pollutes the environment and creates an unhealthy atmosphere for living beings. So, demand for clean energy is essential for saving the environment. So, more and more research is going on renewable energy systems like solar energy and wind energy, which are pollution-free and environmentally friendly. But storage of solar and wind energy and the cost of adoption are the major challenges of renewable energy. Hence, researchers are focusing on the adoption of hydrogen energy. This energy adoption not only reduces emissions of harmful pollutants like So_x and NO_x but also resolves the issues with air quality. Hydrogen fuel cells during burning do not produce any hazardous gas or never emit carbon dioxide gas, they only produce water during burning, which is non hazardous.Even production of hydrogen gas occurs in the electrolysis process with the help of renewable energies like wind and solar energy, so no carbon emission occurs during the production process. Hence, hydrogen energy is widely accepted all over the world as a non-polluting energy source. Many developing and developed countries are accepting hydrogen energy and started making power stations and generating hydrogen fuel. Still, many countries have lack of infrastructure, costs and technology to adopt hydrogen energy as the source of energy.

Hydrogen has a poor volumetric energy density, making it difficult to transport and store. Large storage volumes or high-pressure tanks are required for the safe and efficient development of methods for hydrogen distribution and storage. Worldwide, hydrogen energy is regarded as a clean energy source that will lessen environmental pollution and have no effect on climate change. Another term for hydrogen energy is "potential solution"—it refers to lowering greenhouse gas emissions. Because hydrogen fuel cell vehicles (FCVs) use hydrogen to produce electricity through a chemical process with oxygen and water vapor as a byproduct, they can be considered a sustainable energy alternative. throughout the process of producing hydrogen. Hydrogen has the potential to revolutionize both the grid system and automobiles.

Since hydrogen can be produced in a variety of ways and is widely available, it is becoming a popular green energy source with significant interest from the general public, businesses, and governments because of its potential as a fuel. Research on hydrogen and the development of novel materials for fuel cells, sensors, transmission, and storage are receiving significant funding. Given the difficulties in accident prevention, ensuring hydrogen facilities that are trustworthy and safe is essential. Resolving material compatibility problems in hydrogen systems is still a top priority. Outlooks for technology and long-term planning are guided by challenges, road-maps, and scenarios. Strategic visions take into account both socioeconomic and ecological aspects while coordinating policies and operations. The potential for transforming our energy environment in a sustainable way exists when the promise of hydrogen combines with material advancement. Developing shared future visions to anticipate the possible advantages of this new technology is important.

To further comprehend the relationship between materials and hydrogen dynamics, the interaction of materials and composites with hydrogen is explored, especially in relation to pipelines, infrastructure, and hydrogen transmission. The investigation also covers the phenomenon of embrittlement in storage and transit, along with a thorough analysis of the innovations and challenges inherent in hydrogen fuel cells.

Hydrogen finds widespread application in process industries as a viable substitute for fossil fuels. The transportation sector is preparing for the usage of fuel cells, which primarily use hydrogen as a fuel. Regardless of the industry or use, hydrogen handling and storage safety are essential. Compressed gas hydrogen is frequently stored in metal cylinders. But in these situations, hydrogen embrittlement presents a problem that must be resolved. Embrittlement is a safety risk because it causes the metal cylinders that hold the hydrogen gas to deteriorate. Because hydrogen embrittlement is more likely to occur in high-strength steels, the phenomenon's susceptibility rises with steel strength. Better storage materials and the modification of current ones are necessary for safe hydrogen storage systems.

The adoption of hydrogen energy has many challenges. In this chapter, an effort is made to study and find challenges/ barriers to adopting hydrogen energy, Then WASPAS method is used to prioritize the challenges such that necessary technologies can be adopted or policies can be framed to mitigate the adverse effect of the barriers.

BACKGROUND

As per the report of the World Health Organization (2018), it has been-reported that the air pollution and climate change occur due to the use of coal, oil and natural gas in industries. By using coal, oil and natural gas human health is badly affected. Endo et al. (2021) have reported that hydrogen energy has gained its popularity in recent years. Hydrogen energy is a clean energy source can act as a substitute of fossil fuels (Singh et al., 2020). Climate change and global warming are the changes brought by the use of fossil fuels (I.e., oil, coal, and natural gas etc.).The gravimetric density of fossil fuels about seven times lesser than that of hydrogen fuels. Hence, Yang et al.(2010) have stated that, in future, hydrogen energy will be the most important energy source, by improving/ solving the issues/ problems of generating hydrogen energy. Murray et al. (2009) have discussed the major problems of hydrogen energy are transportation and storage. Baykara (2018) has explained that hydrogen energy has no negative impact on environment or earth. Farrell et al. (2003) have stated that hydrogen will be the main source of energy. It will modify the transportation sector by replacing oil. Ewan and Allen (2005) have stated that about 40-45% of hydrogen production comes from gas-steam reform, about a third from oil reform, a fifth from coal reform and about 4% from water electrolysis.

It is predicted by European Commission. High Level Group for Hydrogen (International Renewable Energy Agency, 2019) that by 2040, 35% of cars in Europe will use hydrogen as their fuel, replacing oil (European Commission, 2003).

RESEARCH METHODOLOGY

A study is carried on by extensive literature review on adoption challenges of hydrogen energy as a source of energy and a set of 22 questionnaires is designed. The survey is conducted by circulating it among academicians, research scholars and energy experts. Around 100 questions circulated on yes and no. questions (yes=1 no=0) and 70% response received then statistical analysis followed by the WASPAS (Weight aggregated sum and product assessment method is used to prioritize the challenges of hydrogen energy.

Table 1. Question

Question	Yes=1	No=0
Production of hydrogen by electrolysis method is very costly		
For production, storage, distribution, and dispensing, good infrastructure and facilities are required		
For storing high pressure tank requirements, which are technically difficult		
For transportation of hydrogen is requires different kinds of pipes and transportation systems as highly flammable		
A new innovative durable fuel cell is essential		
Advance technology and more research is required for **hydrogen production**		
Large-scale energy demands require significant investment and scaling		
Green hydrogen production is very costly		
Infrastructure development is essential for establishing production facilities, distribution networks, refueling stations, and storage solutions, which requires high investment		
Fuel cell vehicles are in their initial stage and very costly		
Infrastructure hydrogen fuel needs to be developed		
The development of a hydrogen distribution network has challenges like cost, safety etc.		
Lack of refueling station for hydrogen fuel		
Safety standards to be developed for use of hydrogen fuels		
supporting policies for use of hydrogen fuel incentives is required		
Environmental regulations stop fossil fuel use but not support use of hydrogen fuel.		
Production of hydrogen energy from fossil fuels emits carbon, so contradicts the environmental policy		

continued on following page

Table 1. Continued

Question	Yes=1	No=0
For production of hydrogen, high use of water resources may create a scarcity of water		
Hydrogen generated from renewable energy also leaves a carbon footprint.		
Hydrogen must compete with well-established energy sources like fossil fuels		
Public awareness and adoption e of hydrogen as fuel is essential		
policies supporting hydrogen infrastructure development, research, and deployment are crucial		
Awareness of hydrogen energy benefits is required.		
Integrating hydrogen with existing energy systems is a challenge		
Public perception and regulatory concerns around safety need to be addressed.		

RESULTS AND DISCUSSION

The factor analysis is conducted after getting the responses of adoption challenges of hydrogen energy under seven dimensions (I.e., technical challenges, economic challenges, infrastructure challenges, environmental challenges, policy challenges, social acceptance challenges and integration challenges).Only 21 items selected had Cronbach's alpha value more than 0.5, during statistical analysis done in Minitab 17.The items below 0.5 values are discarded.Table.2 shows the result of factor analysis. Then the WASPAS Method is used to prioritize the challenges.

Table 2. Factor analysis of challenges

Dimensions	Question	Factor
Technical challenges	Production of hydrogen by electrolysis method is very costly	0.532
	For production, storage, distribution, and dispensing, good infrastructure and facilities are required	0.527
	For storing high pressure tank requirements, which are technically difficult	0.514
	For transportation of hydrogen is requires different kinds of pipes and transportation systems as highly flammable	0.509
	A new innovative durable fuel cell is essential	0.538
Economic challenge	Green hydrogen production is very costly	0.542
	Infrastructure development is essential for establishing production facilities, distribution networks, refueling stations, and storage solutions, which requires high investment	0.503
	Fuel cell vehicles are in their initial stage and very costly	0.519

continued on following page

Table 2. Continued

Dimensions	Question	Factor
Infrastructure challenge	Infrastructure for hydrogen fuel needs to be developed	0.625
	Development of hydrogen distribution network has challenges like cost, safety etc.	0.611
	Lack of refueling station for hydrogen fuel	0.567
Policy challenges	Safety standards to be developed for use of hydrogen fuels	0.549
	supporting policies for use of hydrogen fuel incentives is required	0.533
	Environmental regulations stop fossil fuel use but not support use of hydrogen fuel.	0.510
Environmental challenges	Production of hydrogen energy from fossil fuels emits carbon, so contradicts the environmental policy	0.713
	For production of hydrogen high use of water resources may create scarcity of water	0.584
	Hydrogen generated from renewable energy also leaves carbon footprint.	0.557
Sociatal acceptance challenges	Hydrogen must compete with well-established energy sources like fossil fuels	0.562
	policies supporting hydrogen infrastructure development, research, and deployment are crucial	0.527
Integration challenges	Integration and acceptance of hydrogen energy	0.634
	Integrating hydrogen with existing energy systems is a challenge	0.572

Then the WASPAS method is implemented for ranking purposes. The WASPAS method is a combination of two methods: weighted sum and weighted production method. Chakraborty and Zavadskas (2014) have written that the weighted aggregate sum product assessment method is effective MCDM (multi criteria decision making method, as it is very simple but accurate in finding the result.

Step-1 **A** Decision matrix is constructed by considering criteria and alternatives. In Table 3 criteria are taken as challenges of adoption of hydrogen energy and in alternative application of hydrogen energy in the residential sector, commercial sector and industrial sector are taken

Step-2 *Normalization of decision matrix is done* in table.4, all the criteria are considered as beneficial criteria so below equation is used

$$\overline{X}_{ij} = \frac{Xij}{Max\ i\ Xij} \left(\text{for beneficial criteria}\right) \tag{1}$$

Step 3: Relative performances is calculated

All the weights are taken the same value 0.33 as per expert suggestion. Then in the table.5 Q1 and Q2 are calculated by using the below equations

$$\mathbf{Q1}\left(1\right) = \sum\nolimits_{j=1}^{n} XijWj \tag{2}$$

$$\text{Q2(1)} = \prod_{j=1}^{n} \left(\overline{X}_{ij}\right) Wij \tag{3}$$

Step 4: Final performance found by using the below formula

$$Q_i = 0.5\ P_i^{(1)} + 0.5\ P_i^{(2)} \quad (4)$$

And based on Q value, ranking is done

Table 3. Decision matrix

	Sectors		
Challenges	**Residential**	**Commercial**	**Industrial**
Technical challenges	0.875	0.987	0.833
Economic challenges	0.812	0.787	0.666
Infrastructure challenge	0.375	0.937	0.833
Environmental challenges	0.575	0.5	0.555
Policy challenges	0.187	0.183	0.889
Sociatal acceptance challenges	0.375	0.474	0.778
Integration challenges	0.473	0.525	0.944

Table 4. Normalized decision matrix

	Sectors		
Challenges	**Residential**	**Commercial**	**Industrial**
Technical challenges	0.886524823	1	0.843971631
Economic challenges	1	0.969211823	0.820197044
Infrastructure challenge	0.400213447	1	0.889007471
Environmental challenges	1	0.869565217	0.965217391
Policy challenges	0.210348706	0.205849269	1
Societal acceptance challenges	0.482005141	0.609254499	1
Integration challenges	0.501059322	0.556144068	1

Table 5. Final rank

Challenges	Q1	Q2	Q	Rank
Technical challenges	0.90106383	0.908711841	0.904887835	3rd
Economic challenges	0.920504926	0.927067119	0.923786023	2nd
Infrastructure challenge	0.755442903	0.711041647	0.733242275	4th
Environmental challenges	0.935478261	0.943834848	0.939656555	Ist
Policy challenges	0.467345332	0.354847237	0.411096284	7th
Sociatal acceptance challenges	0.690115681	0.667406815	0.678761248	6th
Integration challenges	0.678877119	0.655956788	0.667416953	5th

After ranking by WASPAS method, it is seen that environmental challenges is the first ranked challenge, followed by economic challenge which is ranked 2nd and technical challenge ranked third. So, necessary frame work must be developed to mitigate the environmental challenge on priority basis, such that hydrogen energy can be widely accepted. To find linear relationship between these challenges Interpretative structural modelling method is implemented.

ISM is a method is used to find inter relationship between variables finding complex relation between each variables.

If these variables/challenges are considered as I and J variables(row and column variables as per table.6) are both impacts on each other X is used, if I has better impact on j, V is written .If reverse occurs means j impacts on I, A written, if no relation found then O is written in the below table. Then in table 7, binary nos (1 and 0) are used.

In table 7, I to J entry 1 then J to I must be 0.for Jto I entry 1 then I to J must be 0 for A variable.For X, both must be one and For O both must be 0.In table. 8 dependency and driver power of variables calculated by summing rows and column respectively and * mark shows transitivity relation between variables.Table. 9 and Table 10 shows final level of matrix selection.

Table 6. SSIM (self structural matrix)

		Integration challenges	Societal acceptance challenges	Policy challenges	Environmental challenges	Infrastructure challenge	Economic challenge	Technical challenge
	Challenges	7	6	5	4	3	2	1
1	Technical challenges	O	O	A	V	O	V	
2	Economic challenges	V	V	V	V	A		

continued on following page

Table 6. Continued

		Integration challenges	**Societal acceptance challenges**	**Policy challenges**	**Environmental challenges**	**Infrastructure challenge**	**Economic challenge**	**Technical challenge**
3	Infrastructure challenge	O	O	V	V			
4	Environmental challenges	O	V	A				
5	Policy challenges	O	A					
6	Societal acceptance challenges	V						
7	Integration challenges							

Table 7. Initial reachability matrix

	1	2	3	4	5	6	7
1	1	1	0	1	1	1	1
2	1	1	0	1	1	1	1
3	1	1	1	1	1	1	1
4	1	1	0	1	1	1	1
5	0	0	0	1	1	1	0
6	0	1	0	0	1	1	0
7	1	0	0	1	1	1	0

Table 8. Final reachability matrix with transitivity relationship

	1	**2**	**3**	**4**	**5**	**6**	**7**	**Driver**
1	1	1	0	1	1*	1*	1	6
2	1	1	0	1*	1*	1	1	6
3	1	1	1	1	1	1	1	7
4	1	1	0	1	1	1	1	6
5	0	0	0	1	1	1	0	3
6	0	1	0	0	1	1	0	3
7	1	0	0	1	1*	1	0	4
Dependency	5	5	1	6	7	7	4	

Table 9. Iteration 1

Variables	Reachability Set	Antecedent Set	Intersection Set	Level
1	1,2,4,5,6,7	1,2,3,4,7	1,2,4,7	
2	1,2,4,5,6,7	1,2,3,4,6	1,2,4,6	
3	1,2,3,4,5,6,7	3	3	
4	1,2,4,5,6,7	1,2,3,4,5,7	1,2,4,5,7	
5	4,5,6	4,5,6	4,5,6	I
6	2,5,6	1,2,3,4,5,6	2,5,6	I
7	1,4,5,6	1,2,3,4	1,4	

Table 10. Final level

Variables	Level
5,6	I
1, 4,7	II
2	III
3	IV

Figure 1. ISM model

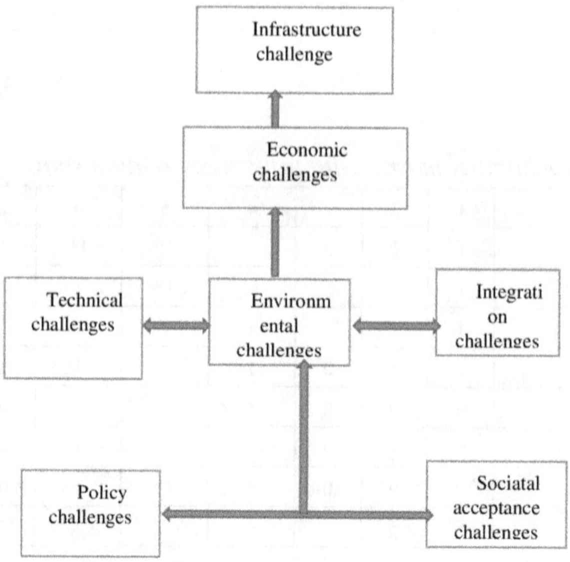

CONCLUSIONS

The adoption of hydrogen energy as a future fuel, needs a lot of investment, infrastructure and advanced technology, which will help to design storage containers, proper transportation systems, fueling systems, fuel cells etc. With all these factors, like technical and economic challenges, most important is sustainable hydrogen energy. It means, the necessary technology must be implemented and innovated which will help reduce the carbon footprint during hydrogen generation, such that no environmental pollution occurs. By mitigating the challenges of hydrogen energy adoption, it will automatically bring awareness among people and it will be widely used in every sector and in every industry, and it can also replace fossil fuels and their adverse impact on the earth. By adopting hydrogen energy, more job opportunities can be created in every field by creating a sustainable future. It will also be able to resolve issues like climate change and improve air quality while enhancing energy security, which will automatically foster economic growth.

REFERENCES

Baykara, S. Z. (2018). Hydrogen: A brief overview on its sources, production and environmental impact. *International Journal of Hydrogen Energy*, *43*(23), 10605–10614. DOI: 10.1016/j.ijhydene.2018.02.022

Chakraborty, S., & Zavadskas, E. K. (2014). Applications of WASPAS method in manufacturing decision making. *Informatica (Vilnius)*, *25*(1), 1–20. DOI: 10.15388/Informatica.2014.01

Endo, N., Goshome, K., Tetsuhiko, M., Segawa, Y., Shimoda, E., & Nozu, T. (2021). Thermal management and power saving operations for improved energy efficiency within a renewable hydrogen energy system utilizing metal hydride hydrogen storage. *International Journal of Hydrogen Energy*, *46*(1), 262–271. DOI: 10.1016/j.ijhydene.2020.10.002

European Commission. (2003). *Hydrogen energy and fuel cells: A vision of our future: Final report of the High Level Group* (Vol. 20719). European Commission, Directorate-General for Research.

Ewan, B. C. R., & Allen, R. W. K. (2005). A figure of merit assessment of the routes to hydrogen. *International Journal of Hydrogen Energy*, *30*(8), 809–819. DOI: 10.1016/j.ijhydene.2005.02.003

Farrell, A. E., Keith, D. W., & Corbett, J. J. (2003). A strategy for introducing hydrogen into transportation. *Energy Policy*, *31*(13), 1357–1367. DOI: 10.1016/S0301-4215(02)00195-7

International Renewable Energy Agency. (2019). *Global energy transformation: A roadmap to 2050 (2019 edition)*. https://www.irena.org/publications/2019/Apr/Global-energy-transformation-A-roadmap-to-2050-2019Edition

Murray, L. J., Dincă, M., & Long, J. R. (2009). Hydrogen storage in metal–organic frameworks. *Chemical Society Reviews*, *38*(5), 1294–1314. DOI: 10.1039/b802256a PMID: 19384439

Singh, R., Altaee, A., & Gautam, S. (2020). Nanomaterials in the advancement of hydrogen energy storage. *Heliyon*, *6*(7), e04487. DOI: 10.1016/j.heliyon.2020.e04487 PMID: 32743097

World Health Organization. (2018). COP24 special report: health and climate change.https://www.energy.gov/eere/articles/hydrogen-clean-flexible-energy-carrier [Access 05/11/2020].

Yang, J., Sudik, A., Wolverton, C., & Siege, D. J. (2010). High capacity hydrogen storage materials: Attributes for automotive applications and techniques for materials discovery. *Chemical Society Reviews*, *39*(2), 656–675. DOI: 10.1039/B802882F PMID: 20111786

Chapter 3
Advancements in Battery and Energy Storage Materials:
Paving the Way for Sustainable Energy Solutions

Surjit Singha
https://orcid.org/0000-0002-5730-8677
Kristu Jayanti College, India

Ranjit Singha
https://orcid.org/0000-0002-3541-8752
Christ University, India

Shruti Jose
https://orcid.org/0000-0003-3139-2505
Christ University, India

Joy Josephine
Good Shepherd English School, Karaikal, India

R. Sreethi Rebeka
https://orcid.org/0009-0008-9011-7048
Mount Carmel College, India

Babitha Nadampalli
Christ University, India

Alphonsa Diana Haokip
https://orcid.org/0000-0003-2578-0114
Don Bosco College, Maram, India

ABSTRACT

Advanced materials are key to battery and energy storage technology improvements, which are a cornerstone of sustainable energy for the future and are the topic of this chapter. It explores advances in solid-state electrolytes, lithium-sulfuric and sodium-ion batteries, nanomaterials and organic compounds, which all have the potential to enhance energy density, cycle life and environmental sustainability. These materials hold great promise, as they may overcome current limitations in

DOI: 10.4018/979-8-3693-9316-1.ch003

battery performance, safety, and cost, the authors say. The chapter also explores the economic and environmental implications of these innovations, spotlighting their role in the global transition to renewable energy. Given ongoing research efforts and favourable policies, next-generation energy storage systems will play an essential role in advancing clean energy technologies in areas from electric vehicles to electric grid storage.

INTRODUCTION

The global energy landscape is changing rapidly as the world seeks to combat climate change and transition to clean sources of energy. The development of energy storage systems is a key part of this transition, as well as exceptionally efficient and safe energy systems that are sustainable. As a result, these technologies are pivotal in balancing supply and demand, promoting grid stability, and improving the resilience of renewables. Batteries, among them, have been instrumental in the incorporation of intermittent sources of energy (solar and wind) and fuelling the burgeoning market for electric vehicles (EV) and portable electronics. However, in addition to batteries, other energy storage technologies, such as supercapacitors, flywheels, and thermal storage systems, offer complementary capabilities that meet different operating requirements.

Among these, lithium-ion (Li-ion), lithium-sulfuric (Li-S), and sodium-ion (Na-ion) are the most prominent in energy storage due to their high energy density and broad applicability. Li-ion batteries dominate the market due to their high energy efficiency and long cycle life (Goodenough & Park, 2013; Manthiram et al., 2014). Worries over resource scarcity, cost and environmental impact, however, contribute to the search for alternatives. Li-S batteries do provide higher theoretical energy densities, requiring a more abundant element, such as sulfur, but suffer from rapid capacity fading. Na-ion batteries offer an attractive low-cost alternative but are considered to have an inferior energy density to Li-ion, compounded by the abundance of raw materials.

On the other hand, supercapacitors have high power density and rapid charge-discharge cycles, which are suitable for applications that require a rapid supply of energy. Flywheel systems are highly efficient and have a long working life that helps in the efficient stabilization of the grid. Thermal energy storage systems, such as phase-change materials and molten salt systems, are beneficial for industrial applications and large-scale integration with renewable energy sources. All these technologies are important for building a sustainable energy infrastructure, and the complementary deployment of these technologies is necessary.

A Timeline of Some Significant Innovations in Energy Storage Lead-acid batteries, the pioneering electrochemical energy storage form, were developed in the mid-19th century. Nickel-cadmium (Ni-Cd) batteries showed up in 1899 as rechargeable options. There were nickel-metal hydride (Ni-MH) batteries in 1967, which had greater energy density than the first nickel-cadmium batteries. Li-ion batteries were commercialized in 1991, which changed the consumer electronics and mobile power demand landscape. Li-S and Na-ion Batteries: Li-ion technology is becoming limited, and the focus transferred in the 2000s toward the development of Li-S and Na-ion batteries. In the last few decades, the development of solid-state electrolytes, nanostructured electrodes, and organic compounds has been researched for improved performance, safety, and sustainability.

Future advancements, such as lithium-sulphur batteries and solid-state electrolytes, will deliver additional enhancements in energy density, environmental sustainability, and cost-effectiveness. New battery materials will also be key to escaping the limitations of traditional battery technology. Solid-state electrolytes also improve safety and allow lithium-metal anodes to be used, which boosts energy density. Electrode performance is enhanced through the use of nanostructured materials like graphene and carbon nanotubes, which increase charge availability. Organic synthesis has become a resourceful alternative, reducing the requirements of rare metals. This fosters a circular economy and mitigates damage to the environment.

The latest advancements in battery materials and their importance in the future of energy storage are the focus of this chapter. It discusses how such materials can increase safety, decrease environmental impact, and improve economic viability. It also considers the broader implications of these innovations, whether they be catalyzing renewable energy systems with scale or sustainable development at large. Through examination of current academic research and future trends, this chapter maps out the way ahead for energy storage, revealing the synergy between modern technology, ecological health, and financial development.

BACKGROUND

The global energy landscape is evolving along with the world's increasing reliance on renewable energy sources like solar, wind, and hydropower. These renewables play a vital role in mitigating climate change and lowering greenhouse gas emissions. However, their inconsistency and variability pose significant challenges to maintaining an uninterrupted and dependable energy supply. To tackle this issue, new energy storage technologies must be developed to store excess energy when production is high and then release energy when production is low and demand is high. Of the technologies, batteries have become a key energy solution thanks to their scalability, flexibility, and long-time energy storage capabilities (Wei et al., 2023).

Although they have gained widespread acceptance, conventional battery technologies, particularly lithium-ion batteries (LIBs), are limited in several aspects. LIBs are limited by weak cycle life, safety, cost of production, and environmental hazards but provide high energy density as well as efficiency. The flammability of liquid electrolytes in LIBs, for example, presents considerable safety hazards in practical scenarios, particularly under extreme operating conditions. Furthermore, dependence on rare and costly resources like cobalt and nickel drives up production expenses and raises ethical and environmental considerations concerning mining methods. This limitation has highlighted the important need to develop next-generation battery technologies that are better suited to replace conventional systems (Manthiram, 2020).

Forthcoming materials innovation is the most essential step along the pathway to solving the high-performing battery technologies. New components from innovations in materials science with improved materials fuel the development of energy storage methods. One notable breakthrough in this area is the development of solid-state electrolytes, which are a replacement for the liquid electrolyte traditionally used in batteries. These solid-state electrolytes are much more thermally stable than their liquid counterparts, making for significantly safer batteries that are far less likely to catch fire. These materials also allow for lithium metal anodes, which have the potential to boost significantly energy density. Nanomaterials similarly transformed electrode design, with high surface area, better conductivity, and better mechanical properties (Xu et al., 2013; Zhao et al., 2019).

One promising innovation in batteries is the lithium-sulphur (Li-S) battery, which uses sulphur as a cathode material. Sulphur is also abundant, inexpensive, and environmentally benign, making it an attractive alternative when compared to traditional cathode counterparts. Li-S batteries can yield energy densities several times greater than those of LIBs. Polysulfide shuttle effect and poor cycle stability are still research problems. Sodium-ion (Na-ion) batteries are also being explored as an alternative for grid-scale applications because sodium is much more abundant and has a lower cost than lithium. Recent progress in the design of electrode and electrolyte materials is engagingly increasing the energy density and cycle life of Na-ion batteries toward commercialization (Raza et al., 2023; Shao et al., 2023).

Advanced materials are also in line with global sustainability goals. We can also reduce the environmental footprint of next-generation batteries by minimizing their dependence on scarce and toxic materials. In addition, enhanced battery efficiency and durability lead to cost-effectiveness throughout the product lifecycle, underlining the economic feasibility of these technologies for a wide range of applications, including EVs and large-scale energy storage systems. The use of organic molecules and recyclable materials contributes even further to their sustainability, which aligns with the world shifting to a circular economy (Wang et al., 2023; Zhao et al., 2019).

All these advancements in battery materials are improving energy storage performance, which in turn has a direct impact on accelerating renewables adoption . A lot more needs to be done in this space. More efficient energy storage systems allow for higher penetration of renewables in the energy grid, reducing fluctuations in renewable energy generation and stabilizing the grid. Empowered by technological breakthroughs in battery innovation, the transportation sector is also transitioning towards electrification to decrease greenhouse gas emissions and facilitate a shift to a low-carbon economy.

Overall, designing novel materials is a game changer for energy storage technologies, overcoming the drawbacks of traditional chemistries and facilitating the worldwide shift towards sustainable energy. Such materials are setting the path for future-era batteries capable of meeting the challenges of a shifting energy landscape through improved energy density, efficiency, safety, and environmental sustainability. The ongoing research into new materials and their incorporation into real-world use cases will play a crucial role in unlocking the optimal use of renewable energy and driving towards a sustainable energy future.

THEORETICAL FRAMEWORK

With our growing dependence on renewable energy sources such as solar, wind, and hydropower, the energy landscape around the world is changing. These so-called renewables, in turn, are key to the fight against climate change and the reduction of greenhouse gases. However, their intermittency and variability present significant challenges in ensuring a continuous and reliable energy supply. To solve this problem, new energy storage technologies are needed that store excess energy when production is high and release energy when production is low and demand is high. Among these technologies, batteries are becoming an essential energy solution due to their scalability, flexibility, and long-time energy storage capacity (Wei et al., 2023).

Though widely welcomed, traditional batteries, especially Li-ion batteries (LIB), are still greatly limited on many fronts. While LIBs offer high energy density and efficiency [11,12], they are hampered by poor cycle life, safety, manufacturing cost and environmental threats. For instance, the flammability of liquid electrolytes in LIBs poses significant safety risks in practical environments, especially under extreme working situations. Also, depending on rare and expensive materials like cobalt and nickel increases production costs, along with ethical and environmental concerns about how these are extracted. This limitation has impetus for the critical need to develop next-generation battery technologies to serve as replacements for conventional systems (Manthiram, 2020).

The most promising solution to the pathway of high-performing battery technologies is the materials innovation that follows. Novel components drive the development of energy storage approaches from advances in materials science with superior materials. A key innovation in this space is the solid-state electrolyte, which substitutes for the liquid electrolyte used in conventional batteries. These solid-state electrolytes are far more thermally stable than liquid ones, resulting in significantly safer batteries that carry a greatly reduced risk of catching fire. These materials also enable lithium metal anodes, which could significantly increase energy density. Nanomaterials also revolutionized electrode design due to high surface area, improved conductivity, and mechanical properties (Xu et al., 2013; Zhao et al., 2019).

One potential battery innovation that stands out is the lithium-sulphur (Li-S) battery, which uses sulphur as the cathode material. Sulphur is also abundant, inexpensive, and environmentally benign, which makes it attractive to conventional cathode counterparts. In principle, Li-S batteries can deliver an energy density several times higher than that of LIBs. The polysulfide shuttle effect and unsatisfactory cycle stability have to be solved systematically. Due to the excess availability of sodium (inexhaustible) and lower cost, sodium-ion (Na-ion) batteries are also being considered as an alternative for grid-scale applications. The development of electrode and electrolyte materials has made exciting strides towards the commercialization of Na-ion batteries, significantly improving their energy density and cycle life (Raza et al., 2023; Shao et al., 2023).

In addition, advanced materials align with the global goals of sustainability. We can also shrink the environmental footprint of next-generation batteries by minimizing their reliance on rare and toxic materials. Furthermore, improved battery performance and longevity deliver economic benefits throughout the product lifecycle, emphasizing the economic viability of the technologies across various applications, such as EVs and large-scale energy storage systems. Their sustainability is further supported by the use of organic molecules and recyclable materials, which follows the tendency of the world that is passing to a circular economy (Wang et al., 2023; Zhao et al., 2019).

All these breakthroughs in battery materials translate into energy storage performance and, consequently, directly drive renewables adoption, although a lot remains to be done in this area. Increased penetration of renewables in the energy grid, in addition to reduced fluctuations of renewable energy generation through efficient energy storage systems, stabilizes the grid. Driven by technological advances in battery technology, the transportation sector is also moving towards electrification to significantly reduce greenhouse gas emissions and help transition to a low-carbon economy.

Generally, the design of new materials could enforce a revolution in energy storage technologies, and the limitation of traditional chemistries could be avoided, which is necessary for the worldwide shift to sustainable power. These materials are paving the way for the batteries of a new era to overcome the challenges of an evolving energy landscape with gainful energy density, efficiency, safety, and environmental sustainability. The constantly expanding field of new materials development and their inevitable adoption into real use cases will be pivotal in enabling the best possible utilization of renewable energy and the journey towards sustainable energy supply.

CONCEPTUAL FRAMEWORK

The premise behind this chapter, as the broader conceptual framework will illustrate, is that developments in battery materials are crucial to overcoming fundamental challenges in energy storage and hastening the transition to renewable energy systems on a global scale. This framework links the characteristics of high-end materials to key performance indicators of batteries, such as energy density, cycle life, safety, environment-friendliness, and cost-efficiency. It underscores the critical importance of materials innovation in identifying next-generation battery technologies and outlines the relationship with the more integrated societal and economic transitions that accompany these transformations.

Main Factors That Dictate Performance of Battery

Energy Density

Energy density—the amount of energy a battery can store in relation to its size or weight—is a key performance metric. Increased energy density means that batteries can store more energy without increasing their physical dimensions, which is especially useful for use cases like electric vehicles (EVs) and portable electronics. The use of advanced materials like solid-state electrolytes and lithium metal anodes can significantly enhance energy density by minimizing the inefficiencies seen with traditional materials. Similarly, lithium-sulphur (Li-S) batteries have also been recognized as potential substitutes for traditional lithium-ion batteries due to their high theoretical energy density.

Cycle Life

Cycle life is essentially how many charge-discharge cycles a battery can go through before it is no longer capable of holding much of a charge. Improvements in materials, including nanostructured electrodes and new electrolyte compositions, are critical to extending cycle life. Nanomaterials, for instance, strengthen electrodes' structural integrity and facilitate charge conductivity, decreasing material degradation in repeated cycles. Advanced cathode and electrolyte materials of Li-S and Na-ion batteries alleviate the polysulfide shuttle effect and dendrite formation, which effectively enhances their cycle stability.

Safety

Safety is the most important aspect of battery technology, especially for EV and grid storage applications, where failure could result in catastrophic consequences. Solid-state electrolytes have a revolutionary design that removes all flammability risks from liquid electrolytes. Because of this, these materials are also inherently more thermally stable and less prone to thermal runaway, even if the polymer structure does not have inherent low flammability. Moreover, the use of organic compounds with enhanced electrochemical stability also increases the safety of the battery.

Environmental Sustainability

The sustainability of batteries is also a rising issue as the substantial impact of material extraction, manufacturing and disposal on the environment forces it. Advanced, organic and recyclable materials can replace scarce and toxic elements, including cobalt and nickel. Thus, Li-S batteries use sulphur, which is plentiful and eco-friendly. Likewise, Na-ion batteries utilize sodium, which is abundant on earth, reducing the environmental and economic costs of the extraction of lithium.

Cost-Efficiency

Battery production price matters for its deployment in many industries. Highly energy-dense, long-cycle life advanced materials significantly reduce total ownership costs. Next-generation batteries that use inexpensive and widely available materials — such as sodium and sulfur — are even cheaper to produce, making them more accessible to large-scale uses.

ADVANCED MATERIALS FOR ENERGY STORAGE SYSTEMS

This framework highlights several classes of advanced materials as critical enablers of battery innovation:

Solid-State Electrolytes: These materials enhance safety, energy density, and thermal stability, allowing for the use of lithium metal anodes and minimizing the risk of fires.

Nanomaterial: Nanomaterials have modified the electrical conductivity and improved the structural stability of electrodes, thus enhancing other performance indicators like energy density, cycle life, and charging speed.

Nanotechnology in Energy Storage: With their high surface area-to-volume ratio, nanomaterials provide enormous opportunities to increase the performance of energy storage devices.

Sodium-Ion Batteries: Ground sodium offers a low-cost, sustainable alternative for grid-scale energy storage.

Organic Compounds: Organic usage has less impact on the environment and does not impede recycling, thus benefiting the circular economy.

Connecting Innovation to Transition in Renewable Energy

The inclusion of such advanced materials in battery systems has significant implications for the worldwide transition to green energy. In order to overcome the inherent intermittency of renewable energy sources such as solar and wind, energy storage must be efficient, reliable, and sustainable. For example, next-generation batteries are paving the way for higher penetration of renewables into the energy grid by providing more extended storage periods and lower costs. Moreover, with battery efficiency and safety continuing to grow, EVs are adopted in more market segments, decreasing the dependence on fossil fuels and adding to the decarbonization of ecosystems.

This example of the conceptual framework highlights the linkages between development in material science, the performance measures of batteries and their significance for sustainability and energy transition. If conventional systems were to be complemented by advanced materials, it could be a game-changer for energy storage, heralding a transition to a new energy era of clean energy.

RESEARCH PROBLEM

While advanced materials hold significant promise for improving battery and energy storage technologies, challenges remain in their practical application, scalability, and integration into existing systems. There needs to be more understanding of how these materials can be effectively incorporated into commercial battery designs, particularly regarding long-term performance, cost reduction, and environmental impact. Furthermore, these innovations' economic and policy implications require in-depth exploration to facilitate their widespread adoption.

RESEARCH QUESTION

How can advanced materials such as solid-state electrolytes, lithium-sulfur, sodium-ion batteries, nanomaterials, and organic compounds be integrated into next-generation energy storage systems to improve performance, sustainability, and economic viability?

RESEARCH OBJECTIVES

1. To investigate the role of advanced materials, such as solid-state electrolytes, lithium-sulfur, sodium-ion batteries, nanomaterials, and organic compounds, in enhancing battery performance.
2. To assess the environmental and economic implications of incorporating these advanced materials into energy storage technologies.
3. To explore the challenges and opportunities associated with integrating next-generation materials into existing battery technologies.
4. To identify strategies for overcoming the limitations of current energy storage systems through material innovation.
5. To forecast the potential impact of these innovations on global energy storage solutions and the transition to renewable energy.

THE ROLE OF ADVANCED MATERIALS IN BATTERY TECHNOLOGY

Material innovation is driving materials for battery technology, and it is the key to delivering better performance, safety, and sustainability in energy storage systems. The materials chosen for batteries dictate key metrics like energy density, power output, cycle life, and thermal stability. The last couple of decades have witnessed a significant shift towards the electrification of various sectors, driven by an increasing demand for high-performance batteries to meet the growing energy needs (Raza et al., 2023; Machín et al., 2024).

In all these accomplishments, advanced materials break the barriers in energy storage, such as low ionic conduction, thermal instability, and shortage of resources. With the aid of nanostructured electrodes, they also enhance safety and durability by replacing flammable liquid electrolytes with solid-state electrolytes and improving ion transport and battery lifespan. This evolution supports the thematic scalability of renewable energy, where material innovations are the keys to achieving a sustainable future.

However, lithium-ion batteries (LIBs) are still limited by low energy density, safety concerns, and dependence on scarce materials such as lithium and cobalt. Advice Additionally, the regular cycles of charge and discharge deteriorate the materials, limiting their efficiency and lifespan. Moreover, as are typical battery materials, from extraction and processing to disposal and recycling. We need breakthroughs in materials science to make safer, more efficient, eco-friendly batteries.

Novel materials present game-changing solutions. Solid-state electrolytes contribute to stability, and the use of high-capacity lithium-metal anodes and nanostructured materials (e.g., graphene) improves the performance of electrodes. Organic compounds provide sustainable alternatives that reduce dependence on rare metals. One of these innovations, lithium-sulfur and sodium-ion batteries provide lower-cost, high-energy-density options while being more sustainable. In addition, recycled material contributes to a circular economy and reduces environmental impact. By incorporating these advanced technologies, next-generation batteries will lead the shift towards a low-carbon, energy-secure future in line with global sustainability objectives.

SOLID-STATE ELECTROLYTES: THE FUTURE OF SAFE AND EFFICIENT BATTERIES

Solid-state batteries (SSBs) are a game-changing technology for energy storage as they overcome key shortcomings of conventional lithium-ion batteries by substituting liquid electrolytes for solid-state ones. However, these electrolytes are flammable with the potential to cause thermal runaway or reduced electrical input at high temperatures (e.g. between 60–150°C). These electrolytes are composed of non-combustible materials, such as garnet, perovskite, or NASICON-type compounds, making them substantially non-flammable and thermally stable, which helps mitigate thermal runaway and as well as fire risks. By enabling high-capacity lithium-metal anodes, solid-state batteries provide higher energy density, and their solid electrolytes also suppress dendrite growth to promote long cycle life and enhanced safety. Moreover, they can operate efficiently at high temperatures, making them excellent candidates for use in electric vehicles, aerospace applications and portable electronics (Wang et al., 2023). Ceramic electrolytes have the pros of high thermal and chemical stability but also the cons of brittleness and complicated fabrication methods. A polymer electrolyte made of flexible materials (for example, polyethene oxide (PEO)) shows many advantages, including simple fabrication and compatibility with lithium-metal anodes. However, its ionic conductivity and thermal stability are not sufficiently high. Composite electrolytes combine the benefits of ceramic electrolytes and polymer electrolytes, achieving a trade-off between conductivity, flexibility and mechanical strength, thus emerging as a potentially powerful material for advanced batteries. Objectives: Solid-state batteries are a promising technology due to their potential safety advantages, but they face challenges, including high production costs, interface instability between the electrolytes and electrodes and upscaling from laboratory proof-of-concept to commercial-scale production. Commercialization appears more amenable to innovation as increasingly novel material design strategies combined with new manufacturing techniques and synergy between academia, industry, and government pave the way. As the technology matures, economies of scale and developments in manufacturing are projected to bring down costs, paving the way for solid-state batteries to be a safer, more efficient, and more sustainable choice than conventional systems. These developments hold the potential to transform the energy system and the global energy sector, enabling a more rapid transition to a low-carbon future.

NANOMATERIALS: ENHANCING BATTERY PERFORMANCE

Nanotechnology is revolutionizing battery development, providing answers to long-standing issues in energy storage systems. Nanotechnology takes advantage of phenomena that arise when materials are manipulated at the nanoscale, potentially leading to significant improvements in battery performance (Asamoah et al., 2024; Kumar et al., 2024).

Nanomaterials Applications in Battery Manufacturing Nanomaterials enhance battery performance by increasing surface area, promoting ion and electron transfer, and improving structural stability. These characteristics allow for quicker charge times, greater energy densities, and longer cycle life. Nanotechnology also enables the design of novel electrode and electrolyte materials, which can lead to new battery chemistries, such as lithium-sulphur and sodium-ion systems (Hussein, 2023; Lu et al., 2016).

Nanomaterials are being incorporated into batteries, addressing the need for energy storage technologies with broad applications, from electric vehicles to renewable energy systems to portable electronics.

Using nanomaterials is one of the innovations in battery technologies, which is promising better battery performance. Electrodes, especially in carbon-based nanomaterials like graphene (due to outstanding conductivity, surface area, and mechanical strength), are considered to improve energy density and charge/discharge rates while also improving conductivity and structural stability with carbon nanotubes (CNTs). Metal oxide nanoparticles, including TiO_2 and MnO_2, have been reported to provide great improvements to the capacity, cycling stability, and efficiency of electrodes. Nanostructured silicon: Given its high theoretical capacity, silicon is used in anodes, and Nanostructuring helps overcome volume expansion and degradation. Similarly, nanostructured sulphur in lithium-sulphur batteries tackles issues such as the dissolution of polysulfides and the enhancement of cathode power and capacity retention (Grotto et al.—they are studying quantum dots, which are semiconductor nanocrystals that can optimize charge transport and energy storage. Moreover, the use of nanostructured polymers, and conducting polymer nanomaterials in particular, turns out the potential for lightweight, flexible batteries with enhanced ionic conductivity coupled with superior stability that contrive promising next-generation electrochemical energy storage devices. Effect on Energy Density and Cycle Life (Asamoah & others, 2024; Kumar & others, 2024)

Nanomaterials play an integral role in two major performance indicators: energy density and cycle life. These materials are tuned at a nanoscale since they can increase the energy density of batteries through enhanced electrochemical activity of the electrodes. For example, the development of silicon anodes at the nanoscale and cathodes based on graphene can accommodate a higher volume of ions, allowing for significant increases in storage capacity.

Degradation is tackled with the help of these nanomaterials; they offer structural stability and limit stress on charge-discharge cycles. Some, such as carbon nanotubes and metal oxide nanoparticles, are introduced into the electrodes to improve their mechanical integrity, avoid cracking, and extend battery longevity.

Moreover, the polysulfide migration in lithium-sulphur batteries is inhibited by nanostructured sulphur cathodes, allowing for prolonged cycle stability. With ongoing investigations, developing low-cost, scalable processes for nanomaterial syntheses will also further expedite their implementation in battery systems. The utilization of nanomaterials for improved energy density, charge rates, and longevity ensures their pivotal role in the development of next-generation batteries, providing much-needed energy storage solutions for a sustainable future.

ORGANIC COMPOUNDS IN ENERGY STORAGE

Organic materials and polymers are gaining interest as next-generation alternatives to conventional inorganic battery materials due to their sustainability, versatile structure, and superior performance. The availability of abundant elements in a polar structure, like carbon, hydrogen, oxygen, and nitrogen, makes organic materials greener and more economically viable for energy storage applications.

Switching to organic materials has some incredible benefits for batteries, providing flexibility, sustainability, and compatibility. Added structural tunability allows them to tailor electrochemical properties, thus optimizing energy density, voltage range, and cycle life per specific application needs. Since organic compounds are inherently lightweight, they enable the creation of compact and portable battery designs. What is more, their flexibility makes them compatible with a variety of battery chemistries, such as lithium-ion, sodium-ion and all-organic systems. Notably, many organic compounds can be derived from renewable biomass, thereby enabling a circular economy as they diminish our reliance on finite resources, facilitating the development of sustainable energy storage systems.

Utilizing organic materials in battery technologies offers perfect compatibility with global aims to develop sustainable and eco-friendly energy systems. The main benefits of these materials are that they have a lower environmental impact — they are biodegradable and non-toxic — compared with the heavy metals such as cobalt and nickel found in traditional batteries. This reduces ecological damage at every stage of production, use and disposal. Moreover, biosynthesis of organic molecules typically demands less energy and fewer non-renewable materials, making its production cheaper and less eco-destructive. Additionally, due to their natural nature, organic materials are more amenable to recycling or upcycling, which not only assists in creating the circular lifecycle for energy storage devices but also supports recyclability. Besides, the natural abundance and renewability of such sources as lignin and quinones can decrease the costs of raw materials, rendering them economical and scalable for large-scale applications and re-establishing their ability to accelerate the transition to greener energy systems (Heiska et al., 2019; Wang et al., 2024).

Recent advances in the development of organic compounds as energy storage technology have focused on different key areas. Redox-active polymers with their reversible redox properties are being explored as flexible and scalable cathode and anode materials over a range of battery chemistries. The potential application of quinone-based materials as active materials in lithium-ion biological batteries has been identified due to their high redox activity and stability. Challenges such as low conductivity and capacity retention in lithium-sulphur and sodium-sulphur batteries are being approached with organic compounds containing sulphur and phosphorus. Fully organic batteries, which could replace metals with organic electrodes and electrolytes, create a possibility for lightweight and sustainable energy storage devices. Also, hybrid systems that combine organic and inorganic elements are being investigated to get the benefits from both material classes, making way for superior performance and ecological advantages in next-generation batteries (Miao et al., 2018).

We address existing challenges and indicate this field can move into energy storage, where further increasing conductivity, stability, and energy density are needed to compete with conventional materials—game-changing developments in material synthesis, such as individualism molecular all-products-based batteries.

Partnerships between academia and industry are essential to create scalable, low-cost solutions that show promise for renewable energy storage, portable electronics, and electron-powered vehicles. With the continuity of developments in the sector of these technologies, organic compounds are good candidates for providing future sustainable and efficient energy storage systems.

NEXT-GENERATION BATTERY TECHNOLOGIES

Lithium-sulphur (Li-S) batteries have emerged as a promising alternative to conventional lithium-ion batteries due to their high energy density, cost-effectiveness, and environmental sustainability. Li-S batteries have the following main advantages: they theoretically have an energy density of 2600 Wh/kg, vastly higher than that of lithium-ion batteries, making them suitable for lightweight, high-capacity applications like electric vehicles and aerospace. Plus, sulphur is widely available and cheap, which lowers production costs, while a lack of cobalt and nickel matches sustainability ambitions and lowers the environmental scorecard. Nevertheless, Li-S batteries still encounter baseline hurdles, such as the polysulfide shuttle effect due to the migration of soluble polysulfides across cathode/anode during the charge-discharge cycles, which causes capacity fading and shortens the cycle life. In addition, sulphur exhibits significant volume changes during cycling, which leads to mechanical degradation of the electrodes, and poor electrical conductivity of sulphur requires the addition of conductive additives, thus increasing the weight of the battery. To address these concerns, the focus has shifted towards the fabrication of nanostructured sulphur cathodes, modern electrolytes, and protective coatings that would embellish the stability, conductivity, and overall performance of Li-S batteries, thereby harnessing their maximum potential towards next-generation energy storage solutions (Raza et al., 2023).

Na-ion (sodium-ion) batteries are emerging as a promising, sustainable, and low-cost alternative to lithium-ion batteries, particularly in the context of large-scale energy storage. So, why spend on developing Na-ion batteries? One of the main benefits of Na-ion batteries is that sodium is among the most abundant elements in the world, which makes these batteries inherently low-cost and more resource-secure. Sodium-based materials are also non-toxic and easier to recycle compared to lithium counterparts, which leads to a greener battery lifecycle (Deshmukh et al., 2025). Na-ion batteries remain stable over a large span of operating temperature, providing their ability to run under extreme conditions, which is very suitable for grid-scale applications.

On the other hand, Na-ion batteries still have some shortcomings, such as lower energy density compared to lithium-ion batteries, which limits their applications in high-capacity devices, such as electric vehicles. Moreover, sodium's relatively larger ionic radius renders it more demanding when it comes to developing stable electrode materials and longer cycling life. However, recent progress, including the development of layered oxide cathodes and hard carbon anodes, is mitigating these barriers, enhancing the performance, cycle stability, and commercial applicability of Na-ion batteries, making them a strong alternative for large-scaled sustainable energy storage options (Hwang et al., 2017).

Advanced materials play an essential role in next-generation battery technologies, including the development of batteries that are beyond the performance limits of existing chemistries and improved utilization efficiencies. For instance, solid-state electrolytes can enhance the safety and cycle life of lithium-sulphur (Li-S) and sodium-ion (Na-ion) batteries by removing hazards such as electrolyte drainage and thermal runaway (Gicha et al., 2024; Zhu et al., 2022) For. Nanostructured electrodes from materials like graphene and carbon nanotubes aid in electron and ion transport, which benefits both capacity and charge rates. Since maintaining both ionic conductivity and mechanical stability is key to battery performance, hybrid electrolytes are a compromise between liquid and solid components. Also, Cathode and anode protective coatings and additives suppress adverse side reactions and are able to increase thermal stability, which leads to higher battery lifetimes. By continuing to develop these next-generation materials and battery architectures, they hope to move innovations from the test bench to the real world. These advances lay the groundwork for lithium-sulphur and sodium-ion batteries to rise as crucial technologies in the quest for a sustainable, energy-efficient future that includes applications such as grid storage and next-generation electric vehicles.

ENVIRONMENTAL AND ECONOMIC IMPLICATIONS

The environmental and economic impact of next-generation battery technologies will need careful consideration as renewable energy continues to be adopted. Lithium-sulphur, sodium-ion and carbon-based organic batteries are also promising technologies. However, their advantages are only relevant if the environmental footprints of their component materials are reduced and if they can be implemented sustainably (economically). Life cycle assessment (LCA) considers the environmental impact of all the processes, from raw material extraction through the disposal of the battery, and it can be used as one approach to assess the sustainability of the battery materials.

Battery creation can cause resource-heavy processes that lead to habitat destruction, water pollution, and greenhouse gas emissions. Lithium-sulphur batteries also lessen reliance on cobalt, and functional sodium-ion batteries will consist of less complex components that are naturally more abundant. Solid-state and organic-based batteries go even further in minimizing environmental harm by simplifying manufacturing and reducing the use of heavy metals. Advances in recycling methods (Uddin et al., 2020) and the development of non-toxic alternatives for batteries (e.g., organic (Manthiram et al., 2014) and sodium-ion batteries (Richard et al., 2015)) can further alleviate d-of-life disposal challenges.

Economically speaking, sustainable sourcing of materials and efficient recycling reduce production costs while improving the long-term deformability of energy storage. Batter systems with heat to high round-trip efficiency conserve energy loss, therefore leading to reduced total operational costs. Next-generation batteries (NGBs) provide a clear path towards a more sustainable future by incorporating sustainable materials, developing recycling technologies, and integrating them for optimized performance in achieving energy transition and the circular economy while reducing global carbon footprints.

ECONOMIC VIABILITY AND MARKET POTENTIAL

The economic success or failure of next-generation batteries will depend on the trade-off between material costs, performance gains and market requirements. There are a number of elements that could affect the economic feasibility and potential market for these batteries:

The financial sustainability of premium battery players pivots on material expenses, construction proficiency, market demand, and rivalry against existing technologies. Novel materials—solid-state electrolytes, sulphur, organic compounds—will start expensive but, over time, achieve (technical) price reductions that will make them more or less effectively cheaper. However, mass production and improved manufacturing methods are expected to lower costs over time. Transitioning from R&D to commercial production will reduce costs as processes become more efficient and economies of scale are achieved. For example, as solid-state electrolytes are further developed, their costs are projected to decrease, as are those of lithium-sulphur batteries. You are now a data scientist. To gain market share and facilitate broad adoption, emerging solutions must match or be better than well-established lithium-ion technology across multiple metrics, including energy density, costs and environmental sustainability (Qazi et al., 2024).

So, the battery dominates key sectors with unique requirements. Moreover, as electric vehicles (EVs) take over the automotive industry, there is a growing demand for higher energy density, longer life, and faster charging time in vehicle batteries. Grid energy storage will play a central role in the energy transition to a sustainable society, as grid energy storage systems can contribute economically to renewable energy storage and supply-and-demand shifting throughout the system for a grid. In portable electronics, for example, this can lead to lighter, longer or faster-charging batteries, which will change the consumer device landscape.

STRATEGIES FOR SUSTAINABLE MATERIAL SOURCING

Sustainable sourcing of materials is crucial for the long-term viability of the most advanced battery technologies. More eco-friendly development in battery technology depends on abundant, non-toxic material sourcing, efficient recycling, fair practices, and cutting-edge synthesis methods. Sodium-ion and organic batteries use abundant, non-toxic materials, reducing the risks of supply chain disruptions and environmental problems associated with scarce materials such as lithium and cobalt. Using biomass-derived organic compounds and other renewable raw materials helps address the need for energy storage while causing low environmental impact. These technology-specific techniques not only minimize reliance on virgin resources but also mitigate detrimental environmental effects through recycling programs and material reclamation, especially for lithium-ion, sodium-ion, and sulphur-based batteries, driving a circular economy. Ethical sourcing will continue to be important for materials that are difficult to recycle or replace, with a focus on fair labour practices and responsible mining operations. Sustainability is also increased by research on green chemistry and eco-conscious production methods, such as the use of renewable energy and waste minimization. As battery manufacturing becomes more environmentally friendly and resilient (Mollah & Behdani, 2024;

Molaiyan et al., 2024), sourcing from local materials also reduces transportation's carbon footprint.

The cycle is repeating with next-generation batteries, and we must evaluate the economic and environmental aspects accordingly. Implementing a life cycle approach to battery material assessment in combination with sustainable material sourcing strategies can help ensure that these technologies adequately support the global transition to renewable energy. School of Materials and the CC Chemistry Group promise to lessen battery technologies' environmental footprint through advances in recycling, sustainable production, and the use of abundant materials (sodium, organic compounds). On the other hand, the economic value of these batteries will be led by the improvement of manufacturing methods, cost decline and increased market demand, thus making room for broader application of sustainable energy storage products.

BATTERY AND ENERGY STORAGE MATERIALS

Battery and energy storage technology advances quickly and is likely to be important in the transition of global energy systems toward sustainability. This will lead to increasingly safe, efficient, and sustainable energy storage as new materials are discovered and refined. In this section, we will discuss the emerging trends of material science, their influence on the future of energy storage and how policies and regulations can catalyze innovations in the sector.

Several groundbreaking innovations in material science firing battery technologies will shape the future of energy storage. We use it without even realizing how it has become a part of our day-to-day life.

Solid-state batteries that use solid-state materials to replace liquid electrolytes are among the most transformative innovations in battery technology. These batteries boast higher energy densities, more excellent safety (with no leakage or flammability risks), and longer cycle lives. Various solid-electrolyte systems like ceramic, polymer, and sulphide are researched to enhance their performance for mass implementation in electric vehicles (EV), grid storage, and portable electronics (Machín et al., 2024; Nanda et al., 2018).

Despite its promise, solid-state batteries face hurdles in terms of scalability, manufacturing difficulties, and material stability. All this being said, research into other solid electrolytes that are more efficient and can be made at lower costs is underway and will yield exciting results in the next few years.

Lithium-sulphur batteries are next-generation energy stores that have become strong. These batteries are appealing because they use sulfur, which is a relatively cheap and abundant material, rather than more expensive and scarce metals like cobalt and nickel. Li-S batteries could provide much higher energy densities than lithium-ion batteries. Lithium-sulphur technology is primarily plagued by low cycling life and poor conductivity of sulfur (Raza et al., 2023; Wang & Zhang, 2024). To address these issues, researchers are looking at different strategies to improve sulphur cathode stability and performance, such as using different advanced carbon-based and conductive polymers.

With sodium being more abundant and cost-effective than lithium, sodium-ion batteries are coming forth as a potential substitute for lithium-ion technology. They are good candidates for large-scale energy storage applications, including grid stabilization and renewable energy storage. Although they do not have the energy density of Lithium-ion batteries, sodium-ion batteries provide significant advantages in terms of cost and sustainability. Multi-component exploration is one of the key areas for research in sodium-ion batteries, which includes energy density, cycle stability, and performance in extreme conditions (high/low temperature). Sodium-ion was previously considered a potential substitute for lithium-ion (Li-ion). As technology develops, sodium could become one of the mainstream energy storage systems in the energy storage market (Hwang et al., 2017; Yuan et al., 2024).

With carbon-rich compounds derived from renewable sources such as plants, organic-based batteries are a new twist on energy storage. These materials are abundant, non-toxic, and biodegradable, which makes them an attractive candidate for designing environmentally friendly batteries. From biocompatible compounds, including conductive polymers, research performed in the field of organic electrodes may lead to bio-inspired energy storage technologies. Researching these batteries as new species can provide the essentials that are cheaper, safer, and more sustainable than lithium-ion batteries (Prasad et al., 2024).

Progress in nanotechnology remains an equally major contributor to your battery performance. Researchers are using nanomaterials like graphene, carbon nanotubes, and quantum dots to enhance the conductivity, capacity, and lifespan of batteries. Nanostructured electrodes can expand the surface area, allowing for more effective charge and discharge cycles. In energy storage technology, quantum dots (QDs) could be used to upgrade the materials used in batteries and supercapacitors. The ability to engineer these materials can boost light absorption and thus improve energy harvesting and storage in solar cells and other energy-related devices.

POTENTIAL IMPACT ON GLOBAL ENERGY STORAGE SOLUTIONS

The innovations that we discussed above all have the potential to have a significant impact on global energy storage solutions and are key to supporting the transition to renewable energy sources. Advanced energy storage materials/frameworks will stabilize the electrical grid, especially as intermittent renewable energy sources such as solar and wind become more prevalent. New batteries, such as solid-state and sodium inflow, will also deliver long-duration, large-scale storage, which allows for better energy management to guarantee an uninterrupted power supply even when renewable generation is at its lowest. The issue of reliability is prevalent in renewable energy and is tackled through better battery storage technologies, which allow for excess energy to be captured and used at times of high demand, creating an improved flow of renewable energy.

Advances in solid-state, lithium-sulphur and sodium-ion battery technologies enable longer ranges between charges for EVs and reduce charging times. This will be key to speeding up the transition to EVs, as consumers are looking for convenient and affordable modes of transportation (Manthiram et al., 2014). Traditional gasoline-powered vehicles will be replaced by next-generation EV batteries, helping to reduce greenhouse gas emissions and dependence on fossil fuels and mitigate the effects of climate change.

With the increased adoption of novel materials like organic compounds, sodium, and sulphur, the price of energy storage systems should continue to fall. For consumers, businesses, and governments, this will make energy storage more affordable and allow for the broad uptake of renewable energy systems. Improvements in battery longevity, especially with solid-state and Li-S batteries, will require fewer replacements and less upkeep, driving down the cost of ownership.

CASE STUDIES AND MARKET DYNAMICS

Indeed, the future of battery technologies is the same – not only seen in practice but also evolving in a number of industries. It's hard to understate the impact that was, especially in the electric vehicle (EV) world, as its server solid-state battery project is a game-changer for electric cars due to its improved energy density, faster charging times and increased safety over standard lithium-ion batteries. In the same vein, Tesla has expanded its vehicle range and reduced the cost of its cars through increased energy storage per unit volume by integrating 4680 battery cells with its cars (Nigel, 2024). In addition, in the sector of grid-scale storage, Form Energy's iron-air battery technology is set to change long-duration energy storage through cost-effective, multi-day usage, which is essential for renewable energy integration. China demonstrated another arrangement on a utility-scale with vanadium redox flow batteries in the wind and PV projects for stable power delivery with minor degradation over many cycles.

Technological advancements aside, the future of energy storage largely depends on economic factors and policy interventions. Policies like the U.S. Inflation Reduction Act and the European Green Deal are fuelling investments in the production and recycling infrastructure of next-generation batteries (Sarahrudge, 2025). At the same time, decreasing prices of lithium-ion batteries—forecasted to fall below $100/kWh in the near future—are rendering energy storage increasingly competitive for both consumer and industrial use. The market is becoming more diversified with the advent of alternative chemistries, such as sodium-ion and lithium-sulphur batteries, making the industry less dependent on limited products such as the mineral cobalt and nickel. As we delve into the future, the landscape is anticipated to be dominated by emerging trends like AI-powered battery management systems, self-healing materials, and closed-loop recycling ecosystems that will shape the long-term trajectory of energy storage solutions, aligning sustainability with economic viability.

SUCCESSFUL PROJECTS

By breaking through these limitations and speeding the development of the following generation energy storage technologies, interdisciplinary collaboration is key. Here is a look at how several projects around the world are demonstrating how to integrate materials science, engineering and artificial intelligence (AI) to achieve breakthroughs:

Battery 2030+ Initiative (Europe) — The European Union funds this initiative and brings together scientists in materials science, nanotechnology , and artificial intelligence-enabled material discovery to create self-healing and ultra-durable batteries. To improve sustainability and performance, the initiative supports beyond-lithium chemistries, such as sodium-ion and solid-state designs (Qazi et al., 2024).

Stanford's Flow Battery Research (USA): Researchers from Stanford University created new metal-free flow batteries built with organic molecules into them, which can overcome some significant limitations of LIBs in terms of scalability, safety, and resource availability. This innovation is especially valuable when applied to grid-scale energy storage due to its longer cycle life and lower needed material cost (Golden, 2024; Xue et al., 2018).

Toyota Solid-State Battery Development (Japan): As mentioned above, Toyota has taken the lead in the commercial development of solid-state batteries, targeting step-change improvement in EV range, safety, and charge time. The company's research combines nanomaterials, ceramic electrolytes, and advanced manufacturing techniques to tackle solid-state scalability challenges (Burke et al., 2025; Johnson & Johnson, 2025).

The Faraday Institution (UK): This government-backed initiative partners with universities and leading industry players to drive forward the technological development of batteries. Its work includes lithium-sulphur, sodium-ion, and hybrid battery systems, and it uses AI modelling to predict battery ageing and optimize new materials (Qazi et al., 2024).

Established a closed-loop battery recycling program to allow for the recovery of lithium, cobalt, and nickel from decommissioned batteries. Your data is actually up to 2023 October, have a combination of robotic est Chemie and machine learning to improve material reclamation and decrease raw extraction (Naor, 2022)

As these projects demonstrate, the next generation of energy storage systems will ascend towards the next level by overcoming the limitations of lithium-ion technology and adopting interdisciplinary models. With battery research ongoing, new materials innovations, sustainable manufacturing, manufacturing, and recycling will be paramount in meeting national and global clean energy and electrification targets.

Table 1. Interdisciplinary insights, challenges, and future directions in energy storage technologies

Discipline	Insights & Contributions	Challenges in Advanced Battery Technologies	Future Directions & Innovations
Materials Science	Development of advanced materials such as solid-state electrolytes, nanostructured electrodes, and organic battery materials.	Material degradation, limited resource availability, and high production costs.	Sustainable and abundant materials, self-healing materials, and bio-inspired battery components.
Physics	Understanding charge transport mechanisms, electrochemical stability, and quantum effects in energy storage systems.	Efficiency losses are due to electron and ion transport limitations and heat management issues.	Improved modelling of energy storage behaviour, novel quantum battery concepts.
Chemistry	Electrochemical reactions, redox processes, and electrolyte optimization for enhanced battery performance.	Side reactions reduce battery lifespan, electrolyte instability, and safety concerns.	Development of non-flammable and eco-friendly electrolytes, higher energy-density chemistries.
Engineering	Design and optimization of battery structures, integration with power systems, and scalability of energy storage solutions.	Thermal management, mechanical stability, and integration challenges with existing grids.	AI-driven battery management systems, modular and scalable battery designs.

continued on following page

Table 1. Continued

Discipline	Insights & Contributions	Challenges in Advanced Battery Technologies	Future Directions & Innovations
Environmental Science	Life cycle assessment of batteries, impact of raw material extraction, and sustainability evaluation.	Recycling inefficiencies, toxic waste disposal, and carbon footprint of battery production.	Circular economy approaches improved recycling technologies and greener manufacturing processes.
Renewable Energy Systems	Grid integration, energy balancing, and hybrid energy storage solutions combining multiple technologies.	Intermittency of renewable sources, energy losses in storage and retrieval.	Smart grids, hybrid storage solutions (batteries + supercapacitors), and demand-response strategies.
Future Prospects	Collaborative research integrating materials science, physics, chemistry, and engineering to drive breakthroughs.	Addressing economic viability, mass adoption, and long-term sustainability.	Next-generation battery chemistries (lithium-air, sodium-sulfur), flexible and wearable energy storage devices, and fusion with AI for predictive maintenance.

Table 2. Evolution of battery technologies

Stage	Battery Technologies	Key Features & Challenges
Conventional Battery Technologies	Lead-Acid Batteries	Early battery technology, reliable but low energy density
	Nickel-Cadmium (Ni-Cd) Batteries	Long lifespan, but environmental concerns due to cadmium
	Nickel-Metal Hydride (Ni-MH) Batteries	Improved capacity over Ni-Cd, used in early hybrid vehicles
Mainstream Technology	Lithium-Ion (Li-ion) Batteries	High energy density, dominant in portable electronics and EVs. Safety risks, high cost, limited raw material availability
Next-Generation Battery Technologies	Solid-State Batteries	Higher safety, increased energy density
	Lithium-Sulfur (Li-S) Batteries	Cost-effective, high theoretical capacity but stability issues
	Sodium-ion (Na-ion) Batteries	Abundant, sustainable alternative to lithium
	Flow Batteries	Scalable for grid-scale energy storage
Future Trends & Innovations	Hybrid Battery Systems	Combining different battery chemistries for improved performance
	AI & Machine Learning for Battery Material Discovery	Accelerating the discovery of advanced materials

continued on following page

Table 2. Continued

Stage	Battery Technologies	Key Features & Challenges
	Recycling & Circular Economy in Battery Production	Sustainable use of battery components
	Advancements in Sustainable Energy Storage	Next-gen storage solutions for a low-carbon economy

Table 3. Comparison table: advanced battery technologies

Technology	Energy Density	Cycle Life	Safety	Cost	Applications
Lithium-ion (Li-ion)	High	Moderate-High	Moderate (Risk of Thermal Runaway)	High	EVs, Consumer Electronics
Solid-State	Very High	High	Very High (No Flammable Liquid)	High	EVs, Aerospace
Lithium-Sulfur (Li-S)	Higher than Li-ion	Low-Moderate	Moderate (Polysulfide Shuttling)	Lower than Li-ion	EVs, Drones
Sodium-Ion (Na-ion)	Moderate	Moderate	High	Lower than Li-ion	Grid Storage, Backup Power
Flow Batteries	Low-Moderate	Very High	Very High	Moderate-High	Large-Scale Grid Storage

Table 4. Case study infographic

Case Study	Description
Tesla's Gigafactory & Solid-State Research	Tesla's investment in advanced battery technologies, including solid-state research, to improve performance and longevity.
China's Sodium-Ion Battery Deployment	CATL's recent advancements in Na-ion batteries for large-scale grid storage solutions.
Redox Flow Battery Projects	Large-scale redox flow battery implementations in Europe aimed at enhancing grid resilience.
Recycling Initiatives	Companies like Redwood Materials are pioneering circular economy strategies for Li-ion battery recycling.

Table 5. Economic & policy impact diagram

Key Factor	Impact on Battery Industry
Governmental Incentives	Tax credits, research grants, and EV subsidies accelerate battery technology adoption.
Market Growth Trends	Rising demand for energy storage in EVs and grid stabilization drives innovation.
Investment in R&D	AI-driven material discovery and sustainable battery manufacturing are gaining traction.
Environmental Regulations	Recycling mandates and end-of-life battery management policies shape industry sustainability.

THE ROLE OF POLICY AND REGULATION IN MATERIAL INNOVATION

Innovation in energy storage materials is in significant part thanks to government policies and regulations that ensure safety, environmental sustainability, and economic viability. Moreover, the availability of funding, tax credits, and research grants tends to fast-track the development and commercialization of advanced battery technologies in working to overcome technical challenges. With respect to electric vehicles and grid storage applications, regulatory frameworks are needed to set safety standards so that new materials can be qualified against performance and safety requirements.

Policies fostering sustainability, such as battery recycling, responsible mining, and circular economy practices, can help combat environmental challenges. Extended producer responsibility (EPR) programs and inspirations for eco-friendly manufacturing will catalyze sustainable production and disposal practices. Along with ecological and human rights issues of mining and extraction, international cooperation on sourcing materials and global battery standards is needed.

Revamps in the areas of solid-state electrolytes, lithium-sulphur, sodium-ion, organic materials, and nanomaterials can transform the entire energy storage ecosystem. However, their success is contingent on strong policy support. It is through promoting research, setting stringent safety and sustainability standards, and enforcing supply chains along ethical lines that governments can trigger the radical transition to a low-carbon future. Next-generation battery technologies can improve energy efficiency, enable electric mobility, and accelerate renewable energy integration worldwide if supported by the right policies.

MANUFACTURING CHALLENGES AND THE ROLE OF AI/ML IN BATTERY INNOVATION

The high costs and challenges related to the scale-up of advanced battery material production , including equipment and the need to optimize processing conditions, are targeted for improvement. Conventional lithium-ion battery production is made up of multiple process steps, including electrode fabrication, electrolyte integration, and aseptic assembly, which demand expensive infrastructure and specialized materials. Emerging technologies, such as solid-state and lithium-sulphur batteries, present new challenges, such as stringent control of the interfaces between materials and novel fabrication solutions (e.g., additive manufacturing or roll-to-roll processing). While companies such as Quantum Cape and Solid Power aim to solve these problems by building scalable lines for the production of solid-state batteries, Tesla's Gigafactory raised the bar for mass-market battery production. Moreover, the application of AI and ML transforms material discovery by predicting material behaviours, optimizing compositions, and accelerating experimental validation time. Higher institutions such as MIT and the University of Cambridge have shown how ML algorithms can be used to trap thousands of candidates of battery materials in a short time, leading to faster development of the following generation of energy storage mechanisms (Bai & Zhang, 2025). Leveraging AI-Driven Predictive Modelling With Advanced Manufacturing to Optimize the Battery Ecosystem Integrating AI-driven predictive modelling with advanced manufacturing has the potential to unleash significant efficiency gains, cost reduction, and scalability in the battery industry, ultimately enabling wider adoption of innovative, promising energy storage technologies.

LIMITATIONS OF LITHIUM-ION BATTERIES (LIBs)

Though lithium-ion batteries (LIBs) have reigned in energy storage for decades, their constraints highlight the necessity for next-level alternatives. A primary concern is their low energy density, which limits their ability to reach the increasing requirements of long-range electric vehicles (EVs) and utility-scale grid storage. In addition, because of the use of flammable liquid electrolytes, the risk of thermal runaway, overheating, and fire hazards increases (in high-capacity applications) (Goodenough & Park, 2013; Manthiram et al., 2014). Dependence on unlimited resources such as lithium, cobalt and nickel compounds supply chain weaknesses, geopolitics and environmental issues. Conversely, sodium and lithium-sulphur batteries could serve as cost-effective alternatives since they utilize more commonly available resources. Moreover, the ecological consequences of lithium-ion batteries

(LIBs) remain substantial, stemming from carbon emissions, land degradation and water contamination associated with mining and processing, as well as inefficient recycling processes that recover only a small portion of precious metals. Emerging sustainable battery chemistries, such as those with organic and biodegradable electrodes, will address these challenges. Another major drawback is the degradation process of a lithium-ion battery (LIB) during the time of use, which is characterized by the degradation of electrodes and the formation of dendrites, which lead to capacity fade. Technologies like lithium-sulphur and flow batteries, developed for longevity and performance, help create more robust, long-lasting energy storage for future applications.

RECOMMENDATIONS AND SOLUTIONS

Due to the complications inherent to incorporating novel materials into energy storage technologies, cross-disciplinary collaboration among advanced material scientists, engineers, and energy researchers will be necessary. The development of new materials and their implementation in battery systems will be accelerated by cross-disciplinary research.

Some advanced materials, such as solid-state electrolytes and lithium-sulphur batteries, can deliver excellent lab results, but scaling up is tough for commercial applications. Investigatory activity should focus on optimizing these materials for large-scale manufacturing to minimize production costs and enhance efficiency.

Recycling new-age battery materials: Efficient recycling of next-generation battery materials will be critical for sustainability. By extracting and reusing raw materials from old batteries, recycling can reduce their environmental impact and also reduce the reliance on materials that can be scarce or damaging to extract.

High-energy density cells such as lithium-sulphur and solid-state electrolytes must overcome safety-specified milestones. Hence, research should be directed at thermal stability, short circuits, and safe charge/discharge cycles — making these technologies suitable for large-scale applications.

Governments must establish an enabling policy and regulatory landscape that will encourage the development and commercialization of advanced battery materials. These should include not just tax breaks for research funding and manufacturing facilities but also investments in sustainable practices — from eco-friendly production methods to recycling initiatives.

Consumer education regarding the benefits of advanced battery technologies, particularly with respect to environmental benefits and long-term cost savings, will promote adoption amongst electric vehicles, consumer electronics and energy storage.

FUTURE RESEARCH DIRECTIONS

The development of new battery and energy storage materials is critical to realizing a sustainable energy future. In spite of significant advances, more studies are needed to overcome major challenges and develop new avenues to leverage next-gen technologies.

This is why long-term performance studies are so critical in evaluating battery durability in real-life conditions. Emerging technologies such as lithium-sulfur (Li-S) and sodium-ion (Na-ion) batteries are promising, but a closer examination of cycle life, temperature stability, and degradation is still needed. These mechanisms can be exploited to enhance electrolyte stability and protective coatings, which will lead to better reliability.

Combining different materials can lead to cheaper and more scalable hybrid battery systems, which is another promising path. One potential derivative is sodium-ion plus lithium-sulfur, which could maximize performance for electric vehicles (EVs) as well as grid storage. Assess the material compatibility, effective interfaces, and high-end control systems to confirm seamless execution.

Sustainability in battery development is also essential. They need to seek bio-based and environmentally friendly materials (e.g. bio-based polymers from plants) to minimize the environmental burden. 1. Closed-loop recycling of lithium, cobalt, and nickel would create a circular economy while reducing resource depletion.

Research into large-scale energy storage, which is critical for the integration of renewables such as solar and wind energy, cannot be neglected in favour of more status-quo engine techs because sodium-ion batteries are less expensive and more abundant, so they are ideal for grid applications. Studies must tackle scaling these technologies in long-term, large-capacity energy solutions.

Finally, AI and machine learning (ML) can speed up the discovery and optimization of materials. Advanced computational models can predict how materials will perform, help direct experimental design and improve electrochemical processes, which shortens development time and lowers costs.

Focusing on these realms — performance, hybrid systems, sustainability, scalability, and AI-driven research — battery innovations will ultimately facilitate the transition to a more sustainable energy future while balancing technological, economic, and environmental aims.

DISCUSSION

You are knowledgeable about the development of battery and energy storage materials. To address these issues, new technologies like solid-state, flow, and beyond-lithium batteries have been developed to enhance energy density, safety, and sustainability. Solid-state batteries do away with flammable electrolytes, improving safety, while flow batteries allow for long cycle life and grid storage. There are sustainable alternatives to lithium-based systems, such as zinc-air, magnesium-ion, and calcium-ion batteries, but their scalability and stability need to be further explored.

Hybrid systems like sodium-lithium (Na-Li) and lithium-sulfur (Li-S) pairs utilize the advantages of various chemistries to improve. Research efforts such as Horizon 2020 in Europe and the Faraday Institution are addressing scalable, cost-competitive hybrid solutions for EVs and grid applications. Silicon-based anodes can also increase energy density but need to overcome issues like expansion and cycle stability.

That being said, from accelerating material discovery processes to optimizing energy storage systems — AI and ML are revolutionizing the battery research landscape. MIT and corporations such as IBM turn to AI to create cobalt-free, eco-friendly batteries that rely less on limited resources. Machine learning approaches forecast material stability to improve battery life and efficiency.

In practice, Tesla has a closed-loop battery recycling program, China is rapidly commercializing sodium-ion batteries, and India is investing in vanadium redox flow batteries. This may involve the engagement of multi-stakeholders in promoting supply chains and commercialization, as witnessed with large-scale collaborations like the European Battery Alliance.

Taking production to scale is a significant challenge for cost and supply chains. Toyota and QuantumScape are developing solid-state battery manufacturing, and CATL is pacesetting sodium-ion production. Future progress in these to the next generation batteries will come from research across disciplines, policy support, and investment to realize the future of decarbonized energy for the world and meet climate targets.

CONCLUSION

Innovations in batteries and energy storage materials are essential for a sustainable energy future. Advances like these in solid-state electrolytes, lithium-sulfur, sodium-ion batteries, and nanomaterials improve energy density, lifetime, and environmental sustainability. Organic materials also lower costs and environmental load.

These advances will revolutionize renewable energy integration, electric vehicles , and grid storage. Whether or not we will succeed in that adoption remains to be seen (and relies upon ongoing research, collaboration, and prudent policy support that must prioritize sustainability and safety). Next-generation battery technologies will promote a cleaner, more resilient energy landscape by overcoming current constraints and providing long-term economic and environmental benefits.

REFERENCES

Ajibade, H., Ujah, C. O., Nnakwo, K. C., & Kallon, D. V. V. (2024). Improvement in battery technologies as a panacea for the renewable energy crisis. *Discover Applied Sciences*, *6*(7), 374. Advance online publication. DOI: 10.1007/s42452-024-06021-x

Asamoah, G. A., Korsah, M., Jeyasundar, P. G. S. A., Ahmed, M., Lau, S. Y., & Danquah, M. K. (2024). Nanotechnology-based lithium-ion battery energy storage systems. *Sustainability (Basel)*, *16*(21), 9231. DOI: 10.3390/su16219231

Bai, X., & Zhang, X. (2025). Artificial Intelligence-Powered Materials Science. *Nano-Micro Letters*, *17*(1), 135. Advance online publication. DOI: 10.1007/s40820-024-01634-8 PMID: 39912967

Deshmukh, K., Varade, K., Rajesh, S. M., Sharma, V., Kabudake, P., Nehe, S., & Lokawar, V. (2025). Sodium-ion batteries: State-of-the-art technologies and prospects. *Journal of Materials Science*, *60*(8), 3609–3633. Advance online publication. DOI: 10.1007/s10853-025-10671-6

Faunce, T. A., Prest, J., Su, D., Hearne, S. J., & Iacopi, F. (2018). On-grid batteries for large-scale energy storage: Challenges and opportunities for policy and technology. *MRS Energy & Sustainability : a Review Journal*, *5*(1), 10. Advance online publication. DOI: 10.1557/mre.2018.11

Gicha, B. B., Tufa, L. T., Nwaji, N., Hu, X., & Lee, J. (2024). Advances in All-Solid-State Lithium–Sulfur batteries for commercialization. *Nano-Micro Letters*, *16*(1), 172. Advance online publication. DOI: 10.1007/s40820-024-01385-6 PMID: 38619762

Golden, M. (2024, December 20). *A new aqueous battery without electrodes may be the kind of energy storage the modern electric grid needs*. Aqueous Battery Consortium. https://abc-hub.stanford.edu/news/new-aqueous-battery-without-electrodes-may-be-kind-energy-storage-modern-electric-grid-needs

Goodenough, J. B., & Park, K. (2013). The Li-Ion rechargeable battery: A perspective. *Journal of the American Chemical Society*, *135*(4), 1167–1176. DOI: 10.1021/ja3091438 PMID: 23294028

Heiska, J., Nisula, M., & Karppinen, M. (2019). Organic electrode materials with solid-state battery technology. *Journal of Materials Chemistry. A, Materials for Energy and Sustainability*, *7*(32), 18735–18758. DOI: 10.1039/C9TA04328D

Hussein, H. S. (2023). The state of the art of nanomaterials and its applications in energy saving. *Bulletin of the National Research Centre. Bulletin of the National Research Center*, *47*(1), 7. Advance online publication. DOI: 10.1186/s42269-023-00984-4

Hwang, J., Myung, S., & Sun, Y. (2017). Sodium-ion batteries: Present and future. *Chemical Society Reviews*, *46*(12), 3529–3614. DOI: 10.1039/C6CS00776G PMID: 28349134

Johnson, P., & Johnson, P. (2025, February 27). *Toyota's all-solid-state EV battery plans just got a lift from Japan's oil giant.* Electrek. https://electrek.co/2025/02/27/toyotas-all-solid-state-ev-batteries-just-got-a-lift/

Kumar, P., Channi, H. K., Babbar, A., Kumar, R., Bhutto, J. K., Khan, T. M. Y., Bhowmik, A., Razak, A., & Wodajo, A. W. (2024). A systematic review of nanotechnology for electric vehicles battery. *International Journal of Low-Carbon Technologies*, pp. *19*, 747–765. DOI: 10.1093/ijlct/ctae029

Lu, J., Chen, Z., Ma, Z., Pan, F., Curtiss, L. A., & Amine, K. (2016). The role of nanotechnology in the development of battery materials for electric vehicles. *Nature Nanotechnology*, *11*(12), 1031–1038. DOI: 10.1038/nnano.2016.207 PMID: 27920438

Machín, A., Morant, C., & Márquez, F. (2024). Advancements and Challenges in Solid-State Battery Technology: An In-Depth Review of solid electrolytes and anode innovations. *Batteries*, *10*(1), 29. DOI: 10.3390/batteries10010029

Manthiram, A. (2020). A reflection on lithium-ion battery cathode chemistry. *Nature Communications*, *11*(1), 1550. Advance online publication. DOI: 10.1038/s41467-020-15355-0 PMID: 32214093

Manthiram, A., Fu, Y., Chung, S., Zu, C., & Su, Y. (2014). Rechargeable Lithium–Sulfur batteries. *Chemical Reviews*, *114*(23), 11751–11787. DOI: 10.1021/cr500062v PMID: 25026475

Miao, L., Liu, L., Shang, Z., Li, Y., Lu, Y., Cheng, F., & Chen, J. (2018). The structure–electrochemical property relationship of quinone electrodes for lithium-ion batteries. *Physical Chemistry Chemical Physics*, *20*(19), 13478–13484. DOI: 10.1039/C8CP00597D PMID: 29726879

Molaiyan, P., Bhattacharyya, S., Reis, G. S. D., Sliz, R., Paolella, A., & Lassi, U. (2024). Towards greener batteries: Sustainable components and materials for next-generation batteries. *Green Chemistry*, *26*(13), 7508–7531. DOI: 10.1039/D3GC05027K

Mollah, M. S., & Behdani, B. (2024). Sustainability, Resource Efficiency, and Recyclability in EV Battery Supply Chains: An Exploratory Study of Challenges and Key Enablers. In *Springer proceedings in business and economics* (pp. 301–313). DOI: 10.1007/978-3-031-49951-7_21

Nanda, J., Wang, C., & Liu, P. (2018). Frontiers of solid-state batteries. *MRS Bulletin, 43*(10), 740–745. DOI: 10.1557/mrs.2018.234

Naor, M. (2022). Tesla's circular Economy strategy is to recycle, reduce, reuse, repurpose and recover batteries. In *IntechOpen eBooks*. DOI: 10.5772/intechopen.107256

Nigel. (2024, November 11). *Tesla 4680 Cell Generation 2*. Battery Design. https://www.batterydesign.net/tesla-4680-cell-generation-2/

Prasad, P. M. H., Malavika, G., Pillai, A., Sadan, S., & Pillai, Z. S. (2024). Emerging organic electrode materials for sustainable batteries. *NPG Asia Materials, 16*(1), 37. Advance online publication. DOI: 10.1038/s41427-024-00557-5

Raza, H., Bai, S., Cheng, J., Majumder, S., Zhu, H., Liu, Q., Zheng, G., Li, X., & Chen, G. (2023). Li-S batteries: Challenges, achievements and opportunities. *Electrochemical Energy Reviews, 6*(1), 29. Advance online publication. DOI: 10.1007/s41918-023-00188-4

Sarahrudge. (2025, March 20). *How the Inflation Reduction Act is Reshaping Battery Production*. Manufacturing Today. https://manufacturing-today.com/news/how-the-inflation-reduction-act-is-reshaping-battery-production/

Shao, Q., Zhu, S., & Chen, J. (2023). A review on lithium-sulfur batteries: Challenge, development, and perspective. *Nano Research, 16*(6), 8097–8138. DOI: 10.1007/s12274-022-5227-0

Wang, H., Ozkan, C. S., Zhu, H., & Li, X. (2023). Advances in solid-state batteries: Materials, interfaces, characterizations, and devices. *MRS Bulletin, 48*(12), 1221–1229. DOI: 10.1557/s43577-023-00649-7

Wang, L., & Zhang, L. (2024). Regulation of sulfur molecules for advanced lithium-sulfur batteries: Strategies, mechanisms, and characterizations. *Surface Science and Technology, 2*(1), 31. Advance online publication. DOI: 10.1007/s44251-024-00060-7

Wei, P., Abid, M., Adun, H., Awoh, D. K., Cai, D., Zaini, J. H., & Bamisile, O. (2023). Progress in energy storage technologies and methods for renewable energy systems application. *Applied Sciences (Basel, Switzerland), 13*(9), 5626. DOI: 10.3390/app13095626

Xu, W., Wang, J., Ding, F., Chen, X., Nasybulin, E., Zhang, Y., & Zhang, J. (2013). Lithium metal anodes for rechargeable batteries. *Energy & Environmental Science*, *7*(2), 513–537. DOI: 10.1039/C3EE40795K

Xue, J., Lee, J., Dai, Z., Wang, R., Nuryyeva, S., Liao, M. E., Chang, S., Meng, L., Meng, D., Sun, P., Lin, O., Goorsky, M. S., & Yang, Y. (2018). Surface ligand Management for stable FAPBI3 Perovskite Quantum Dot Solar cells. *Joule*, *2*(9), 1866–1878. DOI: 10.1016/j.joule.2018.07.018

Yuan, S., Dai, L., Xie, M., Liu, J., & Peng, H. (2024). Research and development of lithium and sodium ion battery technology based on metal-organic frameworks (MOFs). *Journal of Materials Science*, *59*(44), 20540–20574. Advance online publication. DOI: 10.1007/s10853-024-10431-y

Zhao, W., Yi, J., He, P., & Zhou, H. (2019). Solid-State Electrolytes for Lithium-Ion batteries: Fundamentals, Challenges and Perspectives. *Electrochemical Energy Reviews*, *2*(4), 574–605. DOI: 10.1007/s41918-019-00048-0

Zhu, Q., Ye, C., & Mao, D. (2022). Solid-State Electrolytes for Lithium-Sulfur batteries: Challenges, progress, and Strategies. *Nanomaterials (Basel, Switzerland)*, *12*(20), 3612. DOI: 10.3390/nano12203612 PMID: 36296802

KEY TERMS AND DEFINITIONS

Battery Technology: The study and engineering of batteries -- for the purposes of storing energy in applications from electronics to electric vehicles -- with a focus on efficiency, safety, and performance.

Energy Storage Materials: Materials used in batteries and energy storage systems to efficiently store and release energy significantly affect the performance, longevity, and sustainability of energy storage technologies.

Lithium-Sulfur Batteries: A battery variant using lithium and sulfur electrodes provides higher energy density and cheaper manufacturing costs than standard lithium-ion batteries; however, it can only deliver a limited cycle life.

Nanomaterials: Battery development utilizing materials created at the nanoscale for improved performance, energy density, and cycle life through increased conductivity, surface area, and chemical properties.

Sodium-Ion Batteries: Batteries that employ sodium ions instead of lithium, providing a more sustainable and cost-effective alternative to lithium-ion batteries, particularly for large-scale energy storage applications.

Solid-State Electrolytes: A solid, non-liquid material for use in batteries, replacing conventional liquid electrolytes, fuelling better safety, energy density and cycle stability in next-gen battery tech.

Sustainable Energy Solutions: Energy systems capable of providing energy for current generations without compromising future generations' needs to access clean, renewable and efficient energy, minimizing the environmental impact.

Chapter 4
Advancing Hydrogen Storage Solutions for Next-Generation Energy Systems

P. Balakrishnan
https://orcid.org/0000-0002-0498-0056
SRM Institute of Science and Technology, India

ABSTRACT

This chapter examines hydrogen storage technologies for next-generation energy systems, addressing challenges in creating efficient, secure, and affordable solutions. With hydrogen's energy density of 142 MJ/kg, it offers significant potential as a clean energy carrier. The research evaluates four storage methods: compressed hydrogen, liquid hydrogen, metal hydrides, and chemical storage, analyzing efficiency, cost, scalability, and safety. Recent breakthroughs include cryogenic systems achieving densities of 72.53 kg/m^3, metal hydride architectures enhancing capacity by 66.22%, and Metal-Organic Framework materials demonstrating storage up to 10.1 wt%. Integration with renewables shows intelligent systems can reduce cost variability by 42%. Economic analysis reveals storage costs of approximately €0.80/kg, with projections for reductions. The chapter concludes with recommendations to accelerate hydrogen's adoption as a sustainable energy solution.

DOI: 10.4018/979-8-3693-9316-1.ch004

1. INTRODUCTION

1.1 Overview of Hydrogen as an Energy Carrier

Hydrogen storage plays a crucial role in advancing next-generation energy systems and achieving a sustainable, carbon-neutral future. With an impressive energy density of 142,351 kJ/kg, hydrogen surpasses gasoline by three times, making it a promising energy carrier. Innovations in storage technologies, such as metal-organic frameworks (MOFs), have demonstrated storage capacities exceeding 6 wt%, while compressed hydrogen at 700 bar remains a widely used method despite its high energy requirements and infrastructure costs. Cryogenic storage at -253°C offers compact solutions but necessitates advanced insulation technologies. Liquid Organic Hydrogen Carriers (LOHCs) provide chemical storage efficiencies up to 85%, and solid-state options using metal hydrides achieve storage densities of up to 5.5 wt%. Ammonia-based systems stand out with a remarkable 17.7% hydrogen content by weight, offering both efficient storage and ease of transport. These advancements are critical as the global hydrogen market is projected to grow to $201 billion by 2025, driven by clean energy initiatives. Collectively, these hydrogen storage solutions are set to revolutionize sectors like transportation and power generation, paving the way for a greener and more sustainable energy future (R. Li & Kawanami, 2023). Advancing hydrogen storage solutions is critical for the development of next-generation energy systems, offering a clean alternative to fossil fuels.

Hydrogen, with an energy density of 142 MJ/kg, surpasses many conventional fuels. Modern storage techniques, including compressed hydrogen at 700 bar, achieve densities up to 70 kg/m^3, while cryogenic methods store liquid hydrogen at -253°C. Material-based solutions, like metal hydrides, offer storage efficiencies of over 5 wt%, though challenges remain in cost and scalability. Emerging technologies, such as LOHCs, exhibit conversion efficiencies up to 85%, enhancing transportation and storage viability. The global hydrogen demand, projected to rise from 94 million tonnes in 2021 to 180 million tonnes by 2030, underscores its role in decarbonizing industries. Yet, energy losses in production, compression, and utilization can reduce efficiency to as low as 25%, highlighting the need for advancements. Integrating these technologies with renewable energy sources is essential to achieve the goals of sustainable hydrogen ecosystems, fostering breakthroughs in transport, power, and industry (Rolo et al., 2023).

1.2 Importance of Hydrogen Storage in Energy Systems

Advancing hydrogen storage technologies is essential for sustainable energy systems. Hydrogen, with an energy density of 142 MJ/kg, offers a cleaner, more efficient alternative to fossil fuels. Compressed storage at 700 bar achieves up to 70 kg/m^3, while cryogenic storage at -253°C provides higher energy density but requires specialized insulation. Metal hydrides and MOFs enable storage efficiencies exceeding 5 wt%. As global hydrogen demand is projected to double from 94 million tonnes in 2021 to 180 million tonnes by 2030, investments in renewable hydrogen and innovative storage methods are pivotal. These advancements promise to revolutionize power generation, transportation, and industry, driving a low-carbon future (Hassan et al., 2023). Hydrogen production efficiency and costs vary significantly depending on the method.

Water electrolysis achieves efficiencies of 55-80% but costs \$4.15–\$10.30 per kg, whereas pyrolysis offers 35-50% efficiency at a lower cost of \$1.59–\$2.20 per kg. In 2022, global hydrogen production primarily relied on fossil fuels, with 62% derived from natural gas and 21% from unabated coal, while only 0.7% came from low-emission sources. For storage, metal hydrides provide a high volumetric density of 120 MJ/kg, making them suitable for various applications, but their system costs remain high due to material and containment requirements (Kamran & Turzyński, 2024).

1.3 Research Methodology

This study employs a comprehensive approach, combining literature review, computational simulations, experimental analysis, and optimization techniques to evaluate advancements in hydrogen storage technologies. The methodology ensures a holistic understanding of the technical, economic, and environmental aspects of hydrogen as a sustainable energy carrier.

- Conducted systematic literature review to extract key performance metrics (e.g., storage capacities of 70 kg/m^3 at 700 bar, SEC of 5.87 kWh/kgH_2 for cryogenic systems).
- Used Aspen HYSYS and genetic algorithms for optimizing storage configurations, achieving exergy efficiency up to 52.66%.
- Performed experimental validation of novel materials like ZIF-67 composites (1.53 wt%) and cryogenic insulation systems (reducing material layers by 64.2%).

- Modelled hybrid energy systems integrating hydrogen storage with renewable energy, achieving an 80% renewable fraction and annual system cost of $603,537.85.
- Evaluated economic and environmental impacts through LCA, identifying underground hydrogen storage as the most sustainable option with a preference score of 0.487.
- Analysed policy frameworks and proposed strategic enhancements to support large-scale adoption of hydrogen storage technologies.

2. HYDROGEN STORAGE TECHNOLOGIES

Hydrogen storage technologies are essential for enabling the use of hydrogen as a clean energy carrier. Figure 1, showcases a hybrid approach to hydrogen storage, integrating liquid hydrogen storage and compressed hydrogen storage. Liquid hydrogen is delivered via trailers and stored in cryogenic vessels, maintaining extremely low temperatures. It is then converted to gaseous form using an evaporator and compressed for storage in high-pressure tanks. This dual-stage method ensures efficient storage, transportation, and dispensing of hydrogen, addressing diverse application needs in hydrogen energy systems. Such technologies play a pivotal role in advancing hydrogen as a sustainable energy carrier.

Figure 1. Schematic of liquid-to-compressed hydrogen storage and delivery system

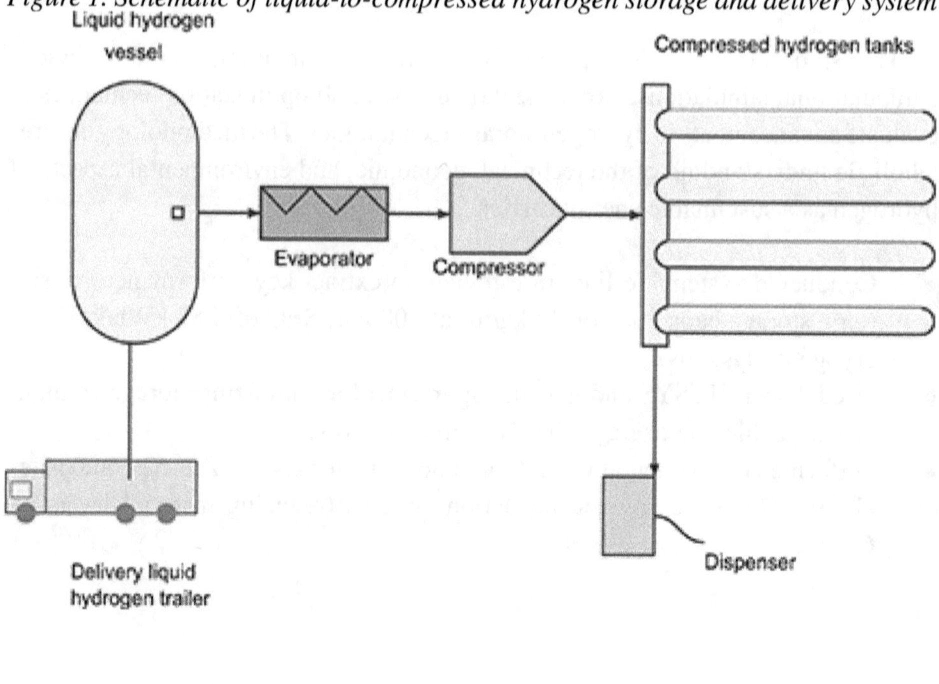

Figure 2 provides an overview of hydrogen storage technologies, categorized into four primary methods: compressed storage, liquid storage, metal hydrides, and chemical storage. Each method highlights specific considerations, mechanisms, or advancements, such as design challenges, safety issues, materials innovation, and release mechanisms. These are briefly discussed below.

Figure 2. Overview of hydrogen storage technologies and their features

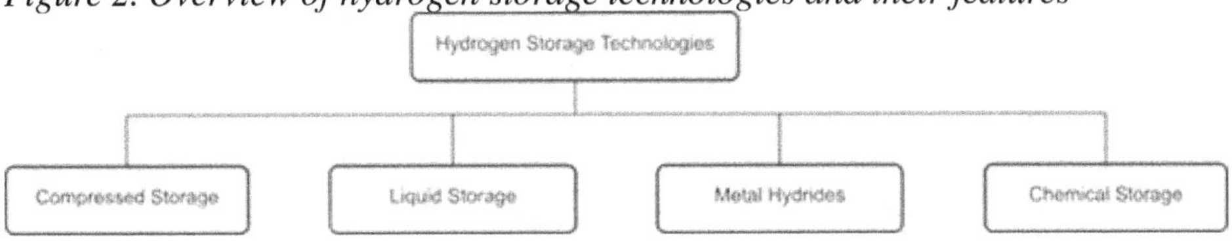

2.1 Compressed Hydrogen Storage

2.1.1 Design and Engineering Considerations

The benefits of self-pressurization in liquid hydrogen (LH2) tanks during maritime transport, achieving boil-off reductions of 31% at 5 kPaG and up to 91% at 60 kPaG. Utilizing an analytical model developed in Matlab, the research evaluated perlite insulation's thermal performance and found very low loss storage feasible in large double-wall tanks for short journeys. However, further investigation is required to understand the effects of ship motion on phase changes within the tanks (Wang et al., 2024). A cryogenic supercritical hydrogen storage system that compresses and cools hydrogen to 70 K and 30 MPa, achieving a storage density of 72.53 kg/m^3 without phase change. Aspen HYSYS software and a genetic algorithm optimize 29 variables, focusing on minimizing the specific energy consumption (SEC). Neon, identified as the most effective expansion cycle working gas, achieves an SEC of 5.87 kWh/kgH$_2$ and an exergy efficiency of 49.74%. The theoretical minimum SEC is 2.92 kWh/kgH$_2$, highlighting the system's efficiency potential. This approach demonstrates significant advancements in hydrogen storage for energy applications (Xu et al., 2024a). A novel cryogenic supercritical hydrogen cooling process using a helium expansion cycle with liquid nitrogen pre-cooling, achieving hydrogen storage at 10 MPa, −235.15 °C, and a density of 63.36 g/L. Optimized through

Aspen HYSYS and a genetic algorithm, the system achieves a SEC of 5.432 kWh/ kgH_2 and a figure of merit (FOM) of 43.88%.

Thermodynamic analysis shows a total exergy loss of 289.69 kW, demonstrating efficiency improvements over conventional liquefaction processes. This innovative approach addresses gaps in cooling technologies for supercritical hydrogen storage and offers significant advancements for the hydrogen industry (Song et al., 2023). Challenges for hydrogen storage include understanding the impact of ship motion on phase changes in maritime storage and advancing efficient cooling processes for cryogenic systems. Continued innovation is vital to unlocking hydrogen's potential as a sustainable energy carrier.

2.1.2 Recent Developments and Innovations

The development of metal hydride hydrogen compressors (MHHC) using AB5-based $La_{0.5}Ce_{0.5}Ni_{5-x}Al_x$ and AB2-based $Ti_{1-y}Zr_yMn_{0.8}Cr_{1.2}$ alloys. In single-stage compressors, these alloys achieved hydrogen pressures of 117 and 132 atm at 100 °C, and both reached 200 atm at 140 °C. The double-stage MHHC demonstrated a pressure increase from 50 to 245 atm by raising the temperature from 25 to 100 °C. The $Ti_{0.9}Zr_{0.1}Mn_{0.8}Cr_{1.2}$ alloy exhibited superior kinetics and discharging pressure, highlighting its potential for efficient hydrogen compression. These innovations contribute to more compact and efficient hydrogen storage systems (Chen et al., 2024). An isobaric compressed hydrogen energy storage system integrating pumped hydro storage and high-pressure proton exchange membrane water electrolyser technology. The system maintains constant pressure in the gas chamber, eliminating compressors and cooling units, and uses insulated reservoirs to minimize heat loss. It achieves a round-trip electrical efficiency of 35.39%, thermal efficiency of 81.03%, and exergy efficiency of 37.89%. The energy storage density is measured at 53.46 kWh/m^3, showcasing its compact and efficient energy storage potential. These innovations represent a significant advancement in hydrogen storage technologies (Cao et al., 2024).

While significant progress has been made, challenges like high costs, limited storage density, and inadequate infrastructure remain barriers to hydrogen's widespread adoption. Overcoming these hurdles will require interdisciplinary collaboration, technological innovation, and strong policy support to unlock hydrogen's full potential as a clean and sustainable energy carrier.

2.2 Liquid Hydrogen Storage

2.2.1 Cryogenic Storage Techniques

Cryogenic storage for liquid hydrogen, emphasizing minimizing heat flux using multiple insulation layers in a vacuum environment. For large-scale tanks, a medium vacuum (100–10^{-1} Pa) reduces maintenance costs while maintaining insulation efficiency. A numerical model demonstrated that optimal placement of a vapor-cooling shield on solid insulation reduced multilayer insulation layers by 64.2% at 5 Pa vacuum pressure. A total insulation thickness reduction of 51.4% was achieved at 1 Pa for a 4200 m^3 hydrogen storage tank. These findings highlight significant advancements in thermal insulation performance for efficient hydrogen storage (Kang et al., 2024).

A cryogenic supercritical hydrogen storage method that achieves a density of 70.74 kg/m^3, comparable to liquid hydrogen, addressing the challenges of low density and evaporation in traditional storage. Using dual parallel mixed refrigerant (DPMR) and dual cascade mixed refrigerant (DCMR) cycles, the processes were optimized via genetic algorithms. The DPMR achieved a specific energy consumption (SEC) of 6.422 kWh/kgH_2 and an exergy efficiency of 52.66%, while DCMR showed an SEC of 6.872 kWh/kgH_2 and an exergy efficiency of 49.24%. These advancements highlight the feasibility and energy efficiency of supercritical hydrogen storage systems over conventional methods (Xu et al., 2023). A cryogenic supercritical hydrogen storage system that cools hydrogen to 70 K and pressurizes it to 30 MPa, achieving a storage density of 72.53 kg/m^3 without phase change. Aspen HYSYS simulations and genetic algorithm optimization minimized the specific energy consumption (SEC), which was 5.87 kWh/kgH_2 for neon, the most efficient working gas. The exergy efficiency of the system reached 49.74%, showcasing its energy efficiency. A theoretical minimum SEC of 2.92 kWh/kgH_2 was calculated, highlighting the system's potential. This approach demonstrates significant advancements in high-density hydrogen storage technology (Xu et al., 2024b). While these techniques represent notable progress in liquid hydrogen storage, challenges persist, including the need to maintain uniform temperature distribution and effectively manage pressure fluctuations. Ensuring the dynamic thermal performance of storage tanks, especially during transitions from gas-liquid stratification to supercritical states, remains essential for achieving both safety and operational efficiency.

2.2.2 Safety and Efficiency Challenges

Liquid hydrogen storage poses challenges like flammability risks from its low boiling point and high energy demands for cryogenic liquefaction, leading to inefficiencies. Maintaining low temperatures with robust insulation adds significant costs, and inadequate infrastructure complicates safe transportation and storage. Advancements in high-pressure storage and cryogenic liquefaction technologies have mitigated some issues, but low storage density and high costs persist. Emerging solutions, such as metal–organic frameworks and nanomaterials, offer potential improvements. Collaboration, policy support, and ongoing research are essential to make liquid hydrogen a viable clean energy carrier (Xie et al., 2024). Challenges in liquid hydrogen storage, including high SEC, low exergy efficiency, and boil-off losses, which increase costs and safety risks. Cost-efficient liquefaction systems should achieve SEC below 6 kWh/kgLH2, efficiency above 40%, production rates over 100 tonnes/day, and investment costs of 1–2 $/kgLH2. Methods like absorption cooling cycles, mixed refrigerant systems, and LNG cold recovery can significantly reduce costs and improve efficiency.

Utilizing low-temperature waste heat for precooling offers a more efficient alternative to traditional electricity-based systems. The paper emphasizes integrating renewable energy and optimization algorithms for future advancements in hydrogen liquefaction technologies (Ghorbani et al., 2023). While liquid hydrogen storage presents significant challenges, advancements in research and technology provide hope for viable solutions. Innovations like metal-organic frameworks and nanomaterials, coupled with renewable energy integration, have the potential to improve both efficiency and safety in hydrogen storage. However, realizing these benefits will demand considerable investment and robust policy support to unlock their full potential in shaping a sustainable hydrogen economy.

2.3 Metal Hydrides

2.3.1 Mechanisms of Hydrogen Absorption and Release

The mechanisms of hydrogen absorption and release in metal hydrides are influenced by thermodynamic and kinetic barriers, such as sluggish absorption/desorption rates and high dehydrogenation temperatures. Nano structuring techniques, particularly nanoconfinement, enhance performance by improving host–guest interactions at nanointerfaces. This approach enables tailored strategies for optimizing hydrogen sorption reactions, addressing specific challenges of different metal hydrides. For instance, nanoconfined hydride composites exhibit modified thermodynamic and kinetic characteristics, leading to improved efficiency. These advancements are essential for progressing hydrogen storage technologies and supporting the hydrogen economy (Cho et al., 2023). Hydrogen absorption and release in metal hydrides involve atomic-level interactions where hydrogen dissociates and diffuses into the metal lattice, forming hydrides. Transition metal substitutions and nano-sized structures enhance hydrating and dehydrating capabilities, improving efficiency.

Materials like magnesium, Mg_2Ni, and rare earth metals are explored for their transformative impact on hydrogen storage. Challenges include activation energy barriers and oxide substitutions, which influence performance. These advancements position metal hydrides as a critical component for efficient hydrogen storage in the energy landscape (Nivedhitha et al., 2024). While these strategies offer potential, practical applications of metal hydrides for hydrogen storage still encounter challenges like limited reversibility, hydrogen losses, and slow reaction kinetics. Addressing these issues requires deeper exploration of the underlying mechanisms and the development of efficient, cost-effective solutions. Such advancements are crucial for the widespread adoption of hydrogen storage technologies.

2.3.2 Recent Advances in Metal Hydride Materials

A novel gyroid-structured canister design for metal hydride (MH) hydrogen storage, increasing hydrogen storage capacity by 66.22% compared to previous designs. Topology optimization enhances reactor capacity by 49%, maintaining bending strength up to 5000 N. Adjusting chamber offsets to 2 mm expands storage capacity by 22% while keeping charging time under 4500 seconds. The combined methods improve storage density while addressing limitations of MH materials in mobility applications. This innovative approach showcases significant advancements in MH hydrogen storage design (Lesmana & Aziz, 2024). A novel metal hydride reactor design integrated with honeycomb fins and phase change material (PCM) to enhance hydrogen storage efficiency. This approach improves heat transfer, reducing hydrogen absorption and desorption times by approximately 86.4 and 81.1%, respectively. The reactor achieves a hydrogen storage capacity of 1.326 wt%, with gravimetric and volumetric densities of 0.411% and 14.76 kg/m^3, respectively. Simulations highlight the benefits of sandwich PCM integration for thermal management. The findings showcase significant advancements in hydrogen storage technologies through improved energy efficiency and reaction rates (Dong et al., 2024). While these advancements hold significant potential, challenges persist in meeting the required gravimetric and volumetric storage densities for commercial viability. To ensure practical implementation in hydrogen energy systems, it is crucial to balance the integration of innovative materials and reactor designs with cost-effectiveness and scalability. These factors will determine the feasibility of transitioning from research to widespread industrial application.

2.4 Chemical Hydrogen Storage

2.4.1 Chemical Reactions for Hydrogen Release

A novel two-step strategy to synthesize high-dispersion gold-palladium alloy nanodots (AuPd alloy NDs), achieving nearly 100% formic acid conversion at room temperature with a remarkable turnover frequency of 9049 h^{-1}. The engineered Au3Pd7/CN0.25 catalyst also enables CO_2 hydrogenation back to formic acid, yielding 90.8% under mild conditions. Detailed studies using isotope labelling and infrared spectroscopy reveal a five-step reaction mechanism involving critical *HCOO species and C—H bond cleavage. This work highlights the potential for constructing a complete CO_2-mediated hydrogen release-storage system. Such advancements showcase the synergy between microstructure engineering and catalytic performance optimization (Luo et al., 2024). The electrochemical release of hydrogen from hydrogen boride (HB) sheets, achieving a Faradaic efficiency exceeding 90% without requiring additional proton sources.

Hydrogen release occurs at an onset potential of -0.445 V versus Ag/Ag+, which is more favourable than other proton sources like water or formic acid. Long-term experiments reveal nearly 100% of the hydrogen capacity of HB sheets can be released, highlighting their efficiency. The process is economical and viable, positioning HB sheets as safe, lightweight hydrogen carriers. These findings support the potential of HB sheets in sustainable hydrogen storage and energy applications (Kawamura et al., 2024). While these approaches hold significant promise, challenges persist in enhancing the efficiency and scalability of hydrogen release systems. Developing catalysts and materials capable of operating under mild conditions with high efficiency is essential for real-world applications. Moreover, integrating these advanced systems into existing energy infrastructures demands further research and innovation to ensure seamless adoption and functionality.

2.4.2 Current Research on Ammonia and Other Carriers

MoC/NFC catalyst is synthesized via liquid-phase plasma, achieving a maximum ammonia yield of 115 $\mu g\ h^{-1}\ mg^{-1}_{cat.}$ with a Faradaic efficiency of 1.15% at -0.7 V vs RHE in 0.1 M KOH electrolyte. Pyridinic- and pyrrolic-N atoms near carbon pores acted as active sites for N_2 adsorption, promoting N_2 triple bond cleavage. Fluorine doping further enhanced N_2 activation due to its high electronegativity, improving catalytic performance. This innovative approach addresses challenges in electrochemical ammonia synthesis under ambient conditions. The findings advance sustainable NH_3 production for renewable hydrogen storage solutions (Park et al., 2024). Ammonia decomposition is an efficient method for hydrogen production, emphasizing its role as a sustainable hydrogen carrier. It reviews various catalysts, including Ru-based, Ni-based, and carbide/nitride catalysts, showcasing their performance and catalytic mechanisms.

Challenges like low efficiency and high costs are identified, necessitating continued research to overcome these barriers. The study underscores the potential of ammonia decomposition to support hydrogen energy storage in clean energy transitions. Advancements in catalyst development remain key to enhancing this technology's efficiency and scalability (X. Li et al., 2024). While ammonia offers significant potential as a hydrogen storage solution, challenges persist in areas like catalyst optimization, improving reaction kinetics, and seamless system integration. Advancing these aspects through ongoing research and technological innovation is essential to unlocking ammonia's full capability in supporting the hydrogen economy.

Table 1 provides a comprehensive comparison of the four primary hydrogen storage methods compressed hydrogen, liquid hydrogen, metal hydrides, and chemical storage. It highlights their main features into their suitability for various applications in the hydrogen energy landscape.

Table 1. Comparison of hydrogen storage technologies

Aspect	Compressed Hydrogen	Liquid Hydrogen	Metal Hydrides	Chemical Storage
Energy Density	Moderate	High	Moderate	High
Storage Conditions	High pressure (350–700 bar)	Cryogenic (-253°C)	Ambient to moderate temperatures	Ambient conditions
Safety	Requires robust containment	Boil-off risks, insulation critical	Safe, low risk of leakage	Safe, but requires catalysts for release
Energy Efficiency	High energy for compression	High energy for liquefaction	Moderate energy for absorption/desorption	Moderate energy for release
Scalability	Suitable for small to medium-scale	Suitable for large-scale	Suitable for small to medium-scale	Suitable for medium to large-scale
Cost	Moderate (tanks, compression)	High (liquefaction, insulation)	Moderate to high (material costs)	Moderate to high (catalyst, infrastructure)
Applications	Transportation, fuel cells	Aerospace, large-scale storage	Portable devices, stationary storage	Industrial, large-scale storage

3. RECENT ADVANCES IN HYDROGEN STORAGE

3.1 Innovations in Materials Science for Hydrogen Storage

ZIF-67-PA12 composites synthesized through an ex-situ method, significantly enhancing hydrogen storage capacities. ZIF-67-PA12-30(p) achieved a storage capacity of 1.53 wt.%, compared to 0.38 wt.% for pure PA12, showcasing the material's efficiency. The ex-situ approach allows high ZIF-67 loadings (up to 60 wt.%) and addresses challenges in traditional in situ methods, such as limited porosity and processability. Advantages include improved repeatability, cost-efficiency, and ease of production, making these composites practical for energy storage. These findings highlight advancements in integrating MOFs with polymers for hydrogen storage technologies (Shang et al., 2024). Hydrophobic mesoporous silica as a host material for hydrogen storage, enabling the formation of clathrate hydrates at pressures approximately 20% lower than bulk systems. Nano-confinement within the silica alters water properties, making hydrogen storage feasible under milder conditions. In-situ inelastic neutron scattering (INS) and neutron diffraction (ND) confirmed the stability of hydrogen clathrate hydrates within the porous structure. These findings highlight a safe and efficient method for hydrogen storage, addressing challenges like high energy costs and low hydrogen density. The approach advances the potential for practical hydrogen storage applications (Ciocarlan et al., 2024). While these innovations offer significant promise, challenges like ensuring long-term stability, reducing costs, and achieving scalability remain obstacles to widespread adoption. Addressing these issues requires sustained interdisciplinary research and technological advancements. Collaborative efforts across fields are essential to overcome these barriers and unlock the full potential of hydrogen as a clean and sustainable energy source.

3.2 Breakthroughs in Nanotechnology and Porous Materials

Silicon nanostructures (SiNSs), such as silicon nanowires and quantum dots, as promising materials for solid-state hydrogen storage, offering a high storage capacity of up to 6 wt.%. These materials address challenges like surface oxidation and reversibility issues common in traditional methods. Catalysts play a vital role in enhancing hydrogen desorption and improving efficiency, making SiNSs key to advancing hydrogen energy storage. This innovation is particularly impactful in automotive applications, supporting the transition to sustainable energy solutions (Saeed et al., 2024).

High-pressure compaction of the UiO-66 metal-organic framework (MOF) increases its volumetric hydrogen storage capacity to 40 g/L while maintaining its gravimetric uptake of 6.5 wt.%. Co-pelletization was used to create UiO-66/nanofibre monoliths, further enhancing the deliverable hydrogen storage capacity by 30%. This also identified an optimal activation temperature of ≤ 150°C for UiO-66 to ensure efficient hydrogen storage performance. These advancements in MOF-based materials are pivotal for sustainable energy applications, providing scalable solutions for hydrogen storage (Langmi et al., 2024). While these advancements are promising, significant challenges persist in the commercialization of hydrogen storage technologies. Material stability, cost, and scalability remain key hurdles that must be addressed to unlock the full potential of nanotechnology and porous materials. Furthermore, the development of sustainable and economically viable storage solutions is critical for enabling the widespread adoption of hydrogen as a clean and reliable energy source. Continued innovation and interdisciplinary collaboration will be essential to overcoming these barriers.

3.3 Advances in Hybrid Storage Systems

The optimal configuration for a hybrid energy system, including a 589.58 kW photovoltaic (PV) system, a 664 kW wind turbine (WT), a 675-kW supercapacitor, and a 1000 kWh battery bank. This setup achieves an 80% renewable energy fraction and reduces the annual system cost to $603,537.85. Among six optimization algorithms evaluated, the CGO algorithm demonstrated superior efficiency, achieving solutions with fewer iterations. The system maintains a competitive levelized cost of electricity (LCOE) at $0.2380 per kWh. These advancements underscore the importance of integrated storage solutions in enhancing renewable energy performance (Fatih Guven et al., 2024). A hybrid system integrating advanced adiabatic compressed air energy storage with a reverse osmosis system, achieving an 11.3% increase in generated power during discharge and a 1.2% improvement in round-trip efficiency compared to separate systems. Reducing the compressed air tank capacity from 1200 to 800 m^3 further increases the efficiency difference to 5.6%. The system demonstrates enhanced energy storage efficiency and freshwater production capabilities. An optimal compression ratio enables a maximum round-trip efficiency of 61.9%. Exergy analysis identifies the compressed air tank and turbines as the most exergy-destructive components (Javaheri & Shafiei Ghazani, 2023). While these advances highlight the promise of hybrid storage systems, significant challenges persist, including high costs, scalability issues, and integration with current infrastructure. Addressing these hurdles requires sustained research and development to unlock the full potential of these technologies. By overcoming these barriers, hybrid storage systems can play a crucial role in advancing sustainable energy solutions.

4. INTEGRATION WITH RENEWABLE ENERGY SYSTEMS

Figure 3 illustrates how renewable energy sources such as wind turbines, solar panels, hydroelectric dams, and nuclear energy integrate seamlessly with an electrolyser to produce hydrogen. Electricity from these renewable sources powers the electrolyser, which uses water to generate hydrogen through electrolysis. The produced hydrogen is then distributed to various applications, including fuel cells, transportation, and energy storage, enabling a sustainable energy ecosystem. This integration supports decarbonization efforts and maximizes the utilization of renewable energy resources.

Figure 3. Integration of renewable energy systems with hydrogen production and utilization

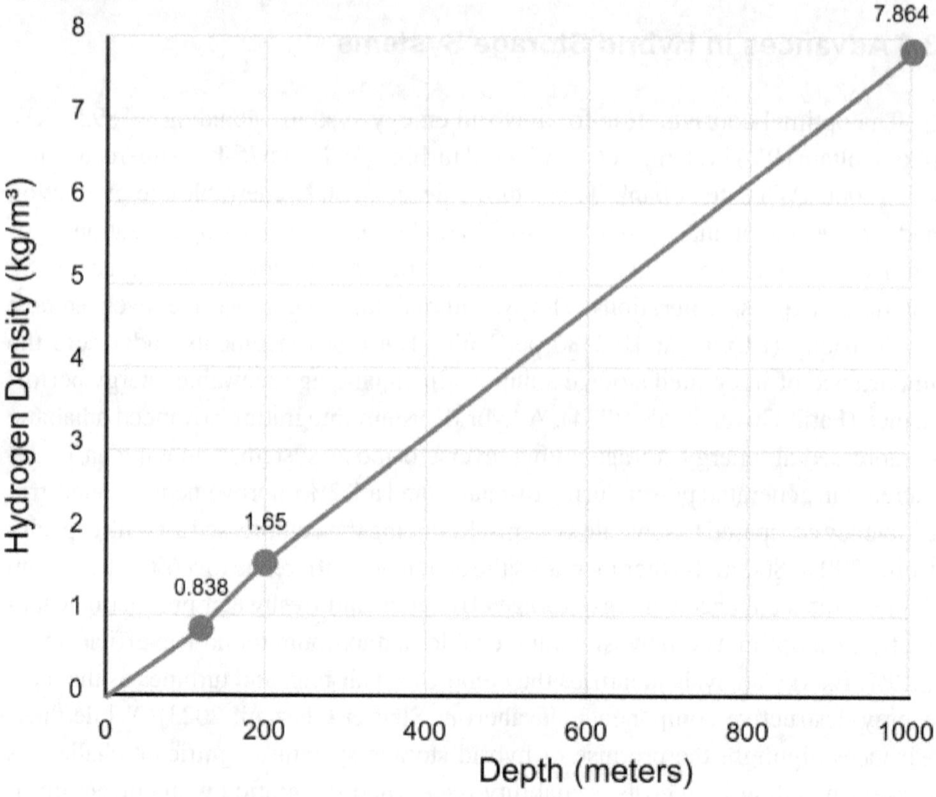

4.1 Role of Hydrogen Storage in Balancing Renewable Energy Supply

Hydrogen energy storage (HES) effectively balances power systems when renewable energy (RE) generation reaches 44.33–54.78% of total power generation, significantly reducing system costs. By 2060, RE is expected to account for 63.12–91.1% of power generation in China, with HES contributing 3.58–18.02% of electricity consumption. Provinces with abundant wind resources may see a 36.43% higher demand for HES compared to solar-dominated regions. The study utilizes a simulation model to analyse HES pathways, addressing cost and scalability challenges in RE integration. This highlights HES's pivotal role in achieving China's carbon neutrality goals (Zhang et al., 2024). Hydrogen storage is vital for balancing renewable energy supply, addressing the variability of renewable generation, and ensuring a reliable power supply over weeks and months. It enables long-duration energy storage (LDES), allowing excess energy from peak periods to be stored and used during low generation times. This highlights hydrogen's role in reducing zero-emission system costs while enhancing energy self-sufficiency. Prioritizing domestic hydrogen production and strategic imports can mitigate risks from fluctuating battery and renewable energy costs. This approach supports decarbonization and informs policies for a sustainable energy future. While hydrogen storage offers significant benefits, challenges like advancing technology, ensuring economic viability, and developing adequate infrastructure persist. Integrating hydrogen storage with other energy systems demands careful planning to optimize its advantages and mitigate potential risks. Addressing these issues is essential to fully harness hydrogen's potential as a clean and reliable energy solution.

4.2 Case Studies on Coupling Hydrogen Production with Wind and Solar Energy

Green hydrogen production probabilities using data from a 3 MWp photovoltaic system and a 3.45 MW wind turbine. It employs the Metalog probability distributions to model power generation and hydrogen output, integrating energy from both sources. By interpolating time series data, the combined outputs provide a robust model for estimating hydrogen production probabilities. This approach incorporates geographic and hardware contexts, supporting preliminary assessments of large-scale hydrogen production. The findings demonstrate a practical application of data-driven analysis in renewable energy integration (Dudziak et al., 2024). Hydrogen production for a ceramics factory using solar photovoltaics and wind energy, applying a multipurpose profit-maximizing model implemented in GAMS. Results show that selling hydrogen to the gas grid is more profitable in low-gas-price scenarios, while self-consumption is advantageous when gas prices are high, reducing reliance on natural gas. Electricity self-consumption is crucial for project profitability, as the model highlights its significant impact on economic outcomes. This approach provides valuable insights into optimizing hydrogen production and usage under varying market conditions (Sousa et al., 2024). While these case studies highlight the promise of integrating hydrogen production with wind and solar energy, challenges like geographic variability, fluctuating market conditions, and seamless technological integration persist. Overcoming these hurdles necessitates comprehensive planning and advanced optimization strategies to fully harness the potential of renewable hydrogen systems and ensure their long-term viability.

5. ECONOMIC AND ENVIRONMENTAL CONSIDERATIONS

5.1 Cost Analysis of Hydrogen Storage Solutions

The levelized cost of underground hydrogen storage (LCHS) for an Austrian project is 0.80 €/kg, with costs divided among facilities (0.33 €/kg), wells (0.09 €/kg), cushion gas (0.23 €/kg), and operational expenses (0.16 €/kg). By integrating renewable energy sources like wind and photovoltaics with hydrogen generation and storage, seasonal energy storage becomes feasible and cost-effective. Optimizing the renewable energy source (RES)-to-electrolyser capacity ratio can lower hydrogen production costs, with LCOH ranging from 6.26 €/kg to 2.61 €/kg when electricity sales revenues are included. Process modelling highlights the importance of adjusting RES sizing to market dynamics, further enhancing cost savings. These findings emphasize the economic viability of renewable hydrogen systems for addressing energy supply-demand mismatches (Clemens et al., 2024). Hydrogen storage is the most cost-effective long-term solution for energy storage, outperforming pumped hydro, compressed air, and batteries in a 100% renewable energy scenario. It demonstrates that gas turbines powered by 100% hydrogen are more economical at 0.207 €/kWh compared to fuel cells at 0.284 €/kWh. The research emphasizes the role of hydrogen in managing power fluctuations through the Power-to-Power (P2P) process, involving conversion to hydrogen using PEM electrolysers and storage in salt caverns. This approach supports decarbonization and efficient energy management. Hydrogen's competitive levelized cost of storage makes it a viable option for seasonal energy storage (Martínez de León et al., 2023). While hydrogen storage offers a promising pathway for long-term energy solutions, its implementation faces challenges related to site-specific performance and economic feasibility. Costs can fluctuate significantly depending on the chosen storage technology and regional characteristics, such as geological conditions and energy market dynamics. Addressing these variations requires targeted advancements in storage materials, infrastructure, and system integration. Continued innovation and investment are essential to reduce costs and ensure hydrogen's competitiveness as a scalable energy storage solution.

5.2 Life Cycle Assessment (LCA) of Hydrogen Storage Technologies

Life cycle assessments of hydrogen production technologies, focusing on alkaline water electrolysis (AWE), proton exchange membrane water electrolysis (PEMWE), solid oxide electrolysis (SOEC), and anion exchange membrane water electrolysis (AEMWE). It highlights the environmental impacts and recent advancements in these methods. PEMWE, when integrated with renewable energy, stands out for its rapid response and efficiency, reducing carbon emissions. The study provides comparative insights into operational efficiencies and scalability of each technology. It emphasizes advancements in membrane technology and electrocatalysts for sustainable hydrogen production (Shaya & Glöser-Chahoud, 2024). The systematic review analysed 32 studies on hydrogen production and utilization technologies from 2010 to 2022, highlighting gaps in applying prospective life cycle assessment (LCA) methodologies. Most studies relied on attributional approaches, lacking forward-looking elements in life cycle inventory (LCI) modelling. Common issues include limited use of prospective life cycle impact assessment (LCIA) methods, inadequate data quality evaluation, and insufficient sensitivity analysis. These findings stress the need for robust methodologies for low-technology readiness level hydrogen technologies. The review emphasizes improving LCA practices to enhance the reliability of environmental assessments (Martinez et al., 2024). Harmonized LCA methodologies and comprehensive assessments are essential to accurately evaluate the sustainability of hydrogen technologies and identify areas for improvement.

5.3 Environmental Impacts and Sustainability Metrics

Compressed Hydrogen Gas (CHG), Cryogenic Liquid Hydrogen (CLH), Metal Hydride (MH), and Underground Hydrogen (UH) is evaluated using the Interval-Valued Intuitionistic Fuzzy Analytic Hierarchy Process (IVIF-AHP) to assess sustainability across economic, environmental, social, and technical dimensions. CHG emerged as the most promising option with a preference score of 0.487, followed by UH at 0.453, while MH scored the lowest at 0.301. The evaluation criteria were decomposed into sub-criteria for a comprehensive assessment, revealing the trade-offs among these options. This approach highlights the need for robust multicriteria tools to guide sustainable hydrogen storage technology decisions (Acar et al., 2023).

Hydrogen's potential as a sustainable fuel due to its renewable nature and low emissions during combustion. It emphasizes the importance of developing innovative nanomaterials for solid-state hydrogen storage to address challenges in secure and efficient storage. This focuses on enhancing physisorption uptake and engineering advancements to improve storage capacity under practical conditions. By advancing economically viable hydrogen storage solutions, the research aims to reduce greenhouse gas emissions and mitigate environmental pollution. These developments are critical for the commercialization of hydrogen as an alternative fuel for automobiles (Attia et al., 2024). Hydrogen storage technologies offer promising opportunities for sustainable energy solutions, but scaling them up to meet global energy demands remains a significant challenge. Advancing these technologies requires ongoing research, innovation, and engineering breakthroughs. Equally important is robust policy support to drive investments and create favourable conditions for large-scale adoption. Addressing these challenges is essential to unlocking the full potential of hydrogen as a reliable and sustainable energy carrier.

6. REGULATORY FRAMEWORK AND POLICY LANDSCAPE

6.1 Current Policies Supporting Hydrogen Development

The need for transitional regulation mechanisms to support Power-to-Power (P2P) systems, enabling seasonal storage of renewable energy surplus as hydrogen. It emphasizes public policies to create market opportunities, enhance energy surplus valuation, and reduce dependency, promoting decarbonization of the electric power sector. A case study outlook between (2030–2040) in Portugal demonstrates the feasibility of storing renewable energy surplus as hydrogen and converting it back into electricity, ensuring generation adequacy. The study integrates long-term adequacy assessments with regulatory evaluations, showcasing the economic viability of these solutions. Proposed strategies aim to ensure energy security while supporting hydrogen's role as a sustainable energy carrier (Henrique Santos et al., 2024). Large-scale green hydrogen storage and transportation methods, including compressed hydrogen, liquid hydrogen, blending with natural gas pipelines, and ammonia as a carrier. It highlights global projects and policies supporting these technologies, emphasizing the need for technical and economic evaluations, such as using salt caverns for storage.

Costs for storage and transportation are projected to decrease over time due to technological advancements and economies of scale. The study stresses optimizing efficiency and leveraging existing infrastructure to overcome challenges in commercialization. These efforts are essential for a sustainable and viable green hydrogen industry (Ma et al., 2023). While these policies play a crucial role in advancing hydrogen storage technologies, challenges persist, including the development of comprehensive infrastructure and the need to optimize storage and transportation efficiency. Addressing these issues requires continued interdisciplinary collaboration and sustained policy support to unlock hydrogen's full potential as a clean and sustainable energy carrier.

6.2 Recommendations for Policy Enhancements to Foster Innovation

The need for ongoing research, technological advancements, and cross-sector collaboration to maximize hydrogen's eco-friendly and low-carbon potential. It emphasizes creating supportive regulatory frameworks, fostering partnerships, and incentivizing investment in hydrogen technologies. Hydrogen's versatility in transportation, manufacturing, and energy storage is underscored, with recommendations to address production and integration challenges. The method evaluates of hydrogen production from renewable and fossil fuel sources, focusing on their technological, economic, and environmental impacts. Public awareness and education are highlighted as crucial for driving innovation and adoption in hydrogen energy systems (Sikiru et al., 2024).

The critical role of green and blue hydrogen in reducing greenhouse gas emissions and advancing a sustainable energy future. It emphasizes the need for scalable production methods, such as solar-thermochemical and photo-electrochemical processes, to achieve cost-effective green hydrogen. The study underscores the importance of policy support, market incentives, and technological advancements in promoting hydrogen as a key energy vector. Addressing challenges like resource availability and geopolitical factors is crucial to establishing a global hydrogen economy. International cooperation and renewable energy expansion are identified as essential for fostering a low-carbon future (Jaradat et al., 2024). While these recommendations offer a promising pathway to advance hydrogen storage technologies, it is essential to balance innovation with environmental sustainability and economic feasibility. Policymakers must carefully ensure that progress in hydrogen storage does not unintentionally cause environmental harm or worsen resource scarcity. Moreover, continuous evaluation and adaptation of policies will be critical to addressing new challenges and leveraging opportunities as the hydrogen energy landscape evolves.

7. CHALLENGES AND BARRIERS TO ADOPTION

7.1 Technical Challenges in Scaling Up Hydrogen Storage Solutions

The transformative potential of hydrogen energy systems, with the capability to reduce global CO_2 emissions by up to 830 million tonnes annually due to hydrogen's high energy density of 120 MJ/kg. Current challenges include electrolysis inefficiencies (60–80%) and high production costs around \$5 per kilogram. Strategic advancements are needed to lower costs below \$2 per kilogram and improve efficiencies above 80%. Additionally, addressing hydrogen storage challenges, requiring pressures up to 700 bar or temperatures below −253°C, necessitates advanced materials and infrastructure improvements. These developments are crucial for maximizing hydrogen's role as a sustainable energy solution (Sadeq et al., 2024). The optimization challenges in designing large-scale hydrogen storage pipeline networks by introducing a mixed integer nonlinear optimization model and a novel hybrid genetic algorithm (HGA). This algorithm, combining the Modified Feasible Directions Method (MFDM) and Genetic Algorithm Theory (GAT), achieves superior performance, reducing convergence costs to 232.44×10^4 Yuan in continuous space and 212.16×10^4 Yuan in discrete space. The study highlights that wellhead temperature significantly impacts investment costs, while ambient temperature influences iteration curves. These findings highlight the technical challenges and optimization needs for scaling hydrogen storage solutions effectively (Liu et al., 2024). While these challenges are substantial, they open pathways for innovation and collaboration. Overcoming these hurdles requires interdisciplinary efforts, robust policy support, and continued research. Integrating renewable energy sources with hydrogen production and storage systems can enhance efficiency and sustainability, paving the way for hydrogen to play a pivotal role in the transition to a low-carbon future.

7.2 Market Barriers and Public Perception Issues

The design of a large-scale transportable liquid hydrogen export terminal with a daily production capacity of 120 tons and a storage capacity of 75,000 m^3, powered entirely by renewable energy sources. It identifies key technical challenges, including scaling up single-stack electrolysis systems, improving liquefaction processes, and enhancing vacuum-insulated liquid hydrogen storage systems. These advancements are essential to reduce boil-off losses and ensure structural integrity under variable conditions. The study proposes potential solutions based on current technology readiness while emphasizing the need for future developments. Addressing these challenges is critical for enabling efficient large-scale hydrogen exportation (Kim et al., 2024). While these challenges are substantial, ongoing research and advancements in technology provide promising solutions. Integrating machine learning with computational chemistry can accelerate the development of advanced hydrogen storage materials, improving efficiency and reducing costs. Policy support and interdisciplinary collaboration remain vital to overcoming technical hurdles and ensuring scalability. These efforts collectively facilitate the transition toward a hydrogen-based energy system. The combined approach of innovation and strategic planning paves the way for a sustainable energy future.

8. FUTURE DIRECTIONS AND RESEARCH OPPORTUNITIES

8.1 Emerging Trends in Hydrogen Storage Research

Emerging hydrogen storage research focuses on biomass-derived porous carbon materials, valued for their cost-effectiveness, high specific surface area, and structural diversity. These include activated porous carbons and heteroatom-doped variants, achieving enhanced hydrogen uptake capacities. Plant- and animal-based biomass are explored as raw materials, offering sustainable and abundant sources. However, further research is essential to address technological challenges and optimize their potential in the hydrogen economy. This advancement promises safe, efficient storage solutions while leveraging renewable resources for sustainability (Elyasi et al., 2024). Emerging trends in hydrogen storage research highlight LOHCs as a viable solution, compatible with existing gasoline infrastructure and addressing limitations of traditional methods. The main systems like methylcyclohexane/toluene and H-18-dibenzyltoluene/dibenzyltoluene show promise for large-scale applications.

Multifunctional heterogeneous catalysts, including monometallic, bimetallic, and pincer complexes, are being developed to enhance efficiency in reversible dehydrogenation and hydrogenation processes. These advancements aim to mitigate the challenges posed by the endothermic nature of these reactions, improving cost-effectiveness and energy efficiency (Jangir & Jagirdar, 2024). While these advancements are promising, challenges persist in scaling up hydrogen storage technologies, managing costs, and seamlessly integrating them with existing infrastructure. Overcoming these barriers requires sustained research, innovation, and development efforts. By addressing these issues, hydrogen's potential as a sustainable and versatile energy carrier can be fully realized, paving the way for a cleaner energy future.

8.2 Potential Future Technologies for Enhanced Storage Solutions

The potential of metal-organic frameworks (MOFs) for hydrogen storage, focusing on innovative design strategies like supramolecular Building Layers and MOF Pillaring Design. Notable examples include she-MOF-1, with a hydrogen storage capacity of 12.60 wt % at 77 K and 100 bars, and MOF-5, featuring a BET surface area of 3512 m^2/g. Fluorination significantly enhances storage, as seen in Co-FINA-1 with 1.97 wt % at 77 K and 1 bar. Advanced computational methods, such as Density Functional Theory (DFT) and Grand Canonical Monte Carlo (GCMC) simulations, are crucial for optimizing MOF designs. Despite progress, challenges like safety and practical conditions require further research for scalable applications (Yuvaraj et al., 2024). A modified zeolite material with tailored pores and surface chemistry, enabling hydrogen storage as hydrates at significantly reduced pressures of 8–10 bar, two orders of magnitude lower than current materials. This advancement addresses key challenges like high operating pressures and slow formation kinetics in hydrogen storage systems. The material enhances hydrogen solubility and facilitates fast charging and ambient temperature discharging, improving efficiency. This novel approach sets the stage for further optimization in storage efficiency and kinetics. The study highlights its potential to advance sustainable hydrogen storage solutions for the hydrogen economy (Firuznia et al., 2023). While these technologies demonstrate significant potential, challenges persist, including the practicality of storing hydrogen at extremely low temperatures and high pressures, as required by some metal-organic frameworks. The seamless integration of these materials into existing infrastructure and ensuring their economic feasibility remain crucial hurdles. However, with ongoing research and advancements, these innovative approaches could revolutionize hydrogen storage, paving the way for more efficient and sustainable energy systems.

Figure 4 highlights the emerging trends such as biomass-derived porous carbon materials and liquid organic hydrogen carriers, along with future technologies like metal-organic frameworks and advanced computational methods. It also identifies critical challenges, including scaling up technologies and managing costs, emphasizing the balance between innovation and feasibility in hydrogen storage development.

Figure 4. Emerging trends, future technologies, and challenges in hydrogen storage materials

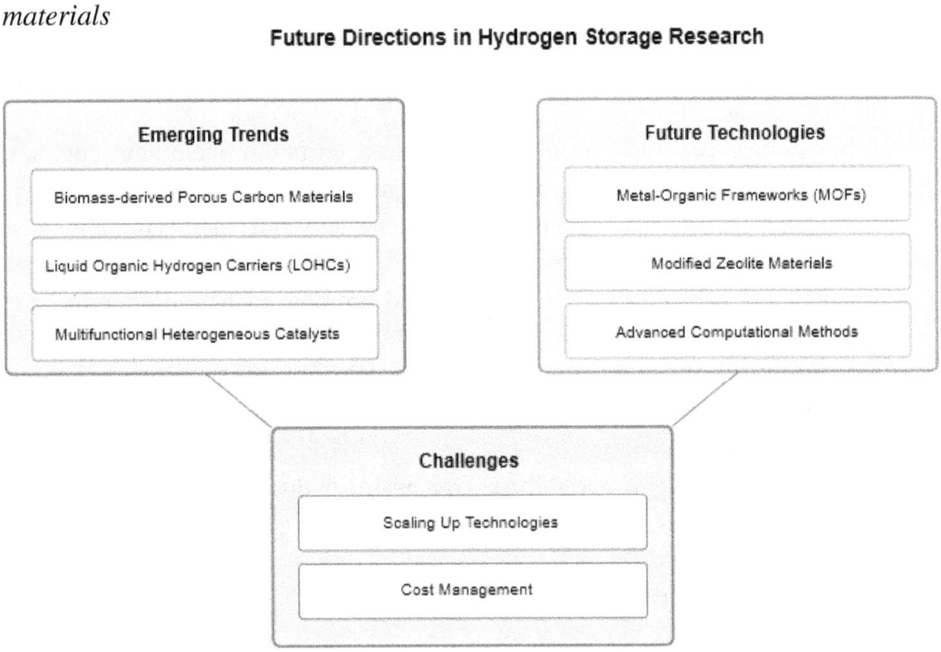

9. CONCLUSIONS AND RECOMMENDATIONS

9.1 Conclusions

Hydrogen storage technologies play a pivotal role in creating sustainable and carbon-neutral energy systems. Advancements in storage methods, materials, and integration with renewable energy sources highlight hydrogen's potential, yet significant challenges in cost, scalability, and infrastructure remain to be addressed.

- Hydrogen offers an energy density of 142 MJ/kg, surpassing conventional fuels, making it an ideal energy carrier.
- Innovations like metal hydrides (over 5 wt% storage efficiency) and MOFs (up to 12.6 wt% at 77 K and 100 bars) have advanced storage capabilities.
- Cryogenic hydrogen storage systems have achieved a storage density of 72.53 kg/m^3, with improved insulation reducing energy losses by over 60%.
- Chemical storage solutions, such as LOHCs, demonstrate efficiencies of up to 85%, making them suitable for large-scale hydrogen transportation.

- Hybrid energy systems combining hydrogen with solar and wind energy have achieved an 80% renewable fraction and competitive electricity costs.
- Significant challenges remain, including high production costs ($4.15–$10.30/kg for electrolysis), storage inefficiencies, and boil-off losses in cryogenic systems.
- Infrastructure development for hydrogen transportation, including pipelines and export terminals, needs to scale up significantly.
- Public perception and awareness of hydrogen safety and benefits require more targeted outreach and education.
- Policies and subsidies to reduce costs and promote hydrogen adoption are essential for accelerating the transition to hydrogen-based energy systems.
- Advanced computational tools like Aspen HYSYS and genetic algorithms have significantly optimized storage processes, reducing energy consumption and improving efficiency.
- Addressing geographic variability in renewable hydrogen production can further enhance global adoption and sustainability.

9.2 Recommendations

- Increase research funding for advanced hydrogen storage materials, including nanotechnology, MOFs, and chemical hydrogen carriers.
- Develop international standards for hydrogen storage, transportation, and safety to streamline global adoption.
- Incentivize the development of renewable hydrogen production methods, such as water electrolysis, by providing subsidies and tax benefits.
- Expand infrastructure for hydrogen production, storage, and distribution, including pipelines and large-scale export terminals.
- Promote collaboration between governments, industries, and academia to drive innovation and scale-up technologies.
- Leverage public-private partnerships to accelerate investment in hydrogen ecosystems and reduce dependency on fossil fuels.
- Enhance integration of hydrogen storage systems with renewable energy grids to stabilize supply and demand.
- Implement educational campaigns to increase public trust in hydrogen technologies and highlight their role in achieving a low-carbon future.
- Focus on reducing the cost of hydrogen production to below $2/kg to make it competitive with traditional energy sources.
- Explore the use of waste heat and renewable energy sources to improve hydrogen liquefaction and storage efficiencies.

REFERENCES

Acar, C., Haktanır, E., Temur, G. T., & Beskese, A. (2023). Sustainable stationary hydrogen storage application selection with interval-valued intuitionistic fuzzy AHP. *International Journal of Hydrogen Energy*, *49*, 619–634. DOI: 10.1016/j.ijhydene.2023.10.081

Attia, N. F., Elashery, S. E. A., Nour, M. A., Policicchio, A., Agostino, R. G., Abd-Ellah, M., Jiang, S., & Oh, H. (2024). Recent advances in sustainable and efficient hydrogen storage nanomaterials. *Journal of Energy Storage*, *100*, 113519–113519. DOI: 10.1016/j.est.2024.113519

Cao, R., Li, W., Chen, Z., & Li, Y. (2024). Development and assessment of a novel isobaric compressed hydrogen energy storage system integrated with pumped hydro storage and high-pressure proton exchange membrane water electrolyzer. *Energy*, *294*, 130798. Advance online publication. DOI: 10.1016/j.energy.2024.130798

Chen, Y. H., Shen, C. C., Hsu, C. W., & Lee, S. L. (2024). Development of a metal hydride hydrogen compressor with high-pressure hydrogen storage alloys. *Materials Chemistry and Physics*, *314*, 128917. Advance online publication. DOI: 10.1016/j.matchemphys.2024.128917

Cho, Y. J., Cho, H., & Cho, E. S. (2023). Nanointerface Engineering of Metal Hydrides for Advanced Hydrogen Storage. *Chemistry of Materials*, *35*(2), 366–385. DOI: 10.1021/acs.chemmater.2c02628

Ciocarlan, R. G., Farrando-Perez, J., Arenas-Esteban, D., Houlleberghs, M., Daemen, L. L., Cheng, Y., Ramirez-Cuesta, A. J., Breynaert, E., Martens, J., Bals, S., Silvestre-Albero, J., & Cool, P. (2024). Tuneable mesoporous silica material for hydrogen storage application via nano-confined clathrate hydrate construction. *Nature Communications*, *15*(1), 8697. DOI: 10.1038/s41467-024-52893-3 PMID: 39379386

Clemens, T., Hunyadi-Gall, M., Lunzer, A., Arekhov, V., Datler, M., & Gauer, A. (2024). Wind–Photovoltaic–Electrolyzer-Underground Hydrogen Storage System for Cost-Effective Seasonal Energy Storage. *Energies*, *17*(22), 5696–5696. DOI: 10.3390/en17225696

Dong, X., Zhao, H., Li, H., Fucucci, G., Zheng, Q., Zhao, H., & Pu, J. (2024). A novel design of a metal hydride reactor integrated with phase change material for H2 storage. *Applied Energy*, *367*, 123321. Advance online publication. DOI: 10.1016/j.apenergy.2024.123321

Dudziak, A., Małek, A., Marciniak, A., Caban, J., & Seńko, J. (2024). Probabilistic Analysis of Green Hydrogen Production from a Mix of Solar and Wind Energy. *Energies*, *17*(17), 4387. Advance online publication. DOI: 10.3390/en17174387

Elyasi, S., Saha, S., Hameed, N., Mahon, P. J., Juodkazis, S., & Salim, N. (2024). Emerging trends in biomass-derived porous carbon materials for hydrogen storage. *International Journal of Hydrogen Energy*, *62*, 272–306. DOI: 10.1016/j.ijhydene.2024.02.337

Fatih Guven, A., Abdelaziz, A. Y., Mahmoud Samy, M., & Barakat, S. (2024). Optimizing energy Dynamics: A comprehensive analysis of hybrid energy storage systems integrating battery banks and supercapacitors. *Energy Conversion and Management*, *312*, 118560. Advance online publication. DOI: 10.1016/j.enconman.2024.118560

Firuznia, R., Abutalib, A., Hakimian, A., Nazifi, S., Huang, Z., Lee, T. R., Rimer, J. D., & Ghasemi, H. (2023). High-capacity hydrogen storage through molecularly restructured and confined hydrogen hydrates. *Materials Today Physics*, *38*, 101248. Advance online publication. DOI: 10.1016/j.mtphys.2023.101248

Ghorbani, B., Zendehboudi, S., Saady, N. M. C., Duan, X., & Albayati, T. M. (2023). Strategies To Improve the Performance of Hydrogen Storage Systems by Liquefaction Methods: A Comprehensive Review. *ACS Omega*, *8*(21), 18358–18399. DOI: 10.1021/acsomega.3c01072 PMID: 37273600

Hassan, Q., Sameen, A. Z., Salman, H. M., Jaszczur, M., & Al-Jiboory, A. K. (2023). Hydrogen energy future: Advancements in storage technologies and implications for sustainability. *Journal of Energy Storage*, *72*, 108404. DOI: 10.1016/j.est.2023.108404

Henrique Santos, B., Peças Lopes, J., Carvalho, L., Matos, M., & Alves, I. (2024). Public policies to foster green hydrogen seasonal storage: Portuguese study case model until 2040. *Energy Strategy Reviews*, *52*, 101354. Advance online publication. DOI: 10.1016/j.esr.2024.101354

Jangir, J., & Jagirdar, B. R. (2024). Unveiling the Potential of Heterogeneous Systems for Reversible Hydrogen Storage in Liquid Organic Hydrogen Carriers. *ChemSusChem*, •••, 202402018–e202402018. DOI: 10.1002/cssc.202402018 PMID: 39417356

Jaradat, M., Almashaileh, S., Bendea, C., Juaidi, A., Bendea, G., & Bungau, T. (2024). Green Hydrogen in Focus: A Review of Production Technologies, Policy Impact, and Market Developments. *Energies*, *17*(16), 3992–3992. DOI: 10.3390/en17163992

Javaheri, M., & Shafiei Ghazani, A. (2023). Energy and exergy analysis of a novel advanced adiabatic compressed air energy storage hybridized with reverse osmosis system. *Journal of Energy Storage*, *73*, 109250. Advance online publication. DOI: 10.1016/j.est.2023.109250

Kamran, M., & Turzyński, M. (2024). Exploring hydrogen energy systems: A comprehensive review of technologies, applications, prevailing trends, and associated challenges. *Journal of Energy Storage*, *96*, 112601. DOI: 10.1016/j.est.2024.112601

Kang, D. H., An, J. H., & Lee, C. J. (2024). Numerical modeling and optimization of thermal insulation for liquid hydrogen storage tanks. *Energy*, *291*, 130143. Advance online publication. DOI: 10.1016/j.energy.2023.130143

Kawamura, S., Yamaguchi, A., Miyazaki, K., Ito, S., Watanabe, N., Hamada, I., Kondo, T., & Miyauchi, M. (2024). Electrolytic Hydrogen Release from Hydrogen Boride Sheets. *Small*, *20*(27), e2310239–e2310239. DOI: 10.1002/smll.202310239 PMID: 38299473

Kim, J., Vu, H. T. T., Kim, S., Choi, M., Lee, E., Park, K., Kim, J., Park, H., Choi, B., Han, J., Kim, H., Choi, J., & Chang, D. (2024). Technical feasibility of large-scale transportable liquid hydrogen export terminal. *International Journal of Hydrogen Energy*, *66*, 499–511. DOI: 10.1016/j.ijhydene.2024.03.343

Langmi, H. W., Musyoka, N. M., Kemmegne-Mbouguen, J. C., Kowenje, C., Kengara, F., & Mokaya, R. (2024). Capacity building in porous materials research for sustainable energy applications. *Interface Focus*, *14*(4), 20230067. Advance online publication. DOI: 10.1098/rsfs.2023.0067 PMID: 39129854

Lesmana, L. A., & Aziz, M. (2024). Strategic Integration of Metamaterials Properties and Topology Optimization of Gyroid Metal Hydride Reactor for High-Density Hydrogen Storage. *Energy*, *308*, 133018–133018. DOI: 10.1016/j.energy.2024.133018

Li, R., & Kawanami, H. (2023). A Recent Review of Primary Hydrogen Carriers, Hydrogen Production Methods, and Applications. *Catalysts 2023, Vol. 13, Page 562, 13*(3), 562. DOI: 10.3390/catal13030562

Li, X., Wang, Q., Wu, S., Hu, Z., & Bai, J. (2024). Critical upstream technologies for hydrogen energy industry: Research progress on ammonia decomposition catalysts. *Sustainable Chemistry and Pharmacy*, *38*, 101492. Advance online publication. DOI: 10.1016/j.scp.2024.101492

Liu, S., Zhou, J., Liang, G., Du, P., Li, Z., & Li, C. (2024). Optimizing large-scale hydrogen storage: A novel hybrid genetic algorithm approach for efficient pipeline network design. *International Journal of Hydrogen Energy*, *66*, 430–444. DOI: 10.1016/j.ijhydene.2024.04.098

Luo, R., Qiao, P., Zeng, M., Deng, X., Wang, H., Hao, W., Fan, J., Bi, Q., Li, G., & Cao, Y. (2024). CO2-Mediated Hydrogen Energy Release-Storage Enabled by High-Dispersion Gold-Palladium Alloy Nanodots. *Small*. Advance online publication. DOI: 10.1002/smll.202407578 PMID: 39568215

Ma, N., Zhao, W., Wang, W., Li, X., & Zhou, H. (2023). Large scale of green hydrogen storage: Opportunities and challenges. *International Journal of Hydrogen Energy*, *50*, 379–396. DOI: 10.1016/j.ijhydene.2023.09.021

Martinez, G. E., Degens, R., Espadas-Aldana, G., Costa, D., & Cardellini, G. (2024). Prospective Life Cycle Assessment of Hydrogen: A Systematic Review of Methodological Choices. *Energies*, *17*(17), 4297–4297. DOI: 10.3390/en17174297

Martínez de León, C., Ríos, C., Molina, P., & Brey, J. J. (2023). Levelized Cost of Storage (LCOS) for a hydrogen system. *International Journal of Hydrogen Energy*, *52*, 1274–1284. DOI: 10.1016/j.ijhydene.2023.07.239

Nivedhitha, K. S., Beena, T., Banapurmath, N. R., Umarfarooq, M. A., Ramasamy, V., Soudagar, M. E. M., & Ağbulut, Ü. (2024). Advances in hydrogen storage with metal hydrides: Mechanisms, materials, and challenges. *International Journal of Hydrogen Energy*, *61*, 1259–1273. DOI: 10.1016/j.ijhydene.2024.02.335

Park, C. H., Kim, U. J., Choi, J. H., & Lee, S. H. (2024). Synergistic N/F Dual-Doped MoC/C Catalyst Synthesized via Liquid Phase Plasma for Sustainable Ammonia Production. *ACS Applied Materials & Interfaces*, *16*(46), 63540–63552. Advance online publication. DOI: 10.1021/acsami.4c12535 PMID: 39505495

Rolo, I., Costa, V. A. F., & Brito, F. P. (2023). Hydrogen-Based Energy Systems: Current Technology Development Status, Opportunities and Challenges. *Energies 2024, Vol. 17, Page 180*, *17*(1), 180. DOI: 10.3390/en17010180

Sadeq, A. M., Homod, R. Z., Hussein, A. K., Togun, H., Mahmoodi, A., Isleem, H. F., Patil, A. R., & Moghaddam, A. H. (2024). Hydrogen energy systems: Technologies, trends, and future prospects. *The Science of the Total Environment*, *939*, 173622–173622. DOI: 10.1016/j.scitotenv.2024.173622 PMID: 38821273

Saeed, M., Marwani, H. M., Shahzad, U., Asiri, A. M., & Rahman, M. M. (2024). Nanoscale silicon porous materials for efficient hydrogen storage application. *Journal of Energy Storage*, *81*, 110418. Advance online publication. DOI: 10.1016/j.est.2024.110418

Shang, C., Liu, Y., Ghita, O., Lachman, N., Wang, D., & Tian, M. (2024). Integrating Metal–Organic Frameworks and Polyamide 12 for Advanced Hydrogen Storage Through Powder Bed Fusion. *Energies*, *17*(21), 5430–5430. DOI: 10.3390/en17215430

Shaya, N., & Glöser-Chahoud, S. (2024). A Review of Life Cycle Assessment (LCA) Studies for Hydrogen Production Technologies through Water Electrolysis: Recent Advances. *Energies*, *17*(16), 3968–3968. DOI: 10.3390/en17163968

Sikiru, S., Oladosu, T. L., Amosa, T. I., Olutoki, J. O., Ansari, M. N. M., Abioye, K. J., Rehman, Z. U., & Soleimani, H. (2024). Hydrogen-powered horizons: Transformative technologies in clean energy generation, distribution, and storage for sustainable innovation. *International Journal of Hydrogen Energy*, *56*, 1152–1182. DOI: 10.1016/j.ijhydene.2023.12.186

Song, Z., Xu, J., & Chen, X. (2023). Design and optimization of a high-density cryogenic supercritical hydrogen storage system based on helium expansion cycle. *International Journal of Hydrogen Energy*, *49*, 1401–1418. DOI: 10.1016/j.ijhydene.2023.07.298

Sousa, J., Azevedo, I., Camus, C., Mendes, L., Viveiros, C., & Barata, F. (2024). Decarbonizing Hard-to-Abate Sectors with Renewable Hydrogen: A Real Case Application to the Ceramics Industry. *Energies*, *17*(15), 3661–3661. DOI: 10.3390/en17153661

Wang, J., Webley, P. A., & Hughes, T. J. (2024). Thermodynamic modelling of pressurised storage and transportation of liquid hydrogen for maritime export. *International Journal of Hydrogen Energy*, *62*, 1273–1285. DOI: 10.1016/j.ijhydene.2024.02.285

Xie, Z., Jin, Q., Su, G., & Lu, W. (2024). A Review of Hydrogen Storage and Transportation: Progresses and Challenges. *Energies*, *17*(16), 4070–4070. DOI: 10.3390/en17164070

Xu, J., Huang, F., Chen, X., Yang, Q., & Zhang, H. (2024a). Design and optimization of cryogenic supercritical hydrogen storage system coupled with mixed refrigerant and gas expansion cycle. *Applied Thermal Engineering*, *243*, 122606. Advance online publication. DOI: 10.1016/j.applthermaleng.2024.122606

Xu, J., Huang, F., Chen, X., Yang, Q., & Zhang, H. (2024b). Design and optimization of cryogenic supercritical hydrogen storage system coupled with mixed refrigerant and gas expansion cycle. *Applied Thermal Engineering*, *243*, 122606. Advance online publication. DOI: 10.1016/j.applthermaleng.2024.122606

Xu, J., Song, Z., Chen, X., & Yang, Q. (2023). Design and optimization of high-density cryogenic supercritical hydrogen storage systems integrating with dual mixed refrigerant cycles. *Energy*, *290*, 130124. Advance online publication. DOI: 10.1016/j.energy.2023.130124

Yuvaraj, A. R., Jayarama, A., Sharma, D., Nagarkar, S. S., Duttagupta, S. P., & Pinto, R. (2024). Role of metal-organic framework in hydrogen gas storage: A critical review. *International Journal of Hydrogen Energy*, *59*, 1434–1458. DOI: 10.1016/j.ijhydene.2024.02.060

Zhang, R., Zhao, X., Liu, Y., & Zhang, Q. (2024). Development pathway and influencing factors of hydrogen energy storage accommodating renewable energy growth. *Journal of Energy Storage*, *87*, 111441. Advance online publication. DOI: 10.1016/j.est.2024.111441

KEY TERMS AND DEFINITIONS

Compressed Hydrogen Storage: The process of storing hydrogen gas under high pressure (up to 700 bar) in specialized tanks.

Cryogenic Storage: A hydrogen storage method that involves cooling hydrogen to extremely low temperatures (-253°C) to convert it into a liquid state for compact storage.

Decarbonization: The reduction of carbon dioxide (CO_2) emissions by transitioning to cleaner energy solutions like hydrogen.

Electrolysis: A method of producing hydrogen by splitting water (H_2O) into hydrogen (H_2) and oxygen (O_2) using electricity.

Energy Density: The amount of energy stored in a system or material, measured in terms of energy per unit mass or volume (e.g., MJ/kg).

Hydrogen Storage: The process of storing hydrogen for energy applications, involving methods like compressed gas, liquid hydrogen, and metal hydrides.

Liquid Organic Hydrogen Carriers (LOHCs): Chemicals that can absorb and release hydrogen, offering a safe and efficient way to store and transport hydrogen.

Metal-Organic Frameworks (MOFs): Porous materials used for hydrogen storage, capable of capturing large amounts of hydrogen due to their high surface area.

Renewable Hydrogen: Hydrogen produced using renewable energy sources like wind, solar, and hydroelectric power, resulting in a sustainable and clean energy carrier.

Scalability: The ability to scale up hydrogen storage technologies for large-scale industrial and commercial applications.

Chapter 5
AI-Driven Approaches for Next-Gen Solar Cells

Pooja Dehankar
Ajeenkya D.Y. Patil School of Engineering, India

Susanta Das
https://orcid.org/0000-0002-9314-3988
Ajeenkya D.Y. Patil University, India

ABSTRACT

AI plays a significant role in enhancing solar cell performance and has implications for sustainability and renewable energy integration. Additionally, it investigates how ML & AI may be applied to optimize energy production, boost solar cell efficiency, and facilitate intelligent grid integration. This chapter focuses on using AI-driven algorithms for solar cell material design and development. It shows how AI could accelerate the search for and development of materials with increased stability and effectiveness. It also explores the application of AI in developing solar cell technologies. The chapter discusses AI-powered control systems and prediction models for tracking and preserving solar cell performance. To improve the overall dependability of solar energy systems, it emphasizes the function of AI algorithms in identifying anomalies, predicting energy usage, and maximizing energy production. The chapter talks about the challenges of manufacturing and implementing solar cells. It draws attention to the ways AI could boost the sustainability of solar energy capacity.

DOI: 10.4018/979-8-3693-9316-1.ch005

INTRODUCTION

The Sun is a fantastic source of energy and plays a significant part in many countries' energy generation mix. Specifically, photovoltaic (PV) technology is an established, dependable, and tested way to turn the Sun's enormous energy into electrical power. Thus, the Sun offers a free energy source that may be captured and transformed into electrical power through the use of photovoltaic (PV) technology. One advantage of PV technology is its scalability and modularity. As a result, it may be swiftly installed in a variety of settings. These sites might range from residential and commercial buildings to traditional ground installations. However, the cost of solar electricity compared to other energy sources is one of the obstacles to its wider adoption. To assist lower the cost of solar electricity and create new solar cells that can produce more electricity per unit area, innovation in the creation of new materials and solar cell layouts is crucial. However, artificial intelligence is already showing up as another useful method to assist in achieving these goals because of improvements in computing speed and capacity. There are three primary areas in which AI can be used in the PV industry: weather data forecasting and modeling, solar cell basic modeling, and photovoltaic system sizing (Ghannam et al., 2019).

Power networks are evolving into next-generation systems due to their increased efficiency, environmental friendliness, and energy-smartness. These innovative systems, which are frequently referred to as Smart Energy Integration (SEI), combine numerous cutting-edge technologies to improve the efficiency of the production, distribution, and use of energy. Energy flow was unidirectional and output was regulated in the previous power grid. Energy storage, smart gadgets, and local green energy sources are all being handled by this evolving dynamic, two-way communication system (Gyawali & Patil, 2024).

Smart Energy Integration uses AI, the Internet of Things (IoT), and sophisticated monitoring to create a flexible and sensitive energy environment. The primary goal is to increase the resilience and dependability of electricity networks while promoting sustainability and reducing their environmental effect. Because it enables real-time monitoring, control, and optimization of energy flows, SEI differs from conventional grids. This enables businesses and consumers to make informed decisions regarding how much energy to generate and consume (Zhao & Lee, 2021).

Two renewable energy sources that are very significant to SEI are wind and solar power, which offer dispersed clean power generation. Combining these sources can be challenging since they appear and disappear at different periods. However, new technology that can anticipate when green energy will be accessible and save excess energy for later use, such as prediction analytics and energy storage solutions, can help with these issues. This not only maintains power but also reduces our reliance on fossil fuels, contributing to a more sustainable and healthy energy landscape (Gupta et al., 2020).

In Industry 4.0, maximizing solar energy efficiency necessitates an automated system driven by artificial intelligence. In order to determine the most effective energy sources for the demands of the sector and modify them appropriately, the AI-powered system uses clever algorithms that learn from each task it is given. The system optimizes solar energy for the industrial machinery, saves the energy in a reserve storage system, and uses sensors to analyze the conditions of the solar cells and their surroundings in order to boost solar energy efficiency. Through continuous monitoring and energy-setting adjustments, the AI-powered system effectively optimizes the efficiency of solar energy produced for industrial applications (Boobalan et al., 2023).

Solar radiation can be transformed into thermal and electrical energy using photovoltaic thermal (PV/T) solar systems. While solar thermal collectors collect excess energy and use it to heat cooling fluids and remove residual heat from photovoltaic cells, solar cells use solar radiation to generate electricity. To improve total system efficiency, the PV/T cooling system's architecture must be optimized (Shaban et al., 2024).

The combination of solar cells, radio wave technology, and AI is an intriguing prospect in the quickly developing fields of sustainable technology and commercial innovation. The complex relationships between these technologies and the significant business ramifications are examined. Applications for energy production, wireless communication, and the Internet of Things are examined in relation to the integration of solar cells and radio wave technologies. Predictive maintenance, load forecasting, and AI-driven forecasts are regarded as essential tools for companies looking to optimize energy efficiency and cost reduction (Smith, 2024).

From leveraging AI to maximize solar energy generation to leveraging 5G technology and the internet of things (IoT) to transform wireless communication, the convergence of these dynamic sectors has ushered in a new era of opportunities. Verizon's 5G-enabled smart cities and Sun Power's AI-enhanced solar panels are two successful instances of how these technologies' convergence creates the perfect atmosphere for entrepreneurship. This combination presents challenging issues in addition to revolutionary possibilities. These include legislative challenges, ethical AI issues, and the efficient management of electromagnetic spectrum. To properly manage this dynamic environment, stakeholders must encourage interdisciplinary research, flexible regulatory frameworks, international collaboration, and STEM education (Harry, 2023).

BACKGROUND

AI techniques have improved the sustainability, efficiency, and dependability of renewable energy systems (RES). The potential of AI-driven optimization techniques to transform RES in a number of areas, from resource assessment to system management and maintenance, has been the subject of a growing corpus of research in recent years. The need to implement creative solutions that can optimize the use of renewable resources while minimizing environmental effects and guaranteeing economic viability is important given the acceleration of climate change and the rise in energy demand. AI integration has emerged as a possible optimization path in the field of RES. The combination of AI and renewable energy technology offers a fresh strategy with enormous promise for improving system sustainability, dependability, and efficiency (Ukoba et al., 2024).

Significant progress is being made in the field of perovskite-based materials for energy applications because to AI. Materials based on perovskites have drawn a lot of interest in energy applications because of their exceptional optoelectronic qualities and adaptable makeup. Perovskite LEDs are perfect for display technologies because of their great color purity and adjustable emission. Perovskites are important components of sustainable energy systems, and research is being done to improve their performance in spite of issues like stability and scalability. By speeding up the search for new materials, streamlining production procedures, improving performance and stability, and encouraging sustainability (Alam & Prasad, 2025).

SOLAR PHOTOVOLTAIC TECHNOLOGY

One effective and adaptable way to turn the Sun's enormous energy into electrical power is through solar photovoltaic (PV) technology. To guarantee that small, lightweight, and flexible electronic gadgets run for extended periods of time with less battery usage, innovation in the creation of novel materials and solar cell structures is needed. Growing interest in effective energy-harvesting technologies has corresponded with recent advancements in biomedical wearable and implantable devices. Rechargeable batteries are the main source of energy for these devices. Furthermore, methods for ML and AI are hailed as revolutionary advances in energy harvesting, particularly in solar energy materials (Bhatti et al., 2022).

Recent advancements in technology, lower material costs, and government backing for the production of power from renewable sources have all contributed to the rapid growth of solar PV technology. Around the world, photovoltaics is a key component of using solar energy to produce power. The photovoltaic industry is one of the fastest expanding sectors at the moment, with a global market size of about 23.5 GW in 2010 and an annual growth rate of 35–40%. One crucial factor in establishing this technology in the market is the solar cell's efficiency (Tyagi et al., 2013).

New potential for the use of renewable energy resources are being created by growing energy demand, global environmental concerns, and continual advancements in renewable energy technologies. Of all the renewable energy sources available today, solar energy is the most plentiful, limitless, and pure. About 1.8 × 1011 MW of solar electricity gets intercepted by the earth, which is significantly more than the current pace of energy consumption. One of the best methods to use solar energy is via photovoltaic technology (Parida et al., 2011).

MODELING AND FORECASTING SOLAR RADIATION DATA USING ARTIFICIAL INTELLIGENCE

Statistics on solar radiation are recognized as reliable and accessible renewable energy sources. Models for forecasting and estimating these meteorological data must consequently be developed. These statistics are crucial for the sizing of PV systems. Daily clearness index modeling, insolation forecasting, and modeling of daily, monthly, and hourly solar radiation are all examples of how AI approaches are applied to solar radiation data. AI techniques are emphasized for their increasing application in data analysis, providing a viable substitute for more conventional statistical methods in many scientific domains. This is an alternative method of avoiding the aforementioned issues (Mellit, 2008).

Solar power forecasting is a crucial tool for grid operators to dispatchably manage the generated energy, especially in home microgrids. The intermittent generation of solar power cannot be reliably predicted by deterministic approaches since they rely on distinct sets of inputs and outputs. In order to anticipate solar power, AI tactics and stochastic methods are used. The power produced by the home smart microgrid is then forecast using the weather model that was created. To verify the model's accuracy, the NN-based model is used to forecast the solar irradiance and generated solar power in the home microgrid (Sabzehgar et al., 2020).

ARTIFICIAL INTELLIGENCE IN SOLAR SYSTEM MANAGEMENT AND OPERATION

As clean and sustainable energy sources have become more popular, solar systems have gained attention and are now at the center of many energy-related conversations. In order to optimize and improve solar systems both architecturally and productionally, the creation of integrated systems has recently become a groundbreaking method. These systems have been substantially impacted by AI in the areas of operations and management. The best performance and least amount of downtime have been made possible by AI's ability to help detect any issues and inefficiencies in solar panels, inverters, and other components early on. In order to apply AI for data analysis, modern solar systems come with integrated monitoring systems. The presentation will cover functional aspects of solar systems, such as anomaly detection and malfunctions. In order to maintain a steady and dependable solar energy source in the face of fluctuating unstable environmental patterns, adaptive control solutions for parameter adjustment for dynamic systems must be examined. It should be mentioned that AI offers benefits in terms of energy distribution optimization and grid integration (Ghoshal & Tripathy, 2024).

The development of new, high-performing materials like narrow band gap donor polymers and non-fullerene acceptors highlights the consistent rise in record efficiencies in organic solar cells over the past few years. Due to their prohibitively high requirements for human workforce time and resources, the current catalog of organic photovoltaic materials is becoming too large to evaluate using traditional experimentation methodologies, which slows the advancement of organic photovoltaic technology. In order to take advantage of their naturally fast exploration speeds and hasten the discovery of new materials, high-throughput experimental and computational approaches are therefore encouraged (Rodríguez-Martínez et al., 2021).

AI Technologies in Solar Applications

Neural networks were employed by Moustris et al. (2008) to provide hourly global and diffuse sun irradiance statistics at representative Greek locales. For the regions under investigation, a very good agreement with a satisfactory result is obtained between hourly solar irradiance values taken from pyranometer measurements and global and diffuse solar irradiance data sets obtained by NNs (when trained with other easily accessible weather and geographical parameters like air temperature, sunshine duration, cloud cover, latitude, etc.). Complete solar irradiance time-series sets could be produced with a reasonably high degree of precision in situations where solar data is lacking or in regions where weather stations do not measure and/or record solar data.

Using global solar radiation and other meteorological information as inputs, Soares et al. (2004) employed artificial neural networks to estimate hourly values of diffuse solar radiation at a surface in Sao-Paulo City, Brazil. It was discovered that the neural-network performance is enhanced when atmospheric long-wave radiation is used as input. However, long-wave radiation, which serves as a stand-in for cloud-cover data at the regional level, is more significant than conventional meteorological factors like air temperature and atmospheric pressure. An unbiased assessment has demonstrated that neural network synthetic series outperform correlation models in reproducing diffuse solar radiation.

AI in Power Generation

By improving efficiency, integrating renewable energy more effectively, streamlining grid management, and enabling predictive maintenance, artificial intelligence is transforming power generation and paving the way for a more dependable and sustainable energy future (Hamdan et al., 2024). AI algorithms use sensor, smart meter, and historical consumption data to identify and react to grid malfunctions or defects in real-time, reducing downtime and guaranteeing a steady supply of electricity (Koshy et al., 2021). AI can analyze data from several sources to optimize energy distribution, decrease downtime, and reduce energy losses (Onwusinkwue et al., 2024). By anticipating equipment failures, AI can minimize interruptions and enable prompt maintenance (Rojek et al., 2023). Power distribution is optimized by AI-powered adaptive control systems using real-time data, increasing dependability and efficiency (Dhiman & Kadam, 2024).

Artificial Intelligence to Boost Solar Cell Efficiency

The performance of inexpensive photovoltaic (PV) cells in small electrical devices can be improved with the use of ML and AI techniques. The creation of more economical and efficient solar cells is, in fact, expected to be greatly aided by machine learning. Many machine learning techniques, including deep neural networks (DNNs), Bayesian optimization (BO), and Gaussian process regression (GPR), which have been successful in increasing the stability, efficiency, and cost of solar cells. According to the research, the most viable approach for creating affordable solar cells is GPR in conjunction with BO. These methods can greatly accelerate the development of novel PV materials and structures while improving the stability and efficiency of inexpensive solar cells (Bhatti et al., 2023).

The potential of PV technology to produce power directly from sunshine has generated a lot of attention. ML is a method that uses known data to teach a computer how to do a certain activity. It has several applications and has recently gained popularity as a research issue because of the quick collection of data and the development of computer technology. Though it began slowly, the use of machine learning techniques in solar cell design and manufacturing has recently accelerated dramatically. According to a thorough review of the literature, machine learning approaches can successfully support all of the main facets of solar cell research and development (Li et al., 2019).

The high energy efficiency, longer system longevity, and lower cost of materials of ceramic electrochemical cells (CECs) make them potential technologies for clean and efficient energy conversion and storage. Widespread adoption and continued development in the sector are hampered by persistent difficulties in finding appropriate materials with the required qualities, such as high ionic and electronic conductivity, thermal stability, and chemical compatibility. The materials discovery cycle in CECs can be improved and accelerated by AI using data-driven modeling, simulation, and optimization tools (Bello et al., 2024).

The ability of photovoltaic technology to generate electricity directly from sunshine draws researchers from both industry and academia. The dye-sensitized solar cell has become the most popular of all the photovoltaic technologies because of its flexible electrolyte, dye, substrate, and catalyst options and inexpensive manufacturing costs. The materials' optical, electrical, and structural characteristics affect a solar cell's power conversion efficiency. However, determining the best materials for the production of an effective solar cell through laboratory research takes a lot of time, money, and human labor. To forecast the performance of solar cells, the researchers used machine learning approaches because of their demonstrated potential in pattern matching and computer vision (Tomar et al., 2022).

Recent developments in solar PV technology have accelerated the shift to more sustainable energy sources by greatly improving the materials, applications, and efficiency of solar energy systems. High-efficiency photovoltaic cell development is one of the main advances in solar PV technology. Significant gains in conversion efficiency have resulted from advancements in cell design and manufacturing techniques. Under focused sunlight, for example, multi-junction solar cells—which stack several semiconductor materials—have demonstrated efficiencies of above 40%. Furthermore, developments in bifacial solar cells—which absorb light from both sides—and passivated emitter rear cell (PERC) technology have increased the efficiency of conventional silicon-based PV cells. Additionally, materials science has been essential to the development of solar PV technologies. New materials like perovskite solar cells, which have the potential to be very efficient and inexpensive to produce, have completely changed the field. Flexible and simple to fabricate, perovskite materials enable lightweight and versatile solar panels. The range of applications for PV panels is also being expanded by developments in organic photovoltaics (OPVs) and quantum dot solar cells, which are helping to create flexible, semi-transparent, and aesthetically beautiful PV panels. Because of these developments, the use of solar PV technology has expanded dramatically. The smooth integration of solar energy systems into architectural designs has been made possible by advancements in solar panel integration into building materials, such as solar windows and photovoltaic roofing tiles (Ogundipe et al., 2024).

The development of PV cell technology is evidence of human resourcefulness and the unwavering search for sustainable energy sources. Pioneering developments in materials and production techniques have characterized the development of PV cells from the first days of solar energy exploration to the complex systems of today. Crystalline silicon, a substance that has remained popular because of its shown durability and efficiency, was used in the early stages of solar cell research. Beyond merely increasing efficiency, material developments in solar cell technology are significant. Solar energy appears as a ray of hope in light of growing environmental concerns and the need for sustainable energy sources worldwide. Improvements in materials, technology, and panel efficiency have all contributed to the significant drop in solar energy generation costs over the last ten years (Jaiswal et al., 2021).

A key component in addressing the global transition to sustainable energy sources is solar energy. Two well-known clean energy sources are fuel cells and PV systems, however their inconsistent and fluctuating power generation patterns prevent them from being widely used. a unique strategy that integrates AI techniques to enhance hybrids' (SPV-FCH) performance. With the help of nature renewable energy, which includes fuel cells' steady contribution, the synergy seeks to develop a more dependable, continuous power generation system. AI algorithm integration provides an intelligent control mechanism that maximizes the hybrid system's performance, overcoming variations in solar irradiation and the dynamic nature of energy consumption. Predictive analytics and machine learning algorithms are used by the AI-enabled control system to estimate trends in energy use, weather, and solar irradiance patterns. The system may dynamically modify both parts, maximizing their performance for optimal energy output, efficiency, and overall system reliability by utilizing real-time data and previous trends. Additionally, the AI system makes proactive maintenance and problem detection possible, which increases the hybrid system's overall longevity and robustness. The AI controller improves its predictions and control schemes through ongoing learning and adaptation, guaranteeing peak performance in a range of environmental circumstances (Soni et al., 2023).

TRENDS IN AI-ENHANCED SOLAR CELLS

Rapid advancements in artificial intelligence are producing new technologies that have the potential to completely transform renewable energy systems (Feng et al., 2024). In energy management and grid operations, quantum machine learning algorithms are becoming a powerful tool for resolving challenging optimization issues. Compared to classical algorithms, these algorithms analyze enormous volumes of data and explore solution spaces much quicker by utilizing the concepts of quantum computing (Yang et al., 2024).

The use of federated learning in distributed energy systems is another exciting advancement. More resilient and flexible AI systems for renewable energy management are made possible by this method, which permits cooperative model training across several energy assets without jeopardizing data privacy. In order to satisfy the demand for interpretability and transparency in AI-driven decision-making processes, explainable AI (XAI) solutions are becoming more popular in the energy sector (Jiang et al., 2024).

With 5G networks, edge computing—where AI algorithms analyze data locally at the network edge—is expected to become more popular. Solar-powered radio wave devices will be the main source of power for these edge computing nodes, ensuring real-time processing and responsiveness. 5G's massive IoT connectivity enables a multitude of low-power devices to be connected. For these IoT devices to operate effectively and have longer battery lives, solar-powered AI-enhanced radio wave technologies will be necessary (Ali, 2024).

This 5G capability allows for the creation of virtual networks with unique characteristics to serve a range of applications. AI will assist in managing and optimizing network slices so that they can be customized to meet the various demands of various organizations and industries. Startups and industry initiatives are paying more and more attention to the convergence of solar cell, radio wave, and artificial intelligence technologies. These newcomers are encouraging innovation by pushing the boundaries of what is currently possible (Bhardwaj et al., 2023).

CHALLENGES OF MANUFACTURING AND IMPLEMENTING SOLAR CELLS

A major factor in the enormously sustainable and renewable worldwide energy transmission is the yearly rise in the world's energy consumption as well as the associated environmental problems and concerns. Out of all the renewable energy sources, solar energy systems have garnered the most attention in the past ten years. But even renewable energy can have some negative effects on the environment, therefore more care and appropriate precautions need to be taken (Rabaia et al., 2021).

One of the biggest obstacles to the market adoption of solar PV technology has been the high cost of power from these panels. However, proponents contend that many of the advantages of solar PV are missed by conventional analysis. Some of those benefits take place in the electrical markets, while others fall under the category of externalities related to the environment and security (Borenstein, 2008).

A smart energy production system called building integrated photovoltaics (BIPV) integrates solar PV panels into windows, roofs, facades, and shading structures. The development of BIPV products has been going on for the last 30 years, but compared to traditional rack-mounted solar PV, their practical applications have been gradual. One of the primary causes is that stakeholders have not fully comprehended the technological obstacles, which extend from the design stage to the commissioning and maintenance stages (Yang, 2015).

Access to a plentiful supply of inexpensive energy is essential to today's quality of life. Energy from non-fossil sources is necessary for a sustainable future; it should ideally be dependable, safe, adaptable, inexpensive, and infinite. In many cases, the intermittent and fluctuating nature of renewables' output limits their adoption. This necessitates using energy-storage technologies in practice. Prospects for usage in centralized and distributed energy supply, as well as in electric and hybrid electric cars, are assessed for the following possible storage technologies: pumped-hydro, flywheels, hydrogen (for use in fuel cells), and batteries (Dell & Rand, 2001).

Even with the advancements, there are still obstacles in the way of completely incorporating renewable energy into manufacturing processes. Obstacles still exist because of things like energy storage, grid integration, and the sporadic nature of some renewable sources. The viability of renewable energy for manufacturing is expected to be significantly improved by further research and development initiatives aimed at resolving these issues. The manufacturing sector's embrace of renewable energy has progressed from cautious investigation to calculated application. This change is a reflection of a larger trend toward sustainability that is being fueled by regulatory support, economic factors, and technology advancement. The industry's dedication to sustainable manufacturing methods is anticipated to grow as it continues to negotiate the benefits and challenges presented by renewable energy, supporting international efforts to tackle climate change (Usman et al., 2024).

The power generation environment has changed dramatically in recent years, shifting from centralized power plants to decentralized power systems. Significant developments in grid architecture have prompted this change, bringing in the idea of distributed generation (DG), which is now an essential part of electrical power systems, particularly in distribution networks (DNs). Reducing power losses and guaranteeing dependable power generation require DG integration. However, as decentralized technologies become more widely used, power systems face new difficulties, including determining the best places to put DG, setting up safety devices, controlling voltage, and dealing with power quality (PQ) problems. These complications are increased by the unpredictability and variability of renewable energy sources (RES), such wind and solar electricity, necessitating careful consideration and strategic planning (Saxena et al., 2024).

Federal funds for enhanced workforce training for U.S. solar installers are among the many federal, state, and municipal expenditures in the solar industry that have resulted from the potential economic benefits of the expanding renewable energy sector. To assess the number and makeup of the workforce required to meet the demand for solar electricity, there are still gaps in the data (Friedman et al., 2011).

This field is expanding quickly, emphasizing transparency, bias prevention, and responsible AI development. These processes are crucial for building trust and settling ethical conundrums. Alongside the convergence of these technologies, scalability and sustainability will be prioritized (Babu et al., 2023).

Sustainability will continue to be a primary goal. Innovations in energy efficiency, renewable energy production, and responsible manufacturing will help create a more sustainable future. To meet the needs of expanding cities and expanding populations, integrated systems must be scalable. Widespread use of these technologies requires scalable solutions (jin Kim, 2024).

Ensuring the robustness and dependability of interconnected systems is essential. Strong disaster recovery plans, redundancy, and cyber security measures are necessary to avoid disruptions. Governments, corporations, and international organizations will need to work together to set global standards, manage regulatory concerns, and ensure that technological advancements benefit everybody. Notwithstanding various challenges, a promising and transformative future is presented by the combination of commerce, artificial intelligence, radio wave communication, and solar cell technology. These technologies working together have the power to fundamentally alter the way we connect, produce energy, and conduct business. Prioritizing sustainability, scalability, and ethical innovation holds promise for creating a more technologically sophisticated, energy-efficient, and connected world in the future (Sujatha et al., 2022; Husnain et al., 2023) .

APPLICATION OF AI IN THE DEVELOPMENT OF SOLAR CELL TECHNOLOGIES

An ever-more-important part of the development of perovskite-based materials for energy applications is AI. Powerful tools for speeding up the identification, optimization, and use of perovskite materials have been made possible by the quick development of AI approaches, especially in ML and data-driven modeling. AI's capacity to process enormous datasets, spot intricate patterns, and make very accurate predictions about material properties is revolutionizing how scientists tackle the difficulties involved in creating perovskite-based energy solutions like solar cells, LEDs, and energy storage devices (Lye et al., 2023).

Beyond solar energy and lighting, perovskite materials are being investigated for energy storage, specifically in supercapacitors and lithium-ion batteries. The adjustable redox characteristics and strong ionic conductivity of perovskites make them viable options for battery electrode materials. Because perovskites can store and release charge quickly, they can help supercapacitors achieve high capacitance and energy density (Hu et al., 2023).

By drastically cutting down on the time and resources needed to find novel perovskite compositions with desirable features, AI and ML algorithms are transforming the materials discovery process. Conventional techniques for characterizing and synthesizing materials are frequently costly and time-consuming. AI-driven methods allow researchers to explore large chemical spaces and forecast the performance of new perovskite compounds prior to experimental confirmation. Examples of these methods include predictive modeling and high-throughput virtual screening. For instance, researchers can forecast important material features like bandgap, stability, and charge-carrier mobility using machine learning models trained on existing information. This helps direct the synthesis of new perovskites with ideal attributes for certain energy applications. The use of neural networks and genetic algorithms to forecast and optimize the composition of perovskite materials for improved solar cell performance is a noteworthy application of AI in this field. These models are able to determine the best elemental and processing combinations that optimize stability and efficiency. One of the main obstacles to the commercialization of perovskite solar cells has already been overcome by these methods, which have produced novel perovskite formulations with enhanced resistance to environmental influences (Glück & Bein, 2020).

REAL-WORLD IMPLEMENTATION

The best proof of concept for the commercial applications of solar cells, radio waves, and AI comes from case studies and actual applications. These examples demonstrate how these integrated technologies are more than just theoretical concepts; they are workable solutions that foster productivity, sustainability, and innovation in a range of industries (Najumunnisa et al., 2023).

One of the most well-known applications of this connection is smart grids. These grids are characterized by their utilization of advanced technologies, including wireless networking, artificial intelligence, and Internet of Things devices, to optimize energy usage and delivery. Pacific Gas and Electric (PG&E), a major utility company with its headquarters located in California, has implemented a sophisticated smart grid technology. They use artificial intelligence algorithms to forecast energy use, optimize energy distribution, and integrate energy from multiple sources, including solar panels. Lower energy waste, greater dependability, and cost savings have resulted for PG&E and its customers (Blunt, 2022).

Panel orientations and tracking systems can be modified by AI-driven analytics to optimize efficiency and sunshine capture (Ohalete et al., 2023). Particularly in times of low sunshine, artificial intelligence (AI) can optimize energy storage systems (batteries) to guarantee optimal use of stored energy (Song et al., 2024). By forecasting energy output and demand, AI can assist in more efficiently integrating solar energy into the electrical grid, guaranteeing a steady and dependable power supply (Wen et al., 2024).

AI-enabled systems can track the performance of solar panels, identify inefficiencies, and anticipate possible breakdowns, enabling preventative maintenance and lowering the need for expensive repairs (Kumar et al., 2024).

Practical Applications

AI has had a truly revolutionary effect on the solar energy sector. Artificial Intelligence has become a disruptive force in numerous industries. By harnessing the potential of AI, researchers and engineers have found new ways to enhance the efficiency, affordability, and functionality of PV technology. The integration of AI into solar energy represents a paradigm change in the future of renewable energy production. One of the main areas where AI has made a big contribution is the improvement of solar panel efficiency (Mohammad & Mahjabeen, 2023).

The various elements that influence solar panel performance may be properly measured and analyzed thanks to advanced data analytics and artificial intelligence systems. These variables include things like incident light intensity, temperature, and shading effects. By continuously monitoring and assessing these factors in real-time, AI systems may dynamically alter the solar panels' working parameters to boost their output. Both the energy production of solar panels and the overall dependability and performance stability of solar energy systems are increased by this AI-driven optimization (Wen et al., 2024).

Regulatory and Policy Frameworks

The need for AI explainability has also been brought to light by the European Union's general data protection regulation (GDPR), which promotes the "right to explanation," which requires that decisions made by AI that have an impact on people or enterprises be understandable and justified (Lipu et al., 2019). In an effort to tackle the black-box issue, scholars have created Explainable AI (XAI) frameworks that improve the interpretability of AI-powered power system decision-making (Farzaneh et al., 2021). In order to help operators understand which parameters—such as voltage fluctuations, frequency anomalies, or current distortions—had the greatest impact on an AI model's prediction, shapley additive explanations (SHAP), one of the most popular methods for AI explainability, assigns importance scores to input features (Adnan et al., 2021). Local interpretable modelagnostic explanations (LIME) is a further method that produces simplified surrogate models that interpretably imitate deep learning behavior (Serban & Lytras, 2020). AI-driven power grid monitoring has effectively used LIME, which offers comprehensible explanations of fault categorization outcomes (Antonopoulos et al., 2020).

Case Studies

Google's AI division, DeepMind, partnered with the UK's National Grid in 2018 to maximize wind energy production efficiency. DeepMind predicted wind patterns by analyzing previous turbine data and weather projections using machine learning techniques. Compared to conventional techniques, the National Grid was able to schedule electricity production more effectively thanks to the AI system's precise forecasts, which increased energy output by 20% (Masood et al., 2025).

A project that used AI for precise solar forecasting was carried out by IBM Research Ireland. To forecast solar power generation, the system combined sophisticated meteorological modeling with machine learning techniques. The AI model helped to lessen the intermittent nature of solar energy by enhancing forecasting accuracy, which in turn allowed for improved grid management and integration of solar power (Shahzad et al., 2025).

AI is being used in an autonomous microgrid project in South Australia by SIMEC Zen Energy to maximize the performance of a variety of energy sources, such as solar, wind, and energy storage. In order to maximize the microgrid's resilience and efficiency, the AI system continuously evaluates data on weather, equipment state, and energy demand to make decisions on energy dispatch and storage in real time (Ukoba et al., 2024).

CONCLUSION

There is a lot of promise for climate change mitigation and sustainable energy solutions with the installation of next-generation solar panels in cities. Significant determinants of adoption and performance include higher income levels, increased environmental consciousness, and favorable opinions about the advantages of technology. Policymakers ought to create focused educational initiatives and subsidies that take socioeconomic and environmental aspects into account. The development and promotion of effective solar technology with an emphasis on long-term advantages should be the primary focus of industry stakeholders. Stakeholders may help energy efficiency and sustainable urban development activities by better promoting the usage of modern solar panels by being aware of these important criteria.

Smart Energy Integration's transition to Next-Gen Power Networks is a significant step toward a more intelligent, sustainable, and effective energy future. Numerous facets of this paradigm shift were examined in this comprehensive study, including the interplay between smart grids and renewable energy sources, as well as the potential applications of artificial intelligence and energy storage technologies to improve network performance.

The relationship between smart grids, renewable energy, energy storage, and artificial intelligence is not merely a theoretical concept; it is a reality that could occur and will significantly impact the operation of the energy infrastructure in our planet. Adopting these new concepts is not only essential, but it also presents an opportunity to build a better, more accessible future where energy is directly connected to people's desires and objectives worldwide.

With developments in photovoltaic power plants, battery materials, and antenna technologies, the power industry experiences a paradigm shift. A sustainable and connected energy future is made possible by advanced RF/millimeter-wave antennas, next-generation electric cars, and solar energy systems that are optimized.

In the operational stage of perovskite-based energy devices, artificial intelligence is also very important. Machine learning-driven predictive maintenance models can track the functionality of installed LEDs or perovskite solar panels and anticipate possible malfunctions before they happen. By tracking performance indicators, identifying anomalies, and forecasting maintenance requirements, these models evaluate data from sensors built into the devices, increasing their operational longevity and decreasing downtime.

REFERENCES

Adnan, S. M., Hamdan, A., & Alareeni, B. (2020, November). Artificial intelligence for public sector: chatbots as a customer service representative. In *International Conference on Business and Technology* (pp. 164-173). Cham: Springer International Publishing.

Alam, T., & Prasad, A. (2025). Artificial intelligence in Perovskite-Based materials for energy applications. *LatIA*, *3*, 125–125. DOI: 10.62486/latia2025125

Ali, M. (n.d). Integrating solar cell technology, radio waves, and AI for a transformative IT business ecosystem: A comprehensive review. *International Journal of Multidisciplinary Sciences and Arts, 3*(2), 263-269.

Antonopoulos, I., Robu, V., Couraud, B., Kirli, D., Norbu, S., Kiprakis, A., Flynn, D., Elizondo-Gonzalez, S., & Wattam, S. (2020). Artificial intelligence and machine learning approaches to energy demand-side response: A systematic review. *Renewable & Sustainable Energy Reviews*, *130*, 109899. DOI: 10.1016/j.rser.2020.109899

Babu, K. V., Sudipta, D., Sree, G. N. J., Almawgani, A. H., Islam, T., & Alhawari, A. R. (2023). Deep learning assisted fractal slotted substrate MIMO antenna with characteristic mode analysis (CMA) for Sub-6 GHz n78 5 G NR applications: Design, optimization and experimental validation. *Physica Scripta*, *98*(11), 115526. DOI: 10.1088/1402-4896/ad00e6

Bello, I. T., Taiwo, R., Esan, O. C., Adegoke, A. H., Ijaola, A. O., Li, Z., Zhao, S., Wang, C., Shao, Z., & Ni, M. (2024). AI-enabled materials discovery for advanced ceramic electrochemical cells. *Energy and AI*, *15*, 100317. DOI: 10.1016/j.egyai.2023.100317

Bhardwaj, S., Malik, P. K., Islam, T., Gehlot, A., Das, S., & Asha, S. (2023). A Printed Monopole Antenna for Next Generation Internet of Things: Narrow Band Internet of Things (Nb-IoT). *Progress in Electromagnetics Research C. Pier C*, *138*, 117–129. DOI: 10.2528/PIERC23090202

Bhatti, S., Manzoor, H. U., Michel, B., Bonilla, R. S., Abrams, R., Zoha, A., Hussain, S., & Ghannam, R. (2022). Machine learning for accelerating the discovery of high performance low-cost solar cells: a systematic review. *arXiv preprint arXiv:2212.13893.*

Bhatti, S., Manzoor, H. U., Michel, B., Bonilla, R. S., Abrams, R., Zoha, A., Hussain, S., & Ghannam, R. (2023). Revolutionizing low-cost solar cells with machine learning: A systematic review of optimization techniques. *Advanced Energy and Sustainability Research*, *4*(10), 2300004. DOI: 10.1002/aesr.202300004

Blunt, K. (2022). *California Burning: The Fall of Pacific Gas and Electric—and What It Means for America's Power Grid*. Penguin.

Boobalan, S., Lakshmi, T. K., Ghate, S. N., Haqqani, M. H., & Jaiswal, S. (2023). Enhancing Power Efficiency in 4IR Solar Plants through AI-Powered Energy Optimization. *Engineering Proceedings*, *59*(1), 87.

Borenstein, S. (2008). The market value and cost of solar photovoltaic electricity production.

Dell, R. M., & Rand, D. A. J. (2001). Energy storage—A key technology for global energy sustainability. *Journal of Power Sources*, *100*(1-2), 2–17. DOI: 10.1016/S0378-7753(01)00894-1

Dhiman, O., & Kadam, P. S. (2024). AI-Driven Adaptive Control Systems for Power Distribution. *Research Journal of Computer Systems and Engineering*, *5*(1), 71–82.

Farzaneh, H., Malehmirchegini, L., Bejan, A., Afolabi, T., Mulumba, A., & Daka, P. P. (2021). Artificial intelligence evolution in smart buildings for energy efficiency. *Applied Sciences (Basel, Switzerland)*, *11*(2), 763. DOI: 10.3390/app11020763

Feng, Y., Qi, Y., Li, H., Wang, X., & Tian, J. (2024, July). Leveraging federated learning and edge computing for recommendation systems within cloud computing networks. In *Third International Symposium on Computer Applications and Information Systems (ISCAIS 2024)* (Vol. 13210, pp. 279-287). SPIE. DOI: 10.1117/12.3034773

Friedman, B., Jordan, P., & Carrese, J. (2011). *Solar installation labor market analysis* (No. NREL/TP-6A20-49339). National Renewable Energy Lab.(NREL), Golden, CO (United States).

Ghannam, R., Klaine, P. V., & Imran, M. (2019). Artificial intelligence for photovoltaic systems. *Solar Photovoltaic Power Plants: Advanced Control and Optimization Techniques*, 121-142.

Ghoshal, N., & Tripathy, B. K. (2024). Artificial Intelligence Applied to the Management and Operation of Solar Systems. *Biomass and Solar-Powered Sustainable Digital Cities*, 317-338.

Glück, N., & Bein, T. (2020). Prospects of lead-free perovskite-inspired materials for photovoltaic applications. *Energy & Environmental Science*, *13*(12), 4691–4716. DOI: 10.1039/D0EE01651A

Gupta, V., Konda, S. R., Kumar, R., & Panigrahi, B. K. (2020). Electric vehicle driver response evaluation in multiaggregator charging management with EV routing. *IEEE Transactions on Industry Applications*, *56*(6), 6914–6924. DOI: 10.1109/TIA.2020.3017563

Gyawali, Y. P., & Patil, S. S. (2024). Next-Gen Power Networks: A Comprehensive Study on Smart Energy Integration. *Acta Energetica*, (01), 35–45.

Hamdan, A., Ibekwe, K. I., Ilojianya, V. I., Sonko, S., & Etukudoh, E. A. (2024). AI in renewable energy: A review of predictive maintenance and energy optimization. *International Journal of Science and Research Archive*, *11*(1), 718–729. DOI: 10.30574/ijsra.2024.11.1.0112

Harry, A. (2023). Synergizing Solar Cell Technology, Radio Waves, AI, and Business: A Comprehensive Review. *International Journal of Multidisciplinary Sciences and Arts*, *2*(2), 169–177. DOI: 10.47709/ijmdsa.v2i2.2995

Hu, S., Smith, J. A., Snaith, H. J., & Wakamiya, A. (2023). Prospects for tin-containing halide perovskite photovoltaics. *Precision Chemistry*, *1*(2), 69–82. DOI: 10.1021/prechem.3c00018 PMID: 37124243

Husnain, A., Rasool, S., Saeed, A., & Hussain, H. K. (2023). Revolutionizing pharmaceutical research: Harnessing machine learning for a paradigm shift in drug discovery. *International Journal of Multidisciplinary Sciences and Arts*, *2*(4), 149–157. DOI: 10.47709/ijmdsa.v2i2.2897

Jaiswal, D., Mittal, M., & Mittal, V. (2021). A Review on Solar PV Cell and Its Evolution. *Latest Trends in Renewable Energy Technologies: Select Proceedings of NCRESE 2020*, 303-313.

Jiang, W., Qian, K., Fan, C., Ding, W., & Li, Z. (2024). Applications of generative AI-based financial robot advisors as investment consultants. *Applied and Computational Engineering*, *77*(1), 265–270. DOI: 10.54254/2755-2721/67/2024MA0057

jin Kim, C. (n.d.). Synergizing solar cell technology, radio wave communication, AI integration, and business opportunities: A comprehensive review. *International Journal of Multidisciplinary Sciences and Arts, 3*(2), 252-262.

Koshy, S., Rahul, S., Sunitha, R., & Cheriyan, E. P. (2021). Smart grid–based big data analytics using machine learning and artificial intelligence: A survey. *Artif. Intell. Internet Things Renew. Energy Syst*, *12*, 241–278. DOI: 10.1515/9783110714043-008

Kumar, R. K., Prabhakaran, S., Moorthy, S. K., Shreeram, S., & Veeramani, R. (2024, March). AI-Enabled Robotic Maintenance System for Solar Panel Health Monitoring. In *2024 10th International Conference on Advanced Computing and Communication Systems (ICACCS)* (Vol. 1, pp. 2185-2189). IEEE. DOI: 10.1109/ICACCS60874.2024.10716831

Li, F., Peng, X., Wang, Z., Zhou, Y., Wu, Y., Jiang, M., & Xu, M. (2019). Machine learning (ML)-assisted design and fabrication for solar cells. *Energy & Environmental Materials*, *2*(4), 280–291. DOI: 10.1002/eem2.12049

Lipu, M. H., Miah, M. S., Jamal, T., Rahman, T., Ansari, S., Rahman, M. S., Ashique, R. H., Shihavuddin, A. S. M., & Shakib, M. N. (2023). Artificial intelligence approaches for advanced battery management system in electric vehicle applications: A statistical analysis towards future research opportunities. *Vehicles (Basel)*, *6*(1), 22–70. DOI: 10.3390/vehicles6010002

Lye, Y. E., Chan, K. Y., & Ng, Z. N. (2023). A review on the progress, challenges, and performances of tin-based perovskite solar cells. *Nanomaterials (Basel, Switzerland)*, *13*(3), 585. DOI: 10.3390/nano13030585 PMID: 36770546

Masood, A., Ahmed, U., Hassan, S. Z., Khan, A. R., & Mahmood, A. (2025). Economic Value Creation of Artificial Intelligence in Supporting Variable Renewable Energy Integration: A Systematic Review of Power System Applications.

Mellit, A. (2008). Artificial Intelligence technique for modelling and forecasting of solar radiation data: a review. *International Journal of Artificial intelligence and soft computing, 1*(1), 52-76.

Mohammad, A., & Mahjabeen, F. (2023). Revolutionizing solar energy: The impact of artificial intelligence on photovoltaic systems. *International Journal of Multidisciplinary Sciences and Arts*, *2*(3), 591856. DOI: 10.47709/ijmdsa.v2i1.2599

Moustris, K., Paliatsos, A. G., Bloutsos, A., Nikolaidis, K., Koronaki, I., & Kavadias, K. (2008). Use of neural networks for the creation of hourly global and diffuse solar irradiance data at representative locations in Greece. *Renewable Energy*, *33*(5), 928–932. DOI: 10.1016/j.renene.2007.09.028

Najumunnisa, M., Sastry, A. S. C. S., Madhav, B. T. P., Islam, T., & Das, S. (2023). Compact and innovative microstrip patch antenna with enhanced microwave circuit performance for RFID applications.

Ogundipe, O. B., Okwandu, A. C., & Abdulwaheed, S. A. (2024). Recent advances in solar photovoltaic technologies: Efficiency, materials, and applications. *GSC Advanced Research and Reviews*, *20*(1), 159–175. DOI: 10.30574/gscarr.2024.20.1.0259

Ohalete, N. C., Aderibigbe, A. O., Ani, E. C., Ohenhen, P. E., Daraojimba, D. O., & Odulaja, B. A. (2023). AI-driven solutions in renewable energy: A review of data science applications in solar and wind energy optimization. *World Journal of Advanced Research and Reviews*, *20*(3), 401–417. DOI: 10.30574/wjarr.2023.20.3.2433

Onwusinkwue, S., Osasona, F., Ahmad, I. A. I., Anyanwu, A. C., Dawodu, S. O., Obi, O. C., & Hamdan, A. (2024). Artificial intelligence (AI) in renewable energy: A review of predictive maintenance and energy optimization. *World Journal of Advanced Research and Reviews*, *21*(1), 2487–2499. DOI: 10.30574/wjarr.2024.21.1.0347

Parida, B., Iniyan, S., & Goic, R. (2011). A review of solar photovoltaic technologies. *Renewable & Sustainable Energy Reviews*, *15*(3), 1625–1636. DOI: 10.1016/j.rser.2010.11.032

Rabaia, M. K. H., Abdelkareem, M. A., Sayed, E. T., Elsaid, K., Chae, K. J., Wilberforce, T., & Olabi, A. G. (2021). Environmental impacts of solar energy systems: A review. *The Science of the Total Environment*, *754*, 141989. DOI: 10.1016/j.scitotenv.2020.141989 PMID: 32920388

Rodríguez-Martínez, X., Pascual-San-José, E., & Campoy-Quiles, M. (2021). Accelerating organic solar cell material's discovery: High-throughput screening and big data. *Energy & Environmental Science*, *14*(6), 3301–3322. DOI: 10.1039/D1EE00559F PMID: 34211582

Rojek, I., Jasiulewicz-Kaczmarek, M., Piechowski, M., & Mikołajewski, D. (2023). An artificial intelligence approach for improving maintenance to supervise machine failures and support their repair. *Applied Sciences (Basel, Switzerland)*, *13*(8), 4971. DOI: 10.3390/app13084971

Sabzehgar, R., Amirhosseini, D. Z., & Rasouli, M. (2020). Solar power forecast for a residential smart microgrid based on numerical weather predictions using artificial intelligence methods. *Journal of Building Engineering*, *32*, 101629. DOI: 10.1016/j.jobe.2020.101629

Saxena, V., Manna, S., Rajput, S. K., Kumar, P., Sharma, B., Alsharif, M. H., & Kim, M. K. (2024). Navigating the complexities of distributed generation: Integration, challenges, and solutions. *Energy Reports*, *12*, 3302–3322. DOI: 10.1016/j.egyr.2024.09.017

Şerban, A. C., & Lytras, M. D. (2020). Artificial intelligence for smart renewable energy sector in europe—Smart energy infrastructures for next generation smart cities. *IEEE Access : Practical Innovations, Open Solutions*, *8*, 77364–77377. DOI: 10.1109/ACCESS.2020.2990123

Shaban, W. M., Kabeel, A. E., Attia, M. E. H., & Talaat, F. M. (2024). Optimizing photovoltaic thermal solar systems efficiency through advanced artificial intelligence driven thermal management techniques. *Applied Thermal Engineering*, *247*, 123029. DOI: 10.1016/j.applthermaleng.2024.123029

Shahzad, T., Nawaz, S., & Tabassum, M. (2025). Renewable energy deployment and guidelines for responsible AI adoption. In *Explainable Artificial Intelligence and Solar Energy Integration* (pp. 363–392). IGI Global.

Smith, K. (2024). Solar cell integration with radio wave ai for enhanced business sustainability. *BULLET: Jurnal Multidisiplin Ilmu*, *3*(1), 65–75.

Soares, J., Oliveira, A. P., Boznar, M. Z., Mlakar, P., Escobedo, J. F., & Machado, A. J. (2004). Modelling hourly diffuse solar-radiation in the city of Sao Paulo using a neural-network technique. *Applied Energy*, *79*(2), 201–214. DOI: 10.1016/j.apenergy.2003.11.004

Song, H., Liu, C., Amani, A. M., Gu, M., Jalili, M., Meegahapola, L., ... & Dickeson, G. (2024). Smart optimization in battery energy storage systems: An overview. *Energy and AI*, 100378.

Soni, P., Dave, V., & Repalle, N. B. (2023, December). AI Based Performance Boost in Solar PV Fuel Cell Hybrids. In *International Conference on Renewable Energy, Green Computing, and Sustainable Development* (pp. 3-16). Cham: Springer Nature Switzerland.

Sujatha, N., Valli, L. N., Prema, A., Rathiha, S. K., & Raja, V. (2022). Initial centroid selection for K-means clustering algorithm using the statistical method. *International Journal of Science and Research Archive*, *7*(02), 474–478. DOI: 10.30574/ijsra.2022.7.2.0309

Tomar, N., Rani, G., Dhaka, V. S., & Surolia, P. K. (2022). Role of artificial neural networks in predicting design and efficiency of dye sensitized solar cells. *International Journal of Energy Research*, *46*(9), 11556–11573. DOI: 10.1002/er.7959

Tyagi, V. V., Rahim, N. A., Rahim, N. A., Jeyraj, A., & Selvaraj, L. (2013). Progress in solar PV technology: Research and achievement. *Renewable & Sustainable Energy Reviews*, *20*, 443–461. DOI: 10.1016/j.rser.2012.09.028

Ukoba, K., Olatunji, K. O., Adeoye, E., Jen, T. C., & Madyira, D. M. (2024). Optimizing renewable energy systems through artificial intelligence: Review and future prospects. *Energy & Environment*, *35*(7), 3833–3879. DOI: 10.1177/0958305X241256293

Usman, F. O., Ani, E. C., Ebirim, W., Montero, D. J. P., Olu-lawal, K. A., & Ninduwezuor-Ehiobu, N. (2024). Integrating renewable energy solutions in the manufacturing industry: challenges and opportunities: a review. *Engineering Science & Technology Journal*, *5*(3), 674–703. DOI: 10.51594/estj.v5i3.865

Wen, X., Shen, Q., Zheng, W., & Zhang, H. (2024). AI-driven solar energy generation and smart grid integration: A holistic approach to enhancing renewable energy efficiency. *Academia Nexus Journal, 3*(2).

Yang, P., Chen, Z., Su, G., Lei, H., & Wang, B. (2024). Enhancing traffic flow monitoring with machine learning integration on cloud data warehousing. *Applied and Computational Engineering*, *77*(1), 238–244. DOI: 10.54254/2755-2721/67/2024MA0058

Yang, R. J. (2015). Overcoming technical barriers and risks in the application of building integrated photovoltaics (BIPV): Hardware and software strategies. *Automation in Construction*, *51*, 92–102. DOI: 10.1016/j.autcon.2014.12.005

Zhao, Z., & Lee, C. K. (2021). Dynamic pricing for EV charging stations: A deep reinforcement learning approach. *IEEE Transactions on Transportation Electrification*, *8*(2), 2456–2468. DOI: 10.1109/TTE.2021.3139674

KEY TERMS AND DEFINITIONS

General Data Protection Regulation: The General Data Protection Regulation (GDPR), a European Union regulation on information privacy, aims to give individuals more control over their personal data and sets guidelines for organizations on collecting, storing, and managing that data, regardless of where they are located if they target or collect data related to people in the EU.

Perovskite Solar Cells: Perovskite solar cells are a promising, emerging photovoltaic technology that utilizes perovskite materials as the light-harvesting active layer, offering high efficiency and low production costs.

Photovoltaic (PV) Technology: Photovoltaic (PV) technology, or solar panels, converts sunlight directly into electricity using semiconductor materials, offering a clean and renewable energy source.

Renewable Energy Systems: Renewable energy systems harness naturally replenishing resources like solar, wind, water, and biomass to produce energy, offering a sustainable and environmentally friendly alternative to fossil fuels.

Smart Energy Integration: Smart energy integration involves using technology to optimize energy systems, including renewable energy sources, smart grids, and energy-efficient devices, for greater efficiency, reliability, and sustainability.

Chapter 6
Artificial Intelligence in Renewable Energy: Transforming Energy Storage and Harvesting Solutions

C. V. Suresh Babu
https://orcid.org/0000-0002-8474-2882
Hindustan Institute of Technology and Science, India

A. Mohamed Mohideen
https://orcid.org/0009-0004-5188-1116
Hindustan Institute of Technology and Science, India

K. Akash Kannan
https://orcid.org/0009-0000-3176-9160
Hindustan Institute of Technology and Science, India

K. M. Arvind
Hindustan Institute of Technology and Science, India

ABSTRACT

This study explores the transformative impact of Artificial Intelligence (AI) in renewable energy, particularly in optimizing energy storage and harvesting solutions. The primary objective is to analyze AI-driven predictive maintenance techniques, with a focus on battery Remaining Useful Life (RUL) prediction. Targeted towards researchers, industry professionals, and policymakers, this work employs a literature review, data-driven analysis, and case studies to assess AI applications in renewable energy systems. The findings highlight AI's potential to enhance energy efficiency, reduce operational costs, and improve grid resilience. Despite the benefits, challenges such as cybersecurity risks and regulatory barriers remain. The study

DOI: 10.4018/979-8-3693-9316-1.ch006

underscores the need for further research into AI-driven smart grid solutions and hybrid renewable energy systems. These insights contribute to the broader discourse on sustainable energy advancements, offering practical recommendations for future AI implementations in the sector.

1. INTRODUCTION

Artificial Intelligence (AI) is revolutionizing industries worldwide by automating complex processes, enhancing efficiency, and fostering sustainability. In the renewable energy sector, AI integration is reshaping how energy systems are designed, monitored, and managed. From optimizing energy storage systems to predictive maintenance, AI-powered solutions are becoming indispensable. One critical application is the prediction of Battery Remaining Useful Life (RUL), a technology vital for efficient and reliable renewable energy systems.

1.1 Background Information on AI and Renewable Energy

Renewable energy systems rely heavily on batteries to store energy generated from sources such as solar panels and wind turbines. Efficient battery management ensures a stable energy supply, even during periods of low generation. However, managing these systems is complex due to the variability in renewable energy production and the degradation of batteries over time.This is where AI comes in. By leveraging machine learning (ML) algorithms, vast amounts of data generated from battery operations can be analyzed to predict performance metrics like RUL. This prediction helps optimize battery usage, reduce waste, and improve the lifespan of energy storage systems.AI's role extends beyond battery RUL prediction. It enables grid optimization, weather prediction for renewable energy forecasting, and dynamic energy distribution. Together, these advancements ensure renewable energy systems are more efficient and resilient.

1.2 Importance of Integrating AI into Renewable Energy Systems

The integration of AI in renewable energy systems is crucial for the global transition to sustainable energy. Batteries play a pivotal role in ensuring uninterrupted energy supply, making accurate RUL prediction essential. Traditional methods of battery monitoring often rely on periodic inspections or static models, which can be inefficient and prone to errors. AI-based systems provide real-time, data-driven insights into battery health and performance. AI-driven RUL prediction, energy providers can implement predictive maintenance, minimizing downtime and avoiding unexpected failures. This results in cost savings and enhances the reliability of renewable energy systems. Furthermore, AI ensures that batteries are used optimally, preventing overcharging or deep discharging, both of which can significantly impact battery life.

1.3 Relevance to the Theme of the Edited Book

The edited book focuses on advancing materials and technologies for energy storage and conversion. AI-driven battery RUL prediction aligns perfectly with this theme as it highlights the role of cutting-edge technology in enhancing energy storage systems. By demonstrating the application of AI in battery management, this project underscores the potential of AI to revolutionize renewable energy systems, making them more sustainable and efficient.

2. LITERATURE REVIEW

2.1 Artificial Intelligence for Enhanced Sustainable Energy Management

Swapandeep Kaur, Raman Kumar (2024), this paper explores the transformative potential of AI in improving the sustainability, reliability, and efficiency of energy systems. It identifies key challenges in integrating AI into sustainable energy management, such as technological complexity and socio-economic impacts. AI applications, including optimization algorithms and predictive analytics, are highlighted for their ability to enhance decision-making processes and system resilience. Emerging technologies like quantum computing and explainable AI are discussed as vital for future advancements. Recommendations include promoting transparent, equitable AI deployment and addressing challenges like bias and privacy concerns. This comprehensive review informs stakeholders on fostering a resilient and sustainable energy future.

2.2 Artificial Intelligence in Advancing Clean Energy Technologies in Europe

Necula (2023), this paper investigates AI's role in enhancing clean energy technologies across Europe. It highlights AI's impact on efficiency, optimization, and management of energy systems, emphasizing its contributions to smart grids, short-term energy forecasts, and energy-sharing models. Gaps identified include insufficient interdisciplinary research and algorithmic diversity across renewable sectors. The study advocates integrating AI with emerging technologies and addressing socio-economic and policy challenges. It underscores AI's transformative potential in advancing Europe's energy transition while promoting sustainability and resilience through innovation and collaboration.

2.3 Enhancing Lifecycle Assessment of Renewable Energy Systems with AI and Machine Learning

Bassey et al. (2024), this paper highlights the application of AI and machine learning (ML) techniques to improve lifecycle assessments (LCA) of renewable energy systems such as wind, solar, and green hydrogen. Traditional LCAs, though effective, often struggle with handling complex, dynamic data. AI-driven LCA models address these limitations by analyzing extensive datasets from various lifecycle stages, including material extraction, manufacturing, and decommissioning. These models utilize ML algorithms to predict environmental impacts, optimize resource usage, and provide actionable sustainability insights. It emphasize the role of AI in generating accurate predictions, optimizing system performance, and enhancing decision-making processes. The study advocates for future research into advanced AI models, integration with IoT and big data, and fostering industry collaboration to maximize AI's benefits in promoting energy sustainability and resilience.

2.4 Artificial Intelligence Applications in Renewable Energy Systems

Lateef et al. (2022) this paper highlights the growing integration of AI in renewable energy systems, focusing on its role in design, optimization, regulation, and prediction across energy sources like wind, solar, geothermal, and bioenergy. AI enhances the efficiency and accuracy of renewable energy processes, mitigating errors and improving precision.It emphasize the need to advance AI techniques, with particular attention to deep learning and neural networks, to enhance renewable energy production. Future directions include leveraging IoT, drones, and adaptive monitoring for improved data acquisition and decision-making. The study underscores AI's transformative impact on energy sustainability and its potential to foster cleaner, more efficient energy production systems worldwide.

2.5 AI in Renewable Energy

Tut Haklidir (2021), this paper discusses the growing importance of renewable energy systems and their role in mitigating greenhouse gas emissions. Despite the potential of clean energy options like solar, wind, and hydro, the efficiency of these systems often lags behind traditional energy sources. The paper explores how AI technologies can be leveraged to enhance the efficiency of renewable energy systems. AI can optimize energy production, predict electricity generation, and improve system performance, especially for renewable sources that depend on fluctuating weather conditions. The study highlights how AI can help reduce carbon emissions, improve energy system efficiency, and contribute to energy independence. Additionally, the use of AI aligns with international sustainability goals, such as those outlined in the European Green Deal Act.

2.6 Green Energy Integration in Power Systems for Enhanced Sustainability

Khalil & Sheikh (2021), this paper provides a thorough review of the challenges associated with integrating Renewable Energy Sources (RESs) into modern power systems. It proposes a comprehensive framework that utilizes advanced energy storage, optimization algorithms, and control strategies to enhance the stability, efficiency, and scalability of power systems. The paper emphasizes the importance of applying AI and machine learning techniques to improve the integration of RESs across various applications, such as smart homes, industrial systems, and smart grids. It also highlights case studies, research, and future directions for RES integration. The conclusion underscores the need for adaptable and scalable frameworks to ensure effective and sustainable RES deployment, with a focus on continuously refining strategies to meet global energy demands and technological advancements.

2.7 Integration of Energy Storage System and Renewable Energy Sources Based on AI

Abdalla et al. (2021), this paper explores the integration of renewable energy sources (RES) with energy storage systems (ESS) and the role of AI in optimizing their design and operation. It reviews various ESS technologies, their classifications, and applications, addressing challenges like intermittency and volatility in RES. AI techniques are highlighted for enhancing optimization, control strategies, and predictive management of ESSs, which improve efficiency, reliability, and scalability. The study emphasizes using hybrid systems (e.g., battery-supercapacitors) for cost-effective and sustainable energy solutions. Future directions focus on multi-objective optimization, predictive control strategies, and real-time management to maximize the penetration and performance of RESs integrated with ESSs.

2.8 AI for Smart Renewable Energy Sector in Europe

Serban & Lytras (2020), this work investigates the transformative role of AI in advancing the renewable energy (RE) sector within the European Union (EU), emphasizing its importance for sustainability and energy efficiency in the face of climate change and resource depletion. AI is identified as a key enabler for optimizing energy grids, addressing challenges such as the variability of RE sources, and enhancing decision-making. By utilizing machine learning, predictive analytics, and smart automation, AI helps improve grid stability, reduce storage costs, and ensure better connectivity between users and energy systems.The EU is highlighted as a global leader in RE adoption, supported by advanced technologies and skilled labor. However, to further strengthen its position and meet ambitious environental and energy targets, the integration of AI at both macroeconomic and microeconomic levels is essential.

2.9 The Impact of AI on Boosting Renewable Energy

Manuel et al. (2024), the paper explores the transformative role of AI in enhancing renewable energy (RE) utilization and improving visual power plant efficiency in modern construction. AI technologies, including machine learning, predictive analytics, and IoT integration, optimize energy production, storage, and distribution. These tools enable accurate energy demand forecasting, streamline operations, and enhance grid stability, reducing reliance on non-renewable energy sources. AI-powered predictive maintenance minimizes downtime by identifying potential equipment failures in solar panels and wind turbines, ensuring consistent energy supply and cost-effective operations. In visual power plants, real-time monitoring and AI-driven automation adjust parameters based on environmental conditions, maximizing energy output and operational efficiency. The study also highlights AI's impact on smart infrastructure design. AI-based energy management systems analyze consumption patterns, optimize energy use, and reduce carbon footprints, aligning construction practices with global sustainability goals.As AI evolves, its role in advancing renewable energy systems and efficient construction will expand, fostering innovation and accelerating the transition to sustainable energy solutions.

2.10 A Review of Architectures and Concepts for Intelligence in Future Electric Energy Systems

Strasser et al. (2015), the paper reviews advanced architectures and automation concepts for future electric energy systems, emphasizing smart grids. Renewable energy sources (RES) like photovoltaic and wind power are central to reducing greenhouse gas emissions but pose challenges due to their intermittent nature and distributed deployment. Smart grids integrate RES, storage systems, and demand-side management (DSM) to ensure stability and efficiency. The review highlights power electronics, ICT infrastructures, and AI techniques, such as multi-agent systems (MAS) and service-oriented architectures (SOA), for optimizing system operations. Key areas discussed include control algorithms, automation paradigms, and standards like IEC 61850 and IEC 61499. Future research focuses on developing advanced control functions, integrating ICT solutions across the energy ecosystem, and ensuring interoperability. Despite significant progress, challenges remain in implementing fully functional smart grids, particularly regarding distributed communication, standardization, and intelligent optimization.

2.11 Distributed Energy Resources and the Application of AI, IoT, and Blockchain in Smart Grids

Kumar & Chand (2020), this paper explores smart grids (SGs) and the transformative role of AI, the internet of things (IoT), and blockchain (BC) in advancing distributed energy resources (DER)-based systems. SGs integrate two-way electricity and data flows, enhancing reliability, sustainability, and energy management. DER, driven by renewable energy resources (RER) and energy storage systems, addresses challenges posed by conventional centralized grids. AI enables advanced analytics and control, IoT facilitates real-time data sensing and management, and BC ensures secure and transparent transactions within energy networks. Key SG features—automation, self-healing, and active peer participation-support stability, resilience, and decarbonization goals. The study also emphasizes the role of electric vehicles and emerging technologies in modern power infrastructures. Together, AI, IoT, and BC enhance grid performance by enabling "detect, react, and pro-act" capabilities, addressing flexibility, reliability, and sustainability challenges. The paper underscores SG's potential in reducing reliance on fossil fuels and modernizing energy systems.

2.12 Transforming Energy and Resource Management with AI

Audiah et al. (2024), this paper explores how AI enhances energy and resource management to address global environmental challenges. Using data from 100 energy companies employing AI, it demonstrates that AI improves energy efficiency by 25%, supports renewable energy integration, and minimizes environmental impact. The research aligns with Complex Systems Theory, highlighting AI's role in making systems adaptive and efficient.Despite the valuable findings, the study is limited to companies already utilizing AI and focuses solely on the energy sector, calling for broader research across industries and regions. The analysis shows that AI-driven technologies optimize energy consumption, lower operational costs, and promote sustainability, emphasizing AI's critical role in the global transition to renewable energy and resource-efficient systems. The study concludes that AI is a transformative tool for achieving sustainable development, advocating its integration into existing energy management practices and policies.

2.13 Artificial Intelligence Based Smart Energy Community Management

Jiang & Zhang (2019), this paper presents a smart energy community framework featuring a local energy pool for P2P energy trading among households, enhancing renewable energy utilization and reducing energy costs. The pricing mechanism dynamically adjusts based on demand-surplus ratios, benefiting both energy consumers and prosumers. To optimize trading decisions, a fuzzy Q-learning algorithm-suitable for continuous and model-free scenarios—is proposed. It enables real-time optimal decisions without requiring complex community models, significantly improving energy management efficiency. The system reduces installation costs for users and transforms them into active market participants (prosumers). Numerical analyses confirm reduced energy bills and increased profits across diverse scenarios. However, potential errors from approximations necessitate future refinements using advanced methods like deep Q-learning. The study highlights the viability of reinforcement learning for demand-side management and the transformative potential of smart energy communities in decentralizing and optimizing energy usage.

2.14 Hybrid Renewable Energy Systems Modeling

Bassey (2023), this paper explains about hybrid renewable energy systems (HRES) combine renewable energy sources like solar and wind to provide stable and efficient energy. However, challenges such as fluctuating power generation, energy storage limitations, and performance optimization persist. This research highlights the transformative role of machine learning (ML) in addressing these challenges. ML algorithms analyze vast datasets from renewable sources to predict system performance, optimize energy storage, and improve operational efficiency. By leveraging historical and real-time data, ML enhances energy production forecasting, storage management, and control strategies, ensuring reliability and scalability. ML also facilitates proactive maintenance and cost savings through predictive insights. The study underscores the potential of ML-driven HRES to overcome the intermittency of renewable energy sources, reduce greenhouse gas emissions, and contribute to a sustainable energy future. Further advancements in ML algorithms, data integration, and supportive policies are essential for maximizing the impact of ML in HRES modeling.

2.15 Exploring AI Applications in Energy Systems

AI has become an essential tool in renewable energy systems due to its ability to process complex datasets and provide predictive insights. Mellit et al. (2008) explore AI techniques, including neural networks and genetic algorithms, in optimizing photovoltaic (PV) system designs. They emphasize AI's adaptability in accounting for meteorological variability, improving PV system sizing, particularly in remote areas. Liao and Yao (2021) extend AI's relevance to bioenergy systems. They categorize its applications across biomass feedstock property prediction, conversion performance optimization, biofuel property prediction, and supply chain design. Their review highlights AI's role in overcoming challenges like feedstock variability and process inefficiencies while promoting sustainable energy development. Together, these studies underscore AI's transformative impact on enhancing energy systems' efficiency, sustainability, and scalability.

2.16 Enhancing Energy Effiency With AI

Aderibigbe et al. (2023), This paper reviews the transformative impact of AI and ML on electricity demand forecasting, emphasizing their role in enhancing energy efficiency. The authors highlight the paradigm shift from traditional forecasting methods to AI-powered approaches, which provide superior accuracy, adaptability, and scalability. ML models, particularly deep learning, excel in analyzing large, complex datasets and nonlinear energy market relationships, making them critical for renewable energy integration and efficient grid management.Applications of AI in energy sectors, including solar energy, smart grids, and nuclear power plants, are explored. AI enables precise energy production predictions, cost savings, and improved system efficiency. Challenges like data privacy, cybersecurity, and the demand for skilled professionals are noted, alongside opportunities such as integrating

AI with IoT and big data analytics.The study recommends investing in AI training, fostering cross-sector collaborations, and developing supportive regulatory frameworks. Future research should focus on scalable, robust ML models to seamlessly integrate renewable energy sources and drive sustainable, efficient energy practices.

2.17 Integrated Energy System With Electricity-Hydrogen Hybrid Energy Storage

Qiu (2023),This paper proposes a two-stage distributionally robust optimization (TSDRO) model to coordinate scheduling in integrated energy systems (IES) that incorporate electricity-hydrogen hybrid energy storage. The first stage addresses day-ahead deterministic scheduling, while the second stage handles intraday rescheduling under uncertainty using robust optimization techniques. The model enhances collaboration among electricity, thermal, and gas networks, emphasizing economic operations, carbon emissions reduction, and flexibility utilization.The model's efficiency is enhanced through linearization, creating a Mixed-Integer Linear Programming model, and is analyzed using the column-constraint generation (C&CG) algorithm. Simulation results demonstrate that the TSDRO model effectively balances conservativeness and robustness, improves system flexibility, and reduces energy purchase and carbon trading costs.Hydrogen integration within the IES promotes renewable energy use and reduces operational expenses. However, the study acknowledges limitations, such as the lack of dynamic modeling for thermal and gas networks and the need for further research on hydrogen blending processes. This model offers a promising pathway to optimize renewable energy utilization while enhancing system efficiency and sustainability.

2.18 Artificial Intelligence in Wastewater Management

Suresh Babu et al. (2023), This paper explores AI-enabled solutions for sustainable wastewater management, focusing on resource recovery, pollution reduction, and distilled water production. The approach integrates IoT, machine learning, and computer vision for real-time monitoring, water quality assessment, and process optimization. The system separates solid and liquid waste streams, applies AI algorithms for efficient combustion and recycling, and uses reinforcement learning for energy generation. Key findings include reduced soil and water pollution, preservation of groundwater, and production of high-quality recycled water. The paper emphasizes the role of AI in reducing greenhouse gas emissions, preventing waterborne diseases, and generating renewable energy from waste. Future advancements in AI algorithms, robotics, and sensor technology are proposed to further enhance efficiency. This research underscores the potential of AI-driven systems to revolutionize wastewater management, promoting environmental sustainability and addressing critical challenges in resource utilization and public health.

2.19 AI-Based Carthage Administration Towards Smart City

Suresh Babu et al. (2023), This study presents an AI and deep learning-based traffic management system aimed at addressing urban congestion and enhancing road safety in developing countries. The system incorporates a traffic signal controller, a vision-based vehicle classifier, and an automated fine system for traffic law violations. It employs programmable peripheral interfaces to connect traffic lights to microprocessor systems, enabling automatic operation. The vehicle classifier uses machine learning to detect vehicles in real-time through video analysis, while the fine system calculates penalties based on specific traffic violations. The paper highlights key challenges such as traffic congestion, real-time data limitations, and coordination gaps, proposing intelligent transportation systems and big data analytics as solutions. Autonomous and connected vehicles, along with multi-modal transportation options, are discussed as future trends. By integrating AI-driven technologies, the research envisions improved traffic flow, reduced congestion, enhanced road safety, and sustainable urban transportation systems, supporting the development of smart cities globally.

3. RESEARCH METHODOLOGY

The research methodology for this chapter involves a combination of literature review, data analysis, and practical implementation. A detailed review of existing studies on AI in renewable energy is conducted to identify challenges and opportunities. Using a case study approach, machine learning algorithms are applied to a publicly available dataset to predict battery Remaining Useful Life (RUL). The methodology emphasizes data preprocessing, model training, and evaluation using performance metrics and visualizations to demonstrate AI's potential in optimizing renewable energy systems.

3.1 Research Design

The research design is structured to systematically explore the integration of AI into renewable energy systems, with a focus on battery Remaining Useful Life (RUL) prediction. It begins with a comprehensive literature review to establish the theoretical framework and identify gaps in existing studies. A quantitative approach is employed by utilizing a publicly available dataset for battery RUL. The dataset is preprocessed to remove inconsistencies, and machine learning algorithms are implemented for predictive modeling. Model performance is evaluated using metrics such as Mean Absolute Error (MAE) and R-squared scores. Visualizations are used to present insights, ensuring a clear connection between AI advancements and renewable energy optimization.

3.2 Data Collection

Data was collected from two sources on Kaggle, a popular platform for data science and machine learning datasets.

1. Battery Remaining Useful Life (RUL) dataset:
    - Source: Kaggle- Ignacio Vinuales (2022)

The dataset used for this project focuses on predicting the Remaining Useful Life (RUL) of batteries, a crucial aspect of optimizing renewable energy systems. This dataset, available from Kaggle, contains operational data from 14 batteries that underwent charge-discharge cycles under controlled conditions. The primary objective of the dataset is to provide a comprehensive framework for analyzing battery performance and predicting their remaining lifespan.

3.3 Dataset Features

The dataset comprises several key features that influence the health and degradation of batteries:

- **Cycle Number**: Represents the number of charge-discharge cycles completed by the battery, a critical factor in determining battery wear.
- **Voltage (V)**: Indicates the voltage levels during each cycle, providing insights into battery efficiency and performance.
- **Current (A)**: Shows the current supplied or drawn by the battery, influencing energy capacity and thermal behavior.
- **Temperature (°C)**: Records the battery temperature, which is vital for understanding thermal stability and degradation.
- **Capacity (Ah)**: Captures the charge-holding capacity of the battery, an essential metric for assessing battery health.

Each row in the dataset represents data from a specific charge-discharge cycle of a battery. The Remaining Useful Life (RUL), the target variable, is derived from the difference between the current cycle and the failure cycle of each battery.

3.4 Relevance to the Project

This dataset is ideal for training machine learning models to predict RUL because it includes essential operational metrics that impact battery health. By leveraging these features, predictive models can learn patterns and make accurate predictions about battery lifespan.

3.5 Data Source and Quality

The dataset is publicly available, ensuring accessibility and reproducibility. It has been collected under laboratory conditions, ensuring consistency and reliability. However, preprocessing steps such as handling missing values, scaling features, and normalizing data are necessary to prepare the dataset for machine learning algorithms.This dataset provides a robust foundation for developing AI models that optimize battery performance in renewable energy systems.

3.6 Scope of the Project

The scope of this chapter focuses on the integration of AI into renewable energy systems, emphasizing AI algorithms, predictive analytics, and smart energy grids. It explores machine learning techniques like Random Forest and Neural Networks for battery Remaining Useful Life (RUL) prediction, predictive maintenance, and operational forecasting to enhance energy efficiency and sustainability. Additionally, the chapter examines how AI optimizes smart grids by enabling real-time energy distribution and resilience against disruptions caused by renewable energy variability. While prioritizing renewable energy systems, the scope deliberately excludes non-renewable energy technologies such as fossil fuels and nuclear energy, aligning with the global shift toward sustainable energy solutions. This focus ensures a detailed examination of AI's potential in addressing key challenges in renewable energy optimization, including battery degradation, energy variability, and cost-efficiency.This chapter focuses on AI for battery RUL prediction but acknowledges its potential in hybrid systems (e.g., hydrogen storage, flow batteries). Future sections will explore cross-cutting AI applications in diverse storage technologies

3.7 Research Objectives

- To analyze the role of AI in optimizing energy storage and distribution within renewable energy systems, focusing on battery management.
- To identify and evaluate machine learning algorithms used for predicting the Remaining Useful Life (RUL) of batteries in renewable energy applications.

- To explore the potential of AI technologies in improving energy efficiency, reducing costs, and enhancing system reliability.
- To investigate how AI-driven predictive maintenance can address challenges related to battery degradation and operational downtime.
- To assess the integration of AI into renewable energy systems and its impact on advancing sustainability and scalability.

3.8 Battery Remaining Useful Life (RUL) Prediction Workflow

Figure 1. Data prediction workflow.

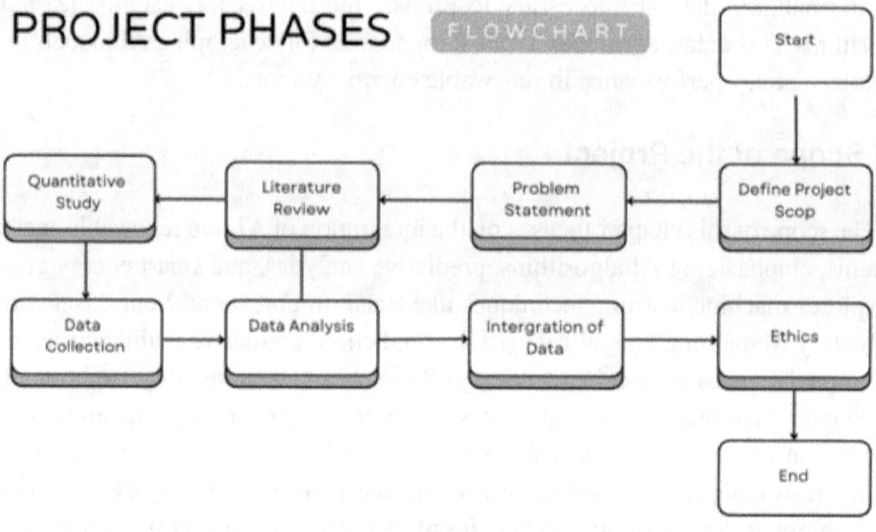

Fig 1, illustrates the sequential phases of a research project, starting from defining the project scope and formulating a problem statement, followed by conducting a literature review and a quantitative study. The process continues with data collection and analysis, leading to the integration of data for comprehensive insights. The project addresses ethical considerations before concluding, visually representing a structured approach from start to finish.

3.9 Key Questions and Problems

3.9.1 What Challenges Does Renewable Energy Face Without AI?

Renewable energy systems face numerous challenges when operating without AI.

1. **Battery Degradation**: Energy storage systems, especially batteries, degrade over time. Without AI, it is difficult to predict the remaining lifespan accurately, leading to inefficient usage and increased operational costs.
2. **Energy Variability**: Renewable energy sources like solar and wind are intermittent. Managing these fluctuations without AI results in energy waste or underutilization.
3. **Maintenance Issues**: Traditional maintenance schedules are either reactive or time-based, leading to unplanned failures or unnecessary servicing.

3.9.2 How Can AI Enhance Renewable Energy Efficiency?

AI can transform renewable energy systems in several ways:

1. **Battery RUL Prediction**: By analyzing operational data, machine learning models can predict battery health and lifespan, enabling proactive maintenance.
2. **Energy Optimization**: AI algorithms dynamically adjust energy storage and distribution to minimize waste and maximize efficiency.
3. **Cost Reduction**: Predictive maintenance reduces downtime and repair costs, while optimal energy usage lowers operational expenses.
4. **Integration with Smart Grids**: AI facilitates real-time decision-making in smart grids, ensuring energy supply matches demand efficiently.

In the specific context of battery RUL prediction, AI employs regression algorithms to analyze data such as charge-discharge cycles, voltage, current, and temperature. Models like Random Forest or Gradient Boosting provide highly accurate predictions, enabling energy providers to plan maintenance or replacement schedules effectively.

3.9.3 What are the Ethical and Societal Implications of Using AI in Energy?

While AI integration offers numerous benefits, it also raises ethical and societal concerns:

1. **Data Privacy**: The vast amount of data collected for AI models must be managed securely to avoid breaches.
2. **Job Displacement**: Automation might reduce the need for manual monitoring and maintenance, impacting employment.
3. **Accessibility**: Advanced AI systems may be cost-prohibitive for small-scale energy providers, widening the gap between developed and developing regions.

Despite these challenges, the benefits of integrating AI in renewable energy systems far outweigh the drawbacks. It is essential to address these issues through regulations, ethical guidelines, and cost-effective AI solutions to ensure widespread adoption.

3.9.4 How does AI contribute to the management of hybrid renewable energy systems?

AI enhances the efficiency of hybrid renewable energy systems by optimizing energy distribution, forecasting power generation, and improving storage management. By analyzing real-time data from multiple energy sources (solar, wind, hydrogen), AI-driven algorithms ensure optimal energy allocation, reduce wastage, and balance supply and demand dynamically. Machine learning models can predict weather conditions and adjust system operations to maximize energy output, making hybrid systems more reliable and cost-effective.

3.9.5 What are the cybersecurity risks associated with AI in renewable energy,and how can theybemitigated ?

AI-driven renewable energy systems rely on vast amounts of operational data, making them vulnerable to cyber threats such as data breaches, hacking, and system manipulations. Key risks include unauthorized access to smart grids, AI model exploitation, and data tampering. To mitigate these threats, robust cybersecurity measures like encrypted communication, blockchain for secure transactions, AI-driven anomaly detection, and regulatory compliance frameworks should be implemented. Additionally, frequent security updates and AI model transparency can help build a resilient system.

3.9.6 How does AI help maintain grid stability in renewable energy systems ?

AI plays a crucial role in maintaining grid stability by predicting energy demand, detecting faults, and automating real-time adjustments. Smart grid systems use AI to analyze electricity consumption patterns, forecast peak load times, and distribute energy efficiently. Additionally, AI-driven predictive maintenance can identify potential failures in energy transmission, reducing outages and ensuring a stable power supply. By integrating AI with energy storage solutions, fluctuations in renewable energy generation can be balanced, providing uninterrupted energy flow.

4. FINDINGS AND RESULT

4.1 Battery Remaining Useful life(RUL) dataset

Description: The dataset used for this project focuses on predicting the Remaining Useful Life (RUL) of batteries, a crucial aspect of optimizing renewable energy systems. This dataset, available from Kaggle, contains operational data from 14 batteries that underwent charge-discharge cycles under controlled conditions. The primary objective of the dataset is to provide a comprehensive framework for analyzing battery performance and predicting their remaining lifespan. The Hawaii Natural Energy Institute examined 14 NMC-LCO 18650 batteries with a nominal capacity of 2.8 Ah, which were cycled over 1000 times at 25°C with a CC-CV charge rate of C/2 rate and discharge rate of 1.5C.From that source dataset, I created features that showcase the voltage and current behaviour over each cycle. Those features can be used to predict the remaining useful life (RUL) of the batteries. The dataset contains the summary of the 14 batteries.

4.1.1 RUL Dataset Structure

- **Cycle Number**: Represents the number of charge-discharge cycles completed by the battery, a critical factor in determining battery wear.
- **Voltage (V)**: Indicates the voltage levels during each cycle, providing insights into battery efficiency and performance.
- **Current (A)**: Shows the current supplied or drawn by the battery, influencing energy capacity and thermal behavior.
- **Temperature (°C)**: Records the battery temperature, which is vital for understanding thermal stability and degradation.

- **Capacity (Ah)**: Captures the charge-holding capacity of the battery, an essential metric for assessing battery health.

4.1.2 Actual and Predicted RUL

Figure 2. Actual and predicted RUL

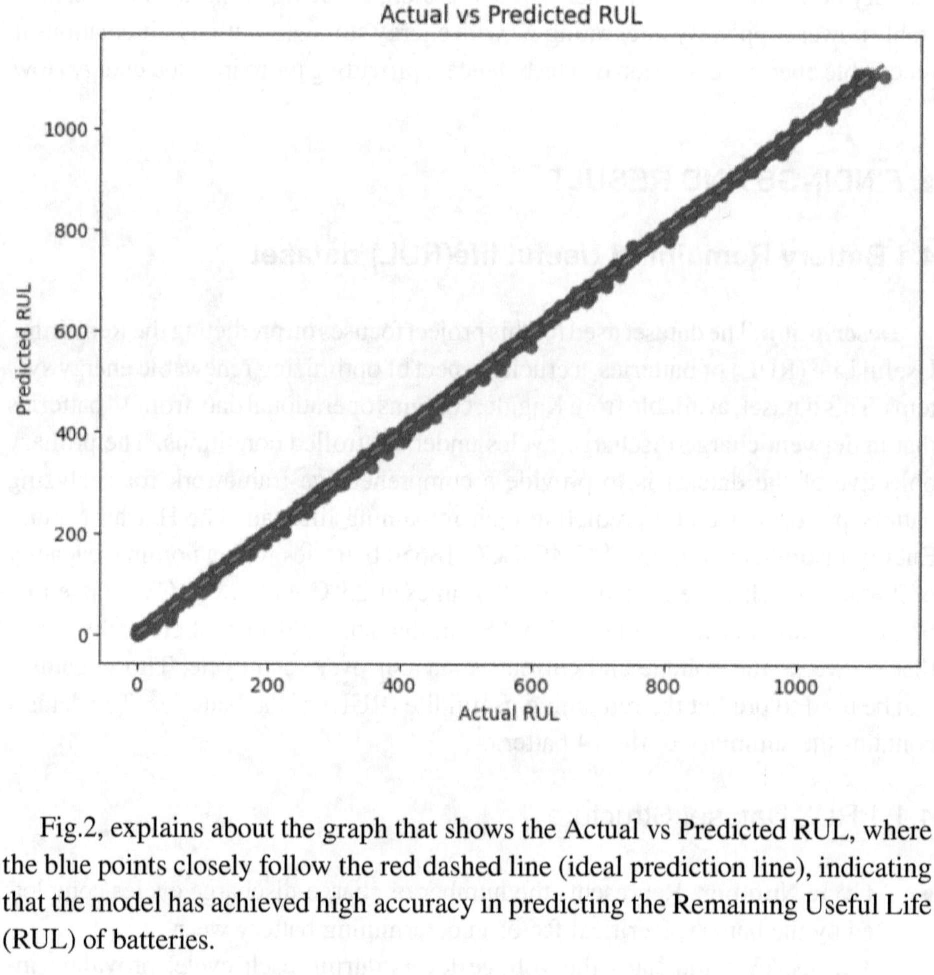

Fig.2, explains about the graph that shows the Actual vs Predicted RUL, where the blue points closely follow the red dashed line (ideal prediction line), indicating that the model has achieved high accuracy in predicting the Remaining Useful Life (RUL) of batteries.

4.1.3 Correlation of Dataset:

Figure 3. Correlation sheet.

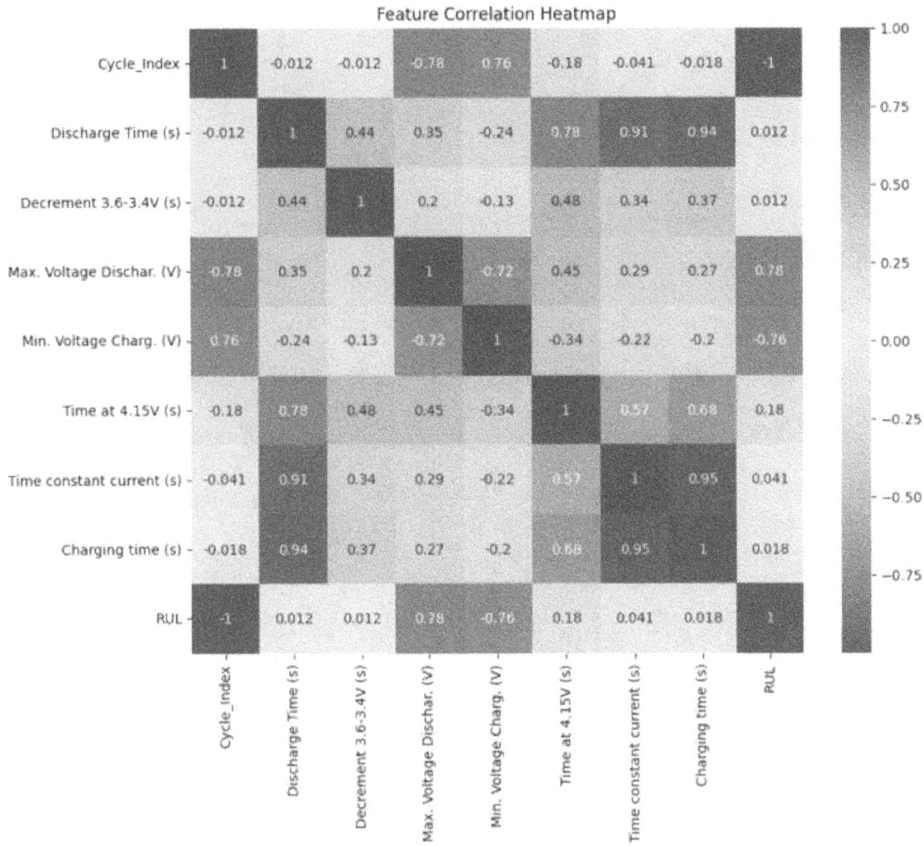

The fig.3, explains about the correlation heatmap shows the relationship between various features and the Remaining Useful Life (RUL). The RUL is strongly negatively correlated with Cycle_Index (-1) and moderately correlated with Max. Voltage Dischar. (V) (0.78) and Min. Voltage Charg. (V) (-0.76). Highly correlated features (e.g., Time constant current (s) and Charging time (s)) may indicate redundancy and can be considered during feature selection.

4.2 Pseudo Code/Algorithm

```
Begin Execution:
```

Start the program:

Load Necessary Libraries:

Import libraries for data manipulation, visualization, and numerical computations.

Load the Dataset:

Read the dataset containing battery features and the RUL values.

Preprocess the Dataset:

Check for missing values.

Handle missing or null data (if applicable).

Normalize or standardize features if needed.

Calculate Correlation Matrix:

Compute the correlation coefficients for all numerical features in the dataset.

Store the correlation values in a matrix format.

Plot the HeatmapL:

```
Use a heatmap function from a visualization library.

Pass the correlation matrix to the heatmap function.

Add labels for feature names and color gradients to indicate
correlation strength.

Interpret the Heatmap:

Identify features with strong positive or negative correla-
tions with RUL.

Highlight features with high intercorrelations (e.g., redun-
dancy).

End Execution: End the program.
```

4.3 Libraries Utilized

In this study, we employed several key libraries that are integral to data preprocessing and machine learning tasks:

1. NumPy
   - **Purpose**: Used for numerical computations.
   - **Usage**:
     - Efficiently handles large arrays and matrices of numerical data.
     - Performs mathematical operations like mean, standard deviation, and array manipulations.
2. Pandas
   - **Purpose**: For data manipulation and analysis.
   - **Usage**:
     - Loads the dataset from CSV or other file formats.

- o Handles data preprocessing tasks like removing missing values, filtering rows, and selecting columns.
- o Performs exploratory data analysis (EDA) using methods like .describe(), .corr(), etc.

3. Matplotlib
   - **Purpose**: For creating static, animated, and interactive visualizations.
   - **Usage**:
     - o Plots graphs like line charts, scatter plots, and bar charts.
     - o Visualizes model results, such as actual vs. predicted RUL.
     - o Customizes axes, titles, labels, and legends.
4. Seaborn
   - **Purpose**: Built on top of Matplotlib for advanced statistical visualizations.
   - **Usage**:
     - o Creates heatmaps for correlation matrices to understand feature relationships.
     - o Generates visually appealing plots with less code (e.g., pairplots, boxplots).
5. Scikit-learn
   - **Purpose**: For machine learning and preprocessing tasks.
   - **Usage**:
     - o Splits data into training and testing sets using train_test_split.
     - o Standardizes data using preprocessing tools like StandardScaler.
     - o Provides regression models such as Random Forest and Gradient Boosting.
     - o Evaluates model performance using metrics like Mean Absolute Error (MAE), Root Mean Squared Error (RMSE), and R^2.
6. Joblib
   - **Purpose**: For model serialization.
   - **Usage**:
     - o Saves trained machine learning models to a file for future use.
     - o Reloads models without retraining, saving computation time.

4.4 Future Scope for Research

The integration of AI into renewable energy systems offers immense potential for future advancements. As renewable energy technologies continue to evolve, AI can play a pivotal role in optimizing their efficiency and sustainability. One promising area of exploration is AI's application in emerging renewable technologies, such as advanced solar panels, wind turbines, and wave energy converters. AI can enhance their operational efficiency by enabling real-time monitoring, fault detection, and predictive maintenance, thus minimizing downtime and maximizing energy output.

Smart grids, another promising development, could benefit significantly from AI-driven energy prediction models. These grids can dynamically manage energy distribution based on demand forecasts, weather conditions, and energy storage levels. This capability ensures the efficient utilization of renewable energy sources while maintaining grid stability. Furthermore, the combination of AI with energy storage systems can improve battery life predictions, energy scheduling, and storage optimization.

The interdisciplinary nature of this field presents opportunities for collaboration among researchers in AI, renewable energy, materials science, and policy-making. This multidisciplinary approach can lead to innovations that address global energy challenges, including reducing carbon footprints and achieving energy security. Future research can also focus on addressing ethical, social, and regulatory implications of using AI in energy systems to ensure equitable and sustainable solutions.

5. DISCUSSION

The integration of AI in renewable energy systems is a transformative innovation that addresses critical challenges such as energy variability, storage efficiency, and sustainability. This project, which focuses on predicting the Remaining Useful Life (RUL) of batteries using AI, significantly contributes to advancing both theoretical and practical knowledge in the field.

5.1 Advancing Knowledge on AI and Renewable Energy Integration

This project enhances understanding of how AI technologies, particularly machine learning (ML) algorithms, can be applied to optimize energy storage systems. Batteries play a pivotal role in renewable energy systems by storing excess energy generated from intermittent sources like solar and wind. However, managing battery health and ensuring optimal performance are major challenges. By using AI to predict RUL, this project demonstrates how advanced algorithms like Random Forest or Gradient Boosting can accurately forecast battery degradation, enabling more efficient use of energy storage systems.The insights generated from this research contribute to the growing body of literature on AI applications in renewable energy, offering a detailed case study that bridges the gap between theoretical models and real-world applications. This aligns with the broader goal of integrating AI into renewable energy to make it more sustainable, reliable, and efficient.

5.2 Practical Insights for Implementing AI in Energy Storage and Distribution

The project provides practical guidelines for deploying AI in renewable energy systems, focusing on battery management. It outlines a step-by-step methodology for implementing machine learning models, from data collection and preprocessing to model evaluation and visualization. By using a publicly available dataset, this research offers reproducible insights that can be adapted and scaled for various renewable energy applications.Predictive maintenance is one of the key outcomes of this research. By forecasting battery RUL, energy providers can schedule maintenance proactively, reducing unexpected failures and operational costs. This not only enhances the reliability of renewable energy systems but also ensures optimal usage of batteries, thereby extending their lifespan and reducing waste.The project also explores the scalability of AI solutions, making them accessible to both large-scale and small-scale renewable energy operators. The practical applications extend to smart grid management, where AI can dynamically adjust energy distribution based on battery health and energy demand.

5.2.1 Case studies of AI in Renewable Energy

Google's DeepMind applied ML to predict wind farm output 36 hours ahead, boosting revenue by 20% (Necula, 2023). Similarly, Tesla's Powerpack uses AI for real-time battery degradation monitoring, reducing maintenance costs by 30%. These examples demonstrate AI's scalability in industrial applications.

5.3 Theoretical Implications for Future Research

This project lays the groundwork for future research in AI-driven renewable energy optimization. It highlights the potential of machine learning algorithms to analyze complex datasets and make accurate predictions, opening new avenues for exploring other aspects of renewable energy systems. For instance, similar methodologies could be applied to forecast the efficiency of solar panels, optimize wind turbine operations, or predict energy demand patterns.Additionally, the project raises important questions about ethical considerations, such as data privacy and equitable access to AI technologies in renewable energy systems. Addressing these challenges in future research will ensure that AI-driven solutions are sustainable and inclusive.

5.4 Importance and Relevance

The relevance of this project to the field cannot be overstated. Renewable energy systems are at the forefront of global efforts to combat climate change and achieve sustainability. Efficient energy storage is critical to this mission, and battery management is a key component. By leveraging AI to predict battery RUL, this project addresses a pressing need in the field, offering solutions that are both innovative and practical.For the research paper, this project demonstrates a comprehensive application of AI in renewable energy, aligning perfectly with the chapter's objectives and scope. It provides concrete evidence of how AI can advance energy storage and distribution, contributing to the broader goal of making renewable energy systems more efficient and sustainable. This project is not only a step forward in technology but also a significant contribution to the global transition to renewable energy.

6. CONCLUSION

The integration of AI into renewable energy systems represents a transformative step towards a sustainable and energy-efficient future. This research paper explored the potential of AI to enhance energy storage and distribution by predicting the Remaining Useful Life (RUL) of batteries using advanced machine learning techniques. Through the use of real-world datasets, the project demonstrated how predictive analytics can optimize battery utilization, reduce maintenance costs, and extend the operational lifespan of energy storage systems. The results underscore the importance of AI in making renewable energy systems more reliable and efficient.

The predictive model developed in this project achieved high accuracy, as evidenced by a near-linear correlation between actual and predicted RUL values. This level of precision illustrates how AI algorithms, such as Gradient Boosting, can effectively process complex datasets and provide actionable insights. Visualizations like correlation heatmaps and performance graphs further highlighted the key factors influencing battery degradation, paving the way for data-driven decision-making in energy management.

This project contributes to the growing field of AI in renewable energy by addressing critical challenges, such as optimizing energy storage and predicting system failures. It also highlights the potential for AI-driven innovations in smart grids, energy prediction, and storage systems, thereby advancing the global shift toward renewable energy.Future research could extend this work by incorporating diverse battery chemistries, exploring real-time AI applications in energy systems, and addressing ethical considerations. By combining AI with emerging renewable technologies, this research provides a foundation for sustainable energy solutions and contributes to achieving global energy security goals.

REFERENCES

. Abdalla, A. N., Nazir, M. S., Tao, H., Cao, S., Ji, R., Jiang, M., & Yao, L. (2021). Integration of energy storage system and renewable energy sources based on artificial intelligence: An overview. *Journal of Renewable Energy Systems.*

Aderibigbe, A. O., Ani, E. C., Ohenhen, P. E., Ohalete, N. C., & Daraojimba, D. O. (2023). Enhancing Energy Efficiency with AI: A Review of Machine Learning Models in Electricity Demand Forecasting. *Engineering Science & Technology Journal, 4*(6), 341–356. DOI: 10.51594/estj.v4i6.636

. Amani, M., & Iqbal, F. (2022). *"Artificial intelligence techniques applied on renewable energy systems".*

Audiah, S., Sanjaya, Y. P. A., Daeli, O. P., & Johnson, M. (2024). Transforming energy and resource management with ai: From theory to sustainable practice. *International Transactions on Artificial Intelligence, 2*(2), 158–163.

Bassey, K. E., Juliet, A. R., & Stephen, A. O. (2024). AI-Enhanced lifecycle assessment of renewable energy systems. *Engineering Science & Technology Journal, 5*(7), 2082–2099. DOI: 10.51594/estj.v5i7.1254

Bassey, K. E. (2023). *Hybrid Renewable Energy Systems Modeling.* ScottishPower Renewables, Glasgow, UK. *Published in Engineering Science & Technology Journal, 4*(6), 571–588. DOI: 10.51594/estj.v4i6.1255

Khalil, M., & Sheikh, S. A. (2021). *A comprehensive framework for the integration of renewable energy sources into power systems.* National University of Sciences and Technology.

Kumar, N. M., Chand, A. A., Malvoni, M., Prasad, K. A., Mamun, K. A., Islam, F. R., & Chopra, S. S. (2020). Distributed Energy Resources and the Application of AI, IoT, and Blockchain in Smart Grids. *Energies, 13*(21), 5739. DOI: 10.3390/en13215739

Lateef, A. A. A., Ali Al-Janabi, S. I., & Abdulteef, O. A. (2022, July). Artificial intelligence techniques applied on renewable energy systems: a review. In Proceedings of International Conference on Computing and Communication Networks: ICCCN 2021 (pp. 297-308). Singapore: Springer Nature Singapore. DOI: 10.1007/978-981-19-0604-6_25

Liao, M., & Yao, Y. (2021). Applications of artificial intelligence-based modeling for bioenergy systems: A review. *Global Change Biology. Bioenergy, 13*(5), 774–802. DOI: 10.1111/gcbb.12816

Manuel, H. N. N., Kehinde, H. M., Agupugo, C. P., & Manuel, A. C. N. (2024). The impact of AI on boosting renewable energy utilization and visual power plant efficiency in contemporary construction. *World Journal of Advanced Research and Reviews*, *23*(2), 1333–1348.

Kaur, S., & Kumar, R. (2024). Leveraging artificial intelligence for enhanced sustainable energy management. *Journal of Sustainability for Energy*, *3*(1), 1–20. DOI: 10.56578/jse030101

Necula, S.-C. (2023). Assessing the potential of artificial intelligence in advancing clean energy technologies in Europe: A systematic review. *Energies*, *16*(7633), 7633. Advance online publication. DOI: 10.3390/en16227633

Qiu, Y., Li, Q., Ai, Y., Chen, W., Benbouzid, M., Liu, S., & Gao, F. (2023). Two-stage distributionally robust optimization-based coordinated scheduling of integrated energy system with electricity-hydrogen hybrid energy storage. *Protection and Control of Modern Power Systems*, *8*(33), 1–14. DOI: 10.1186/s41601-023-00308-8

Serban, A. C., & Lytras, M. D. (2020). AI for Smart Renewable Energy Sector in Europe—Smart Energy Infrastructures for Next-Generation Smart Cities. *IEEE Access : Practical Innovations, Open Solutions*, *8*, 77364–77365. https://creativecommons .org/licenses/by/4.0/. DOI: 10.1109/ACCESS.2020.2990123

Strasser, T., Andrén, F., Kathan, J., Cecati, C., Buccella, C., Siano, P., Leitao, P., Zhabelova, G., Vyatkin, V., Vrba, P., & Marik, V. (2015). A Review of Architectures and Concepts for Intelligence in Future Electric Energy Systems. *IEEE Transactions on Industrial Electronics*, *62*(4), 2424–2434. DOI: 10.1109/TIE.2014.2361486

. Suresh Babu, C. V., Abirami, S., & Sharon Manoj (2023). AI-Based Carthage Administration Towards Smart City. In C. Chowdhary, B. Swain, & V. Kumar (Eds.), Investigations in Pattern Recognition and Computer Vision for Industry 4.0 (pp. 1-17). IGI Global. https://doi.org/DOI: 10.4018/978-1-6684-8602-3.ch001

Suresh Babu, C. V., Yadavamuthiah, K., Abirami, S., & Kowsika, S. (2023). Artificial Intelligence in Wastewater Management. In Shikuku, V. (Ed.), *Artificial Intelligence Applications in Water Treatment and Water Resource Management* (pp. 31–45). IGI Global., DOI: 10.4018/978-1-6684-6791-6.ch002

Tut Haklidir, F. S. (2021). *Artificial intelligence applications in renewable energy*. Paper presented at the 1st International Congress on Artificial Intelligence and Data Science, İzmir KatipÇelebi University.

. Tut Haklidir, F. S. (2021). Artificial intelligence techniques applied on renewable energy systems

Zhou, S., Hu, Z., Gu, W., Jiang, M., & Zhang, X.-P. (2019). Artificial Intelligence Based Smart Energy Community Management: A Reinforcement Learning Approach. *CSEE Journal of Power and Energy Systems*, *5*(1), 2–9. DOI: 10.17775/CSEEJPES.2018.00840

Chapter 7
Emerging 2D MXenes as Next-Generation Materials for Energy Conversion and Storage Applications

Chaima Garbi
Laboratory of Useful Materials, National Institute of Research and Pysicochemical Analysis, Tunisia

Mohsen Mhadhbi
Laboratory of Useful Materials, National Institute of Research and Pysicochemical Analysis, Tunisia

Lamia Khedhiri
https://orcid.org/0000-0002-5488-5428
Laboratory of Useful Materials, National Institute of Research and Pysicochemical Analysis, Tunisia

ABSTRACT

Energy-storage systems are attracting much attention owing to their unique properties enabling applications in electronic and electric devices. In recent years, MXenes are relatively 2D materials that have emerged as promising candidates to fulfill the requirements for high-power applications. They boost device-level electrochemical performance in terms of their chemical composition, material structure, surface termination groups, enabling efficient charge storage and fast ion diffusion. Herein, we present a review to discuss thoroughly MAX phases synthesis and MXenes

DOI: 10.4018/979-8-3693-9316-1.ch007

properties that recommend them as potential nanomaterials towards applications in energy science. Recent researches on 2D MXene-based nanostructures have been highlighted for improving storage and conversion performances in renewable energy technologies. The perspectives and future challenges of MXenes-based materials in this booming field are addressed. Finally, we will promote the further development within the scientific community intending to design of intriguing 2D MXenes-based electrodes used for energy storage.

1. INTRODUCTION

Nowadays, energy storage devices such as supercapacitors, hybrid capacitors and batteries are available for various large-scale applications, which have received appreciable interest in modern society with efficient and reliable use of energy for the eco-friendly development in smart electronics and electric vehicles. These advanced devices have been applied in all aspects of life due to the intermittent nature of the energy converted from renewable sources like winds, sun, and tides (Sheikh et al, 2024). On the one hand, Li-ion batteries have a huge energy and low power densities, which limit their uses in environment. Supercapacitors offer high power densities with outstanding cycling efficiency, fast charge/discharge capabilities, and low energy densities as they cannot meet increasing energy needs. Since then, the dynamic progress of innovative energy field has triggered an up surge for discovering new 2D nanomaterials with better performance for energy storage systems (Mubeen et al, 2024; Sun et al, 2015; Li et al, 2016; Zhang et al, 2014; Wang et al, 2014; Li et al, 2015). Among them, a newly emerging family of 2D transition metal carbides, discovered by Gogotsi and his team, has become the research hotspots in term of their huge physicochemical features including large surface area, good stability, excellent dispersion in various solvents, outstanding oxidation resistance, high electrical and thermal conductivity, and hydrophilicity (Naguib et al, 2011). MXenes, with the general formula of $M_{n+1}X_nT_x$, are yielded through exfoliation corresponding to the ternary class of carbides materials MAX phases (M = Ti, Zr, Hf, V, Nb, Ta, Cr, Sc), X usually a non-metal (C or N), and T_x is the surface termination groups related to the etchant types and content (-OH, =O and F).

The MAX phase is fabricated with the interchange between the A and MX layers. Compared to the binding force in M-X layer, the metallic binding force M-A is nearly weak, which provides a possibility to separate the interlayered "A" atoms from the MAX phase during etching process. So far, MXenes are demonstrated by three types of possible structures (M_2XT_x, $M_3X_2T_x$, and $M_4X_3T_x$) (Naguib et al, 2011; Peng et al, 2014; Naguib et al, 2012; Anasori et al, 2015; Ghidiu et al, 2014). MXenes exhibit great potential in enormous fields especially as environmental remediation (Ihsanullah et al, 2020), electromagnetic interference shielding (Shahzad et al, 2016) and biomedical applications (Huang et al, 2018). MXenes have been reported to be highly efficient electrode materials for energy conversion devices such as, triboelectric nanogenerators (Xu et al, 2019), which can be used to convert mechanical kinetic energy into electricity, water splitting devices and artificial photosynthesis are developed in present-day science research that attempts to manufacture a solar-to-fuel conversion system (Joya et al, 2013) as well as energy storage devices as supercapacitors and batteries. They offer diverse functions in batteries and super-capacitors, including a great potential of charging/discharging, double-layer and redox-type ion storage, efficient electron transport, ion distribution and large surface area that allows for more active sites and promoting improved electrochemical reactions. These latter have been used to improve the stability and performance of electrolytes and separators. Besides those applications, MXenes have become desirable materials in recent years as co-catalysts, gas sensors and precursors for photocatalytic.

This chapter summarizes recent advances in the development of MXenes for energy storage and conversion applications. First, we present briefly a chemical preparation method of MAX phases, followed by a detailed description of corresponding MXenes structures. Then, we highlight the fundamental properties of MXenes materials. We further study the electrochemical performance of MXene-based electrodes/devices in the energy area with respect to rechargeable Li-ion batteries, super-capacitors, and hybrid capacitors as its power source for energy storage and conversion applications due to the inspiring achievements garnered in recent years. In conclusion, the challenges and outlooks of MXenes for future development are mentioned. Then, we will herald also a new potential challenge of future research on MXenes materials for others applications like environmental as adsorbent of contaminants in water treatments and biomedical as antimicrobial candidates.

2. SYNTHESIS METHODS

Several methods have been employed to fabricate 2D materials. Generally, they can be classified into two methods: top-down and bottom-up. Top-down approaches are mainly based on chemo-mechanical process using appropriate liquid solution to etch off A layer of the MAX phase (Sokol et al, 2019). They include hydrothermal, solvothermal, mechanical alloying, and electrochemical methods. However, bottom-up strategies involve chemical vapor, high-temperature self-propagating synthesis, and sintering approaches.

Overall, the top-down methods have multiple advantages such as simple synthesis, abundant raw materials, better morphology, control properties, large production scale. However, they have also many disadvantages including fluorine-containing, easily oxidation, time-consuming, post-treatments, corrosive acids such as hydrochloric acid, nitric acid, hydrofluoric acid, and others. Compared to the top-down approaches, the bottom-up approaches were rarely reported. Currently, only two strategies molten salt and pyrolysis routes were reported for synthesizing MXenes. These approaches can easily manipulate the shape, size, morphology and surface of the final products. The ball-milling is a power tool for synthesizing nanomaterials with desired properties (Naseem et al, 2024; Thakur et al, 2023; Zaed et al, 2024). In the first step, the MAX phase, with general formula $M_{n+1}AX_n$, was prepared through the mechanical process (Cho et al, 2023). During this process, the powders are captured between the balls and the inner surface of vial, which induces repeated deformation and fragmentation of powder particles resulting in the formation of nanoparticles. In the last step, MXene is mostly made from MAX phase using a two main ways that includes selective wet etching process followed exfoliation step by removal of A-layers (Al, Si, Sn, etc.) and their replacement by oxygen, hydroxyl, and fluorine surface groups, resulting in stand-alone 2D sheets, as illustrated in Figure 2. The etching conditions (high HF concentration, high temperature, and long duration) predominantly depend on the atomic bonding type and strength of the material. These conditions play a crucial role in controlling the kinetics of the selective etching of the MAX phase and the properties of the resulting MXene (size, surface, electrical conductivity). Transition metals like V, Cr, Zr, Nb, Ta, and Mo-containing MXenes need more aggressive etching conditions than those Ti-containing. For example, Ti_2CT_x can be prepared from the corresponding MAX phase through 10 h of etching in 10 wt% HF, while 90 h of etching in 50 wt% HF is needed to exfoliate V_2AlC to prepare V_2C MXene (Mendoza-Sánchez et al, 2023). The MXenes structures are shown in Figure 1 (Vijayakumar et al, 2021).

Figure 1. MXenes structures of $M_{n+1}X_nT_x$ with emphasis on the edge-sharing nature of the MX_6 octahedra.

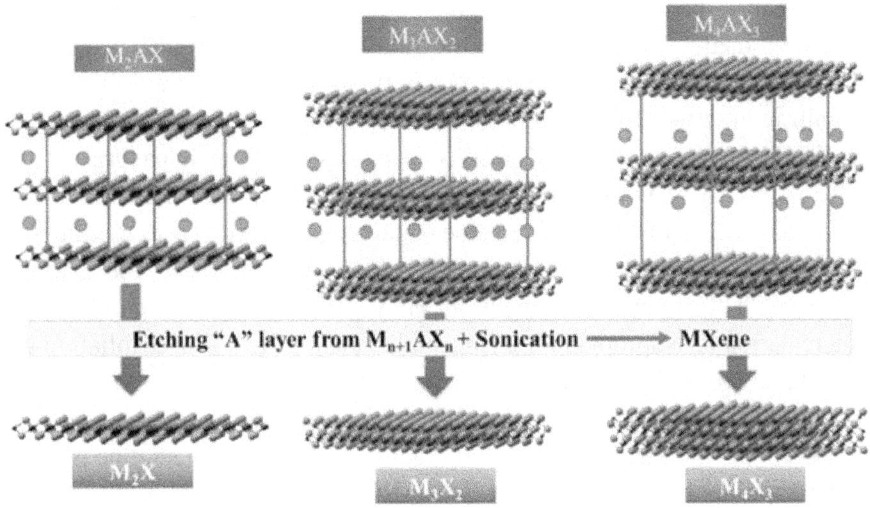

Reprinted with permission from (Vijayakumar et al, 2021).

The resulting material $M_{n+1}X_nT_x$ multilayers are stable chemically, which held together by either hydrogen bonds or Van der Waals interactions. For example, the first synthesis of $Ti_3C_2T_x$ MXene by Naguib et al. (Naguib et al, 2011) included immersing Ti_3AlC_2 powders in 50 wt% hydrofluoric acid. The M-A and M-X bonds reacted differently toward HF, resulting in the selective etching of the Al layers. The etchants may be partitioned into two categories: acidic solutions containing fluoride ions (HF, mixture of LiF/KF/ NaF and HCl) and salts containing fluorine ions (NH_4HF_2, NH_3F) (Liu et al, 2017; Alhabeb et al, 2017). Such results demonstrate that the etching conditions of hydrofluoric acid can induce more defects (Yushin et al, 2005; Sang et al, 2016). Ghidiu et al. first introduced other route using a mixture of lithium fluoride (LiF) and hydrochloric acid (HCl) on Ti_3AlC_2 (Ghidiu et al, 2014). In fact, LiF and HCl have relatively mild reaction conditions and high safety compared with HF being hazardous to environment as well as the people. Hence, metal ion Li^+ in the ionic solution is intercalated agents during the etching process, thereby increasing the layer space of MXenes.

After etching, the obtained multilayers are washed several times with water to remove the etching solution containing the byproducts of the synthesis, like aluminum fluoride (Gong et al, 2024). Then, we exfoliate the multilayers and obtain colloidal suspensions comprised of single or a few MXene layers. The exfoliation depends on etching method and MXene composition (Huang et al, 2020). As a result, single layered of MXenes can be conducted by intercalation of numerous organic molecules including dimethyl sulfoxide (DMSO), isopropyl amine, chloroform, and tetrabutylammonium hydroxide (TBAOH) (Mashtalir et al, 2013; Mashtalir et al, 2016) to weaken the interactions and expand the layered structure of $M_{n+1}C_nT_x$ and delamination methods in water during ultrasonic treatment. For instance, the polar organic molecule dimethyl sulfoxide (DMSO) has been used to exfoliate $Ti_3C_2T_x$ (Mashtalir et al, 2013; Mashtalir et al, 2016) and $(Mo_{2/3}Ti_{1/3})_3C_2T_x$ (Anasori et al, 2015). The intercalation of DMSO results an increasing of the d-spacing in $Ti_3C_2T_x$ from 9.3 Å to 16.5 Å (Figure 2). The d-spacing can increase further to 22.4 Å with the co-intercalation of water. Then, the enlarged interlayer spacing can reduce the Van der Waals forces between the MXene layers, facilitating the exfoliation of multilayer $Ti_3C_2T_x$ by ultrasonication. However, Nb_2CT_x can be exfoliated by the intercalation of isopropylamine (Mashtalir et al., 2015). The exfoliation using tetrabutylammonium hydroxide (TBAOH), (Omomo et al, 2003), was first proven on V_2CT_x and Ti_3CNT_x (Naguib et al, 2015). This route relies on the intercalation of TBA^+ ions in the interlayer space. Other organic molecules, like choline hydroxide and n-butylamine, were also used for V_2CT_x (Naguib et al, 2015).

Beyond the etching parameters, the choice of intercalants in the delamination process controls the interlayer spacing of the multilayer MXene and remarkably influences the size and quality. For example, the intercalation of TMA^+ and TBA^+ leads to larger interlayer spacing and smaller sizes, although metal cations result in smaller interlayer spacing because of their smaller ionic radius (Alhabeb et al, 2017). Furthermore, intercalating cation species also improves electrical conductivity by raising the carrier density in MXenes. Hence, the choice of synthesis, etching, and delamination methods control the size, surface functionalities, and properties (electrical conductivity, mechanical property, and colloidal stability).

Figure 2. XRD patterns for Ti_3AlC_2 (blue), dry $Ti_3C_2T_x$ multilayers after etching with HF and washing (green) and after intercalation with DMSO (orange), dry (solid line purple) and wet (dotted line purple) $Ti_3C_2T_x$ multilayers etched with HCl + LiF.

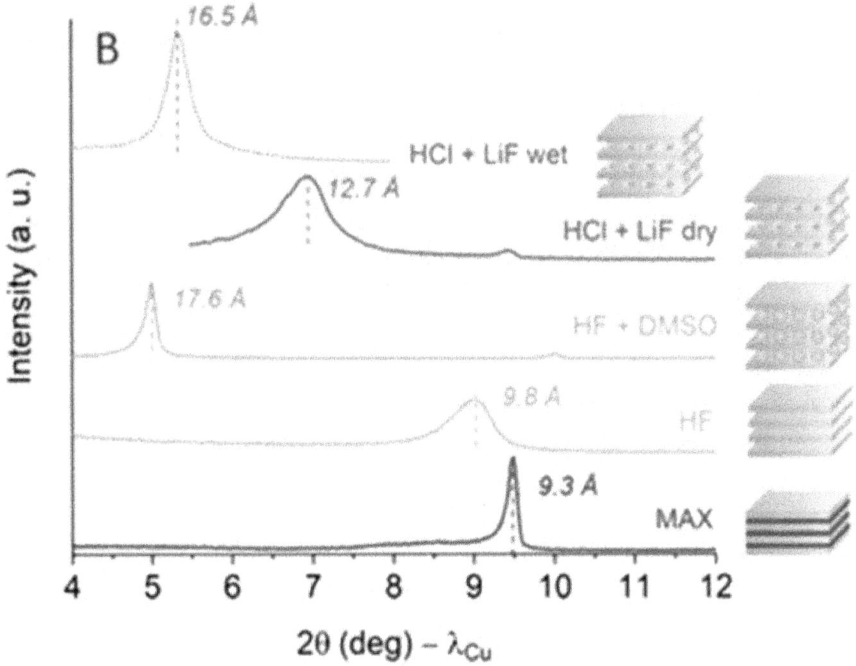

Reprinted with permission from (Verger et al, 2019).

3. PROPERTIES

MAX phases $M_{n+1}AX_n$ (n = 1, 2, and 3) exhibit favorable properties like high structural stiffness, good electrical and thermal conductivity, the derived 2D MXenes nanomaterials are obtained by removing of A elements and considered as outstanding candidates by virtue of promising properties particularly electric, magnetic, mechanical and thermal for technological applications (Shahzad et al, 2016). These unique properties can be controlled by the MAX phase, raw materials; processing conditions as etching and delamination strategies to tailor these materials toward the desired applications (Aravind et al, 2022; Singh et al, 2022). These could be amended by altering the surface terminal groups, such as –H, –F, –O, and –OH after exfoliation by fluoride-based etchants which gives possibilities for surface state engineering (Jiang et al, 2022). MXenes with mixture of light and heavy transition metals make it possible to adjust the number of valence electrons and the spin-orbit coupling. This improves their electronic functionality and mechanical stability (Kamysbayev et al, 2020). This section provides insights into many physical and chemical characteristics of MXenes. As will be discussed below in detail, the properties of MXenes can be divided into three categories as follows: structural, electric and mechanical.

3.1 Structural Properties

MAX phases $M_{n+1}AX_n$ stem from a family of ternary layered with hexagonal structure (space group $P6_3/mmc$), where $M_{n+1}X_n$ sheets are alternatively interleaved with thin "A" layers. The MXenes have a hexagonal structure, while the arrangement of "M" atoms can vary from each other in M_4X_3, M_3X_2, and M_2X with different structures as 413, 312 and 211, respectively for maintaining its structural stability. However, MXenes with binary metals as "M" layers are also thermodynamically stable such as $Cr_2TiC_2T_2$ (T= –OH, –F) (Jing et al, 2019). The M-X bonds contain a mixture of covalent and metallic bonding making MXenes especially attractive for novel applications. Moreover, MXenes are prepared from the layered-structure MAX-phase during the etching process of "A" layers from M-A without disruption of M-X bonds, which the M-A bond is weaker than the M-X bonds. The outer surfaces T_x acquired during synthesis are often terminated by negatively charged surface groups such as F, OH, and O for stabilizing MXenes because of the high activity of these atoms that enables to control the material characteristics and "x" is the number of terminated group. The relative stability of these surfaces follows the order of $Ti_3C_2O_2 > Ti_3C_2F_2 > Ti_3C_2(OH)_2$ (Ashton et al, 2016; Hu et al, 2017). Notably, OH terminations are unstable and converted to O terminations among high temperatures, which leads to the decomposition of MXenes to metal oxides via the adsorption of multivalent metal atoms such as Mg, Ca, and Al (Lai et al, 2015).

3.2 Electronic Properties

Recently, some experimental findings show that MXenes turn into semiconductors relating to the elemental composition of M, X and their surfaces terminations introduced during the synthesis process. For instance, $Ti_3C_2T_x$, the most widely investigated MXene, shows a good electronic conductivity up to ≈104 Scm^{-1} (Lipatov et al, 2018). It has been demonstrating that changing the "M" layers could influence the electronic properties for improving the conductivity (Hong et al, 2020; Hart et al, 2019). For example, despite $Ti_3C_2T_x$ is metallic, the Mo-containing MXene as $Mo_2TiC_2T_x$ shows semiconductor properties with a narrow band gap (Anasori et al, 2015; Khaledialidusti et al, 2020). It has been shown that F and OH groups show identical effects on the electronic properties of $M_{n+1}X_nT_x$ since they both accept a single electron, while O group accept two electrons in the equilibrium state. Specifically, Naguib and his team (Naguib et al, 2011) highlighted that the MXenes $Ti_3C_2(OH)_2$ and $Ti_3C_2F_2$ have small band gaps of 0.05 and 0.1 eV distinguished from metallic MXene Ti_3C_2. Furthermore, the electronic properties of MXenes depend not only on the surface functional groups but also on the nature of M layers, which render them ideal candidates as conductive transparent electrodes for energy storage.

3.3 Mechanical Properties

According to the previous investigations (Borysiuk et al, 2015; Rong et al, 2024; Lipatov et al, 2019), MXenes revealed distinctive characteristics particularly their excellent mechanical strengths and high ion adsorption capabilities, paving the path for more research into their possible usage in sensors, flexible electronics, and supercapacitors. As a result, MXenes can remarkably enhance the mechanical properties of materials like yield strength, Young's modulus, elongation, and hardness. Strength and elastic properties of MXenes are mainly dependent not only on the surface functional groups but also on the number and thickness of layers. The strength between M-X bonds could be characterized via the bond stiffness. This bond can be used as a relation to the theoretical mechanical properties, such as Young's modulus (Firestein et al, 2020). Figure 3a indicates that the rigidity of M-C tends to increase with an increasing group number of "M" used in transition metal carbides as the d-block order increases (Wyatt et al, 2021). Furthermore, Mn, Mo, and W tend to have the highest stiffness in M-C bonds in the 3d, 4d, and 5d transition metal blocks of MXenes, respectively. For instance, experimental studies have been conducted to $Ti_3C_2T_x$ MXene monolayers highlighting its Young's modulus is 330 GPa (Rong et al, 2024), which is higher than other 2D materials like graphene oxide and MoS_2 (Lipatov et al, 2018).

MXenes mechanical properties have also attracted attention with theoretical calculations predicting the hardness of $Ti_{n+1}C_n$ MXenes (Figure 3c), which reveals that an increasing MXene thickness causes an increase in 2D stiffness. The stiffness in Ti_2C is lower than Ti_3C_2 or Ti_4C_3 due to the higher surface bonding interactions between Ti–Ti. While, the structures as Ti_3C_2 or Ti_4C_3 have more interior layers of Ti-C bonding without the influence of surface Ti–Ti bonding. It also be noted that Young's modulus remarkably increases with decreasing layer thickness. The surface terminated MXenes are flexible in comparison with no surface terminated MXenes. Compared to the hardness bond, it is noted that the -O terminated MXenes exhibits more elastic stiffness than -F and -OH terminated ones (Figure 3e). Besides, surface terminations have been predicted to decrease the elastic modulus as compared to bare MXenes due to an increased monolayer thickness, as shown in Figures 3d and 3e.Therefore, the stiffness and elastic properties help us to identify the effects of chemical components, carbon or nitrogen, and surface functionalities on the fundamental M-X bond of MXenes (Wyatt et al, 2021). These exceptional properties including strong stiffness, high elastic constant and Young's modulus can open a novel avenue for fabricating next-generation 2D nanomaterials devices for energy storage applications.

Figure 3. Theoretical and experimental mechanical behavior of MXenes. a) Calculated bond stiffness between M–X for all early transition. b) Average of all predictions on the mechanical stiffness of bare M_2X MXenes. c) 2D stiffness and elastic modulus of $Ti_{n+1}X_n$, which illustrate the effect of increasing flake thickness (n) on MXenes' mechanical stiffness. d) Predictions on the mechanical 2D stiffness for terminated M_2XT_x MXenes for different surface terminations including $-O_2$, $-F_2$, $-(OH)_2$, and $-Cl_2$. e) Prediction of the elastic modulus of terminated MXenes. f) Effect of Ti and C vacancies on the 3D elastic modulus of $Ti_3C_2O_2$ MXene.

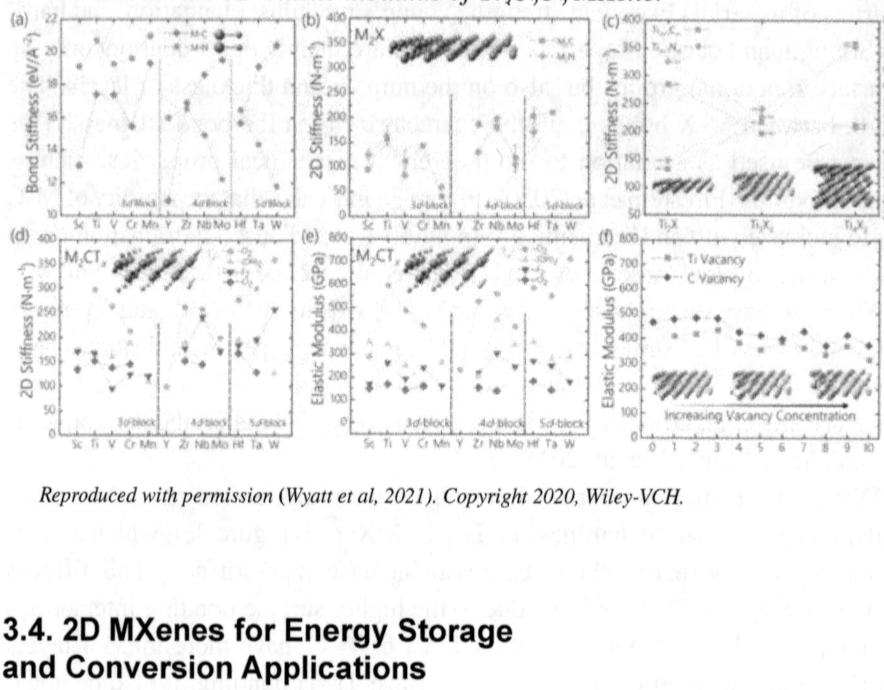

Reproduced with permission (Wyatt et al, 2021). Copyright 2020, Wiley-VCH.

3.4. 2D MXenes for Energy Storage and Conversion Applications

Currently, energy storage devices like rechargeable lithium-ion batteries and supercapacitors have been used in electric vehicles to reduce gas emissions (Abbasi et al, 2011). Furthermore, smart electricity grids demand high energy, high power, and outstanding long-term properties to ensure consistent energy supply. MXene-based materials with high performance for energy storage systems are critical to integrate renewable energy resources into practical applications (Bibi et al, 2024; Ahmad et al, 2024; Kabir et al, 2024). In fact, they have good conductivity, high specific surface area, and a large number of redox sites, which are the great of importance in electrochemical performance. The electrical conductivity of MXenes is able to host ions and provide fast ionic transport, which makes them promising electrodes materials in ionic liquid electrolyte for batteries and supercapacitors. It is very suitable for energy storage and conversion applications (Choudhary et al, 2017; Zhang et al, 2018; Tang et al, 2012). For instance, $Ti_3C_2T_x$ and other MXenes, with high conductivity, hydrophilicity, and the ability to intercalate cations and store charge, have been investigated in recent years. MXenes have been deeply explored and showed promise candidates in numerous other areas such as photo-electrochemical sensing (Tan et al, 2024; Liu et al, 2019) and catalysis (Gao et al, 2017).

3.4.1 Lithium-Ion Batteries

Owing to the state of the art energy storage technology in current society rechargeable lithium-ion batteries (LIBs) are mainly used in industry as energy storage devices for reducing the enormous pressure of the energy crisis and environmental pollution (Rohit et al, 2017). The fundamental mechanism of the LIBs relies on Li-ion insertion/extraction into the positive and negative electrodes during the charge and discharge processes (Lin et al, 2021). In LIBs system, there are three dynamic processes for Li^+ transportation as follows Li^+ motion via the electrolyte accompanied by electron flowing in the external circuit, interfacial ion transfer and chemical diffusion accompanied by electron in the bulk of electrode material (Qin et al, 2023). The diffusion of Li^+ and electron in the electrode material is highly affected in terms of the ionic and electronic conductivity. Although LIBs hold high energy density, good cyclability and high lithium storage capacity, they have low power density (Xu et al, 2023; Ju et al, 2023; Zhang et al, 2024). These latter were controlled by Li^+ intercalation/de-intercalation diffusion of Li^+ inside the compound and its electronic conductivity along with the kinetics of electrochemical processes at the interface.

Furthermore, rechargeable LIBs with graphite anodes are currently the main energy storage devices in the transportation and renewable energy sectors. In fact, graphite is the frequently employed anode component with a low specific capacity of 372 $mAh.g^{-1}$ and deficient rate capability (Tang et al, 2018; Sun et al, 2019). However, this low theoretical capacity of graphite anodes hampers the further development for energy storage in the future. Furthermore, the slow Li ions diffusion rate of 10^{-8} cm^2 s^{-1} in the graphite structure results in a low power density of LIBs. It is crucial therefore to explore alternative electrode materials with better electrochemical performance, which are under intense investigation. In early attempts, conventional 2D materials like graphene, composites, and metal oxides (Zhao et al,

2024; Jafari et al, 2024) have been used as electrodes because of their large surface areas and efficient ion transport between their layers. In rechargeable LIBs, the high rate charge/discharge process could degrade the electrochemical properties. For this reason, it is very important to develop new materials or technologies to improve the electrochemical performance of rechargeable LIBs. Recently, MXenes have also been used as hopeful electrodes, since they can offer intriguing advantages in term of good mechanical stability and outstanding electrochemical properties to acquire high performance batteries (Pang et al, 2018; Tang et al, 2018). 2D transition metal carbides allow fast ion transportation, redox reactions that add a pseudo-capacitive ingredient to the electrical double-layer mode of charge storage and balanced mechanical flexibility at nano-sheet level accommodating large ions for LIBs. Further, MXenes have improved batteries to fulfill high capacity whereas endowing rapid charge/discharge by virtue of its appropriate interlayer spacing. For instance, concerning LIBs, the theoretical lithium diffusion barrier of $Ti_3C_2T_x$ is relatively lower (0.07 eV) than graphite carbon (0.3 V), which provides the ability to support rapidly Li^+ transportation of MXene with high rate performances (Zhang et al, 2019; Ghidiu et al, 2014 ; Wang et al, 2015).

Such evidences (Gopalakrishnan et al, 2024; Kazim et al, 2024; Munonde et al, 2024)) are a result of MXenes intrinsic properties, such as good electrical conductivity and high rate specific capacity, tailoring the electrodes either anode or cathode towards a specific target. As anode materials, MXenes give high lithium/sodium-ion storage capacities and good cycling stability (Naguib et al, 2012). However, as cathode materials, MXenes could provide high specific capacities and rapid redox reactions (Li et al, 2022). Er et al (Er et al, 2014) exhibited that MXene was first tested as a Li-ion battery anode material with encouraging results because of its high specific surface area, weak interlayer forces, open structure, and surface terminations. Additional outlooks like surface termination and interlayer distance are considerable to accommodate LIBSs. Furthermore, the functional groups tend to possess fundamentally different band structure and influence the Li uptake and transport which ultimately impact conductivity of MXenes (Khanal et al, 2023). For example, oxygen termination exhibits the highest theoretical capacity (Xie et al, 2014), whereas hydroxyls and fluorines result in lower capacities as well as lithium-ion transport deterioration (Tang et al, 2012). Chlorides surface groups can give a steric advantage to MXenes by increasing the interlayer spacing and mitigate the detrimental effects of OH and F functional groups (Natu et al, 2023). These could efficiently improve the Li-ions accessibility and afford more active sites, thus enhancing the electrochemical performances of LIBs. Most of the demonstrations of LIBs using MXenes have been provided (Kong et al, 2018; Syamsai et al, 2021; Lu et al, 2019). In particular, a steady-state capacity could be made when Li^+ ions are inserted into MXenes. Besides, Li^+ ions are expected to diffuse rapidly on the surface of MXenes, resulting in a high storage capacity and outstanding high rate capability. Typically, Ti_2CT_x with higher surface energy than $Ti_3C_2T_x$ tends to have a reduced lattice parameter, thus avoiding the intercalation of Li ions.

Although Nb atoms are heavier than Ti atoms, the specific capacity of Nb_2CT_x is higher than Ti_2CT_x. Naguib et al (Naguib et al, 2012; Naguib et al, 2013) reported that both niobium carbides (Nb_2CT_x) and vanadium carbides (V_2CT_x) MXenes have excellent rate performances with high reversible capacities of 170 and 260 mAhg^{-1}, respectively, suggesting fast transport of Li^+ between layers. The electrochemical characteristics of $Ti_3C_2T_x$ were studied (Figure 4) (Sun et al, 2014). $Ti_3C_2T_x$ exhibits linear feature at different current densities, indicating a stable Li-storing behavior and effective charge transfer at the electrode/electrolyte interface. An initial discharge capacity of 265 mAh.g^{-1} and charge capacity of 124 mAh.g^{-1} could be obtained for $Ti_3C_2T_x$, compared with other MXene materials, such as V_4C_3 nanolayers (225 mAh.g^{-1}), exfoliated Ti_2C (110 mAh.g^{-1}), and multilayered $Ti_3C_2T_x$ (123 mAh.g^{-1}).

Figure 4. (a) CV curves of in-Ti_3C_2 from 3.0 V to 0.01 V vs. Li/Li$^+$ at a scan rate of 0.2 mV. s^{-1}; (b) charge–discharge profiles of Ti_3C_2 at 1C; (c) cycling performance of Ti_3C_2 at 1C; (d) first charge–discharge profiles of in-Ti_3C_2 at 0.1C, 1C, 3C and 10C; (e) cycling performance of in-Ti_3C_2 at different rates.

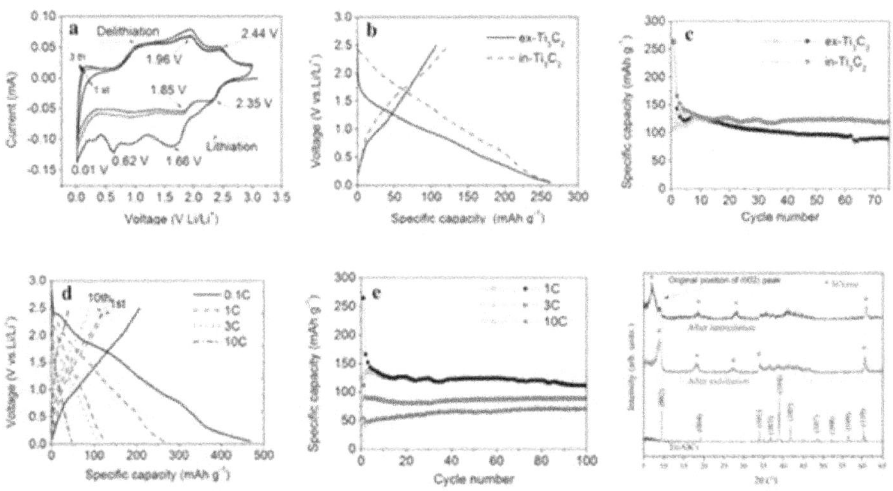

Reprinted by permission from (Sun et al, 2014).

The high chemical diversity of MAX phases encouraged the research of novel MXenes; for this reason, $Mo_2TiC_2T_x$ MXene, prepared by Anasori et al (Anasori et al, 2015), attains a capacity of 269 mAh.g^{-1}, which is higher than $Ti_3C_2T_x$ and Ti_2CT_x with capacity of 123.6 and 110.6 mAh.g^{-1}, respectively (Figure 5). That's why; the electrochemical behavior has been enhanced by integration of Mo atoms in MXenes materials.

Figure 5. Electrochemical performance of Mo_2TiC_2 in LIB (a) Voltage profiles between 0.02 and 3 V vs Li/Li^+ at C/10 rate for the first 10 cycles. (b) Specific lithiation and delithiation capacities versus cycle number at 1 C and C/10 rates. (c) Cyclic voltammograms, at different scan rates, for a freestanding electrode in 1 M H_2SO_4. (d) Capacitance retention test of Mo_2TiC_2 'paper' in 1 M H_2SO_4. Inset shows galvanostatic cycling data collected at a current density of 1 $A.g^{-1}$.

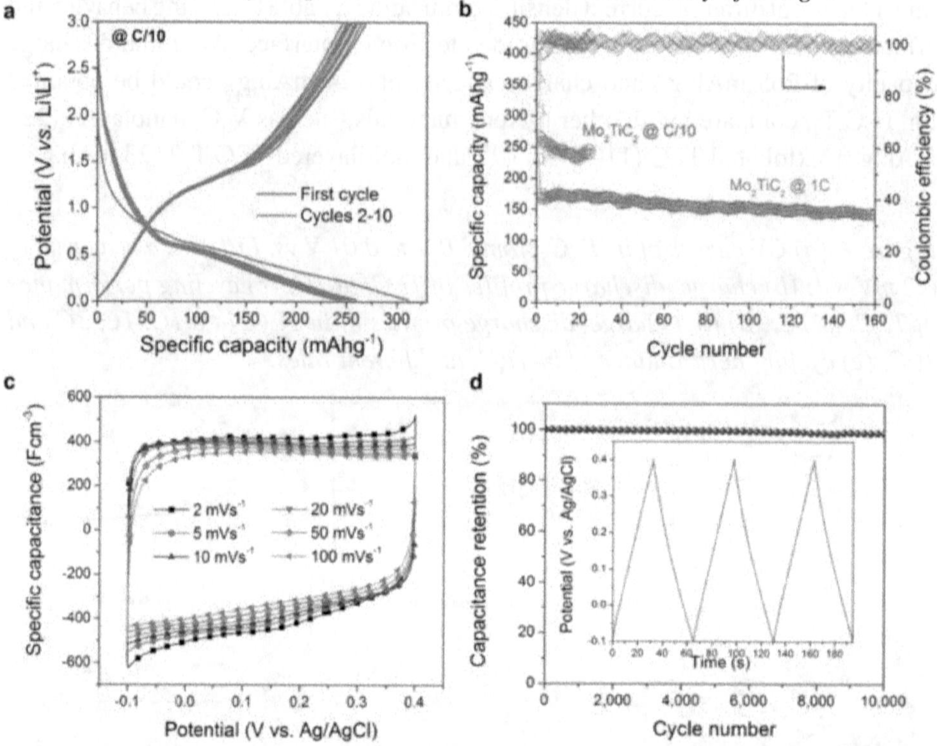

Reproduced with permission from reference (Anasori et al, 2015). Copyright 2015 American Chemical Society.

3.4.2 Supercapacitors

Supercapacitors or electrochemical capacitors devices are suitable to store huge amount of electrical energy and release it in a very short time, making them appropriate for use in enormous potential applications such as electric vehicles, smart electronic devices and energy harvesting systems. They have drawn considerable attention due to their high power density, large capacity, and good cycle stability.

Supercapacitors, which consists of a positive electrode and a negative electrode physically separated by insulating dielectrics, are efficient storage devices that have great potential in several industries in the near future due to their excellent specific power, long life cycle and fast charging–discharging rate. They are able to store and accommodate certain amounts of energy by reversibly adsorbing ions on the electrode surface under an applied potential. According to their charge-storage mechanisms, supercapacitors can be classified as electrochemical double-layer capacitors that mostly dependent on reversible accumulation of ions at the electrode-electrolyte interface to complete electrostatic adsorption and desorption processes, generating a stable potential between the electrodes, providing high power density, stable cycling performance, and long durability but the relatively low energy density with limited discharge time hinders its wide application, as shown in Figure 6 (Liu et al, 2017). However, pseudo-supercapacitors generate faradic process involving redox reactions oxidization and reduction of the electrochemically active reagentsat the electrode surfaces where electrosorption and electrodesorption take place with charge transfer without any bulk phase transformation upon charging/discharging (Augustyn et al, 2014). Alternatively, electrical energy could be stored by converting it into available chemical energy for later use.

Figure 6. (a) Ragone plot comparing the power-energy characteristics and charge/ discharge times of different energy storage devices; (b) Schematic diagram showing fundamental mechanism of electrochemical energy storage devices.

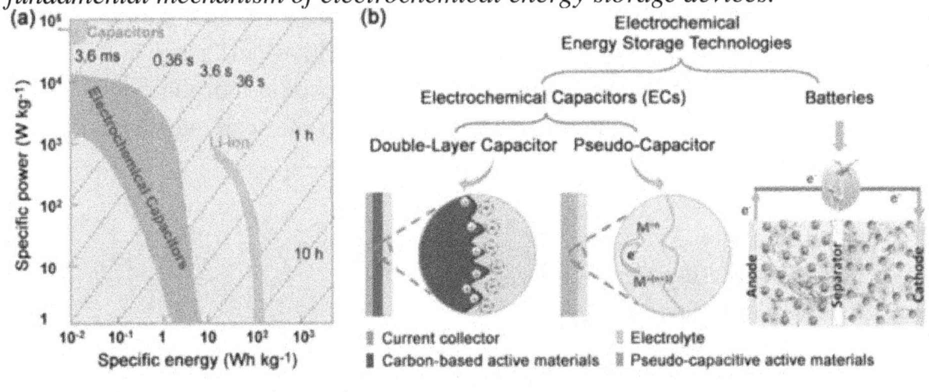

Reproduced by permission from(Liu et al, 2017).

According to Conway et al (Conway et al, 1991; Conway et al, 2003), pseudocapacitance can be ranked into three types as follows underpotential deposition, surface redox system, and intercalation system, as outlined by the schematic diagrams in Figure 7. From underpotential deposition, metal ions form an adsorption monolayer on the surface of another metal with a strong interaction that is very higher than its oxidation-reduction potential (Figure7.b). A classic example of underpotential deposition is Pb on the surface of Au electrodes (Oviedo et al, 2016). Redox pseudocapacitance includes the adsorption of ions on the surface of the electrode material and assisted by faraday charge transfer (Figure7.c). Examples including transition metal oxides as RuO_2 and MnO_2 are the most redox pseudocapacitance materials (Xia et al, 2012; Jiang et al, 2012). Intercalation pseudocapacitance is another faradaic process occurring without a crystallographic phase transition, which arises an intercalation of ions into the tunnels or interlayers of 2D materials (Figure 7.d). Recently, MXenes are introduced as unique host materials for intercalation pseudocapacitors (Attia et al, 2022). High-volume pseudocapacitors have been developed via the intercalation of ions like Li^+, Na^+, K^+, NH_4^+, or Al^{3+} into MXene layers. For example, $Ti_3C_2T_x$ MXene exhibits high volumetric capacitance values because of the changes in Ti oxidation state during the intercalation/de-intercalation processes. At the electrode/electrolyte interface, the ions interact with the electrode material are stored in the active sites of the electrode. Such electrode settings as the porosity, surface area, and conductivity are highly influence the energy density. It is concluded that the quantity of energy stored in a supercapacitor electrode is related to the surface area of the electrode. Furthermore, MXenes are frequently used in supercapacitor applications to assure rapid electron transport; particularly, $Ti_3C_2T_x$ is the most promising MXene to form supercapacitor electrode because of its high volumetric capacitances at high current densities, excellent electrical conductivity, high melting point, good hydrophilicity, and high electrochemical surface. In the application of $Ti_3C_2T_x$ for supercapacitors, the capacitive performance could be affected by different factors, such as the material composition/morphology, surface, electrode configuration, and electrolyte type, as well as the scan rate, current load, and mass ratio.

Supercapacitors
Electric double-layer capacitors
Electric double-layer capacito

Ghidiu et al. (Ghidiu et al, 2014) firstly applied clay-like $Ti_3C_2T_x$ MXene as supercapacitor electrodes in H_2SO_4 solution. A capacitance up to 900 Fcm^{-3} (245 Fg^{-1}) could be obtained at 2 mVs^{-1} without capacitance loss even after 10 000 cycles. Compared to$Ti_3C_2T_x$ with a volumetric capacitance of 350 Fcm^{-3} in KOH electrolyte, the smaller size of H^+ enhanced the electrochemical performance by interlayer spacing and the surface redox processes also contribute to increase the capacitance. The $Ti_3C_2T_x$ delivered outstanding supercapacitor performance in organic electrolytes rather than aqueous electrolyts due to the wider electrochemical stability and the intercalation multivalent cations between layers.

Figure 7. Schematic diagram about charge-storage mechanisms of supercapacitors. a) An electric double layer capacitors (EDLC). Pseudocapacitive materials: b) underpotential deposition, c) redox pseudocapacitor, and d) ion intercalation pseudocapacitor.

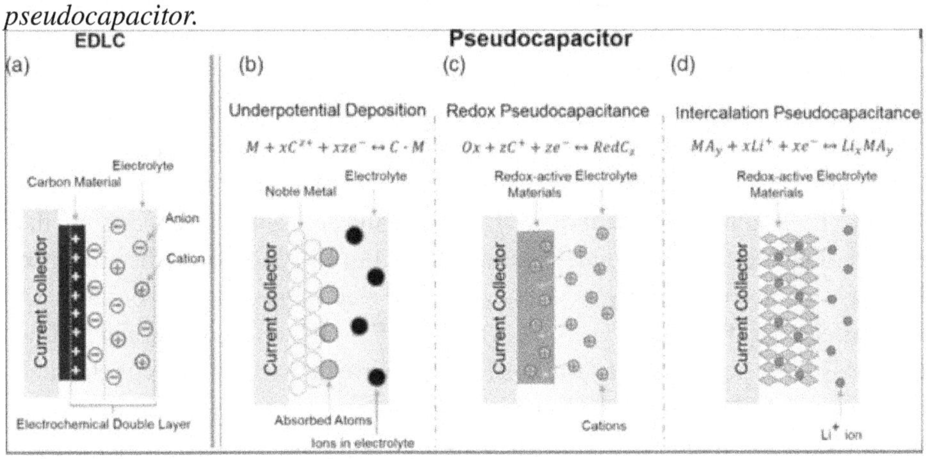

Reproduced by permission from (Shao et al, 2018). Copyright 2018 American Chemical Society.

3.4.3 Hybrid Capacitors

Despite the hybrid capacitors behave the protruding power density, lower energy density compared to batteries makes them hard to satisfy the requirements for energy storage devices (Gao et al, 2023). Therefore, the appearance of metal ion hybrid capacitors is predicted to enlarge energy density without deterioration power density.

Hybrid capacitors are mostly used to serve as advanced energy storage devices for fulfilling both high energy and high power requirements. Hybrid capacitor systems are the promising devises to meet the goal to effectively increase the energy density. From recent contributions (Kurra et al, 2018; Dong et al, 2019; Sun et al, 2022), hybrid capacitors certainly achieve a high energy density and good stability, high power output and ultralong cycle life for offering good storage capacity at higher charge-discharge rates than batteries. They combine the power density related to charging time of supercapacitors with the energy density of batteries, resulting in a device with the ability to charge and discharge energy quickly while also storing a significant quantity of energy. Furthermore, hybrid capacitors possess a combination of two charge storage mechanisms as follows intercalation/deintercalation in a battery and adsorption/desorption in an electric double-layer capacitor. Generally, they contain both non-faradaic electrodes act as cathodes and faradaic electrodes serve as anodes (Kumar et al, 2021; Han et al, 2021). During charging, the anions from electrolyte are adsorbed onto the surface of the cathode, while the cations take part in the redox reactions on the anode. Conversely, during discharging, the anions desorb from the surface of the cathode, and the cations de-intercalate from the anode. Thus, both anions and cations of electrolyte salts take part in the energy storage mechanism of hybrid capacitors, where the anions are involved in the non-faradaic process and the cations take part in the faradaic process.

As a full-cell hybrid electrochemical capacitor, the Lithium-ion capacitor as well-known conventional hybrid capacitors needs a large capacity Li^+ reversible intercalation anode to perform advanced electrochemical energy recovery. MXenes like $Ti_3C_2T_x$, Ti_2CT_x, Nb_2CT_x, and V_2CT_x have totally accessible layer structure that facilitates the intercalation/deintercalation kinetics of Li^+. Moreover, the properties of energy storage, the stability of MXenes, and the electronic transmission speed are highly affected by functional groups (–O, –Cl, –F, –OH) on the surface of MXene layers. Also, Mg-ion hybrid capacitors (MHCs) have an excellent electrochemical performance because of their high energy density, large capacities, and rapid charge transfer kinetics. On the other hand, magnesium is a unique active metal for battery type electrodes in hybrid capacitors, having a volumetric capacity of 3833 $mA.h.cc^{-1}$ (much higher than that of lithium of 2046 $mA.h.cc^{-1}$) (Mandai et al, 2020).

According to the synergistic effect of the highly conductive MXene matrix and porous Fe_2O_3 nanospheres (Tang et al, 2018), the $Ti_3C_2T_x@Fe_2O_3$ anode is expected to exhibit enhanced rate and cycling performance. As illustrated in Figure 8a, a high initial discharge capacity of 1645 $mAhg^{-1}$ and a charge capacity of 1201 $mAhg^{-1}$ are achieved at a current density of 0.1 Ag^{-1}. Figure 8c shows the cyclic voltammetry curves of the $Ti_3C_2T_x@Fe_2O_3$ anode collected at different scan rates from 0.1 to 0.2, 0.3, 0.4, 0.5, 1 and 2 $mV.s^{-1}$, which are compatible with the charge-discharge profiles. Thus, $Ti_3C_2T_x$ MXene exhibits fast transport of electrons, while Fe_2O_3 hollow nanospheres provide large specific capacity, thus synergizing to improve the electrochemical performance electrodes.

Figure 8. Lithium storage performance of the Fe_2O_3, $Ti_3C_2T_x$ and $Ti_3C_2T_x$@Fe_2O_3 electrodes: (a) the first three charge–discharge cycles of the $Ti_3C_2T_x$@Fe_2O_3 hybrid electrode at a current density of 0.1 A g^{-1}. (b) Rate performances of the Fe_2O_3, $Ti_3C_2T_x$ and $Ti_3C_2T_x$@Fe_2O_3 electrodes. (c) CV curves of a $Ti_3C_2T_x$@Fe_2O_3 electrode at different scan rates.

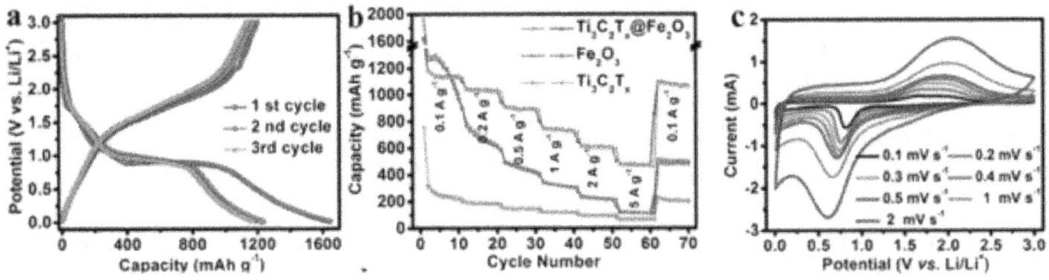

4. CONCLUSION

MXenes have produced immense eagerness in the investigation of 2D materials and emerged as a popular trend towards promise energy storage and conversion applications for research and technological development in current time because of stupendous structural, electronic and mechanical characteristics. In summary, we demonstrate recent advances in the synthesis, structural preparation, and energy applications of MXenes. Recently, 2D metal transition carbides have been prepared by the selective exfoliation and etching of A element from the layered ternary MAX precursors, thus providing larger specific surface area, higher porosity, excellent electronic conductivity, and fast ions transport pathways. Especially, the transition metal carbide core enables conductivity (up to 10 000 S.cm^{-1} for $Ti_3C_2T_x$) for fast electrons transport throughout the electrodes. In this chapter, we highlighted the fundamental considerations and principal mechanisms of Li-ion batteries, supercapacitors and hybrid capacitors, which have made great breakthroughs in the development of energy storage and conversion. The electrochemical performance of these devices is in agreement with some essential parameters like specific capacitance, energy density, and power density. It is particularly important to prepare MXenes electrodes that shown great capacity and stability throughout the charge and discharge processes, suggesting favorable potential for energy storage devices. Finally, we will further discus the challenges and strategies to address the requirements of sustainable energy solutions using MXenes in the enhancement of the effectiveness of energy storage devices and conversion.

REFERENCES

Abbasi, T., Premalatha, M., & Abbasi, S. A. (2011). The return to renewables: Will it help in global warming control. *Renewable & Sustainable Energy Reviews*, *15*(1), 891–894. DOI: 10.1016/j.rser.2010.09.048

Ahmad, F., Asim, M., Mubashar, S., Shahzad, A., Ain, Q. U., Khan, M. A., & Danish, M. (2024). Revolutionizing energy storage: A critical appraisal of MXene-based composites for material designing and efficient performance. *Journal of Energy Storage*, *84*, 110757. DOI: 10.1016/j.est.2024.110757

Alhabeb, M., Maleski, K., Anasori, B., Lelyukh, P., Clark, L., Sin, S., & Gogotsi, Y. (2017). Guidelines for synthesis and processing of two-dimensional titanium carbide ($Ti_3C_2T_x$ MXene). *Chemistry of Materials*, *29*(18), 7633–7644. DOI: 10.1021/acs.chemmater.7b02847

Anasori, B., Xie, Y., Beidaghi, M., Lu, J., Hosler, B. C., Hultman, L., Kent, P. R. C., Gogotsi, Y., & Barsoum, M. W. (2015). Two-dimensional, ordered, double transition metals carbides (MXenes). *ACS Nano*, *9*(10), 9507–9516. DOI: 10.1021/acsnano.5b03591 PMID: 26208121

Aravind, J., Prakash, P., Krithiga, T., Joshua, D., Premkumar, J., Rajamohan, N., Vasseghian, Y., Saravanan, P., & Rajasimman, M. (2022). Methods of synthesis, characteristics, and environmental applications of MXene: A comprehensive review. *Chemosphere*, *286*, 131607. DOI: 10.1016/j.chemosphere.2021.131607 PMID: 34311398

Ashton, M., Mathew, K., Hennig, R. G., & Sinnott, S. B. (2016). Predicted Surface Composition and Thermodynamic Stability of MXenes in Solution. *The Journal of Physical Chemistry. C, Nanomaterials and Interfaces*, *120*(6), 3550–3556. DOI: 10.1021/acs.jpcc.5b11887

Attia, S. Y., Mohamed, S. G., Barakat, Y. F., Hassan, H. H., & Zoubi, W. A. (2022). Supercapacitor electrode materials: Addressing challenges in mechanism and charge storage. *Reviews in Inorganic Chemistry*, *42*(1), 53–88. DOI: 10.1515/revic-2020-0022

Augustyn, V., Simon, P., & Dunn, B., (2014). Pseudocapacitive oxide materials for high-rate electrochemical energy storage.*Energy Environ. Sci. 7*, 1597-1614.

Bibi, F., Soomro, I. A., Hanan, A., Lakhan, M. N., Khan, A., Goraya, N. R., & Zhang, K. (2024). Advances in 2D/2D MXenes-based heterostructures for energy storage/conversion applications. *Journal of Materials Science and Technology*, *202*, 82–118. DOI: 10.1016/j.jmst.2024.03.005

Borysiuk, V. N., Mochalin, V. N., & Gogotsi, Y. (2015). Molecular dynamic study of the mechanical properties of two-dimensional titanium carbides $Ti_{n+1}C_n$ (MXenes). *Nanotechnology*, *26*(26), 265705. DOI: 10.1088/0957-4484/26/26/265705 PMID: 26063115

Cho, I., Selvaraj, A. R., Bak, J., Kim, H., & Prabakar, K. (2023). Mechanochemical Pretreated $M_{n+1}AX_n$ (MAX) Phase to Synthesize 2D-Ti3C2Tx MXene Sheets for High-Performance Supercapacitors. *Nanomaterials (Basel, Switzerland)*, *13*(11), 1741. DOI: 10.3390/nano13111741 PMID: 37299644

Choudhary, N. K., Jin, H., Kim, B., Baek, D. S., Joo, S. H., & Lee, K. (2017). MXene: An emerging two-dimensional material for future energy conversion and storage applications. *Journal of Materials Chemistry. A, Materials for Energy and Sustainability*, *5*(47), 24564–24579. DOI: 10.1039/C7TA09094C

Conway, B. E. (1991). Transition from "supercapacitor" to "battery" behavior in electrochemical energy storage. *Journal of the Electrochemical Society*, *138*(6), 1539–1548. DOI: 10.1149/1.2085829

Conway, B. E., & Pell, W. G. (2003). Double-layer and pseudocapacitance types of electrochemical capacitors and their applications to the development of hybrid devices. *Journal of Solid State Electrochemistry*, *7*(9), 637–644. DOI: 10.1007/s10008-003-0395-7

Dong, L., Yang, W., Yang, W., Li, Y., Wu, W., & Wang, G. (2019). Multivalent metal ion hybrid capacitors: A review with a focus on zinc-ion hybrid capacitors. *Journal of Materials Chemistry. A, Materials for Energy and Sustainability*, *7*(23), 13810–13832. DOI: 10.1039/C9TA02678A

Er, D., Li, J., Naguib, M., Gogotsi, Y., & Shenoy, V. B. (2014). Ti_3C_2 MXene as a high capacity electrode material for metal (Li, Na, K, Ca) ion batteries. *ACS Applied Materials & Interfaces*, *6*(14), 11173–11179. DOI: 10.1021/am501144q PMID: 24979179

Firestein, K. L., von Treifeldt, J. E., Kvashnin, D. G., Fernando, J. F. S., Zhang, C., Kvashnin, A. G., Podryabinkin, E. V., Shapeev, A. V., Siriwardena, D. P., Sorokin, P. B., & Golberg, D. (2020). Young's modulus and tensile strength of Ti_3C_2 MXene nanosheets as revealed by in situ TEM probing, AFM nanomechanical mapping, and theoretical calculations. *Nano Letters*, *20*(8), 5900–5908. DOI: 10.1021/acs.nanolett.0c01861 PMID: 32633975

Gao, G., O'Mullane, A. P., & Du, A. (2017). 2D MXenes: A new family of promising catalysts for the hydrogen evolution reaction. *ACS Catalysis*, *7*(1), 494–500. DOI: 10.1021/acscatal.6b02754

Gao, X., Wu, H., Su, C., Lu, C., Dai, Y., Zhao, S., & He, G. (2023). Recent advances in carbon-based nanomaterials for multivalent-ion hybrid capacitors: A review. *Energy & Environmental Science*, *16*(4), 1364–1383. DOI: 10.1039/D2EE03719J

Ghidiu, M., Lukatskaya, M. R., Zhao, M.-Q., Gogotsi, Y., & Barsoum, M. W. (2014). Conductive two-dimensional titanium carbide "clay" with high volumetric capacitance. *Nature*, *516*(7529), 78–81. DOI: 10.1038/nature13970 PMID: 25470044

Ghidiu, M., Naguib, M., Shi, C., Mashtalir, O., Pan, L. M., Zhang, B., Yang, J., Gogotsi, Y., Billinge, S. J. L., & Barsoum, M. W. (2014). Synthesis and characterization of two-dimensional Nb_4C_3 (MXene). *Chemical Communications (Cambridge)*, *50*(67), 9517–9520. DOI: 10.1039/C4CC03366C

Gong, K., Yin, L., & Zhou, K. (2024), Chemical Exfoliation and Delamination Methods of MXenes. MXene Reinforced Polymer Composites: Fabrication, Characterization and Applications, 39-70.

Gopalakrishnan, M., Kao-ian, W., Aupama, V., Etesami, M., Ganesan, S., Theerthagiri, J., &Kheawhom, S., (2024). Metal-organic framework/mxene heterostructure and its derivatives as electrode materials for rechargeable Zn-based batteries: Design strategies and perspectives. *J. Chem. Eng, 484*, 149624. .DOI: 10.1016/j.cej.2024.149624

Han, C., Wang, X., Peng, J., Xia, Q., Chou, S., Cheng, G., & Li, W. (2021). Recent progress on two-dimensional carbon materials for emerging post-lithium (Na^+, K^+, Zn^{2+}) hybrid supercapacitors. *Polymers*, *13*(13), 2137. DOI: 10.3390/polym13132137 PMID: 34209707

Hart, J. L., Hantanasirisakul, K., Lang, A. C., Anasori, B., Pinto, D., Pivak, Y., Van Omme, J. T., May, S. J., Gogotsi, Y., & Taheri, M. L. (2019). Control of MXenes electronic properties through termination and intercalation. *Nature Communications*, *10*(1), 522. DOI: 10.1038/s41467-018-08169-8 PMID: 30705273

Hong, W., Wyatt, B. C., Nemani, S. K., & Anasori, B. (2020). Double transition-metal MXenes: Atomistic design of two-dimensional carbides and nitrides. *MRS Bulletin*, *45*(10), 850–861. DOI: 10.1557/mrs.2020.251

Hu, T., Li, Z. J., Hu, M. M., Wang, J. M., Hu, Q. M., Li, Q. Z., & Wang, X. H. (2017). Chemical Origin of Termination-Functionalized MXenes: $Ti_3C_2T_2$ as a Case Study. *The Journal of Physical Chemistry. C, Nanomaterials and Interfaces*, *121*(35), 19254–19261. DOI: 10.1021/acs.jpcc.7b05675

Huang, K., Li, Z. J., Lin, J., Han, G., & Huang, P. (2018). Two-dimensional transition metal carbides and nitrides (MXenes) for biomedical applications. *Chemical Society Reviews, 47*(14), 5109–5124. DOI: 10.1039/C7CS00838D PMID: 29667670

Huang, X., & Wu, P. (2020). A facile, high-yield, and freeze-and-thaw-assisted approach to fabricate MXene with plentiful wrinkles and its application in on-chip micro-supercapacitors. *Advanced Functional Materials, 30*(12), 1910048. DOI: 10.1002/adfm.201910048

Ihsanullah, I. (2020). MXenes (two-dimensional metal carbides) as emerging nanomaterials for water purification: Progress, challenges and prospects. *Chemical Engineering Journal, 388*, 124340. DOI: 10.1016/j.cej.2020.124340

Jafari, M., Ganjali, F., Eivazzadeh-Keihan, R., Maleki, A., &Geranmayeh, S. (2024). Recent advances in applications of graphene-layered double hydroxide nanocomposites in supercapacitors and batteries. *Flat Chem, 45,* 100658. DOI: 10.1016/j.flatc.2024.100658

Jiang, H., Li, C., Sun, T., & Ma, J. (2012). A green and high energy density asymmetric supercapacitor based on ultrathin MnO_2 nanostructures and functional mesoporous carbon nanotube electrodes. *Nanoscale, 4*(3), 807–812. DOI: 10.1039/C1NR11542A PMID: 22159343

Jiang, J., Chen, X., Niu, Y., He, X., Hu, Y., & Wang, C. (2022). Advances in flexible sensors with MXene materials. *New Carbon Materials, 37*(2), 303–320. DOI: 10.1016/S1872-5805(22)60589-4

Jing, Z., Wang, H., Feng, X., Xiao, B., Ding, Y., Wu, K., & Cheng, Y. (2019). Superior thermoelectric performance of ordered double transition metal MXenes: $Cr_2TiC_2T_2$ (T=– OH or– F). *The Journal of Physical Chemistry Letters, 10*(19), 5721–5728. DOI: 10.1021/acs.jpclett.9b01827 PMID: 31507188

Joya, K. S., Joya, Y. F., Ocakoglu, K., & Van de Krol, R. (2013). Water-splitting catalysis and solar fuel devices: Artificial leaves on the move. *Angewandte Chemie International Edition, 52*(40), 10426–10437. DOI: 10.1002/anie.201300136 PMID: 23955876

Ju, Z., Xu, X., Zhang, X., Raigama, K. U., & Yu, G. (2023). Towards fast-charging high-energy lithium-ion batteries: From nano-to micro-structuring perspectives. *Chemical Engineering Journal, 454*, 140003. DOI: 10.1016/j.cej.2022.140003

Kabir, L., Qi, J., Wijaya, K., Sagadevan, S., Yoon, C. M., Ullah, K., & Oh, W. C. (2024). Recent advances and latest technologies in energy storage applications based on 2D MXene. *Journal of Energy Storage*, *80*, 110335. DOI: 10.1016/j.est.2023.110335

Kamysbayev, V., Filatov, A. S., Hu, H., Rui, X., Lagunas, F., Wang, D., Klie, R. F., & Talapin, D. V. (2020). Covalent surface modifications and superconductivity of two-dimensional metal carbide MXenes. *Science*, *8311*(6506), 1–11. DOI: 10.1126/science.aba8311 PMID: 32616671

Kazim, S., Huang, C., Hemasiri, N. H., Kulkarni, A., Mathur, S., & Ahmad, S. (2024), MXene-Based Energy Devices: From Progressive to Prospective. *Adv. Funct. Mater 34*, 2315694. https://doi.org/.DOI: 10.1002/adfm.202315694

Khaledialidusti, R., Mishra, A. K., & Barnoush, A. (2020). Atomic defects in monolayer ordered double transition metal carbide ($Mo_2TiC_2T_x$) MXene and CO_2 adsorption. *Journal of Materials Chemistry. C, Materials for Optical and Electronic Devices*, *8*(14), 4771–4779. DOI: 10.1039/C9TC06046D

Khanal, R., & Irle, S. (2023). Effect of surface functional groups on MXene conductivity. *The Journal of Chemical Physics*, *158*(19), 194701. DOI: 10.1063/5.0141589 PMID: 37184011

Kong, F., He, X., Liu, Q., Qi, X., Zheng, Y., Wang, R., & Bai, Y. (2018). Improving the electrochemical properties of MXene Ti_3C_2 multilayer for Li-ion batteries by vacuum calcination. *Electrochimica Acta*, *265*, 140–150. DOI: 10.1016/j.electacta.2018.01.196

Kumar, S., Saeed, G., Zhu, L., Hui, K. N., Kim, N. H., & Lee, J. H. (2021). 0D to 3D carbon-based networks combined with pseudocapacitive electrode material for high energy density supercapacitor: A review. *Chemical Engineering Journal*, *403*, 126352. DOI: 10.1016/j.cej.2020.126352

Kurra, N., Alhabeb, M., Maleski, K., Wang, C. H., Alshareef, H. N., & Gogotsi, Y. (2018). Bistacked titanium carbide (MXene) anodes for hybrid sodium-ion capacitors. *ACS Energy Letters*, *3*(9), 2094–2100. DOI: 10.1021/acsenergylett.8b01062

Lai, S., Jeon, J., Jang, S. K., Xu, J., Choi, Y. J., Park, J. H., & Lee, S. (2015). Surface group modification and carrier transport properties of layered transition metal carbides (Ti_2CT_x, T:–OH,–F and–O). *Nanoscale*, *7*(46), 19390–19396. DOI: 10.1039/C5NR06513E PMID: 26535782

Li, H., Shi, Y., Chiu, M.-H., & Li, L. J. (2015). Emerging energy applications of two-dimensional layered transition metal dichalcogenides. *Nano Energy*, *18*, 293–305. DOI: 10.1016/j.nanoen.2015.10.023

Li, S. L., Tsukagoshi, K., Orgiu, E., & Samorì, P. (2016). Charge transport and mobility engineering in two-dimensional transition metal chalcogenide semiconductors. *Chemical Society Reviews*, *45*(1), 118–151. DOI: 10.1039/C5CS00517E PMID: 26593874

Li, X., Tang, Y., Liu, L., Zhang, Y., Sheng, R., & NuLi, Y. (2022). Ti_3C_2 MXene with pillared structure for hybrid magnesium-lithium batteries cathode material with long cycle life and high rate capability. *Journal of Colloid and Interface Science*, *608*, 2455–2462. DOI: 10.1016/j.jcis.2021.10.175 PMID: 34763892

Lin, X., Khosravinia, K., Hu, X., Li, J., & Lu, W. (2021). Lithium plating mechanism, detection, and mitigation in lithium-ion batteries. *Progress in Energy and Combustion Science*, *87*, 100953. DOI: 10.1016/j.pecs.2021.100953

Lipatov, A., Lu, H., Alhabeb, M., Anasori, B., Gruverman, A., Gogotsi, Y., & Sinitskii, A. (2018). Elastic properties of 2D $Ti_3C_2T_x$ MXene monolayers and bilayers. *Science Advances*, *4*(6), eaat0491. DOI: 10.1126/sciadv.aat0491 PMID: 29922719

Lipatov, A., & Sinitskii, A. (2019). Electronic and mechanical properties of MXenes derived from single-flake measurements. In: Anasori, B., Gogotsi, Y. (eds) 2D Metal Carbides and Nitrides (MXenes). Springer, Cham. 301-325. https://doi.org/DOI: 10.1007/978-3-030-19026-2_16

Liu, F., Zhou, A., Chen, J., Jia, J., Zhou, W., Wang, L., & Hu, Q. (2017). Preparation of Ti_3C_2 and Ti_2C MXenes by fluoride salts etching and methane adsorptive properties. *Applied Surface Science*, *416*, 781–789. DOI: 10.1016/j.apsusc.2017.04.239

Liu, J. (2017). *Graphene-Based Composites for Electrochemical Energy Storage*. Springer. DOI: 10.1007/978-981-10-3388-9

Liu, J., Jiang, X., Zhang, R., Zhang, Y., Wu, L., Lu, W., Li, J., Li, Y., & Zhang, H. (2019). MXene enabled electrochemical microfluidic biosensor: Applications toward multicomponent continuous monitoring in whole blood. *Advanced Functional Materials*, *29*(6), 1807326. DOI: 10.1002/adfm.201807326

Lu, M., Li, H., Han, W., Chen, J., Shi, W., Wang, J., & Zheng, W. (2019). 2D titanium carbide (MXene) electrodes with lower-F surface for high performance lithium-ion batteries. *Journal of Energy Chemistry*, *31*, 148–153. DOI: 10.1016/j.jechem.2018.05.017

Mandai, T. (2020). Critical issues of fluorinated alkoxyborate-based electrolytes in magnesium battery applications. *ACS Applied Materials & Interfaces*, *12*(35), 39135–39144. DOI: 10.1021/acsami.0c09948 PMID: 32805873

Mashtalir, O., Lukatskaya, M. R., Kolesnikov, A. I., Raymundo-Pinero, E., Naguib, M., Barsoum, M., & Gogotsi, Y. (2016). The effect of hydrazine intercalation on the structure and capacitance of 2D titanium carbide (MXene). *Nanoscale*, *8*(17), 9128–9133. DOI: 10.1039/C6NR01462C PMID: 27088300

Mashtalir, O., Lukatskaya, M. R., Zhao, M.-Q., Barsoum, M. W., & Gogotsi, Y. (2015). Amineassisted delamination of Nb_2C MXene for Li-ion energy storage devices. *Advanced Materials*, *27*(23), 3501–3506. DOI: 10.1002/adma.201500604 PMID: 25930685

Mashtalir, O., Naguib, M., Mochalin, V. N., Dall'Agnese, Y., Heon, M., Barsoum, M. W., & Gogotsi, Y. (2013). Intercalation and delamination of layered carbides and carbonitrides. *Nature Communications*, *4*(1), 1716. DOI: 10.1038/ncomms2664 PMID: 23591883

Mendoza-Sánchez, B., Samperio-Niembro, E., Dolotko, O., Bergfeldt, T., Kübel, C., Knapp, M., & Shuck, C. E. (2023). Systematic Study of the Multiple Variables Involved in V_2AlC Acid-Based Etching Processes, a Key Step in MXene Synthesis. *ACS Applied Materials & Interfaces*, *15*(23), 28332–28348. DOI: 10.1021/acsami.3c01671 PMID: 37253453

Mubeen, I., Shah, S., Pervaiz, E., & Miran, W. (2024). The promising frontier for next-generation energy storage and clean energy production: A review on synthesis and applications of MXenes. *Materials Science for Energy Technologies*, *7*, 180–194. DOI: 10.1016/j.mset.2023.10.002

Munonde, T. S., & Raphulu, M. C. (2024). Review on titanium dioxide nanostructured electrode materials for high-performance lithium batteries. *Journal of Energy Storage*, *78*, 110064. DOI: 10.1016/j.est.2023.110064

Naguib, M., Come, J., Dyatkin, B., Presser, V., Taberna, P. L., Simon, P., Barsoum, M. W., & Gogotsi, Y. (2012). MXene: A promising transition metal carbide anode for lithium-ion batteries. *Electrochemistry Communications*, *16*(1), 61–64. DOI: 10.1016/j.elecom.2012.01.002

Naguib, M., Halim, J., Lu, J., Cook, K. M., Hultman, L., Gogotsi, Y., & Barsoum, M. W. (2013). New two-dimensional niobium and vanadium carbides as promising materials for Li-ion batteries. *Journal of the American Chemical Society*, *135*(43), 15966–15969. DOI: 10.1021/ja405735d PMID: 24144164

Naguib, M., Kurtoglu, M., Presser, V., Lu, J., Niu, J., Heon, M., & Barsoum, M. W. (2011). Two-dimensional nanocrystals produced by exfoliation of Ti_3AlC_2. In MXenes Jenny Stanford Publishing. 15-29.

Naguib, M., Kurtoglu, M., Presser, V., Lu, J., Niu, J., Heon, M., Hultman, L., Gogotsi, Y., & Barsoum, M. W. (2011). Two-dimensional nanocrystals: Two-dimensional nanocrystals produced by exfoliation of Ti_3AlC_2. *Advanced Materials*, *37*(37), 4207–4207. DOI: 10.1002/adma.201190147

Naguib, M., Kurtoglu, M., Presser, V., Lu, J., Niu, J., Heon, M., Hultman, L., Gogotsi, Y., & Barsoum, M. W. (2011). Two-Dimensional Nanocrystals Produced by Exfoliation of Ti_3AlC_2. *Advanced Materials*, *23*(37), 4248–4253. DOI: 10.1002/adma.201102306 PMID: 21861270

Naguib, M., Mashtalir, O., Carle, J., Presser, V., Lu, J., Hultman, L., Gogotsi, Y., & Barsoum, M. W. (2012). Two-dimensional transition metal carbides. *ACS Nano*, *6*(2), 1322–1331. DOI: 10.1021/nn204153h PMID: 22279971

Naguib, M., Unocic, R. R., Armstrong, B. L., & Nanda, J. (2015). Large-scale delamination of multi-layers transition metal carbides and carbonitrides MXenes. *Dalton Transactions (Cambridge, England)*, *44*(20), 9353–9358. DOI: 10.1039/C5DT01247C PMID: 25912071

Naseem, K., Ali, Z., Chen, P., Tahir, A., Qin, F., Fayyaz, A., & Javed, M. S. (2024). Supercapacitive behavior and energy storage properties of molybdenum carbide ceramics synthesized via ball milling technique. *Ceramics International*, *50*(6), 9572–9580. DOI: 10.1016/j.ceramint.2023.12.276

Natu, V., & Barsoum, M. W. (2023). MXene Surface Terminations: A Perspective. *The Journal of Physical Chemistry. C, Nanomaterials and Interfaces*, *127*(41), 20197–20206. DOI: 10.1021/acs.jpcc.3c04324

Omomo, Y., Sasaki, T., Wang, L., & Watanabe, M. (2003). Redoxable nanosheet crystallites of MnO_2 derived via delamination of a layered manganese oxide. *Journal of the American Chemical Society*, *125*(12), 3568–3575. DOI: 10.1021/ja021364p PMID: 12643719

Oviedo, O. A., Reinaudi, L., Garcia, S., Leiva, E. P. M., Oviedo, O. A., Reinaudi, L., &Leiva, E. P. M., (2016). Experimental techniques and structure of the under-potential deposition phase. In book Underpotential Deposition: From Fundamentals and Theory to Applications at the Nanoscale, 17-89. DOI: .DOI: 10.1007/978-3-319-24394-8_2

Pang, J., Mendes, R. G., Bachmatiuk, A., Zhao, L., Ta, H. Q., Gemming, T., Liu, H., Liu, Z., & Rummeli, M. H. (2018). Applications of 2D MXenes in energy conversion and storage systems. *Chemical Society Reviews*, *48*(1), 72–133. DOI: 10.1039/C8CS00324F PMID: 30387794

Peng, Q. M., Guo, J. X., Zhang, Q. R., Xiang, J. Y., Liu, B. Z., Zhou, A. G., Liu, R. P., & Tian, Y. J. (2014). Unique lead adsorption behavior of activated hydroxyl group in two-dimensional titanium carbide. *Journal of the American Chemical Society*, *136*(11), 4113–4116. DOI: 10.1021/ja500506k PMID: 24588686

Qin, M., Zeng, Z., Cheng, S., & Xie, J. (2023). Challenges and strategies of formulating low-temperature electrolytes in lithium-ion batteries. *Interdisciplinary Materials*, 2(2), 308–336. DOI: 10.1002/idm2.12077

Rohit, A. K., Devi, K. P., & Rangnekar, S. (2017). An overview of energy storage and its importance in Indian renewable energy sector: Part I—Technologies and comparison. *Journal of Energy Storage*, *13*, 10–23. DOI: 10.1016/j.est.2017.06.005

Rong, C., Su, T., Li, Z., Chu, T., Zhu, M., Yan, Y., & Xuan, F. Z. (2024). Elastic properties and tensile strength of 2D $Ti_3C_2T_x$ MXene monolayers. *Nature Communications*, *15*(1), 1566. DOI: 10.1038/s41467-024-45657-6 PMID: 38378699

Sang, X., Xie, Y., Lin, M. W., Alhabeb, M., Van Aken, K. L., Gogotsi, Y., Kent, P. R., Xiao, K., & Unocic, R. R. (2016). Atomic defects in monolayer titanium carbide ($Ti_3C_2T_x$) MXene. *ACS Nano*, *10*(10), 9193–9200. DOI: 10.1021/acsnano.6b05240 PMID: 27598326

Shahzad, F., Alhabeb, M., Hatter, C. B., Anasori, B., Man Hong, S., Koo, C. M., & Gogotsi, Y. (2016). Electromagnetic interference shielding with 2D transition metal carbides (MXenes). *Science*, *353*(6304), 1137–1140. DOI: 10.1126/science.aag2421 PMID: 27609888

Shao, Y., El-Kady, M. F., Sun, J., Li, Y., Zhang, Q., Zhu, M., & Kaner, R. B. (2018). Design and mechanisms of asymmetric supercapacitors. *Chemical Reviews*, *118*(18), 9233–9280. DOI: 10.1021/acs.chemrev.8b00252 PMID: 30204424

Sheikh, S., &Haghpanah, A. J. (2024). Ionic liquids in green energy storage devices: lithium-ion batteries, supercapacitors, and solar cells. *Monatsh. Chem-Chemical Monthly,*1-17.

Singh, S., Sheoran, K., Mishra, K., Kaur, H., Kumar, A., Saini, V., Vo, D. N., Yazdani, H., & Kumar, V. (2022). Novel synthesis methods and applications of MXene based nanomaterials (MBNs) for hazardous pollutants degradation: Future perspectives. *Chemosphere*, *293*, 133542. DOI: 10.1016/j.chemosphere.2022.133542 PMID: 34999104

Sokol, M., Natu, V., Kota, S., & Barsoum, M. W. (2019). On the Chemical Diversity of the MAX Phases. *Trends in Chemistry*, *1*(2), 210–223. DOI: 10.1016/j.trechm.2019.02.016

Sun, D., Wang, M., Li, Z., Fan, G., Fan, L. Z., & Zhou, A. (2014). Two-dimensional Ti_3C_2 as anode material for Li-ion batteries. *Electrochemistry Communications*, *47*, 80–83. DOI: 10.1016/j.elecom.2014.07.026

Sun, J., Luo, B., & Li, H. (2022). A review on the conventional capacitors, supercapacitors, and emerging hybrid ion capacitors: Past, present, and future. *Advanced Energy and Sustainability Research*, *3*(6), 2100191. DOI: 10.1002/aesr.202100191

Sun, X., Liu, Y., Zhang, J., Hou, L., Sun, J., & Yuan, C. (2019). Facile construction of ultrathin SnO_x nanosheets decorated MXene (Ti_3C_2) nanocomposite towards Li-ion batteries as high performance anode materials. *Electrochimica Acta*, *295*, 237–245. DOI: 10.1016/j.electacta.2018.10.152

Sun, Y., Gao, S., Lei, F., Xiao, C., & Xie, Y. (2015). Ultrathin two-dimensional inorganic materials: New opportunities for solid state nanochemistry. *Accounts of Chemical Research*, *48*(1), 3–12. DOI: 10.1021/ar500164g PMID: 25489751

Syamsai, R., Rodriguez, J. R., Pol, V. G., Van Le, Q., Batoo, K. M., Adil, S. F., & Grace, A. N. (2021). Double transition metal MXene (TixTa4− xC3) 2D materials as anodes for Li-ion batteries. *Scientific Reports*, *11*(1), 688.

Tan, A. Y. S., Awan, H. T. A., Cheng, F., Zhang, M., Tan, M. T., Manickam, S., &Muthoosamy, K., (2024). Recent advances in the use of MXenes for photoelectrochemical sensors. *Chem Eng J.*, *482*, 148774. DOI: 10.1016/j.cej.2024.148774

Tang, H., Hu, Q., Zheng, M., Chi, Y., Qin, X., Pang, H., & Xu, Q. (2018). MXene 2D layered electrode materials for energy storage. *Progress in Natural Science*, *28*(2), 133–147. DOI: 10.1016/j.pnsc.2018.03.003

Tang, Q., Zhou, Z., & Shen, P. (2012). Are MXenes promising anode materials for Li ion batteries? Computational studies on electronic properties and Li storage capability of Ti_3C_2 and $Ti_3C_2X_2$ (X= F, OH) monolayer. *Journal of the American Chemical Society*, *134*(40), 16909–16916. DOI: 10.1021/ja308463r PMID: 22989058

Tang, X., Guo, X., Wu, W., & Wang, G. (2018). 2D metal carbides and nitrides (MXenes) as high-performance electrode materials for lithium-based batteries. *Advanced Energy Materials*, *8*(33), 1801897. DOI: 10.1002/aenm.201801897

Tang, X., Liu, H., Guo, X., Wang, S., Wu, W., Mondal, A. K., Wang, C., & Wang, G. (2018). A novel lithium-ion hybrid capacitor based on an aerogel-like MXene wrapped Fe_2O_3nanosphere anode and a 3D nitrogen sulphur dual-doped porous carbon cathode. *Materials Chemistry Frontiers*, *2*(10), 1811–1821. DOI: 10.1039/C8QM00232K

Thakur, A. B. S., Chandran, N., Davidson, K., Bedford, A., Fang, H., Im, Y., & Anasori, B. (2023). Step-by-step guide for synthesis and delamination of $Ti_3C_2T_x$ MXene. *Small Methods*, *7*(8), 2300030. DOI: 10.1002/smtd.202300030 PMID: 37150839

Verger, L., Xu, C., Natu, V., Cheng, H. M., Ren, W., & Barsoum, M. W. (2019). Overview of the synthesis of MXenes and other ultrathin 2D transition metal carbides and nitrides. *Current Opinion in Solid State and Materials Science*, *23*(3), 149–163. DOI: 10.1016/j.cossms.2019.02.001

Vijayakumar, M., Elsa, G., Nirogi, A., Navaneethan, R., Sankar, A. B., & Karthik, M. (2021). MXenes and their composites for hybrid capacitors and supercapacitors: A critical review. *Emergent Materials*, *4*(3), 655–672. DOI: 10.1007/s42247-021-00204-7

Wang, H., Feng, H., & Li, J. (2014). Graphene and graphene-like layered transition metal dichalcogenides in energy conversion and storage, *Small 10*, 2165-2181.Zhang, X. Y., Huisman, E. H., Gurram, M., Browne, W. R., van Wees, B. J., &Feringa, B. L. (2014).Supramolecular Chemistry on Graphene Field-Effect Transistors. *Small*, *10*, 1735–1740.

Wang, X., Kajiyama, S., Hosomo, E., Oro, S., Moriguchi, I., Okubo, M., & Yamada, A. (2015). Pseudocapacitance of MXene nanosheets for high-power sodium-ion hybrid capacitors. *Nature Communications*, *6*(1), 6544. DOI: 10.1038/ncomms7544 PMID: 25832913

Wyatt, B. C., Rosenkranz, A., & Anasori, B. (2021). 2D MXenes: Tunable mechanical and tribological properties. *Advanced Materials*, *33*(17), 2007973. DOI: 10.1002/adma.202007973 PMID: 33738850

Xia, H., Meng, Y. S., Yuan, G., Cui, C., & Lu, L. (2012). A symmetric RuO_2/RuO_2 supercapacitor operating at 1.6 V by using a neutral aqueous electrolyte. *Electrochemical and Solid-State Letters*, *15*(4), A60. DOI: 10.1149/2.023204esl

Xie, Y., Naguib, M., Mochalin, V. N., Barsoum, M. W., Gogotsi, Y., Yu, X., Nam, K. W., Yang, X. Q., Kolesnikov, A. I., & Kent, P. R. C. (2014). Role of surface structure on Li-ion energy storage capacity of two-dimensional transition-metal carbides. *Journal of the American Chemical Society*, *136*(17), 6385–6394. DOI: 10.1021/ja501520b PMID: 24678996

Xu, G., Li, X., Xia, X., Fu, J., Ding, W., & Zi, Y. (2019). On the force and energy conversion in triboelectric nanogenerators. *Nano Energy*, *59*, 154–161. DOI: 10.1016/j.nanoen.2019.02.035

Xu, J., Cai, X., Cai, S., Shao, Y., Hu, C., Lu, S., & Ding, S. (2023). High-energy lithium-ion batteries: Recent progress and a promising future in applications. *Energy & Environmental Materials*, *6*(5), e12450. DOI: 10.1002/eem2.12450

Yushin, G. N., Hoffman, E. N., Nikitin, A., Ye, H., Barsoum, M. W., & Gogotsi, Y. (2005). Synthesis of nanoporous carbide-derived carbon by chlorination of titanium silicon carbide. *Carbon*, *43*(10), 2075–2082. DOI: 10.1016/j.carbon.2005.03.014

Zaed, M. A., Tan, K. H., Abdullah, N., Saidur, R., Pandey, A. K., & Saleque, A. M. (2024). Cost analysis of MXene for low-cost production, and pinpointing of its economic footprint. *Open Ceramics*, *17*, 100526. DOI: 10.1016/j.oceram.2023.100526

Zhang, P., Wang, D., Zhu, Q., Sun, N., Fu, F., & Xu, B. (2019). Plate-to-layer Bi_2MoO_6/MXene heterostructured anode for lithium-ion batteries. *Nano-Micro Letters*, *11*(1), 81. DOI: 10.1007/s40820-019-0312-y PMID: 34138047

Zhang, X., Zhang, Z., & Zhou, Z. (2018). MXene-based materials for electrochemical energy storage. *Journal of Energy Chemistry*, *27*(1), 73–85. DOI: 10.1016/j.jechem.2017.08.004

Zhang, Y., He, D., Lu, J., Huang, J., Jiang, H., Rong, J., & Chen, H., (2024). Large-scale production of graphene encapsulated silicon nanospheres as flexible anodes for lithium ion batteries. *J. Chem. Eng, 487*, 150564. DOI: 10.1016/j.cej.2024.150564

Zhao, M., Casiraghi, C., & Parvez, K. (2024). Electrochemical exfoliation of 2D materials beyond graphene. *Chemical Society Reviews*, *53*(6), 3036–3064. DOI: 10.1039/D3CS00815K PMID: 38362717

Chapter 8
Exploring Nanomaterials for Enhanced Energy Storage and Conversion

G. Boopathy
https://orcid.org/0000-0003-2515-7277
Vel Tech Rangarajan Dr. Sagunthala R&D Institute of Science and Technology, India

V. Srinivasan
https://orcid.org/0009-0006-2988-6476
Vel Tech Rangarajan Dr. Sagunthala R&D Institute of Science and Technology, India

Balaji Ganesan
https://orcid.org/0000-0001-5799-9423
Hindustan Institute of Technology and Science, India

Sivaprakasam Palani
https://orcid.org/0000-0001-8082-8649
Addis Ababa Science and Technology University, Ethiopia

ABSTRACT

Our pressing need to build better energy storage and conversion systems grows as our world develops faster. Energy storage systems perform better through nanomaterials which have unique physicochemical traits and redefine how these technologies work. This chapter evaluates how nanomaterials strengthen energy technology through product types and develops fabrication processes for particular solutions. Materials at nanoscale show better energy storage because of their huge surface area and flexible design. Nanomaterials provide multiple solutions for storing

DOI: 10.4018/979-8-3693-9316-1.ch008

hydrogen as well as performing electro catalysis and photovoltaics. Our chapter shows that energy technology requires exact engineering along with modeling and environmental-friendly nanomaterial manufacturing to help turn scientific discoveries into real-world applications and next-level energy solutions.

INTRODUCTION

The increase in energy storage and conversion systems is a challenge in relation to the consumption of energy and environmental impacts. As power generation from clean sources like solar and wind gets to the grid, energy storage solutions have become more of a necessity. The existing energy storage systems like the lithium-ion batteries have constraints in abilities including energy density, charge-discharge rates and lifecycle design hence call for improvement of these systems by the use of improved materials. Renewable energy technologies could only be adopted fully if efficient energy storage devices are developed to store and transform energy to sustainable forms.

Nanomaterials hold a big potential towards these energy storage difficulties in view of their high surface area to volume ratios, higher reactivity, and electronically responsive nature. Such characteristics allow the nanomaterials to enhance the rate of ion diffusion and capacity to store charges, and thus suitable to be used in batteries as well as supercapacitors. For example, titanium carbide, also known as MXenes, and graphene layers, derived from 2D nanomaterials, have enormous electrochemical capacity due to the layers' intercalation capability. Further, its application in energy storage devices could increase the energy density and power density hence making the complete systems more effective. Thus, the thermophysical characteristics of nanomaterials increase their reactivity, which is essential for energy conversion processes.

For instance, nanocomposites as catalysts can enhance electrochemistry of systems such as hydrogen production and conversion of carbon dioxide. Furthermore, the likelihood to design the surface of nanomaterials can also enhance their catalytic properties, which make them critically important in creating future energy change platforms. Consequently, the attempt to introduce nano-materials for the purpose of energy storage and conversion not only looks for the deficiency of the traditional systems, but also aims at developing new ways for the sustainable development of energy resources.

The problems of energy storage are complex and are due to the growing use of variable renewable energy sources. The first problem is the supply-demand imbalance that results in energy wastage at instances of excessive generation, for instance during insufficient generation leading to high demands (Osmani, 2024).

Such battery systems currently available are however only partly suitable because they improve neither the capacities, the ratios nor the service life needed for bulk applications (Müller, 2023). For example, market leader lithium-ion batteries have problems associated with its raw materials, risk of fires and thermal runaway, and environmental concerns connected to mining and recycling of materials (Amiri et al., 2022).

Furthermore, traditional energy storage systems as well have a drawback in the sense that their performance fades with use meaning higher costs. These include electrode degradation whereby the active material in the electrode is worn out, decomposition of the electrolyte and formation of film layers that hinder ion movement. Hence the need to look at new materials that not only improve the efficiency of current energy storage technologies but also materials that are sustainable and eco-friendly. One solution to these difficulties is the annexation of nanomaterials in energy storage systems.

Through proper exploitation, scientists are into the synthesis of new materials that can increase the density of energy, rate capability as well as efficiency (Sun et al., 2024). For instance, the application of nanostructured electrodes in batteries is less prone to the loss of surface area that is effectively available for electrochemical reactions, and is, therefore, characterized by better metrics of performance. Moreover, any synthesis method changes the characteristics of nanomaterials and since there are numerous techniques of producing nanomaterials, the properties of these nanomaterials can be tweaked to optimally suit an application, thus solving the problems with conventional energy storage systems.

It is widely known that nanomaterials possess peculiar structural and morphological characteristics, which define a series of useful properties applicable for energy purposes. One of the more distinctive features of nanomaterials is their ability to provide a large surface area for contacting electrolytes and improving the electrochemical reactions kinetics (Zhang et al., 2022). For energy storage devices it is desirable to have a high surface area to volume ratio because that essentially provides more surface area for ions to access and thus faster charges and discharge rates. Apart from the large surface area, there is increased chemical activity resulting from the surface properties of nanomaterials and or quantum confinement effects. For example, the surface defects and active sites of the nanomaterials can strongly affect the catalytic behaviors, which could be widely used in energy transformation (Crowley et al., 2024; Matthews et al., 2023).

The fact that such properties can be engineered through well-defined synthesis techniques increases the versatility of nanomaterials in energy related applications. For instance, an application of transition metal oxides into nanostructured materials can increase their electrochemical characteristics by enhancing the electron transfer site and the ability to store charges. In addition, the multifunctional characteristic of nanomaterials makes them easily incorporated into energy storage and conversion systems. Nanomaterials themselves can be used in any form that is prescribed by the specific use of the supercapacitor or the fuel cell and attempts to broaden their applications in the field of energy. The focus has also learnt itself towards the synthesis and application of hybrid nanostructures, in which various types of nano structures are incorporated together and are also expected to improve the working of energy storage devices by virtue of their properties. This would explain why different methods of utilizing nanomaterials are complex in order to underscore how these nanomaterials are essential in solving the issues related with energy storage and conversion.

TYPES OF NANOMATERIALS FOR ENERGY STORAGE AND CONVERSION

Exploration of nanomaterials for energy storage and conversion has opened up new avenues in energy technologies and their improvement is pointed out in this paper. The different forms of nanomaterials include carbon based nanomaterials, metal oxide nanoparticles, 2D materials and composite nanomaterials have received a lot of attention in light of their distinctive characteristics and applicability. This section will describe each of them with focus on their functions and uses in energy storage and conversion systems.

Carbon-Based Nanomaterials

Carbon based nanomaterials, such as graphene, carbon nanotubes (CNTs) and carbon quantum dots, have been shown to have great potential in the area of energy storage and conversion because of their properties and the possibility of using them on a large scale. These materials have a high electrical conductivity, large surface area, high mechanical strength, and thus can be used in batteries, supercapacitors, energy, etc.

Graphene and Carbon Nanotubes

Graphene is one-atom thick material made of carbon atoms that form a two dimensional crystal lattice and shows exceptionally high electrical, thermal and mechanical conductivities. Due to the relatively large surface area (around 2630 m^2/g) it has a superior charge storing capability and therefore could be suitable for applications in supercapacitors and lithium-ion batteries (Kothandam et al., 2023). Graphene combined with all other carbon nanomaterials (CNMs) including CNTs form composites that show enhanced advantageous characteristics compared to when they are used individually, thus enhancing the performance of energy storage systems (Goyal et al., 2024). For example, graphene-carbon nanotube hybrid aerogels have been fabricated and characterized that have shown improved thermal transport feature and electrochemical characteristics to display the best supercapacitor characteristics. Carbon nanotubes, which have a cylindrical geometry, are also an important energy storage discipline. Owing to their enhanced electrical conductivity and mechanical properties, they can be employed for their conductivity characteristics in electrode materials (Tsai, 2024). A study of the impact of CNT's on the formulation of electrodes has revealed that CNTs could increase the mechanical stability and conductivity of the electrodes so useful in both battery and supercapacitor applications. Moreover, the functionalization of CNTs opens an opportunity to enhance the compatibility of CNTs with different types of electrolytes, and thus, improve the electrochemical characteristics of the composites. Figure 1 depicts the development, properties, and applications of graphene, carbon nanotubes, and fullerenes.

Figure 1. The production, characterization, and uses of graphene, CNTs and fullerenes.

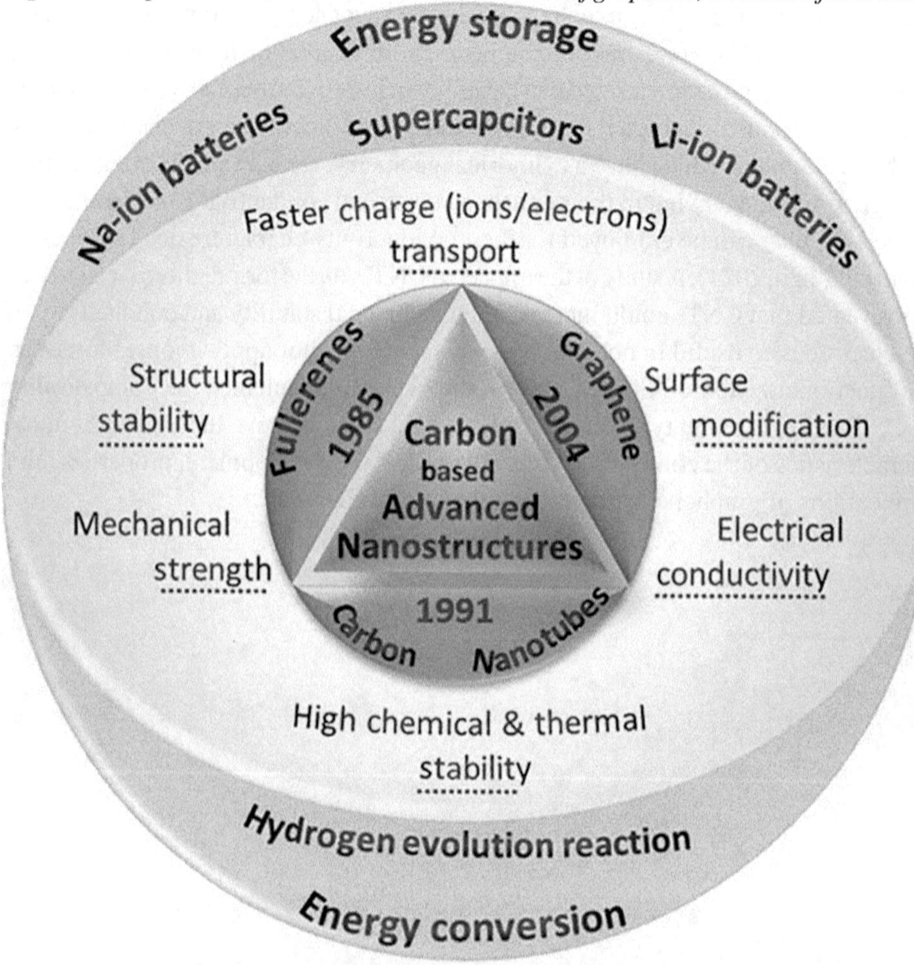

(Image by Kothandam et al., (2023) from "Recent Advances in Carbon-Based Electrodes for Energy Storage and Conversion (https://onlinelibrary.wiley.com/doi/10.1002/advs.202301045)" is licensed under CC BY 4.0.)

Carbon Quantum Dots

Carbon quantum dots (CQDs) are another form of carbon dots that have recently attracted interest for their uses in energy storage and conversion applications. These are zero dimensional nanomaterials and have interesting optical and electronic features, which desirably enable electrochemical applications (Jiang et al., 2022). The synthesis of CQDs does not necessarily require any additional carbon precursors, and thus, it can be derived from biomass rendering the method green. These characteristics, as well as high surface area per unit mass and versatile surface chemistry, make them suitable to be used in supercapacitors and batteries. Past research has shown that incorporating CQDs into composites leads to improvement in the performance of energy storage devices. For instance, in the case of the electrode materials, CQDs have been successfully used to address challenges hence enhancing the improvement of the electrochemical performance in carbon-based electrodes through charge transfer and improved conductivity. This signifies that CQDs can be useful in the construction of future types of energy storage systems with thought and attention given to their integration with other electronics elements.

Sustainable Carbon Sources

Recursive use of carbon related feedstocks in preparation of carbon based nanomaterials has become a rising trend in energy storage and conversion systems. Multiple studies have demonstrated how carbon obtained from lignin or other biomass feedstocks can be useful in development of electrodes for supercapacitors and batteries with high performance. These green materials not only eliminate the related environmental problems that are customary with conventional carbon sources but are also cheaper for the fabrication of emerging high-efficient energy storage gadgets. For example, Lignin–carbon materials exhibit high electrochemical activity to be applied in supercapacitors as it was also presented in Lignin-carbon materials by (Thomas et al., 2022) grade intervention. The beneficial interfacial characteristics of biomass-derived carbon materials including the significantly high specific surface area as well as very high electrical conductivity certainly plays roles in energy storage (Wang et al., 2022). This approach fits well within the current discourse on sustainability in material science and lays the groundwork for green solutions in energy storage. Figure 2 illustrates the process of producing lignin-based carbon particles using a horizontal carbonization furnace.

Figure 2. (a) Schematic showing the preparation of lignin-based carbon particles using the horizontal carbonization furnace. (b) Thermal behavior of and residues obtained for lignins at 1000 °C after the TGA analysis in the N2 atmosphere. (c) Derivative thermogravimetric (DTG) curve for lignins. (d) Carbon yield obtained after carbonization at 1000 °C and 1400 °C for different lignins used in the study.

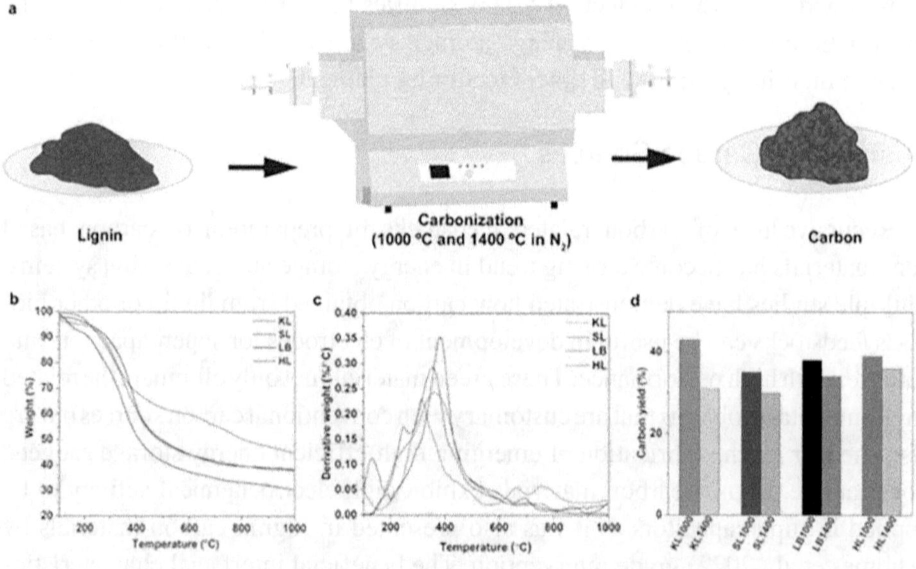

(Image by Thomas et al., (2022) from "Sustainable Carbon Derived from Sulfur-Free Lignins for Functional Electrical and Electrochemical Devices (https://www.mdpi.com/2079-4991/12/20/3630#)" is licensed under CC BY 4.0.)

Metal Oxide Nanoparticles

Among the metal oxide nanoparticles (NPs), Zinc oxide (ZnO) and Titanium-dioxide (TiO_2) nanoparticles have acquired much interest in energy storage and conversion applications mainly because of their favourable characteristics and multiple uses. These materials have high surface area, switchable electronic conductivity and catalytic activity hence applicable in energy conversion and storage devices such as batteries, supercapacitance and photocatalysis applications.

Zinc Oxide Nanoparticles

ZnO NPs have generated considerable interest because of their expected uses in energy storage and conversion systems. ZnO is an n-type semiconductor with a wide bandgap of approximately 3.37 eV, which provides a good electron transport channel and charge separation property (Gharpure et al., 2022). Dye-sensitized solar cells (DSSCs) are one of the best beneficiaries of this property and ZnO nanoparticles are well suited for use as electron transport layers. Silver doping of ZnO can also improve its electrical characteristics and thus increase the efficiency of DSSC by promoting the injection of electrons from the dye molecule into the conduction band of ZnO (Rathnasekara & Hari, 2022). Besides their use in solar cells, ZnO nanoparticles have been investigated for photocatalytic application. (Kaewsaenee et al., 2023) has observed that using ZnO photocatalyst in the presence of ultra-violet light it has high photocatalytic activity that makes it appropriate for environmental and gas treatment as well as water splitting to produce hydrogen. The photocatalytic efficiency of ZnO can be further improved than the nanocomposite with other materials, for example copper oxide (CuO), which can further improve charge carrier separation and minimize the recombination lag (Hafeez et al., 2022). In addition, the method of synthesis plays a crucial role in the physical and electrical characteristics of ZnO nanoparticles. New methods of synthesizing ZnO nanoparticles, using plant extracts, have been purposely to vary the morphology and particle size which in turn improves the performance and stability of the catalyst (Tizazu & Wondimu, 2024). The number of techniques available for the synthesis of ZnO nanoparticles enables a fine tuning of the characteristics of the resulting material for maximising performance in a particular context.

Titanium Dioxide Nanoparticles

Another metal oxide that has received a lot of attention in energy storage and conversion is titanium dioxide that has the chemical formula TiO_2 . Recently, TiO_2 has received immense attention due to its high stability, non-toxicity and highly efficient photocatalytic activities, and hence widely used in photovoltaic solar cell and photocatalytic processes. Among all the polymorphic forms of TiO_2, the anatase phase has been found more suitable for photocatalytic applications owing to its high surface area and suitable electronic structure that promote the separation of photocurrents and thus boosts the photocatalytic activity (Alamgeer et al., 2023). In energy storage applications, titania has been examined especially as an anode material in lithium-ion batteries. Its tolerance of lithium ions during cycling helps it possess high theoretical capacity according to (Alfaawaz et al., 2022). However, when it comes to the factor of electronic conductivity, it is lower in comparison with the results one can achieve when using another material – TiO_2. To eliminate this problem, researchers have consequently applied TiO_2 in composite materials where it has been incorporated with a conductive carbon-based matrix or some other metal oxide to improve the conductivity as well as general electrochemical character of the material.

Synergistic Effects of Metal Oxide Nanocomposites

Herein, the addition of different metal oxides or the incorporation of metal oxides into carbon related materials can result in synergistic or improved properties of nanocomposites for the enhanced electrochemical performances in energy storage and conversion. For instance, the ZnO when incorporated into a polyaniline matrix enhances the electrochemical characteristics of the composite to support photovoltaic application. Likewise, the photocatalytic activity of the $ZnO–TiO_2$ hybrid nanocomposites has been reported to be higher than those of the individual components because of better charge separation and lower rates of electron and hole recombination. The elemental and geometry distribution of the metal oxides determines functionality and performance of the nanocomposites used in various applications. Through fine tuning of the synthesis parameters, the authors can fine tune the resulting properties of these materials to improve the stability, energy density and conductivity of the resulting materials, enabling better performance in the energy storage and conversion technologies.

2D Materials

Layered materials have recently attracted a lot of attention in the energy storage and conversion application because of their distinguishing features as well as excellent electrochemical characteristics. The materials developed here, hence having atomic thickness and high surface area, would enable fast ion conduction and elevated accessible ion surface area for ion adsorption that can be conveniently applied in batteries, supercapacitors, and other energy storage/related technologies.

Transition Metal Dichalcogenides (TMDs)

Among the TMDs candidates, molybdenum disulfide (MoS_2) and Tungsten Diselenide (WSe_2) have attracted large interest for their prospects as anode and cathode materials for lithium-ion and sodium-ion batteries. TMDs have superior theoretical capacities, and excellent electrochemical characteristics; these characteristics are due to the intercalation of ions by layered structures (Kovalska et al., 2023). For example, MoS_2 has been investigated as an anode material where it is revealed glyoxyl excellent cycling stability and high capacities resulting from its ability to manage lithium ions during charge and discharge processes. The TMDs possess other features such as direct band gap, and strong spin orbit coupling which also allows their applicability in optoelectronics apart from energy capture and storage systems (Sharma et al., 2022). Recent discoveries relating to the synthesis of TMDs have triggered the incorporation of TMDs with other materials, which include conductive polymers and carbon materials. Some of these composites can also improve the electrochemical performance since things such as conductivity and structural stability can be affected. For instance, the integration of polypyrrole into MoS_2 nanosheets has been done with a positive effect on the electrochemical performance of lithium-ion batteries, implying the fact that such hybrid structures could be put into use to enhance energy storage characteristics (Kang et al., 2023).

Siligene and Germanene

Other two-dimensional allotropes such as Siligene and germanene of silicon and germanium, respectively, have also been investigated for their use in lithium-ion battery systems. They show reasonable specific capacities and cyclability that points to future high energy devices (Xing et al., 2024). Electrochemical characterization of the anode properties of Siligene and germanene using cycling tests has shown that they could be used in high capacity energy storage devices. An added advantage of these materials is that the electronic properties can be tuned using chemical methods, making the materials suitable for energy storage systems. Figure 3 signifies the Siligene used for Lithium-ion storage.

Figure 3. Electrochemical Decalcification–Exfoliation of Two-Dimensional Siligene, SixGey: Material Characterization and Perspectives for Lithium-Ion Storage.

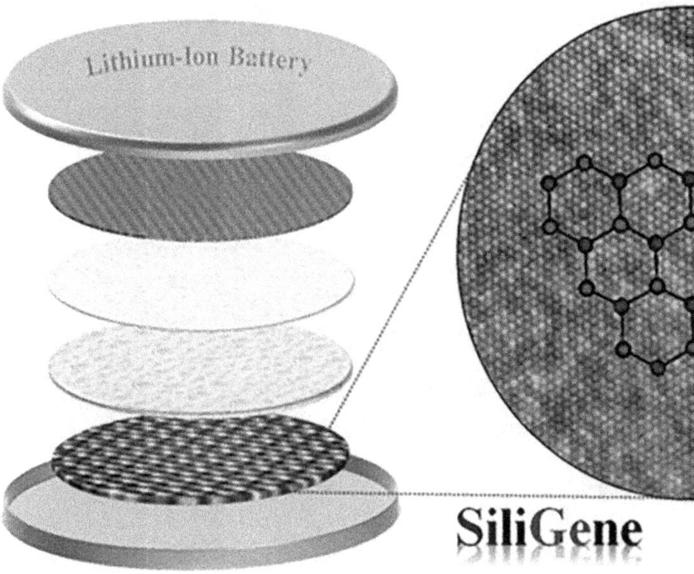

(Image by Kovalska et al., (2023) from "Electrochemical Decalcification–Exfoliation of Two-Dimensional Siligene, SixGey: Material Characterization and Perspectives for Lithium-Ion Storage (https://pubs.acs.org/doi/10.1021/acsnano.3c00658)" is licensed under CC BY 4.0.)

MXenes

MXenes, a family of 2D transition metal carbides and nitrides, have recently attracted much interest for their high conductivity and large surface area to volume ratio rendering them ideal as electrode materials for energy storage. Thus, due to the high density of layers and the presence of hydroxyl groups on the surface of MXenes, it is possible to arrange fast ion transport, which is necessary for high-performance batteries and supercapacitors (Xu et al., 2024). Some research has been conducted to illustrate that MXenes can be used efficiently as anode materials for lithium-ion energy storage devices, proving to possess high capacity, together with outstanding cycling performance (Zheng et al., 2023). The present work on functionalization of MXenes provides an independent approach of tuning the electrochemical properties for improved energy storage potentials for MXenes (Thomas & Cherusseri, 2023). Figure 4 represents the overall schematic diagram of 2D materials applications toward metal anode, electrolyte, and separator design for dendrite-free metal batteries.

Figure 4. Overall schematic diagram of 2D materials applications toward metal anode, electrolyte, and separator design for dendrite-free metal batteries.

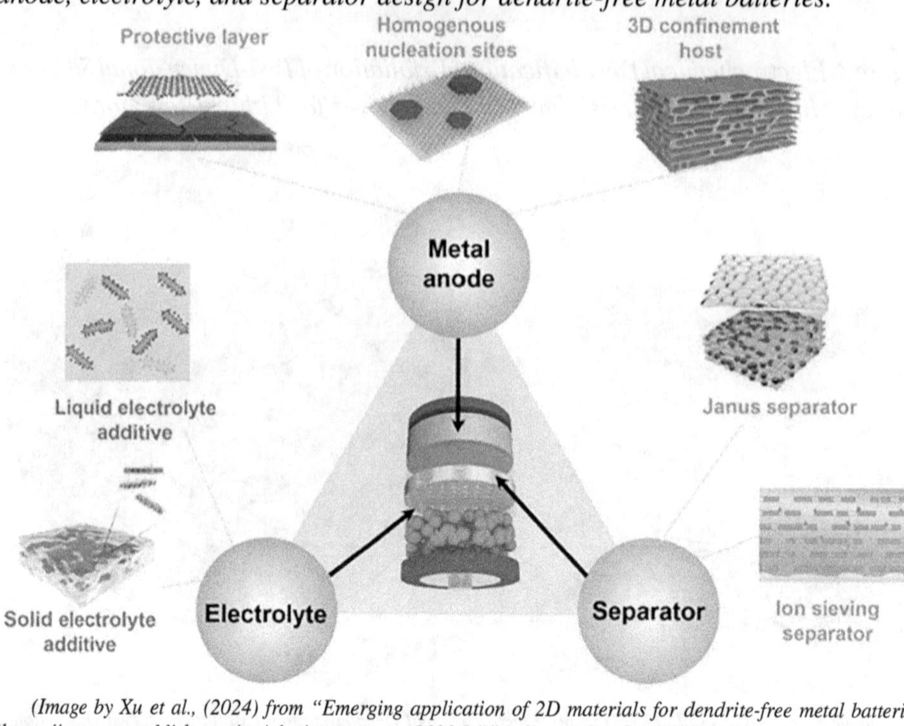

Composite 2D Materials

Staking of dissimilar 2D materials or embedding of 2D materials with other nanostructures has proved advantageous as the performance of energy storage devices is boosted. For instance, incorporating TMDs with graphene or any other forms of carbonic materials results in improved conductibility, structural stability and hence, improved electrochemical characteristics (Nair et al., 2023; Kousik Kumaar et al.,2024). These composite can make an advantage with the characteristic of each component: energy density, charge-discharge rates, and stability will improve synergistically (Thomas, 2024). Furthermore, expanding the new family of 2D materials including boron carbon nitride (BCN), and porous black phosphorus (BP) has introduced more opportunities to work in the aspect of energy storage (Ghosh et al., 2022). These materials possess characteristics that have potential to increase the storage capacity and enhance the electrochemical characteristics of the energy storage systems, which is helpful to the progress of energy storage technology. Figure 5 characterizes the overall schematic diagram of harnessing the unique properties of 2D material toward advanced Zn anodes for high-performance Zinc-Ion Batteries (ZIBs).

Figure 5. Overall schematic diagram of harnessing the unique properties of 2D material toward advanced Zn anodes for high-performance ZIBs.

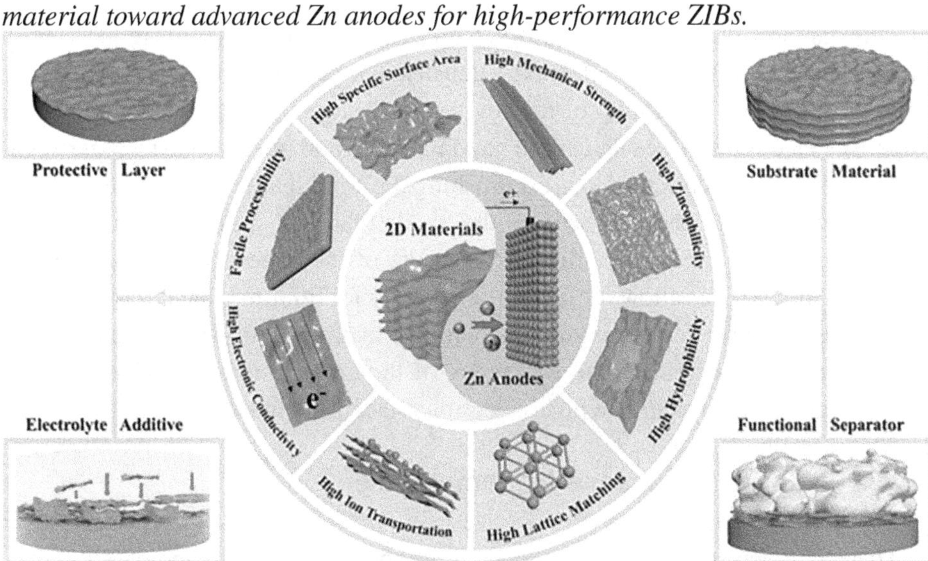

(Image by Zheng et al., (2023) from "2D Materials Boost Advanced Zn Anodes: Principles, Advances, and Challenges (https://link.springer.com/article/10.1007/s40820-023-01021-9)" is licensed under CC BY 4.0.)

Composite Nanomaterials

Heterostructured nanofluids that are formed by mixing at least two kinds of nanomaterials have received considerable interest for improving the energy storage and conversion technologies. These synergistic composites with the improvement of electrochemical performance, stability and conductivity can be obtained by the virtue of the formerly mentioned component attributes. In this section, the authors present an overview of the use of composite nanomaterials in energy-related applications, the methods of their synthesis, their main characteristics, and potential application areas.

Synergistic Effects in Composite Nanomaterials

Some of the nanomaterials used in creating composites may have synergistic properties that improve the efficiency of the structures. For example, the integration of carbonaceous materials, including graphene or CNTs, with metal oxide materials for instance, TiO_2 can greatly enhance the conductivity and electrochemical performances of the integrated composites (Shi et al., 2022). Graphene with high electrical conductivity and large surface area can act as a conductive matrix which can promote electron transfer and metal oxides can deliver high capacity for ion storage (Yang, 2023). There, the synergy between fullerenes and graphene is especially advantageous, as both lithium-ion batteries and supercapacitors require short charge-discharge times and high energy density. The research works done in recent years show that the possibility of improving the electrochemical characteristics of composite electrodes based on carbon nanomaterials and metal oxide nanoparticles exists. For instance, investigations establishing the combination of titanium dioxide (TiO_2) nanoparticles and graphene as the anode material in lithium-ion batteries revealed enhanced cycling stability and capacity of the battery. More specifically, the TiO_2 nanoparticles allow for further active sites for lithium ion intercalation and the graphene matrix increases the conductivity of the electrode. This composite approach not only enhances the efficiency of the energy storage device but also solves some of the problems with conventional electrode materials.

Biomass-Based Composite Nanomaterials

There is renewed interest in producing composite biomass nanomaterials because of their renewability and ability to be processed cheaply. These modified biomass-based nanomaterials can be co-processed with other more conventional nanomaterials to form composites which bring about improvements in performance in energy related uses. For example, the conversion of biomass carbon into metal oxide composites enhances the electrochemical characteristics of the resulting composites for use in supercapacitor and battery applications (Thambiliyagodage et al., 2024; Kousik Kumaar et al., 2024). The production of composite nanomaterials from renewable resources is highly suitable to the current shift in material science towards sustainable development. Furthermore, the theoretical calculations and rational design of biomass-based nanocomposite systems are able to give increased understanding of reaction mechanisms and interdisciplinary interactions between the constituent components, which further improves performance. Restrictions of current and future use of biomass based nanomaterials can also be the integration of artificial intelligence and machine learning in the design process to expand the capability to develop new generation biomass based nanomaterials. Figure 6 denotes the schematic representation of the preparation method and application of biomass-based carbonaceous nanomaterials.

Figure 6. Schematic representation of the preparation method and application of biomass-based carbonaceous nanomaterials.

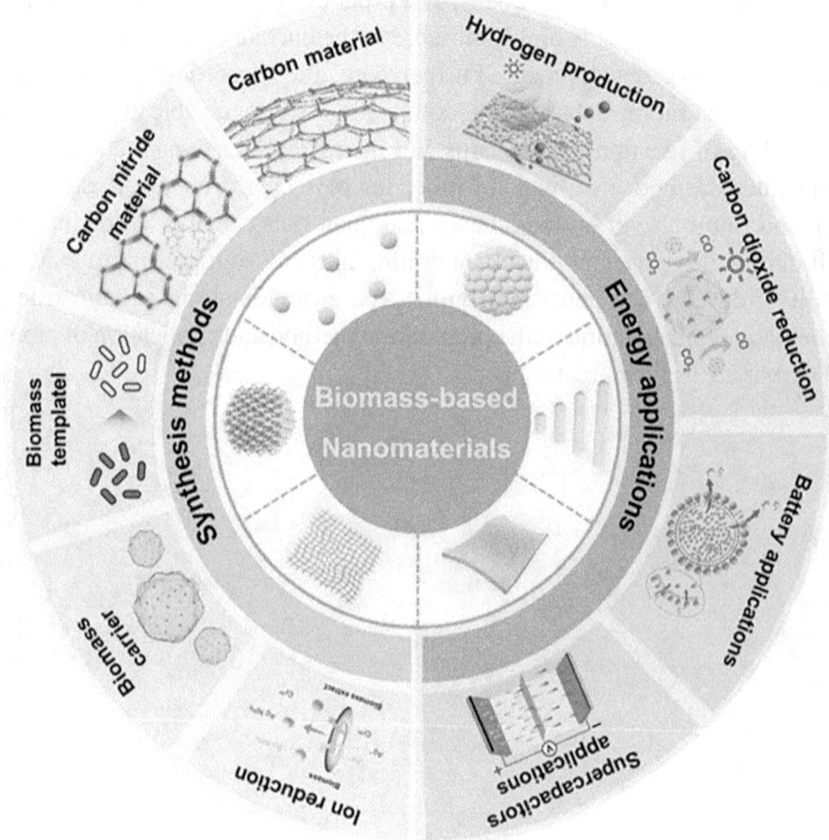

(Image by Yang et al., (2023) from "Application of Biomass-Based Nanomaterials in Energy (https://onlinelibrary .wiley.com/doi/10.1002/aesr.202300141)" is licensed under CC BY 4.0.)

Applications in Energy Storage and Conversion

Composite nanomaterials have been used significantly in numerous energy management systems (EMSs) and power conversion technologies. For example, the incorporation of hybrid nanocomposites in lithium-ion batteries has been shown to considerably enhance the energy density as well as cycling stability (Modugu et al. 2022; Cui, 2024). The use of plural nanomaterial to construct the electrodes thus makes it possible to meet the desired energy storage system requirements through improvement of electrode properties. Apart from batteries, composite nanomaterials are considered for supercapacitors due to their expanded surface area and conductivity that will enhance the capacitive storage of charges. The addition of conductive fillers to polymer matrices including the carbon nanomaterials increases the conductivity of insulating polymer matrices making it possible to fabricate superior structures for supercapacitors (Jing, 2023). Here, the potential of composite nanomaterials is demonstrated with reference to the problems of energy storage and conversion.

ENERGY STORAGE APPLICATIONS OF NANOMATERIALS

Nanomaterials have been used in many applications to improve the performance of the energy storage systems such as batteries, supercapacitors, and hydrogen storage. In this section, nanomaterials applications in lithium-ion batteries (LIBs), advanced battery systems, supercapacitors, and hydrogen storage are described in terms of their impact on the capacity, cycle life, safety, and efficiency aspects. Figure 7 represents the specific energy and specific power plot of energy storage and conversion devices.

Figure 7. Specific energy and specific power plot of energy storage and conversion devices.

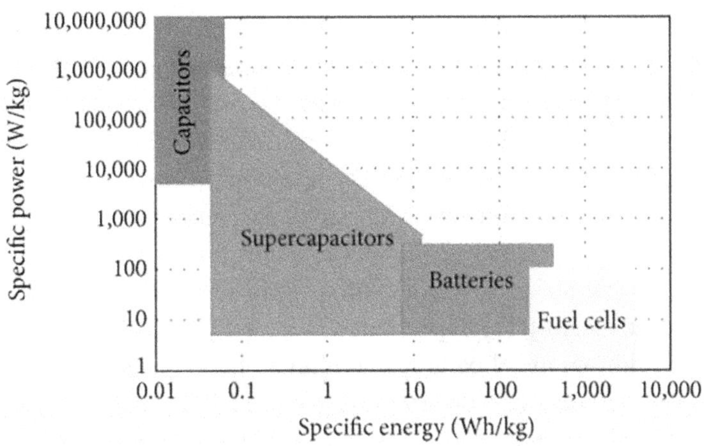

(Image by Saleem et al., (2016) from "Performance Enhancement of Carbon Nanomaterials for Supercapacitors (https://onlinelibrary.wiley.com/doi/10.1155/2016/1537269)" is licensed under CC BY 4.0.)

Batteries

Nanomaterials in general have improved energy storage at various systems and or devices such as battery related systems, supercapacitors and hydrogen storage systems. Nanomaterials for lithium–ion batteries, new generation battery materials, supercapacitors and hydrogen storage systems are also highlighted in this section in terms of capacity, cycle life, safety, and their efficiency.

Lithium-Ion Batteries

LIBs are known to have high capacity, long cycle life, and exhibition of high thermal safety due to the incorporation of nanomaterials. It has been established that using particles and forms of silicon nanoparticles and carbon based nanomaterials enhance the rates of transport and diffusion of lithium ions within electrodes (Tian, 2024). For example, silicon presupposes the theoretical capacity of about 4200 mAh/g and contributes to a growth of the energy density in LIBs. Nonetheless, when discharging, based on the capacity ratio of the anode and cathode, its utilization volume dramatically increases during the cycling process, which is disadvantageous to cycle stability. These volume changes are effectively managed by nanostructuring silicon while increasing mechanical stability and improving cycle life (Liu, 2023). In addition, it was reported that the integration of nanomaterials will further improve the safety of LIBs. For instance, the emerging nanocomposite materials that incorporate flame retardant agents could help to eliminate the risk of thermal runaway, a very sensitive problem affecting lithium-ion batteries (Pan, 2023). Also, it can enhance the electrochemical kinetics of the battery, extending charge and discharge rates that are especially considerable for energy storage technologies for electric vehicles and portable electronics. Other nanomaterial innovations include the use of solid-state electrolytes that make LIBs safer and more efficient. For example, the use of ionic liquid-based electrolyte has presented promoter electrochemical stability and conductivity to lithium metal batteries with high-energy density (Wu et al., 2023; Cui, 2023). These innovations do not only enhance LIBs' safety but also allow the application of lithium metal anodes that can again boost energy density.

Beyond Lithium-Ion Batteries

Nanomaterials also cope with drawbacks of the other potential battery types including sodium-ion and lithium-sulfur batteries. Sodium-ion batteries (SIBs) are emerging as the next suitable option for electric power storage from lithium-ion batteries due to the low cost of sodium. Carbon based composites and transition metal oxides nanostructures have been reported for improved electrochemical performance of SIBs(Zhang et al., 2023). For example, the employment of nanostructured carbon materials enhances sodium-ion electrodes conductivity and cycling stability and makes them comparable to, or even better than, lithium-ion systems. Another candidate is lithiun–sulfur batteries (LSBs) which also have high theoretical energy density. However, because the polysulfides dissolve in the electrolyte, the capacity fades over time. One way around this problem is by using nanomaterials including sulfur-embedded carbon nanofibers in which polysulfides can be coordinated and secured within a conductive matrix which promotes their utilization throughout cycling (Zhang et al., 2024). This approach not only increases the cycle life of LSBs but also increases their overall performance (Khazraji et al., 2022). Current investigations have shown that the use of nanomaterials into LSBs can greatly boost their performance. For example, the authors employed nickel-iron layered double hydroxides (NiFe-LDHs) modified carbon nanotubes as sulfur hosts, and found that incorporation of NiFe-LDHs can enhance the electrochemical performance of LSBs. Such innovations seek to underscore the effectiveness of nanomaterials in dealing with the qualitative factors arising from nonclassical battery systems.

Supercapacitors

Super capacitors or ultracapacitors or electrochemical capacitors have become one of the most important energy storage technologies owing to high power density and ability of charge-discharge and longer cycle life. Incorporation of nanomaterials has been critical in improving the performance of supercapacitor technology. This section gives an overview of the anodic materials on supercapacitors in relation to energy density, capacitance and efficiency. Figure 8 represents the working principle of supercapacitors.

Figure 8. Working principle of supercapacitor.

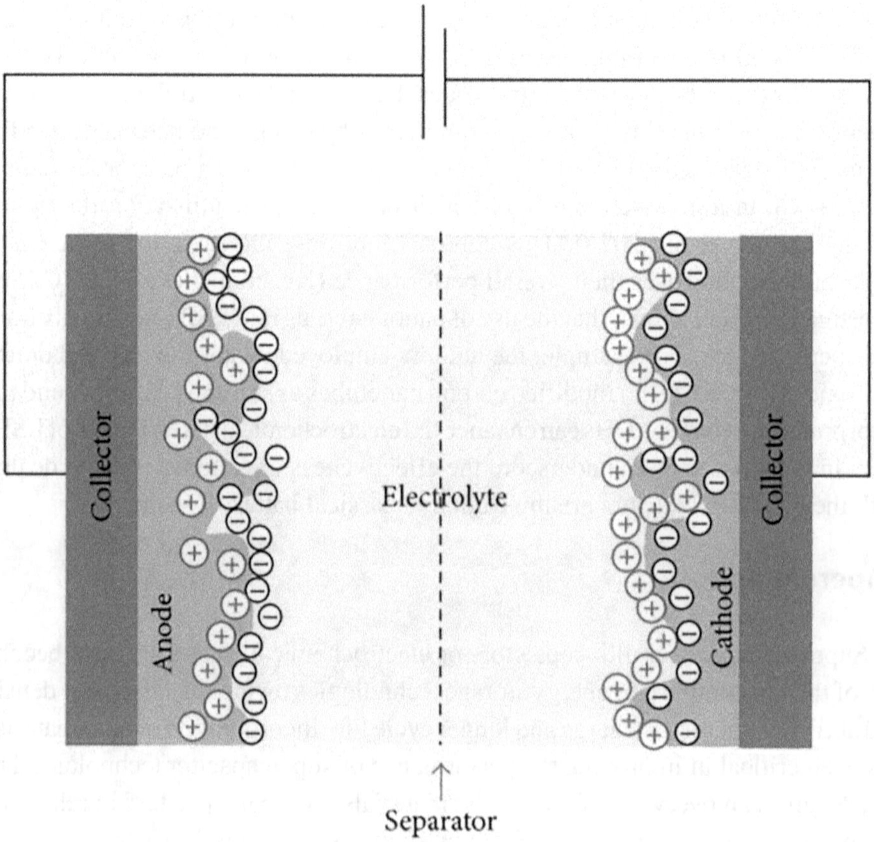

(Image by Saleem et al., (2016) from "Performance Enhancement of Carbon Nanomaterials for Supercapacitors (https://onlinelibrary.wiley.com/doi/10.1155/2016/1537269)" is licensed under CC BY 4.0.)

Contributions of Nanomaterials to Supercapacitors

Carbon-based nanomaterials, transition metal oxide nanoparticles and conducting polymers are at the center of increase and restoration of electrochemical performance of supercapacitors. Carbon materials such as graphene CNTs, and activated carbons are favorites due to their excellent electrical conductivity and high surface area (Jin et al., 2023). For instance, graphene-CNT hybrid aerogels have been reported to be used as electrodes in supercapacitors and can perform optimally because of the high thermal transport behavior and improved charge storage. The functionalization of these carbon-centered nanomaterials can otherwise have a positive impact on their electrochemical performances for energy storage and conversion. Some other transition metal oxides like manganese oxide (MnO_2) and tungsten oxide (WO_3) have also been used because of their pseudocapacitive nature that offers higher capacitance due to faradaic reactions that happen at the electrode-electrolyte interface only (Gowtham et al., 2022; Rehman et al., 2022). The introduction of the nanostructured transition metal oxides in the composite materials can improve the overall performance of the supercapacitors in a wide ranging manner. For instance, the MnO_2 nanocomposite has been effectively applied to enhance the specific capacitance and cycling stability of supercapacitors (GULEN et al., 2023; Wei et al., 2022). Furthermore, conjugated polymers like polypyrrole and polyaniline have been used along with carbons in fabrication of nanocomposite that has high electrochemical activity. These composites can further improve the redox active sites and provide additional capacitance and charge storage ability for the supercapacitors (Ho et al., 2023). Thus, conducting polymers are the most appropriate for creating new improved types of supercapacitors for flexible and wearable technologies (Aydin & Yildiz, 2023).

Performance Enhancements Through Nanostructuring

Nanostructuring of materials comprises an essential step for achieving high performance of supercapacitors. Small particle sizes or high surface areas favored by researchers lead to increased electrochemical activity and better ion diffusion at the electrodes (Zheng et al., 2023, Yun et al., 2022). For example, the synthesis of hierarchical porosity in nanomaterials may contribute to the effective ion transport and improved overall energy density in supercapacitors (De et al., 2023). New developments in fabricating nanostructured materials in the recent past have provided new electrode structures for supercapacitors that immensely improve the electrochemical characteristics of the device. For instance, flower-like hierarchical porous NiO has been synthesized and exhibited superior capacitive performance due to large superficial facets and rich redox active sites. In a similar way, the fabrication of transition metal sulfide nanostructures has been reported to create high capacitance supercapacitors with enhanced energy density and cyclability (Yu, 2024).

Hydrogen Storage

A range of nanomaterials is now viewed as the key solution in the storage of hydrogen for fuel cells and other uses because of the problems experienced in the storage of this gas. Nanostructured materials for hydrogen storage section that contains information on nanomaterials that improves hydrogen storage characteristics: structures, mechanisms, and possible applications. Figure 9 represents the nanomaterials for hydrogen storage.

Figure 9. Nanomaterials for hydrogen storage.

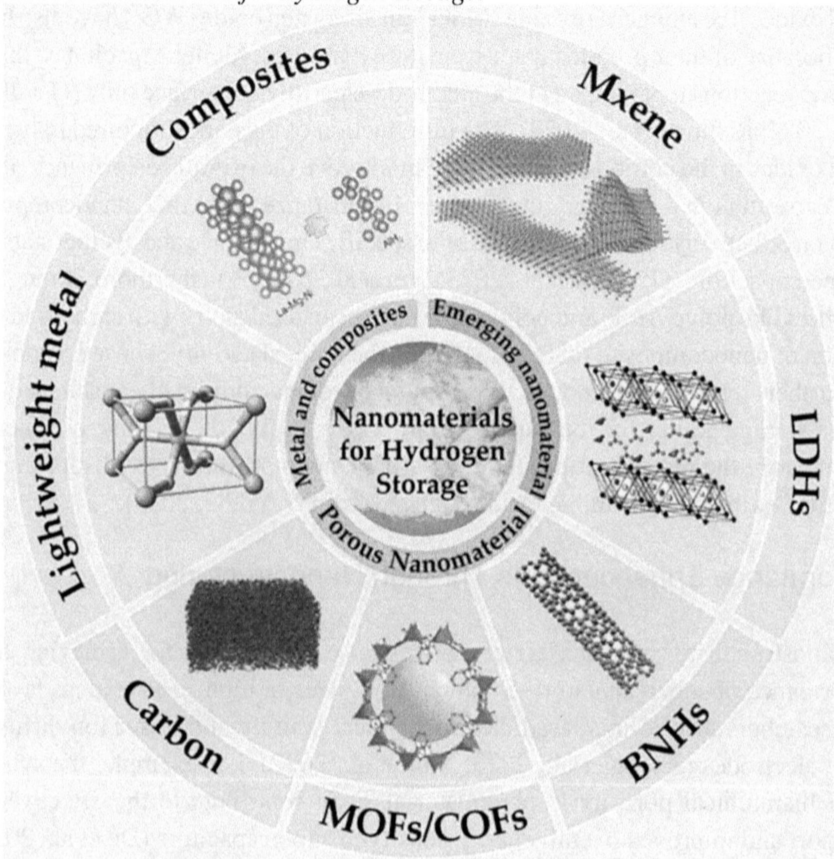

(Image by Xu et al., (2024) from "Advances and Prospects of Nanomaterials for Solid-State Hydrogen Storage (https://www.mdpi.com/2079-4991/14/12/1036#)" is licensed under CC BY 4.0.)

Innovations in Hydrogen Storage

As a result of their large surface to volume ratio and the physical characteristics of nanosized structures, nanomaterials provide capabilities for the adsorption and disintegration of molecular hydrogen and their quick adsorption and release (Xu et al., 2024 ; Shi et al., 2023). For instance, carbon-based nanomaterials like graphene, carbon nanotubes have been the subject of profound investigations for hydrogen storing efficiency. Combination of transition metals into such nanostructures may also increase hydrogen storage capacity by offering synergistic effects that facilitate adsorption (Boateng et al., 2023; Jawhari, 2022). The studies carried out within the last few years have shown that Metal-organic frameworks (MOFs) could also be used in hydrogen storage applications. The internal surface areas of MOFs are large, and their pore size can be adjusted depending on the application so hydrogen storage can be achieved efficiently in MOFs. That is why, due to the possibility to change the chemical and structural characteristics of MOFs, their efficiency in hydrogen storage can be improved, so they remain prospective material for future hydrogen storage systems (Park, 2024).

Nanostructured Materials for Enhanced Performance

Nanostructured materials, for instance, sodium hexa-titanate nanowires decorated with cobalt hydroxide quantum dots as was also depicted above self-assembly nanostructure has been determined to be capable of efficient electro catalytic hydrogen evolution reaction (HER). Such structural features lead to better rate capability due to greater ion diffusion and smaller energy barriers for hydrogen storage. Also, synthesizing Nitrogen-doped reduced graphene oxide nanosheets anchored with palladium (Pd) nanoparticles has shown variable electrochemical hydrogen adsorption and desorption characteristics; again proving that nanomaterials can enhance the hydrogen storage process. The introduction of palladium provides additional hydrogen storage capability to graphene, which is otherwise chemically passive thus eradicating one of the major drawbacks of graphene materials.

Safety and Efficiency in Hydrogen Storage

Nanomaterials also find application in safely storing hydrogen which is one of their biggest advantages of their usage. For instance, the application of carbon nanotube and graphene reinforced composite materials in hydrogen storage tanks and pipelines can help to avoid problems like hydrogen embrittlement. (Shaker, 2024). These composite materials also not only enhance the mechanical properties and service life of storage systems, but also promote the creation of high-sensitivity hydrogen sensors to test leakage and safety. In addition, new developments in the solid-state hydrogen storage materials especially at nanometer scale offer safer and more efficient ways of storing hydrogen. Hydrogen storage materials in solid-state can provide compact storage capabilities as compared to gaseous or liquid-based storage with limitations such as low storage density and flammability of hydrogen (Lü et al., 2023; Wang et al., 2024).

ENERGY CONVERSION APPLICATIONS OF NANOMATERIALS

Nanomaterials have become integral to enhancing the efficiency and stability of energy conversion technologies, particularly in photovoltaic cells and electrocatalysis. This section discusses how nanomaterials are utilized in these applications, focusing on their roles in solar energy conversion, fuel cells, and water splitting for hydrogen production.

Photovoltaic Cells

Nanoparticles have improved the performance and durability of photovoltaic devices especially in the emerging renewable solar power systems. This section reviews the use of different types of nanomaterials in converting Sun's rays into energy; with reference to the application of these nanomaterials in enhancing the capturing and transport of light energy in the solar devices.

Enhancing Efficiency with Nanomaterials

Quantum dot, nanowire and perovskite incorporated structures have been used to enhance the performance of the solar cells. For example, in quantum dots, it is possible to design the possible band gaps that enable efficiency in the absorption of the sun's spectrum to improve the energy conversion rates (Masud et al., 2024). Quantum dot's being able to tune its bandgap makes it possible to engineer many solar cells that have the capability of collecting a wider part of the sunlight therefore increasing the general efficiency. Perovskite solar cells (PSCs) have recently become popular in the photovoltaic systems market because of their high efficiency and low price. Specifically, the integration of novel perovskite quantum dots as nanomaterial into the solar cell has emerged as a research focus for the enhancement of light trapping and charge carrier processes. For example, when MoS_2 quantum dots have been used as an interfacial layer in PSCs, researchers recorded both improved stability and efficiency owing to minimizing energy losses at the interface through this layer (Luo et al. 2022). This shows that nanomaterials are vital in achieving key objectives relating to the development of PSC commercialization barriers. Furthermore, the incorporation of nanostructured conducting polymers including poly (3,4-ethylenedioxythiophene) (PEDOT) has been confirmed to enhance the efficiency of DSSCs. From this study, we see that incorporation of chemically synthesized PEDOT nanomaterials as a counter electrode improved the charge transfer and overall conversion efficiency in solar energy conversion. The absence of toxicity for the solvent used for dispersing these nanomaterials and high donor level also favors the stability and efficiency of the solar cells. Figure 10 shows a schematic diagram of the PSC structure.

Figure 10. A schematic diagram of the PSCs structure.

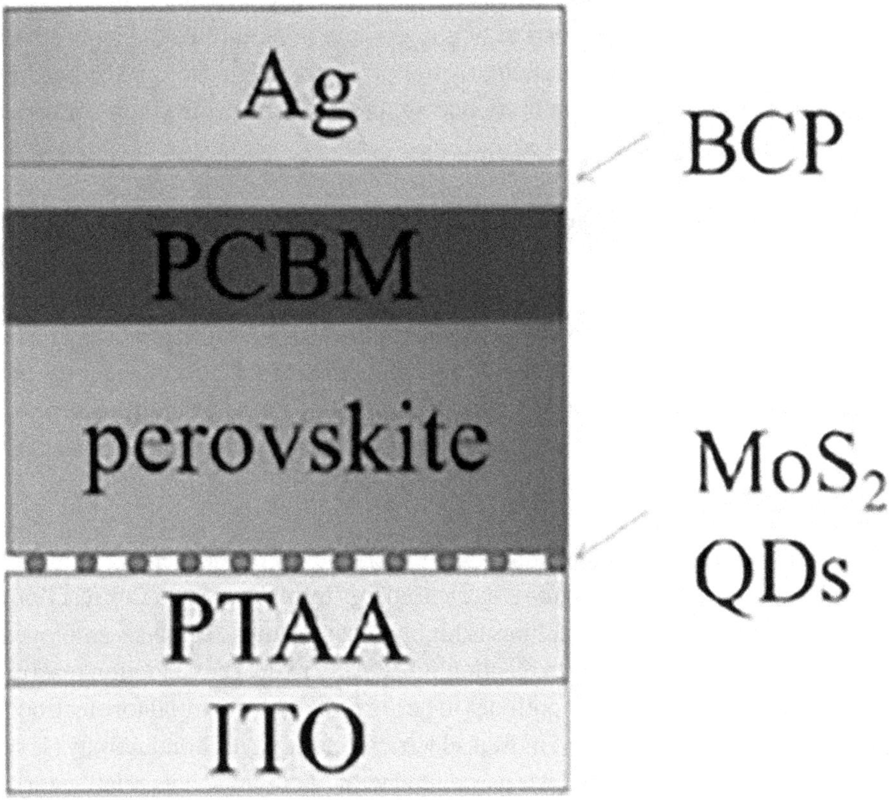

(Image by Luo et al., (2022) from "Enhancing the Efficiency of PSCs through Interface Engineering with MoS_2 Quantum Dots (https://www.mdpi.com/2079-4991/12/17/3079)" is licensed under CC BY 4.0.)

Stability Improvements

One of the major challenges in developing commercial solar cells is the ability of the cells to withstand the various conditions encountered in their application. Nanomaterials at one way act as a protective layer for the solar cell or can improve the mechanical properties of the structure of the solar cell. For instance, the incorporation of graphene oxide into a perovskite solar cell has been found to enhance the resistance to photo deterioration, humidity aging and thermal stress without a concurrent loss of power conversion efficiency. The ability of nanomaterials to improve the efficiency of photovoltaic devices as well as the stability of materials that goes along with it makes such nanomaterials critical in the creation of novel solar energy technologies. Also, the employment of nanostructured materials in fabrication of solar cells could enhance interfacial characteristics, which are vital in reduction of recombination rates. This has shown that if the interface between the different layers present in the solar cells is properly engineered using nanomaterials, the overall efficiency of the transport of charges and the minimized energy losses therein would improve. This is particularly important for the inverted PSCs because they have displayed improved stability and suppressed hysteresis when compared to the typical structure (Isikgor et al., 2022).

Electrocatalysis and Photocatalysis

Nanomaterials are significantly used in the improvement of electrocatalysis and photocatalysis with credit to fuels cells and water splitter for hydrogen production. This section investigates the enhancement of fuel cell performance by nanocatalysts and sustainable hydrogen production by nanomaterials using the water splitting process.

Fuel Cells

Nanocatalysts are essential in enhancing the efficiency of fuel cells that directly convert chemical energy of fuels into electrical energy by undergoing redox reactions. The catalytic activity of these reactions is highly sensitive to the catalysts employed, especially in the oxygen reduction reaction (ORR) and hydrogen oxidation reaction (HOR) reactions. Platinum (Pt) continues to be the preferable material for electrodes prepared for these reactions due to its high electrical and catalytic conductivity (Joshi et al., 2024). Nonetheless, high costs associated with platinum and its relative rarity have stimulated the exploration of other nanocatalysts that would be effective in delivering similar performance at a lower cost. In recent years, the synthesis of new nanomaterials produces bimetallic catalysts, including Pd and Pt alloys, which have higher catalytic activity for ORR and HOR. These nanocatalysts can enhance the electronic properties or the general activity by having useful metallic cooperation (Liu and Su, 2023). It also reveals that the engineering of nanostructured materials for example nanoparticles and nanowires can also enhance the surface area accessible for catalysis and as a result the performance is boosted (Wang, 2023). Furthermore, cheap sources of tran-sition metals and metal oxides along with nitrogen-doped carbon materials are likely substitutes to platinum-group metal catalysts. For instance, cobalt-iron-based catalysts have exhibited remarkable activity of ORR, so they are suitable for fuel cell applications (Modi et al., 2024). These nanomaterials enable the elimination of the cost problem for nanoscale fuel cells and they also help in making fuel cells more sustainable.

They are also central in the fabrication of platinum-based electrocatalysts for the water splitting system, one of the most efficient methods of generating hydrogen. In case of water splitting the faraday efficiency is still dominated by the electrocatalysts involved in the HER and OER. Nanomaterials including TMDs and MOFs have been widely investigated for their high catalytic effectiveness (Cao et al., 2024). For example, TMDs such as molybdenum disulfide (MoS_2) possess high HER activity because of the layered nature that provides charge transfer and hydrogen adsorption as postulated in the work of Cao et al., 2022. The addition of the heteroatoms will also improve the electronic structure of these nanomaterials and increase active site exposure to promote catalytic effectiveness. Photocatalytic water splitting has also been taken up as another important way to split water membranes is that nanometer semiconductor materials can convert solar energy to chemical energy. The synthesis of nanostructured photocatalysts including TiO_2 and ZnO has shown increased light harvesting and better charge separation efficiency and hence increased hydrogen evolution processes (Simpa et al., 2024). Thus, the use of hybrid photocatalysts in which different nanomaterials will be included can help to increase the efficiency of the processes of solar-to-hydrogen conversion, which is considered as potential trends for future studies (Liu et al., 2024; Wang et al., 2022).

FABRICATION AND ENGINEERING OF NANOMATERIALS FOR ENERGY APPLICATIONS

The synthesis and design of nanomaterials are indispensable to the development of energy technologies, especially in energy conversion and storage. This section details three synthesis processes of nanomaterials/ vertically aligned nanostructures (VNNS) namely chemical vapor deposition, the sol-gel method and electro-spinning and the problems facing the scaling up of these techniques for real world uses.

Synthesis Techniques

To incorporate nanomaterials into different industries such as energy storage and conversion, fabrication is of paramount importance. Various synthesis methods have been proposed to synthesize nanomaterials with required characteristics and applications. In this section, techniques and approaches which are used to deposit thin films, such as electrospinning, sol-gel processes, and chemical vapor deposition are described, shedding light on the strengths and uses of each.

Electrospinning

Electro spinning is an effective method of producing nanofibers and nanostructured materials. This technique involves using high voltage to form a charged jet of the polymer solution that undergoes stretching and hardening by the time it gets to the collector (Du et al., 2023; Dai et al., 2023). Electrospinning enables the fabrication of fiber with the largest length to diameter ratio suitable for use in energy storage devices as supercapacitors and batteries (Tan et al., 2024). Electrospun nanofibers targeting a wide range of applications also allow for the formation of versatile structures, such as core-shell fibers, hollow fibers and composite materials. For instance, functional nanoparticles can be incorporated into the electrospun fibers to improve their electrochemical characteristics for applications in sensor and energy storage (Song et al., 2022). Also, new electrospinning technologies and conventional spinning techniques like coaxial and triaxial electrospinning enable the production of functional nanofibrous structures (Salaris et al., 2023; Zhang, 2024).

Sol-Gel Methods

The sol-gel process is also an important method of producing nanomaterials especially metal oxides and hybrid nanocomposites. This method entails the transformation of a solution (sol) into a solid (gel) state so as to enable the development of nanoparticles of desired size and geometry (Gavali, 2023). Sol-gel processing is versatile because it is inexpensive, easily scalable, and can generate high-purity materials at comparatively low temperatures (He et al., 2022). Sol-gel synthesized nanomaterials in energy applications include batteries and supercapacitor electrodes, and photocatalytic systems Bayan et al., 2024. Another aspect that makes sol-gel derived materials particularly attractive is the fact that the composition and structure can be fine-tuned for a particular application, for example, improving the electrochemical performance of energy storage structures (Kumar et al., 2022).

Chemical Vapor Deposition (CVD)

Chemical vapor deposition is one of the most used methods for depositing thin films, fabrication of CNTs and graphene. Chen et al. states that in CVD gaseous precursors are deposited on a substrate to form a solid material whereby there is adequate control of the structure and morphology of the nanomaterials (Chen et al., 2023). This method is especially suitable for the preparation of materials with high crystallinity and uniformity which are so important for applications in photovoltaic cells and fuel cells (Tian, 2022). CVD is not easy to solve for industrial applications because it requires equipment and conditions to be maintained under certain parameters. However, present research is pulling towards improving CVD processes to improve their capacity and affordability (Sharma & James, 2023; Boopathy et al., 2024). Other techniques of CVD like plasma enhanced CVD have also revealed potential in fabricating nanomaterials with considerable properties (Deshmukh et al., 2022).

Challenges in Scaling and Practical Applications

An interesting feature in the studies of nanomaterials is that while they have transitioned from a laboratory synthesis process for various nanomaterials, they are now facing several challenges in the large-scale industrial applications. These are cost, stability, and environmental constraints as well as compliance issues. These challenges are described in this section, and possible ways to promote the more practical and significant use of nanomaterials in diverse forms of energy technologies are elucidated.

Cost

A major problem of developing nanomaterials synthesis processes is the cost of material and processes to be used in the course of producing these nanomaterials. A large number of nanomaterials, especially those containing special metals like platinum or palladium, are costly and restrict the prospects of applications for some technologies, for example, fuel cells (Saleh & Hassan, 2023). Efforts are still being made to look for the new materials and new synthesis techniques to bring the cost down without compromising the performance. For instance, the advances made in non-precious metal catalysis and biomass energy materials have the potential in reducing costs by future years (Sezali et al., 2024). Furthermore, other synthesis techniques that may be useful in the development of the nanoparticles include sol-gel processes and microwave assisted synthesis due to the possibility of their lower cost. These methods always use less energy and fewer toxic substances, which are economically better suitable for mass production (Yao et al., 2024).

Stability

The operational stability of nanomaterials is another major factor of concern. The stability of nanomaterials is of great concern because of the likelihood of them degrading when in use. Most nanomaterials are sensitive to oxidation and degradation of their properties that can be detrimental to their performance in applications such as batteries, super capacitors and fuel cells (Abdulwahab et al., 2023). For example, the application of nanocatalysts in fuel cells can lead to the degradation of the performance because of the leaching of the active phase or the solid-state transformation during the process (Liu, 2023). In response to these stability concerns, authors are conducting scientific searches for such products as protective coatings or stabilizers and composite materials that would improve the stability of nanomaterials. For instance, nanoparticles within a polymer matrix can prolong their life span because of the shield offered by the matrix from environmental impacts (Barhoum et al., 2022; Kishore et al., 2023).

Environmental Impact

Transportation and disposal of a manufactured nanomaterial have a negative impact on the environment. Employment of toxic solvents in synthesis processes or processes that include hazardous chemical substances may have negative impacts on humans and the environment (Harish et al., 2022). Consequently, the research work is expanding towards an eco-friendly one-step synthesis that adopts less hazardous reagents and environmentally friendly reaction pathways. For instance, biomass or waste materials as sources of feedstock in nanomaterial preparation are therefore healthier for the environment and provide viable replacements. However, identifying the environmental impact of nanomaterials in the different phases of their useful life is critical, which is achieved by the help of the life cycle assessment. Scholars are paying much attention to recycling and safe ways for removal of nanomaterials to reduce risks toward environment and human beings (Bae et al., 2022; Ahmed et al., 2022).

Regulatory Challenges

One of the key challenges when incorporating nanomaterials into commoditised products is policy restrictions because of the potential toxicity of the material. Adopting clear set test standards and norms for the safe application of nanomaterials are critical for their applications in various markets (Nkele et al., 2023; Nagappan et al., 2022). Current regulatory guidelines need to be adapted to fit the properties and processes involved with nanotechnology to make certain that nanomaterials are properly assessed for safety qualification and efficacy in comparison with other materials of similar scale. It is strongly recommended that researchers, industrial players, and regulatory agencies come up with standard policies that enhance the use of nanomaterials in energy applications, thus ensuring stakeholders' safety as discussed in (Usman, 2023 ; Ugwuanyi et al., 2024).

MODELING AND SIMULATION OF NANOMATERIALS IN ENERGY SYSTEMS

These factors present the rationale for the modeling and simulation of nanomaterials especially for energy applications. In different theoretical calculations, density functional theory (DFT) & molecular dynamics (MD) is adopted to estimate the characteristics and the behavior of the nanomaterial. These computational techniques are described in this section of the article, as well as a brief explanation of how simulation can help fine-tune nanomaterials for energy storage and conversion applications.

Computational Methods

A mathematical representation of nanomaterials and computational models are useful in predicting their behavior and enhancing their efficiency in energy applications. Quantum chemistry using functional density approximations, MD, and machine learning algorithms are used in determining and modeling the behavior and interactions of nanomaterials. This section presents some of these computational approaches and their utility in energy systems.

Density Functional Theory

This section focuses on the electronic properties of nanomaterials using DFT. They used density functional theory to determine the structure and properties of nanomaterials, their stability and reactivity at the sub-molecular level. This method is of greatest value in studying the behavior of nanomaterials in interacting with the environment, which is the key consideration for catalysis and energy storage. Using DFT, researchers have investigated numerous nanomaterials such as metal nanoparticles, carbons and carbon derivatives, and transition metal dichalcogenides. In the case of graphene and its derivatives, DFT calculations have helped the understanding of its electronic properties, which has aided in development of high performance energy materials (Dahl et al., 2023). Furthermore, DFT can be used to calculate the adsorption energies of corresponding reacting species adsorbed on catalysts, which will assist in the synthesis and optimization of better catalysts for fuel cell and other energy converting applications (Li et al., 2023).

Molecular Dynamics

MD is used to simulate the dynamics of atoms and molecules in nanomaterials to investigate the stability of their structures, as well as their stiffness, and thermal characteristics. This technique is especially useful for observing changes in nanomaterials over time and while exposed to treatments including temperature, stress, etc. In energy applications, MD simulations have addressed itself to study the features of nanomaterials in batteries and supercapacitors such as ion diffusion and charge transport mechanism. For example, new insights of lithium ion transport in nanostructures electrodes may offer important information for increasing the performance of batteries using MD simulations (Baker et al., 2023). Moreover, through integration of MD and DFT, accurate predictions of nanomaterial properties can be achieved to systematically design the appropriate functional material.

Machine Learning Techniques

The application of computationally modeling coupled with machine learning holds new potential for improving nanomaterials. As known, ML algorithms can work with a huge amount of data derived from simulations and find wear patterns and correlations that could be unknown to an analyst. By following this approach, the process of identification of highly performing nanomaterials can be accelerated along with the design process of applications within the energy sector (Xu et al., 2023; Winkler, 2022). For instance, machine learning techniques have been applied to determine the properties of new nanomaterials by analyzing the structure of atomic cells to identify the best composition of the cells when used for a particular application. Moreover, ML can be used in the design of nanomaterials by evaluating their potential in energy storage and conversion systems, which will shorten the design cycle time (Chen & Mou, 2023).

Design Optimization

The perfectibility of nanomaterials is paramount in achieving improved performance of these materials impacting technologies in the energy domain within energy storage as well as energy conversion domains. This optimization process is facilitated by computational methods as well as simulation techniques through which researchers are able to forecast mechanical properties, examine efficiency and ultimately chart development of nanomaterials suited for application in particular uses. In this section, the authors elaborate on the role of the design optimization within the design process and present computational approaches and techniques.

Importance of Design Optimization

Design optimization deals with the systematic alteration of material characteristics and constructs to obtain optimal performance characteristics in energy applications. This process is necessary for producing functional nanomaterials that can be used to fulfill the rising requirements of energy technologies including batteries, supercapacitors, and fuel cell applications. The choice of nanomaterials and their design provide opportunities to increase energy density, electrical power, stability, and, in general, efficiency.

- ***Performance Enhancement:*** Incorporating proper nanomaterials for these applications can enhance many aspects connected with their electrochemical application. For example, the creation of nanostructured electrodes with features such as enlarged surface area and adjustable porosity may improve the transport of ions and capacity of charge accumulation in batteries and supercapacitors (Niu et al., 2023). Likewise, engineering the composition structure and morphology of the electrocatalysts can enhance their efficiency and stability in fuel cell and water splitting technologies.
- ***Cost-Effectiveness:*** Design optimization can also enable the reduction of costs by defining other materials and synthesis methods which can be employed. For example, through optimization strategies the manufacturing cost of fuel cells could be reduced while keeping the systems' performance optimal through the discovery of efficient non-precious metal catalysts (Romano & Johnson, 2022).
- ***Sustainability:*** The application of nanotechnology can improve the sustainability of nanomaterials by decreasing the environmental footprint connected with their synthesis and utilization. Thus, considering environmentally friendly processes and eco-friendly materials, the authors can design and create nanomaterials that will meet the sustainability requirements (S et al., 2023).

Computational Methods for Design Optimization

- ***Machine Learning:*** ML methods have particularly been adopted in the enhancement of nanomaterials because of their efficiency in determining trends in large sets of data that would otherwise be difficult to detect. Structural data of the nanomaterials may be used to estimate the properties of those nanomaterials which makes it easier to determine the right composition and configuration of the nanomaterial that are suitable for certain applications (Varsou &

Sarimveis, 2023). For instance, ML has been used to facilitate the design of porosity in electrodes of batteries to enhance its electrochemical efficiency.

- ***DFT:*** DFT is a highly significant computational tool applied in the study of the electronic structure of nanomaterials. Therefore, DFT can be used to optimize the electronic properties and predict the stability of different structures; as a result, design opportunities can be provided for the creation of new materials for energy storage and conversion applications (Bu et al., 2023). DFT also allows for estimating the adsorption energies of the reactants on the catalyst surface that can be useful for the improvement of electrocatalysts of the fuel cells and water splitting (Ha et al., 2022).
- **MD:** MD simulations embrace understanding of the time response of nanomaterials at the atomic level. Over time, MD can reveal certain structural and mechanical features of nanocomposite based on the position of atoms and molecules (Haas et al., 2024). For practical uses of nanomaterials such as in batteries and supercapacitors and where mechanical strength plays a critical role this information is vital.
- **Inverse Design Approaches:** Inverse design techniques are characterized by defining target properties and then finding structures with those properties with the help of relevant computational algorithms. This approach can therefore go a long way in ensuring that more new nanomaterials with better performance characteristics are developed more frequently. For instance, researchers have integrated inverse design concepts to design nanomaterials with definite catalytic capacity for fuel cells, minimizing the design iterations and reduced experimental trial and error.

FUTURE DIRECTIONS AND EMERGING TRENDS

The field of nanomaterials is advancing ever more quickly due to the ongoing advancements in material properties on the nanoscale, environmentally friendly synthesis techniques, and incorporations into novel technologies. This section explains these trends and their relevance to energy application.

Advancements in Material Performance

The use of nanomaterials is experiencing remarkable improvement adding to its efficiency in energy applications. Current research efforts include continuing improvement in properties as well as materials discovery and synthesis of its constituents. This section describes the recent advancements with advanced predictions of the future of nanomaterial performance based on energy storage and conversion systems.

Innovations in Nanomaterials for Energy Applications

Contemporary research activities have shown that energy conversion efficiency could be greatly enhanced with the help of a variety of two dimensional materials TMDs and graphene. For instance, TMDs have been designed to act as excellent photocatalysts for solar hydrogen peroxide synthesis, an efficient solar material which demonstrates how solar energy can be used to drive chemical reactions (Yang et al., 2024). This constant development in material science makes it easier is a significant key towards improving these 2D photocatalyst materials' efficiency and performance that will make them useful in green chemistry and renewable energy conversion. New successes in lithium-ion batteries energy storage are due to development of new nanomaterials. Innovations have been enhanced through increased research on the microstructure of electrode materials and integration of intelligent battery management systems (BMS) (Boopathy et al., 2023; Zhao et al., 2024). Such innovations demonstrate that material engineering is a crucial factor towards improving performance in energy storage systems.

Sustainable and Green Synthesis of Nanomaterials

Synthesis of nanomaterials in an environmentally friendly manner has been increasingly emphasized due to the demand for energy sustainability. In general, previous synthesis approaches use hazardous reagents and utilise high amounts of energy posing problems on the environment. Biomimetic and biotechnological approaches of nanosynthesis include the use of plant extracts or biomass to synthesize nanomaterials (Solomon et al., 2024;). For instance, biomass derived materials for the synthesis of nanocomposites have presented a grand opportunity to reduce cost of production while improving the performance in energy related applications. Another research focus is the synthesis of sustainable supramolecular polymers targeted at enhancement of the materials' mechanical performance and stability through possible synthesis controlling methods (Wang et al., 2024). This accords not only with approaches to lessen the side effects of nanomaterial production on the environment but also with strategies to further improve the properties of the end products.

Integration with Emerging Technologies

Another promising area in which the utilization of nanomaterials can significantly enhance the developments of energy systems is linked with the integration of the latter with AI and IoT. It might be useful to apply AI to data sets created during the synthesis and characterization of nanomaterials, which can help establish which of these materials and structures are best suited to which tasks (Zhang et al., 2022; Boopathy et al.,2024). Artificial neural networks can be employed for the structure-property relation wherein the behavior of the nanomaterial can be estimated with respect to structure and which can be used to develop efficient materials for energy storage and conversion (Bonthula et al., 2023). Furthermore, many researchers suggest that incorporation of nanomaterials with IoT will enable emergence of highly sophisticated energy management systems that aspects energy consumption in a real time basis. For instance, using nanomaterial-based sensors in energy systems, it is possible to monitor their performance and efficiency, perform preventive maintenance and increase energy levels (Khosravi et al., 2024). This combination of nanomaterials and these emergent technologies is poised to further advance energy application breakthroughs in existing and new ways.

Sustainable and Green Synthesis of Nanomaterials

The preparation of nanomaterials has been done using processes that are both expensive and some of which are hazardous to the environment. But there is a tendency in recent years to use the methods and reagents that are safe for the environment and non-hazardous to health. In this section, the author provides an overview of the latest trends of green synthetic processes as well as their advantages and possible uses in energy sciences.

Green Synthesis Methods

Green synthesis of nanomaterials involves employments of plant extracts, microbes, biopolymers and other bio- degradable materials for the synthesis of nanoparticles. These methods which includes green chemistry synthesis usually demand a lower energy input, fewer toxic chemicals are employed and occupy less space compared to conventional synthesis processes (Jafarzadeh et al., 2023). For example, plant mediated synthesis of metal nanoparticles is growing in popularity because of the presence of various phytochemicals which perform the role of reducing as well as stabilizing agents. Not only does this approach reduce the environmental impact of the method, but it also increases the biocompatibility of the obtained nanoparticles for biomedical purposes (Malini et al., 2022). This trend can be exemplified with the chitosan fungal beads synthesis of silver nanoparticles where the process is environmentally friendly and reusable which enables multiple synthesis without marked degradation of the beads (Hermosilla et al., 2023).

Advantages of Green Synthesis

First of all, these methods show a tendency to be less costly bearing in mind that most of them use easily available natural products and consume less energy. Second, the nanomaterials produced in the process demonstrate higher performance characteristics, including increased stability and biocompatibility which are especially useful in sectors including medicine and environmental protection (Petu et al., 2022). Furthermore, green synthesis techniques can make nanomaterials which are multipurpose and can be easily designed for one use or the other. For instance, carbon nanomaterials produced via green approaches have been applied in energy storage and sensing because of their appealing optical and electrical characteristics (Joseph, 2022).

Challenges and Future Directions

However, there are difficulties on how these green synthesis methods can be applied in large scale in industries. There are concerns such as the ability to replicate the experiment uniformly, compliance to set industry standards and an important concern is whether a green synthesis requires complete understanding of the underlying process, the mechanisms involved (Xu et al., 2023; Xu et al., 2023). Moreover, green synthesis methods are relatively safer but the life cycle of nanomaterials are not fully researched to understand the environmental impacts for sustainability (Wang et al., 2023). Further studies should be directed towards the improvement in yield, selectivity and reliability of the green synthesized processes. If efficient characterization techniques are incorporated, then the benefit of green approaches in fabricating nanomaterials can be further understood to advance energy related technologies and others (Zhou et al., 2023).

Integration with Emerging Technologies

The advent of combining nanomaterials with other technologies is enhancing several fields, especially in energy. The opportunities are realised in the context of AI, the IoT, and advanced sensing systems that this section outlines in relation to smarter energy systems with nanomaterials.

Nanomaterials and Artificial Intelligence

Integration of nanomaterials with AI trends towards improvement of performance in energy oriented applications. The data from the synthesis and characterization of nanomaterials can be used by machine learning algorithms to discover a synthesis of nanomaterials that are ideal for the desired application (Chen et al., 2023). For example, using AI, it is possible to estimate the Nature features of nanomaterials and thus, support the development of materials for energy storage and conversion (Britto et al., 2024). Not to mention, the optimization can also be efficient for creating nanomaterials by using AI optimization techniques that can save time on experiments. Thus, by using AI to predict and analyze the behavior of nanomaterials, the assessment of best performers for energy applications would be faster (Wang et al., 2023). It also enriches the study of these materials and their conjunctions, as well as their functioning in more natural conditions at a faster rate.

Nanomaterials and the Internet of Things (IoT)

The combination of nanomaterials with IoT technologies will also be predictive of hydraulic systems that will assist in monitoring the overall usage of energy in real-time. Wireless sensors used in energy systems allow for the accumulation of useful data for performance and efficiency, as well as for assessment of the state of nanomaterials (Bu et al., 2024). For instance, flexible sensors fabricated from nanomaterials are suitable for sustained health checking within wearables, therefore, providing real-time analysis (Wang et al., 2023). In the same way, nanomaterials can also increase the efficiency and service life of energy storage devices including batteries and supercapacitors. These systems can be developed to incorporate IoT features to help users understand the status of these systems, and make appropriate decisions regarding encouraging energy use or conservation (Jiang et al., 2024). The interaction between nanomaterials and IoT technologies shall continue to fuel innovation in the energy domain and subsequently produce better efficient energy related systems.

Advanced Sensing and Biosensing Applications

These proton exchange membranes (PEMs) are widely used in modern sensor technologies such as enzymatic biosensors and other biosensors in environmental management and health care products. The sort of surface area and its tunability as well as other properties of nanomaterials facilitate the design of highly sensitized and selective sensorable (Sen et al., 2021). For example, nanomaterial based biosensors are used to identify pathogens in water and food through diagnosis which is fast and accurate in enhancing public health (Valenzuela-Amaro et al., 2023). New trends of research appeared during the work, namely paper-based microfluidic devices and smart materials improve the performance of nanomaterial-based sensors. These innovations mean that many analyses can be identified at once, enhancing the effectiveness of tracking systems. The convergence of nanomaterials with enhanced sensing devices is anticipated to create novel works in many sectors such as environmental science, healthcare and food security.

Sustainable Energy Solutions

The integration of nanomaterials is also useful for energy systems under the sustainable development agenda. For instance, the synthesis techniques when manufacturing nanomaterials are environmentally friendly, and product functionality remains high. Furthermore, the safety and performance of renewable energy systems including; solar cells and fuel cells can also be improved through nanomaterials thus supporting the shift to cleaner energies (Ngo et al., 2022). The benefits of nanomaterial applications in altering building structures to be energy efficient is enhanced significantly when combined with new technologies. For example, application of nanomaterials in energy storage systems that improve the efficiency of the system, and IoT for energy efficient systems that improve energy usage patterns (Zhang et al., 2023).

Enhancing the Scope: Addressing Scalability, Economic Viability, Long-Term Stability, and Environmental Impact

Scalability of Nanomaterials for Energy Applications

A big challenge for large scale action of nanomaterial applications is scalability. Given that manufactures of such devices often require stringent synthesis conditions that are difficult to replicate at larger scales, specific synthesis methods such as chemical vapour deposition (CVD) and sol-gel are needed. Besides, the variation between the batches may also influence the uniformity of the performance. Scalability potentials can be provided by the innovative fabrication methods such as roll to roll processing or template assisted synthesis. However, development of such methods for mass production is of utmost importance, while preserving material integrity.

One of the most prominent scalability challenge is the fabrication of 2D materials like transition metal dichalcogenides (TMDs). However, they lack good electrochemical properties that make their large scale production difficult due to complex and time consuming synthesis. To address these challenges the exfoliation and environmentally friendly synthesis techniques that decrease use of toxic reagent need to become an economical high yield production techniques.

Economic Viability of Nanomaterial-Based Energy Solutions

Economic obstacles arise in the way of nanomaterials from being more broadly applied in more energy storage and conversion systems. Significant concerns relate to the costs of raw materials, synthesis, and technologies in use currently. One example is the use of a more affordable alternatives to platinum group metals in fuel cell catalysts and lithium based nanomaterials such as cobalt iron catalysts and sodium ion batteries.

For the economic viability of the industry, nanomaterials should be focused on recycling and reusability strategies. In addition, alloying and nanocrystalline materials can save on financial burdens induced by materials dependence on rare and expensive materials. Another line of policy would be to promote subsidies for green energy technologies based on nanomaterials by reducing initial costs.

Long-Term Stability of Nanomaterials in Energy Devices

Stability of nanomaterials over long periods of time is essential to commercialization of these materials. While numerous nanostructured materials may have excellent initial performance, their structural instability and electrode dissolution, and ion trapping, may lead to collapse over time. One example would be use of silicon-based anodes in lithium ion batteries which have a high theoretical capacity but high volume changes throughout charge and discharge cycling, and electrode fragmentation.

The solution to these problems has been addressed by researchers who are developing nanocomposites with protective coatings or matrix improvements. Encapsulating active material into a robust carbon matrix, using self healing polymers and maximizing the stability of the electrolyte are some of the ways in which stability can be improved. In addition, real time of material degradation and device life time can be integrated with sensors on the devices.(Wang et al., 2023)

Environmental Impact and Sustainability of Nanomaterials

Nanomaterials do offer considerable benefits in the energy application and it is necessary to take a thorough account of their environmental impact. Nanomaterials production commonly uses hazardous chemicals and does so with high energy consumption, which raises related issues about carbon emissions and waste management.

Also, there is a risk that nanoparticles leached out into ecosystems cause harm to environmental and human health.

Lifecycle assessments (LCA) of nanoparticles for nanomaterials helps in thorough evaluation to assess environmental impacts associated with raw material extraction to its disposal. Green synthesis of nanomaterials where carbon sources are derived from biomass and solvent free processes have been an area of initiatives in efforts to improve the sustainability of nanomaterials. Also, recycling and recovering nanomaterials from used batteries and fuel cell promote environmental sustainability by reducing waste and conserving resource.

Lifecycle Analysis and Commercialization Challenges

The lifecycle analysis of nanomaterials is needed to determine its impact in energy systems. Raw material sourcing, manufacturing, operational efficiency and disposal are all considered when the lifecycle assessment is comprehensive. One obstacle of commercialization is there are no standardized testing protocols for the evaluation of nanomaterial. Unlike conventional materials, properties of nanomaterials are varied depending on synthesis method and surface modifications.

Furthermore, there are very few large scale production facilities for the nanomaterial supply chain. In order to enhance commercialization, investments are needed in infrastructure and academics, industry and policymakers work together. However, while addressing these challenges will contribute to broader nanomaterial implementation for advanced energy storage and conversion technologies, they will also have largely achieved these goals in the context of batteries, fuel cells, and solar cells.

CONCLUSION

Review of nanomaterials in energy expansion shows prolific progress and prospective outcomes in numerous fields of energy, storage, conversion and green synthesis. This discussion has revealed the importance of nanomaterials in improving the effectiveness of technologies including Li-ion batteries, super capacitors, fuel cells, and hydrogen generation by water splitting. Due to features such as large surface area, modifiable electronic states, and catalytic ability, very significant enhancements in efficiency, capacity, and stability can be achieved. Various synthesis techniques used in the fabrication of nanomaterials for instance electrospinning, sol-gel methods, chemical vapor deposition among others afford the researchers an opportunity to synthesize materials with properties that fit various uses. Yet there are difficulties with broad application of these methods in terms of cost, stability, environmental acceptability, and legislation. Solving these problems is crucial for the effective commercial application of nanomaterials in energy technologies. However, the combination of nanomaterials with other new technologies like Artificial Intelligence and the Internet of Things presents new possibilities for improving energy systems and improving their efficiency. It means that applications of nanotechnology and these advanced technologies will result in intelligent, effective solutions to enhance energy efficiency in the context of sustainability objectives. However, as the research progresses further, the advent of new and efficient nanostructure and its utilisation in energy systems will greatly pinpoint the precise energy issues across the globe. Extension of the thought to sustainable and green synthesis methods adds to this trend by promoting the sustainable synthesis of nanomaterials without degrading the conditions of the environment. In conclusion, nanomaterials' future in energy related fields is promising as the current and on-going research is expected to pave the way for enhanced energy products that relate to efficiencies, sustainability and affords economic viability. It is crucial to point out that besides researchers, other industry players and regulatory authorities will need to augment their cooperation as they unlock the embedded possibilities of nanomaterials for advancing sustainable energy systems.

REFERENCES

Abdulwahab, K. O., Khan, M. M., & Jennings, J. R. (2023). Doped ceria nanomaterials: Preparation, properties, and uses. *ACS Omega*, *8*(34), 30802–30823. DOI: 10.1021/acsomega.3c01199 PMID: 37663502

Ahmed, S. F., Mofijur, M., Rafa, N., Chowdhury, A. T., Chowdhury, S. N., Nahrin, M., Islam, A. B. M. S., & Ong, H. C. (2022). Green approaches in synthesising nanomaterials for environmental nanobioremediation: Technological advancements, applications, benefits and challenges. *Environmental Research*, *204*, 111967. DOI: 10.1016/j.envres.2021.111967 PMID: 34450159

Alamgeer, T., Tahir, M., Sarker, M. R., Ali, S., Ibraheem, , Hussian, S., Ali, S., Imran Khan, M., Khan, D. N., Ali, R., & Mohd Said, S. (2023). Polyaniline/ZnO Hybrid Nanocomposite: Morphology, Spectroscopy and Optimization of ZnO Concentration for Photovoltaic Applications. *Polymers*, *15*(2), 363. DOI: 10.3390/polym15020363 PMID: 36679244

Alfaawaz, Y. F., Alamri, R., Almohsen, F., Shabab, S., Alhamdan, M. M., Ahdal, K. A., & Abduljabbar, T. (2022). Adhesive bond integrity of experimental zinc oxide nanoparticles incorporated dentin adhesive: An sem, edx, μtbs, and rheometric analysis. *Scanning*, *2022*, 1–9. DOI: 10.1155/2022/3477886 PMID: 36016673

Amiri, M., Dondapati, J. S., Quintal, J., & Chen, A. (2022). Sodium hexa-titanate nanowires modified with cobalt hydroxide quantum dots as an efficient and cost-effective electrocatalyst for hydrogen evolution in alkaline media. ACS Applied Materials &Amp. *Interfaces*, *14*(35), 40021–40030. DOI: 10.1021/acsami.2c11310 PMID: 36006793

Aydin, C., & Yildiz, O. E. (2023). Synthesis, morphological and structural properties of graphene oxide based hybrid supercapacitive go/sn nanocomposites with different solvents. *Journal of Physical Chemistry and Functional Materials*, *6*(2), 86–91. DOI: 10.54565/jphcfum.1393954

Bae, C. W., Chinnamani, M. V., Lee, E. H., & Lee, N. (2022). Stretchable non-enzymatic fuel cell-based sensor patch integrated with thread-embedded microfluidics for self-powered wearable glucose monitoring. *Advanced Materials Interfaces*, *9*(20), 2200492. Advance online publication. DOI: 10.1002/admi.202200492

Baker, D., Dowling, Q., Park, Y., Gerstenmaier, N., Yang, E., Wargacki, A., & King, N. (2023). Hierarchical design of pseudosymmetric protein nanoparticles. DOI: 10.21203/rs.3.rs-3074553/v1

Barhoum, A., García-Betancourt, M. L., Jeevanandam, J., Hussien, E. A., Mekkawy, S. A., Mostafa, M., & Bechelany, M. (2022). Review on natural, incidental, bioinspired, and engineered nanomaterials: History, definitions, classifications, synthesis, properties, market, toxicities, risks, and regulations. *Nanomaterials (Basel, Switzerland)*, *12*(2), 177. DOI: 10.3390/nano12020177 PMID: 35055196

Boateng, E., van der Zalm, J., Burns, N., Chow, D., Kycia, S., & Chen, A. (2023). Tunable electrochemical hydrogen uptake and release of nitrogen-doped reduced graphene oxide nanosheets decorated with pd nanoparticles. *ACS Applied Nano Materials*, *6*(18), 17311–17323. DOI: 10.1021/acsanm.3c03763

Bonthula, S., Bonthula, S. R., Pothu, R., Srivastava, R. K., Boddula, R., Radwan, A. B., & Al-Qahtani, N. (2023). Recent advances in copper-based materials for sustainable environmental applications. *Sustainable Chemistry*, *4*(3), 246–271. DOI: 10.3390/suschem4030019

Boopathy, G., Balaji, G., Bhoominathan, R., Bejaxhin, A. B. H., & Ramanan, N. (2024). Enhancing Mechanical Characteristics and Cost-Efficiency of Composite Materials Through Hybridization and Nanoparticle Incorporation. In M. Sumesh, J. M. R. S. Tavares, S. C. Vettivel, & M. O. Oliveira (Eds.), 2nd International Conference on Smart Sustainable Materials and Technologies (ICSSMT 2023) (pp. 83–89). Springer Nature Switzerland. DOI: 10.1007/978-3-031-49826-8_11

Boopathy, G., Gurusami, K., Vijayakumar, K. R., & Kumar, M. H. (2024). Nanocoatings in Medicine: Revolutionizing Healthcare Through Precision and Potential. In Garg, R., & Anjum, A. (Eds.), (pp. 271–295). Advances in Chemical and Materials Engineering. IGI Global., DOI: 10.4018/979-8-3693-3136-1.ch009

Boopathy, G., Sathish, S., & Kumar, M. H. (2024). Introduction to precision manufacturing for micro- and nanofabrication. In Kalita, K., Ghadai, R. K., & Davim, J. P. (Eds.), *Microfabrication and Nanofabrication* (pp. 1–24). De Gruyter., DOI: 10.1515/9783111335537-001

Britto, J. S. J., Guan, X., Tran, T. K. A., Lei, Z., Bahadur, R., Patel, V., Zhang, X., Wong, S. L., & Vinu, A. (2024). Emerging multifunctional carbon-nanomaterial-based biosensors for cancer diagnosis. *Small Science*, *4*(3), 2300221. Advance online publication. DOI: 10.1002/smsc.202300221 PMID: 40212699

Bu, Q., Li, P., Xia, Y., Hu, D., Li, W., Shi, D., & Song, K. (2023). Design, synthesis, and biomedical application of multifunctional fluorescent polymer nanomaterials. *Molecules (Basel, Switzerland)*, *28*(9), 3819. DOI: 10.3390/molecules28093819 PMID: 37175229

Bu, Y., Cabulong, R. B., & Kim, B. S. (2024). Plant extract-based liquid phase exfoliation enables one-step green production of two-dimensional heterostructure nanohybrids capable of dramatic improvement in polymer properties. *Green Chemistry, 26*(6), 3488–3506. DOI: 10.1039/D3GC04448C

Cao, L., Chen, J., Pang, J., Qu, H., Liu, J., & Gao, J. (2024). Research progress in enzyme biofuel cells modified using nanomaterials and their implementation as self-powered sensors. *Molecules (Basel, Switzerland), 29*(1), 257. DOI: 10.3390/molecules29010257 PMID: 38202838

Cao, Z., Dong, F., Pan, J., Xia, W., Hu, J., & Xu, X. (2022). Heteroatomic platinum–cobalt synergetic active centers with charge polarization enable superior hydrogen evolution performance in both acid and base media. *ACS Applied Energy Materials, 5*(2), 1496–1504. DOI: 10.1021/acsaem.1c02877

Chen, D., Wang, X., Zhu, Z., Zhan, C., Lin, X., Bu, L., & Huang, B. (2023). Highly selective synthesis of monoclinic-phased platinum–tellurium nanotrepang for direct formic acid oxidation catalysis. *Journal of the American Chemical Society, 145*(28), 15393–15404. DOI: 10.1021/jacs.3c03317 PMID: 37429024

Chen, J., Fei, R., & Xie, Y. (2023). Fabrication, microstructures and sensor applications of highly ordered electrospun nanofibers: A review. *Materials (Basel), 16*(9), 3310. DOI: 10.3390/ma16093310 PMID: 37176192

Chen, J., & Mou, C. (2023). The hie-fdtd method for simulating dispersion media represented by drude, debye, and lorentz models. *Nanomaterials (Basel, Switzerland), 13*(7), 1180. DOI: 10.3390/nano13071180 PMID: 37049274

Chen, Y., Jiang, H., Liu, X., & Wang, X. (2023). Engineered electrochemiluminescence biosensors for monitoring heavy metal ions: Current status and prospects. *Biosensors (Basel), 14*(1), 9. DOI: 10.3390/bios14010009 PMID: 38248386

Chetana, S., Abdah, M. A. A. M., Thakur, V. N., Gowda, M. G., Choudhary, P., Sriramoju, J. B., & Khalid, M. (2023). Progress and prospects of MXene-based hybrid composites for next-generation energy technology. *Journal of the Electrochemical Society, 170*(12), 120530.

Chihai, R., Săracu, A. F., & Ungureanu, C. V. (2022). Green synthesis of metal nanoparticles using microalgae: A review. The Annals of "Dunarea de Jos" University of Galati. Fascicle IX. *Metallurgy and Materials Science, 45*(3), 16–19.

Cui, B., Xiao, Z., Cui, S., Hao, S., Liu, S., Gao, X., & Li, G. (2023). Lithiated phosphoryl cellulose nanocrystals enhance cycling stability and safety of quasi-solid-state lithium metal batteries. *ACS Applied Materials & Interfaces*, *15*(35), 41537–41548. DOI: 10.1021/acsami.3c08559 PMID: 37671463

Cui, Z. (2024). Enhancing electrochemical performance of lithium-ion batteries with diverse nanomaterials. *Journal of Physics: Conference Series*, *2798*(1), 012002. DOI: 10.1088/1742-6596/2798/1/012002

Dahl, J. C., Niblett, S., Cho, Y., Wang, X., Zhang, Y., Chan, E. M., & Alivisatos, A. P. (2023). Scientific machine learning of 2d perovskite nanosheet formation. *Journal of the American Chemical Society*, *145*(42), 23076–23087. DOI: 10.1021/jacs.3c05984 PMID: 37847242

Dai, J., Zhu, F., Murugesan, B., Zhang, Q., Zhou, X., Ni, W., & Cai, Y. (2023). Synergistic v5s8/cos nanoparticles-embedded carbon nanofiber as binder-free flexible anode for sodium-ion batteries. *Ionics*, *30*(1), 217–227. DOI: 10.1007/s11581-023-05268-5

De, S., Acharya, S., Maity, C., & Nayak, G. C. (2023). Boron nitride/ti3c2tx mxene nanosheet/ws2 nanostructure ternary composites for all-solid-state flexible asymmetric supercapacitors. *ACS Applied Nano Materials*, *6*(13), 11175–11186. DOI: 10.1021/acsanm.3c01202

de Haas, R. J., Brunette, N., Goodson, A., Dauparas, J., Yi, S. Y., Yang, E. C., Dowling, Q., Nguyen, H., Kang, A., Bera, A. K., Sankaran, B., de Vries, R., Baker, D., & King, N. P. (2024). Rapid and automated design of two-component protein nanomaterials using proteinmpnn. *Proceedings of the National Academy of Sciences of the United States of America*, *121*(13), e2314646121. Advance online publication. DOI: 10.1073/pnas.2314646121 PMID: 38502697

Deshmukh, S., Kathiresan, M., & Kulandainathan, M. (2022). A review on biopolymer-derived electrospun nanofibers for biomedical and antiviral applications. *Biomaterials Science*, *10*(16), 4424–4442. DOI: 10.1039/D2BM00820C PMID: 35820295

Du, Y., Yang, Z., Kang, S., Yu, D., Chen, X., & Shao, J. P. (2023). A sequential electrospinning of a coaxial and blending process for creating double-layer hybrid films to sense glucose. *Sensors (Basel)*, *23*(7), 3685. DOI: 10.3390/s23073685 PMID: 37050745

Gavali, A. (2023). An overview of the synthesis of nanomaterials. Materials Research Foundations, 19-53. DOI: 10.21741/9781644902370-2

Gharpure, S., Yadwade, R., & Ankamwar, B. (2022). Non-antimicrobial and non-anticancer properties of zno nanoparticles biosynthesized using different plant parts of bixa orellana. *ACS Omega*, *7*(2), 1914–1933. DOI: 10.1021/acsomega.1c05324 PMID: 35071882

Ghosh, A., Pal, S., & Sarkar, P. (2022). Rational design of two-dimensional porous boron phosphide as efficient cathode material for li and na ion batteries: A first-principles study. *The Journal of Physical Chemistry. C, Nanomaterials and Interfaces*, *126*(11), 5092–5100. DOI: 10.1021/acs.jpcc.1c09966

Gowtham, M., N, S. K., Sivakumar, C., & Mohanraj, K. (2022). Nanostructured and nanocomposite tungsten oxide electrodes for electrochemical energy storage: a short review. NanoNEXT, 3(2), 1-7. DOI: 10.54392/nnxt2221

Goyal, M., Singh, K., & Bhatnagar, N. (2024). Applications of nanomaterials for enhanced performance, and sustainability in energy storage devices: A review. *ChemistrySelect*, *9*(22), e202400543. Advance online publication. DOI: 10.1002/slct.202400543

Gulen, M., dunya, H., tas, R., & Ayaz, V. E. (2023). Design of mns@mwcnt nanocomposite cathode for ultra-high efficient supercapacitors. International Conference on Pioneer and Innovative Studies, 1, 395-398. DOI: 10.59287/icpis.862

Ha, M., Hajibabaei, A., Pourasad, S., & Kim, K. S. (2022). Sparse gaussian process regression-based machine learned first-principles force-fields for saturated, olefinic, and aromatic hydrocarbons. *ACS Physical Chemistry Au*, *2*(3), 260–264. DOI: 10.1021/acsphyschemau.1c00058 PMID: 36855568

Hafeez, M., Ghazal, A., Khan, J., Ahmad, P., Khandaker, M. U., Osman, H., & Alamri, S. (2022). Eucalyptus globulus extract-assisted fabrication of copper oxide/zinc oxide nanocomposite for photocatalytic applications. *Crystals*, *12*(8), 1153. DOI: 10.3390/cryst12081153

Haghighat Bayan, M. A., Rinoldi, C., Rybak, D., Zargarian, S. S., Zakrzewska, A., Cegielska, O., Põhako-Palu, K., Zhang, S., Stobnicka-Kupiec, A., Górny, R. L., Nakielski, P., Kogermann, K., De Sio, L., Ding, B., & Pierini, F. (2024). Engineering surgical face masks with photothermal and photodynamic plasmonic nanostructures for enhancing filtration and on-demand pathogen eradication. *Biomaterials Science*, *12*(4), 949–963. DOI: 10.1039/D3BM01125A PMID: 38221844

Harish, V., Tewari, D., Gaur, M., Yadav, A. B., Swaroop, S., Bechelany, M., & Barhoum, A. (2022). Review on nanoparticles and nanostructured materials: Bio-imaging, biosensing, drug delivery, tissue engineering, antimicrobial, and agro-food applications. *Nanomaterials (Basel, Switzerland)*, *12*(3), 457. DOI: 10.3390/nano12030457 PMID: 35159802

He, P., Yang, G., Zhu, D., Kong, H., Corrales-Ureña, Y. R., Ciacchi, L. C., & Wei, G. (2022). Biomolecule-mimetic nanomaterials for photothermal and photodynamic therapy of cancers: Bridging nanobiotechnology and biomedicine. *Journal of Nanobiotechnology*, *20*(1), 483. Advance online publication. DOI: 10.1186/s12951-022-01691-4 PMID: 36384717

Hermosilla, E., Díaz, M., Vera, J., Contreras, M. J., Leal, K., Salazar, R., Barrientos, L., Tortella, G., & Rubilar, O. (2023). Synthesis of antimicrobial chitosan-silver nanoparticles mediated by reusable chitosan fungal beads. *International Journal of Molecular Sciences*, *24*(3), 2318. DOI: 10.3390/ijms24032318 PMID: 36768640

Ho, H., Chu, H., Huang, Y., Tsai, D., & Lee, C. (2023). Polypyrrole-coated copper@ graphene core-shell nanoparticles for supercapacitor application. *Nanotechnology*, *34*(12), 125401. DOI: 10.1088/1361-6528/acad87 PMID: 36542854

Isikgor, F. H., Zhumagali, S., Merino, L. V. T., Bastiani, M. D., McCulloch, I., & Wolf, S. D. (2022). Molecular engineering of contact interfaces for high-performance perovskite solar cells. *Nature Reviews. Materials*, *8*(2), 89–108. DOI: 10.1038/s41578-022-00503-3

Jafarzadeh, S., Nooshkam, M., Zargar, M., Garavand, F., Ghosh, S., Hadidi, M., & Forough, M. (2023). Green synthesis of nanomaterials for smart biopolymer packaging: Challenges and outlooks. *Journal of Nanostructure in Chemistry*, *14*(2), 113–136. DOI: 10.1007/s40097-023-00527-3

Jawhari, A. H. (2022). Novel nanomaterials for hydrogen production and storage: Evaluating the futurity of graphene/graphene composites in hydrogen energy. *Energies*, *15*(23), 9085. DOI: 10.3390/en15239085

Jiang, J., Cui, X., Huang, Y., Yan, D., Wang, B., Yang, Z., Chen, M., Wang, J., Zhang, Y., Liu, G., Zhou, C., Cui, S., Ni, J., Yang, F., & Cui, D. (2024). Advances and prospects in integrated nano-oncology. *Nano Biomedicine and Engineering*, *16*(2), 152–187. DOI: 10.26599/NBE.2024.9290060

Jiang, Z., Guan, L., Xu, X., Wang, E., & Wang, C. (2022). Applications of carbon dots in electrochemical energy storage. *ACS Applied Electronic Materials*, *4*(11), 5144–5164. DOI: 10.1021/acsaelm.2c01152

Jin, X., Zhao, J., Ge, Q., Jiang, N., Liu, M., Cong, H., & Zhang, Y. (2023). Carbon nanomaterials made from supramolecular hybridization of cucurbit[6]uril with phosphomolybdic acid for supercapacitors. *ACS Applied Nano Materials*, *6*(4), 2459–2471. DOI: 10.1021/acsanm.2c04737

Jing, C., Li, T., & Su, H. (2023). Application of carbon nanotechnology in conductive composites. *Applied and Computational Engineering*, *23*(1), 157–161. DOI: 10.54254/2755-2721/23/20230646

Joseph, N., & B, M. (2022). Green synthesized fluorescent nano-carbon derived from indigofera tinctora (l.) leaf extract for sensing of pb2+ ions. *ECS Transactions*, *107*(1), 15255–15262. DOI: 10.1149/10701.15255ecst

Joshi, K., Mistry, K., Tripathi, B., Chandra, P., Shinde, S., Kumar, M., Santola, D., Chokshi, H., & Gurrala, P. K. (2023). Mos2 nanostructures for solar hydrogen generation via membraneless electrochemical water splitting. *ACS Applied Electronic Materials*, *5*(3), 1461–1470. DOI: 10.1021/acsaelm.2c01415

Kaewsaenee, J., Singhaset, M. T., Roongraung, K., Kemacheevakul, P., & Chuangchote, S. (2023). Polymer-assisted coprecipitation synthesized zinc oxide nanoparticles and their uses for green chemical synthesis via photocatalytic glucose conversions. *ACS Omega*, *8*(46), 43664–43673. DOI: 10.1021/acsomega.3c05183 PMID: 38027380

Kang, B., Shin, S., Nam, K., Bae, J., Oh, J., Koo, S., Sohn, H., Park, S.-H., & Shin, W. H. (2023). Exfoliated nbse2 nanosheet@polypyrrole hybrid nanocomposites as a high performance anode of lithium-ion batteries. *Journal of Materials Chemistry. A, Materials for Energy and Sustainability*, *11*(35), 19083–19090. DOI: 10.1039/D3TA01335A

Khazraji, M., Wang, J., & Wei, S. (2022). Recent progress of anode protection in li–s batteries. *Energy Technology (Weinheim)*, *11*(1), 2200944. Advance online publication. DOI: 10.1002/ente.202200944

Khosravi, A., Zarepour, A., Iravani, S., Varma, R. S., & Zarrabi, A. (2024). Sustainable synthesis: Natural processes shaping the nanocircular economy. *Environmental Science. Nano*, *11*(3), 688–707. DOI: 10.1039/D3EN00973D

Kishore, S. C., Perumal, S., Atchudan, R., Alagan, M., Wadaan, M. A., & Baabbad, A. (2023). Recent advanced synthesis strategies for the nanomaterial-modified proton exchange membrane in fuel cells. *Membranes*, *13*(6), 590. DOI: 10.3390/membranes13060590 PMID: 37367794

Kothandam, G., Singh, G., Guan, X., Lee, J. M., Ramadass, K., Joseph, S., Benzigar, M., Karakoti, A., Yi, J., Kumar, P., & Vinu, A. (2023). Recent advances in carbon-based electrodes for energy storage and conversion. *Advancement of Science*, *10*(18), 2301045. Advance online publication. DOI: 10.1002/advs.202301045 PMID: 37096838

Kousik Kumaarr, R., Somasundaram Vinoth, K., Srikanth, H. V., & Boopathy, G. (2023). Influence of the load, sliding speed, and weight fractions on wear of a hybrid aluminium metal matrix composite. Journal of Polymer & Composites, 11(Special Issue 13). https://doi.org/DOI: 10.37591/JoPC

Kovalska, E., Wu, B., Liao, L., Mazánek, V., Marek, I., Lajaunie, L., & Sofer, Z. (2023). Electrochemical decalcification–exfoliation of two-dimensional siligene, sixgey: Material characterization and perspectives for lithium-ion storage. *ACS Nano*, *17*(12), 11374–11383. DOI: 10.1021/acsnano.3c00658 PMID: 37283557

Kumaarr, K. R., Boopathy, G., Bejaxhin, B. H., & Ramanan, N. (2023). Considerations on fractured edge dislocations of mechanical tested hand laid hemp flax GFRP hybrid polymer composites with SEM verifications. Journal of Polymer & Composites, 11(Special Issue 13). https://doi.org/DOI: 10.37591/JoPC

Kumar, N., Thakur, V. N., Muthukkumaran, K., Gajraj, V., Sen, A., Joshi, N. C., & Priyadarshi, N. (2022). Morphological reduction of fe3o4 by a single-step hydrothermal synthesis using 1d mno2 as a template and its supercapacitive behaviour. *CrystEngComm*, *24*(25), 4611–4621. DOI: 10.1039/D2CE00620K

Li, H., Jiao, Y., Davey, K., & Qiao, S. (2023). Data-driven machine learning for understanding surface structures of heterogeneous catalysts. *Angewandte Chemie International Edition*, *62*(9), e202216383. Advance online publication. DOI: 10.1002/anie.202216383 PMID: 36509704

Liu, H. (2023). Review on microfluidic technology based synthesis of fe-based nanoparticles for catalyst in fuel cell. *Academic Journal of Science and Technology*, *7*(2), 98–100. DOI: 10.54097/ajst.v7i2.11950

Liu, J. (2023). Application of nanomaterials in the negative electrode of lithium-ion batteries. *Applied and Computational Engineering*, *24*(1), 246–250. DOI: 10.54254/2755-2721/24/20230716

Liu, L., Jin, L., Xiao, Z., Fang, N., Lin, X., Ji, Y., Wang, Y., Li, Y., Huang, X., & Bu, L. (2024). Heterostructured pt-pbs nanobelt achieves remarkable direct formic acid oxidation catalysis. *Nano Letters*, *24*(26), 8162–8170. DOI: 10.1021/acs.nanolett.4c02133 PMID: 38904300

Lü, Q., Zhang, B., Zhang, L., Zhu, Y., & Gong, W. (2023). Monolayer asc5 as the promising hydrogen storage material for clean energy applications. *Nanomaterials (Basel, Switzerland), 13*(9), 1553. DOI: 10.3390/nano13091553 PMID: 37177096

Luo, Z., Guo, T., Wang, C., Zou, J., Wang, J., Dong, W., Li, J., Zhang, W., Zhang, X., & Zheng, W. (2022). Enhancing the efficiency of perovskite solar cells through interface engineering with mos2 quantum dots. *Nanomaterials (Basel, Switzerland), 12*(17), 3079. DOI: 10.3390/nano12173079 PMID: 36080116

Malini, S., Raj, K., Madhumathy, S., El-Hady, K. M., Islam, S., & Dutta, M. (2022). Bioinspired advances in nanomaterials for sustainable agriculture. Journal of Nanomaterials, 2022(1). DOI: 10.1155/2022/8926133

Masud, A., Aftabuzzaman, M., Zhou, H., Kim, S., Yi, J., Park, S. S., Kim, Y. S., & Kim, H. K. (2024). Chemically synthesized poly(3,4-ethylenedioxythiophene) conducting polymer as a robust electrocatalyst for highly efficient dye-sensitized solar cells. *Nanoscale, 16*(29), 13874–13884. DOI: 10.1039/D4NR00949E PMID: 38990512

Matthews, T., Mbokazi, S. P., Dolla, T. H., Gwebu, S. S., Mugadza, K., Raseruthe, K., Sikeyi, L. L., Adegoke, K. A., Saliu, O. D., Adekunle, A. S., Ndungu, P., & Maxakato, N. W. (2023). Electrocatalyst performances in direct alcohol fuel cells: Defect engineering protocols, electrocatalytic pathways, key parameters for improvement, and breakthroughs on the horizon. *Small Science, 4*(1), 2300057. Advance online publication. DOI: 10.1002/smsc.202300057 PMID: 40212622

Modi, K. H., Pataniya, P. M., & Sumesh, C. K. (2024). 2d monolayer catalysts: Towards efficient water splitting and green hydrogen production. *Chemistry (Weinheim an der Bergstrasse, Germany), 30*(23), e202303978. Advance online publication. DOI: 10.1002/chem.202303978 PMID: 38299695

Modugu, S., Schuster, J., & Shaik, Y. (2022). Synthesis and characterization of carbon fiber nanocomposite using titanium dioxide and silicon carbide nanomaterials. *Journal of Composites Science, 6*(10), 312. DOI: 10.3390/jcs6100312

Müller, H. (2023). Effects of nanomaterials on enhancing the performance of energy storage devices in germany. *European Journal of Physical Sciences, 6*(1), 12–22. DOI: 10.47672/ejps.1443

Mustapha, T., Misni, N., Ithnin, N. R., Daskum, A. M., & Unyah, N. Z. (2022). A review on plants and microorganisms mediated synthesis of silver nanoparticles, role of plants metabolites and applications. *International Journal of Environmental Research and Public Health, 19*(2), 674. DOI: 10.3390/ijerph19020674 PMID: 35055505

Nagappan, S., Duraivel, M., Park, N., Prabakar, K., & Park, K. H. (2022). Implementation of heteroatom-doped nanomaterial/core–shell nanostructure based electrocatalysts for fuel cells and metal-ion/air/sulfur batteries. *Materials Advances*, *3*(15), 6096–6124. DOI: 10.1039/D2MA00390B

Nair, A. K., Silva, C. D., & Amón, C. H. (2023). Enhanced alkali-ion adsorption in strongly bonded two-dimensional tis2/mos2 van der waals heterostructures. *The Journal of Physical Chemistry. C, Nanomaterials and Interfaces*, *127*(20), 9541–9553. DOI: 10.1021/acs.jpcc.3c01819

Ngo, N. M., Tran, H., & Lee, T. R. (2022). Plasmonic nanostars: Systematic review of their synthesis and applications. *ACS Applied Nano Materials*, *5*(10), 14051–14091. DOI: 10.1021/acsanm.2c02533

Niu, Z., Zhao, W., Wu, B., Wang, H., Lin, W., Pinfield, V., & Xuan, J. (2023). Π learning: A performance-informed framework for microstructural electrode design. *Advanced Energy Materials*, *13*(17), 2300244. Advance online publication. DOI: 10.1002/aenm.202300244

Nkele, A. C., Offiah, S. U., Chime, C. P., & Ezema, F. I. (2023). Review on advanced nanomaterials for hydrogen production. *IOP Conference Series. Earth and Environmental Science*, *1178*(1), 012001. DOI: 10.1088/1755-1315/1178/1/012001

Osmani, P. S. (2024). Synthesis and characterization of advanced inorganic nanomaterials for energy storage devices. *Journal for Research in Applied Sciences and Biotechnology*, *3*(1), 122–128. DOI: 10.55544/jrasb.3.1.20

Pan, F. (2023). Application of nano-carbon materials in battery electrode. *Highlights in Science Engineering and Technology*, *43*, 597–601. DOI: 10.54097/hset.v43i.7507

Park, H., Majumdar, S., Zhang, X., Kim, J., & Smit, B. (2024). Inverse design of metal–organic frameworks for direct air capture of co2via deep reinforcement learning. *Digital Discovery*, *3*(4), 728–741. DOI: 10.1039/D4DD00010B

Rathnasekara, R., & Hari, P. (2022). Enhancing the efficiency of dye-sensitized solar cells (dsscs) by nanostructured ag-doped zno electrodes. *ChemistrySelect*, *7*(29), e202200830. Advance online publication. DOI: 10.1002/slct.202200830

Rehman, J., Eid, K., Ali, R., Fan, X., Murtaza, G., Faizan, M., Laref, A., Zheng, W., & Varma, R. S. (2022). Engineering of transition metal sulfide nanostructures as efficient electrodes for high-performance supercapacitors. *ACS Applied Energy Materials*, *5*(6), 6481–6498. DOI: 10.1021/acsaem.1c03937

Romano, G., & Johnson, S. G. (2022). Inverse design in nanoscale heat transport via interpolating interfacial phonon transmission. *Structural and Multidisciplinary Optimization*, *65*(10), 297. Advance online publication. DOI: 10.1007/s00158-022-03392-w

Salaris, V., López, D., Kenny, J. M., & Peponi, L. (2023). Hydrolytic degradation and bioactivity of electrospun pcl-mg-nps fibrous mats. *Molecules (Basel, Switzerland)*, *28*(3), 1001. DOI: 10.3390/molecules28031001 PMID: 36770668

Saleh, H. M., & Hassan, A. I. (2023). Synthesis and characterization of nanomaterials for application in cost-effective electrochemical devices. *Sustainability (Basel)*, *15*(14), 10891. DOI: 10.3390/su151410891

Sezali, N. A. A., Ong, H. L., Villagracia, A. R., & Hoang, T. (2024). Bio-based nanomaterials for energy application: A review. *Vietnam Journal of Chemistry*, *62*(1), 1–12. DOI: 10.1002/vjch.202300158

Shaker, L., Al-Amiery, A. A., & Al-Azzawi, W. K. (2024). Nanomaterials: Paving the way for the hydrogen energy frontier. *Discover Nano*, *19*(1), 3. Advance online publication. DOI: 10.1186/s11671-023-03949-8 PMID: 38169021

Sharma, G. K., & James, N. R. (2023). *Electrospinning: the technique and applications*. Recent Developments in Nanofibers Research., DOI: 10.5772/intechopen.105804

Sharma, U., Karazhanov, S., Jose, R., & Das, S. (2022). Plasmonic hot-electron assisted phase transformation in 2d-mos2 for the hydrogen evolution reaction: Current status and future prospects. *Journal of Materials Chemistry. A, Materials for Energy and Sustainability*, *10*(16), 8626–8655. DOI: 10.1039/D1TA10918A

Shi, M., Hou, Y., Du, X., Guo, J., Li, B., & Li, Y. (2023). 2d/2d-ni–co–b/rgo nanolayers prepared by galvanic replacement for hydrogen generation through nabh4 hydrolysis. *ACS Applied Nano Materials*, *6*(8), 6798–6809. DOI: 10.1021/acsanm.3c00515

Shi, Z., Ge, Y., Yun, Q., & Zhang, H. (2022). Two-dimensional nanomaterial-templated composites. *Accounts of Chemical Research*, *55*(24), 3581–3593. DOI: 10.1021/acs.accounts.2c00579 PMID: 36475610

Simpa, P., Solomon, N. O., Adenekan, O. A., & Obasi, S. C. (2024). Nanotechnology's potential in advancing renewable energy solutions. Engineering Science &Amp. *Technology Journal*, *5*(5), 1695–1710. DOI: 10.51594/estj.v5i5.1137

Solomon, N. O., Simpa, P., Adenekan, O. A., & Obasi, S. C. (2024). Sustainable nanomaterials' role in green supply chains and environmental sustainability. Engineering Science &Amp. *Technology Journal*, *5*(5), 1678–1694. DOI: 10.51594/estj.v5i5.1136

Song, J., Lin, X., Ee, L. Y., Li, S. F. Y., & Huang, M. (2022). A review on electrospinning as versatile supports for diverse nanofibers and their applications in environmental sensing. *Advanced Fiber Materials*, *5*(2), 429–460. DOI: 10.1007/s42765-022-00237-5 PMID: 36530770

Sun, Y., Li, X., & Ren, Z. (2024). Tailoring the ion storage of mxene by aramid nanofibers towards self-standing electrodes for flexible solid-state supercapacitors. *Nanotechnology*, *35*(36), 365403. DOI: 10.1088/1361-6528/ad5728 PMID: 38865983

Tan, N. H., Phu, L. T. N., Van Duc, T., Nhi, D. L. T., Nga, D. T., Le, D. T. T., Khanh, H. Q., & Hoa, N. D. (2024). Synthesis of indium oxide nanofibers using electrospinning method towards gas sensor application. *Vietnam Journal of Chemistry*, *62*(3), 354–360. DOI: 10.1002/vjch.202300348

Thambiliyagodage, C., Jayanetti, M., Ekanayake, G., Mendis, A., Liyanaarchchi, H., & Wijayawardana, S. (2024). Llmenite and biomass composite nanomaterials - a review. *Catalysis Research*, *04*(02), 1–31. DOI: 10.21926/cr.2402006

Thomas, B., Sain, M., & Oksman, K. (2022). Sustainable carbon derived from sulfur-free lignins for functional electrical and electrochemical devices. *Nanomaterials (Basel, Switzerland)*, *12*(20), 3630. DOI: 10.3390/nano12203630 PMID: 36296820

Thomas, S. A., & Cherusseri, J. (2023). Recent advances in synthesis and properties of zirconium-based mxenes for application in rechargeable batteries. *Energy Storage*, *5*(8), e475. Advance online publication. DOI: 10.1002/est2.475

Thomas, S. A., Cherusseri, J., Pallavolu, M. R. N., Rajendran, D., & Kumar, D. (2024). Boron carbon nitride (bcn): An emerging two-dimensional material for rechargeable batteries. Energy &Amp. *Fuels*, *38*(15), 13704–13721. DOI: 10.1021/acs.energyfuels.4c02056

Tian, C. (2024). Application of metal-based nanomaterials in lithium batteries. *Applied and Computational Engineering*, *59*(1), 22–29. DOI: 10.54254/2755-2721/59/20240742

Tian, Z. T. (2022). Advances and prospects of self-assembly of nanomaterials in 1d to 4d via electrospinning. *Key Engineering Materials*, *907*, 59–65. . DOI: 10.4028/www.scientific.net/KEM.907.59

Tizazu, G., & Wondimu, E. (2024). Effect of orange fruit peel extract concentration on the synthesis of zinc oxide nanoparticles. DOI: 10.20944/preprints202405.0630.v2

Tsai, L. (2024). Research progress of carbon nanomaterials for lithium-ion battery anodes. *Applied and Computational Engineering*, *59*(1), 155–160. DOI: 10.54254/2755-2721/59/20240788

Ugwuanyi, E. D., Zamathula, Q. S. N., Dada, M. A., Majemite, M. T., & Obaigbena, A. (2024). The impact of nanomaterials in enhancing wastewater treatment processes: A review. *Magna Scientia Advanced Research and Reviews*, *10*(1), 273–285. DOI: 10.30574/msarr.2024.10.1.0030

Usman, U. (2023). Toxicological effects of nanomaterials on the aquatic biota. Emerging Nanomaterials and Their Impact on Society in the 21st Century, 304-340. DOI: 10.21741/9781644902172-13

Valenzuela-Amaro, H. M., Aguayo-Acosta, A., Meléndez-Sánchez, E. R., de la Rosa, O., Vázquez-Ortega, P. G., Oyervides-Muñoz, M. A., Sosa-Hernández, J. E., & Parra-Saldívar, R. (2023). Emerging applications of nanobiosensors in pathogen detection in water and food. *Biosensors (Basel)*, *13*(10), 922. DOI: 10.3390/bios13100922 PMID: 37887115

Varsou, D., & Sarimveis, H. (2023). Deimos: A novel automated methodology for optimal grouping. application to nanoinformatics case studies. *Molecular Informatics*, *42*(8-9), 2300019. Advance online publication. DOI: 10.1002/minf.202300019 PMID: 37258455

Wang, G., Guan, Y., Wang, Y., Ding, Y., & Yang, L. (2023). Direct laser irradiation and modification of 2d te for development of volatile memristor. *Materials (Basel)*, *16*(2), 738. DOI: 10.3390/ma16020738 PMID: 36676475

Wang, J., Suo, J., Song, Z., Li, W. J., & Wang, Z. (2023). Nanomaterial-based flexible sensors for metaverse and virtual reality applications. *International Journal of Extreme Manufacturing*, *5*(3), 032013. DOI: 10.1088/2631-7990/acded1

Wang, L., Meng, Y., & Wang, X. (2024). Sustainable supramolecular polymers. *ChemPlusChem*, *89*(6), e202300694. Advance online publication. DOI: 10.1002/cplu.202300694 PMID: 38355904

Wang, Y. (2023). Rational design of highperformance m-n-c single atom catalysts. [JMMS]. *Journal of Mineral and Material Science*, *4*(5), 1–3. DOI: 10.54026/JMMS/1074

Wang, Y., Chen, Y., Zhao, H., Li, L., Ju, D., Wang, C., & An, B. (2022). Biomass-derived porous carbon with a good balance between high specific surface area and mesopore volume for supercapacitors. *Nanomaterials (Basel, Switzerland)*, *12*(21), 3804. DOI: 10.3390/nano12213804 PMID: 36364579

Wang, Y., Xue, Y., & Züttel, A. (2024). Nanoscale engineering of solid-state materials for boosting hydrogen storage. *Chemical Society Reviews*, *53*(2), 972–1003. DOI: 10.1039/D3CS00706E PMID: 38111973

Wang, Z., Cao, W., Sun, C., Ji, D., Zhao, K., Gao, G., Xu, X., Liu, Y., Zhang, T., Zhu, J., & Han, J. (2023). Nanodiamond reinforced self-healing and transparent poly(urethane–urea) protective coating for scratch resistance. *International Journal of Smart and Nano Materials*. Advance online publication. DOI: 10.1080/19475411.2023.2289608

Wang, Z., Wang, P., Mao, Q., Tian, W., Xu, Y., Li, X., Wang, L., & Wang, H. (2022). Urchin-like pdos nanostructure for hydrogen evolution electrocatalysis. *Nanotechnology, 33*(32), 325401. DOI: 10.1088/1361-6528/ac6c36 PMID: 35504246

Wei, Y., Wu, Y., Cao, X., Cui, L., Chen, J., Wu, Q., Liu, D., Sun, Z., & Zhang, Q. (2022). Simple controllable fabrication of novel flower-like hierarchical porous nio: Formation mechanism, shape evolution and their application into asymmetric supercapacitors. *ChemElectroChem, 9*(16), e202200462. Advance online publication. DOI: 10.1002/celc.202200462

Winkler, D. A. (2022). Probing the properties of molecules and complex materials using machine learning. *Australian Journal of Chemistry, 75*(11), 906–922. DOI: 10.1071/CH22138

Wu, Z., Yi, Y., Hai, F., Tian, X., Zheng, S., Guo, J., Tang, W., Hua, W., & Li, M. (2023). A metal–organic framework based quasi-solid-state electrolyte enabling continuous ion transport for high-safety and high-energy-density lithium metal batteries. *ACS Applied Materials & Interfaces, 15*(18), 22065–22074. DOI: 10.1021/acsami.3c00988 PMID: 37122124

Xing, Y., Cao, C., Huang, Z., Huang, L., Zhang, H., & Jia, Q. (2024). A first-principles study of 2d single-layer sip as anode materials for lithium-ion batteries and sodium-ion batteries. *Physical Chemistry Chemical Physics, 26*(8), 7072–7082. DOI: 10.1039/D3CP05164A PMID: 38345652

Xu, M., Li, J., Yang, Q., Jiang, L., He, J., He, D., Wang, Y., & Yang, Y. (2023). Morphology-controlled green synthesis of tellurium nanostructures and applications of te/mxene hybrid structures. *Materials Advances, 4*(22), 5668–5673. DOI: 10.1039/D3MA00579H

Xu, M., Xin, Z., Wang, J., Tang, T. W., Li, Y., Li, Y., Kim, T.-H., & Luo, Z. (2024). Emerging application of 2d materials for dendrite-free metal batteries. *Energy Materials, 4*(3), 400066. DOI: 10.20517/energymater.2023.94

Xu, Y., Li, Y., Gao, L., Liu, Y., & Ding, Z. (2024). Advances and prospects of nanomaterials for solid-state hydrogen storage. *Nanomaterials (Basel, Switzerland), 14*(12), 1036. DOI: 10.3390/nano14121036 PMID: 38921912

Xu, Y., Xu, D., Yu, N., Liang, B., Yang, Z., Asif, M. S., Yan, R., & Liu, M. (2023). Machine learning enhanced optical microscopy for the rapid morphology characterization of silver nanoparticles. ACS Applied Materials &Amp. *Interfaces*, *15*(14), 18244–18251. DOI: 10.1021/acsami.3c02448 PMID: 37010900

Xu, Z., Guo, X., Wang, S., Wang, S., Wang, J., Jiang, F., & Canham, L. (2023). Green synthesis of mesoporous and biodegradable iron silicide nanoparticles for photothermal cancer therapy. *Journal of Materials Chemistry. B*, *11*(18), 4065–4075. DOI: 10.1039/D2TB02399G PMID: 37082912

Yang, J., Zeng, X., Tebyetekerwa, M., Wang, Z., Bie, C., Sun, X., Marriam, I., & Zhang, X. (2024). Engineering 2d photocatalysts for solar hydrogen peroxide production. *Advanced Energy Materials*, *14*(23), 2400740. Advance online publication. DOI: 10.1002/aenm.202400740

Yang, Z., Zhang, X., Zhang, J., Liu, H., & Yu, X. (2023). Application of biomass-based nanomaterials in energy. *Advanced Energy and Sustainability Research*, *4*(12), 2300141. Advance online publication. DOI: 10.1002/aesr.202300141

Yao, Y., Zhu, W., Teng, Y., Li, C., & Yang, Y. (2024). Polymer-based fabrication of 2d metallic and ceramic nanomaterials. *Accounts of Materials Research*, *5*(8), 944–957. DOI: 10.1021/accountsmr.4c00122

Yu, R., Lu, X., Lu, Z., & Cao, Y. (2024). Facile solid-state chemical synthesis of comoo4 nanorods for high-performance supercapacitors. *Molecules (Basel, Switzerland)*, *29*(6), 1369. DOI: 10.3390/molecules29061369 PMID: 38543005

Yun, Q., Ge, Y., Chen, B., Li, L., Wa, Q., Long, H., & Zhang, H. (2022). Hybridization of 2d nanomaterials with 3d graphene architectures for electrochemical energy storage and conversion. *Advanced Functional Materials*, *32*(42), 2202319. Advance online publication. DOI: 10.1002/adfm.202202319

Zhang, D., Li, Z., & Liang, C. (2022). Diverse nanomaterials synthesized by laser ablation of pure metals in liquids. Science China Physics, Mechanics &Amp. *Science China. Physics, Mechanics & Astronomy*, *65*(7), 274203. Advance online publication. DOI: 10.1007/s11433-021-1860-x

Zhang, L., Li, R., & Yue, W. (2024). Fabrication of nife-ldhs modified carbon nanotubes as the high-performance sulfur host for lithium–sulfur batteries. *Nanomaterials (Basel, Switzerland)*, *14*(3), 272. DOI: 10.3390/nano14030272 PMID: 38334543

Zhang, L., Wei, C., Gao, L., Lin, M.-F., Eh, A. L.-S., Chen, J., & Li, S. (2024). Recent advances in electrospun nanostructured electrodes in zinc-ion batteries. *Batteries*, *10*(1), 22. DOI: 10.3390/batteries10010022

Zhang, Q., He, X., Wang, L., Wang, J., Dong, L., Xie, Y., & Hao, Y. (2023). A novel sugar-assisted solvothermal method for fef2 nanomaterial and its application in libs. *Materials (Basel)*, *16*(4), 1437. DOI: 10.3390/ma16041437 PMID: 36837064

Zhang, T., Chang, L., Zhang, X., Wan, H., Liu, N., Zhou, L., & Xiao, X. (2022). Simultaneously tuning interlayer spacing and termination of mxenes by lewis-basic halides. *Nature Communications*, *13*(1), 6731. Advance online publication. DOI: 10.1038/s41467-022-34569-y PMID: 36347885

Zhang, Y., Poon, K., Masonsong, G. S. P., Ramaswamy, Y., & Singh, G. (2023). Sustainable nanomaterials for biomedical applications. *Pharmaceutics*, *15*(3), 922. DOI: 10.3390/pharmaceutics15030922 PMID: 36986783

Zhao, A., Hao, G., & Zhang, W. (2024). Innovations and strategies for optimizing lithium-ion battery performance: From material advancements to system management. *Applied and Computational Engineering*, *66*(1), 144–149. DOI: 10.54254/2755-2721/66/20240936

Zheng, J., Yan, B., Li, F., Zhang, Q., Han, J., Zhang, C., & He, S. (2023). Al foil-supported carbon nanosheets as self-supporting electrodes for high areal capacitance supercapacitors. *Molecules (Basel, Switzerland)*, *28*(4), 1831. DOI: 10.3390/molecules28041831 PMID: 36838820

Zheng, S., Zhao, W., Chen, J., Zhao, X., Pan, Z., & Yang, X. (2023). 2D Materials Boost Advanced Zn Anodes: Principles, Advances, and Challenges. *Nano-Micro Letters*, *15*(1), 46. DOI: 10.1007/s40820-023-01021-9 PMID: 36752865

Zhou, X., Shi, Y., Hu, Q., Zhang, S., Zhang, X., & Meng, L. (2023). Discrete element simulation of the relationship between
 composition, itz property, and tensile behavior of
 eco-friendly uhpc matrix. *Materials (Basel)*, *16*(10), 3844. DOI: 10.3390/ma16103844 PMID: 37241473

Chapter 9
Innovative Approaches to Renewable Energy Storage:
Bridging the Gap for a Sustainable Future

Surjit Singha
https://orcid.org/0000-0002-5730-8677
Kristu Jayanti College, India

V. Muthu Ruben
https://orcid.org/0009-0006-7723-8596
Christ University, India

Ranjit Singha
https://orcid.org/0000-0002-3541-8752
Christ University, India

R. Sreethi Rebeka
https://orcid.org/0009-0008-9011-7048
Mount Carmel College, India

Shruti Jose
https://orcid.org/0000-0003-3139-2505
Christ University, India

Alphonsa Diana Haokip
https://orcid.org/0000-0003-2578-0114
Don Bosco College, Maram, India

ABSTRACT

In this chapter, we reviewed works related to renewable energy storage technologies with a focus on innovative solutions that can increase the efficiency, scalability and sustainability of energy systems. This includes advanced materials such as solid-state and lithium-air batteries, hybrid storage systems combining multiple technologies, and decentralised storage supporting grid resilience and energy security. By storing energy generated from renewable sources such as the sun and wind, these innovations seek to reduce carbon footprints and enable a transition to a more sustainable energy future by solving the issue of the intermittency of renewables. The chapter examines the emerging trends and technological developments,

DOI: 10.4018/979-8-3693-9316-1.ch009

highlighting the critical role that next-generation energy storage systems can play in driving the global shift towards renewables and transforming energy infrastructure into a cleaner, more resilient world.

INTRODUCTION

The need to slow climate change is changing the world energy map as countries try to cut greenhouse gas emissions and move to greener energy systems. To hit the target, the study states we must take a significant step away from fossil fuels and switch to environmentally friendly energy sources, like solar and wind. However, the inherent intermittency of these energy sources presents a significant challenge in maintaining a stable energy supply. Energy storage solutions must be developed and deployed in order to unlock the full potential of renewables.

Energy Storage Forms an Essential Link Between the Intermittent Advantage of Renewable Energy Generation and Modern Society Energy Demand This allows excess energy produced in times of high output to be stored and used when production is low, providing grid stability and less dependence on non-renewable backup sources. Currently , energy storage technologies such as lithium-ion batteries, pumped hydro storage, and thermal storage have made tremendous advances. Although they require less energy and lessen the environmental impact, they have limitations in terms of scalability, environmental impact, and resource availability.

To combat these limitations, new energy storage systems are under development that cycle advanced materials, hybrid configurations, and new integration methods. These developments are focused on alleviating the challenges related to technology and economics associated with existing solutions and their environmental consequences. These include the design of batteries with less reliance on critical minerals, the combination of supercapacitors with batteries for better performance , and the use of energy-dense materials such as solid-state electrolytes.

In addition, energy storage technologies are crucial for distributed energy systems, in which energy is generated and utilized on-site, reducing transmission losses and increasing grid reliability. Such systems are key to integrating renewable energy into off-grid and remote individual consumers, democratizing access to clean energy, and developing energy independence.

However, this chapter is about how the storage of renewable energy will look in the future based on innovation. This study attempts to investigate the recent developments in material science, systematic methods, and multiscale integration frameworks to create a clearer picture of how these technologies can revolutionize the theoretical potential and practical viability of renewable energy. It also investigates other aspects and higher key practices, such as enabling the carbon neutrality principle, the territory of sustainable societies and intelligent life, and the promotion of the global energy transition.

In upcoming chapters, we will explore specific innovations in energy storage, present key case studies, and examine the challenges and opportunities in store. Ultimately, these perspectives emphasize the transformative potential of next-generation energy storage technologies in achieving a sustainable and resilient energy future.

BACKGROUND

The transition to renewable energy is happening at an unprecedented pace, with solar and wind power leading the charge for a cleaner, sustainable energy future in response to the growing demand for such sources. These technologies may reduce greenhouse gas emissions and slow climate change. However, the controversy surrounding renewable energy generation is its intermitted (variable) nature. Solar power generation requires sunlight, while wind energy is generated using varying speeds of wind. Therefore, adapting these sources to the current energy grid, which was never intended to deal with intermittent and unpredictable power generation from fossil fuels, can be challenging. Due to the lack of an effective way to store surplus energy generated during peak output times, renewables run the risk of being inefficient and wasteful, as surplus energy cannot be consumed. On the other hand, in times of low generation, there may be insufficient power to satisfy energy demand, leading to blackouts. As demand for energy continues to grow while sources of energy generation are increasingly variable, this discrepancy creates a significant challenge: the need for energy storage solutions that can provide an immediate buffer between the supply and consumption of energy.

This problem can be mitigated with energy storage technologies where surplus energy can be stored during high renewable energy production, and then it can be discharged during periods of low production. These systems provide a more reliable and stable grid, allowing for energy to be consistently delivered to power consumers even when solar and wind resources are not available. Conventional energy storage methods like pumped hydro and lithium-ion batteries have been widely used, but they come with their restrictions. Pumped hydro storage, for instance, needs a particular geography with the right topography, whereas lithium-ion batteries face scalability and ecological or resource limitations. Additionally, the corresponding integration of these storage systems in decentralized energy grids (e.g., microgrids) complicates the resilience and reliability of the energy supply (Kandari et al., 2022; Nikolaos et al., 2023).

The limitations of conventional energy storage technologies have catalyzed considerable research and development over the past few years. Advances in materials science and engineering have resulted in a new generation of energy storage technologies like solid-state batteries, lithium-air batteries, and hybrid storage systems. For instance, solid-state batteries boast higher energy densities and enhanced safety over conventional lithium-ion. Lithium-air batteries offer the prospect of much higher storage capacity, but they are still early in the commercialization phase. Another group of solutions are hybrid storage systems, which combine different storage technologies, such as batteries, supercapacitors, and flywheelsflywheels, to optimize performance by harnessing the benefits of multiple technologies, including quick response times and long-term energy storage (Machín et al., 2024; Manthiram, 2020).

Also on the horizon are decentralized energy storage technologies like microgrids and community-based storage solutions that improve grid resilience and energy security. During emergencies or when the power goes out, microgrids enable colleges or even individual buildings to function independently of the primary grid, giving campuses more self-determination and less reliance on centralized sources of energy. Such decentralization is especially advantageous for underserved, rural or impoverished regions without robust grid connectivity (Hunt et al., 2022; Wu et al., 2021).

Although good progress has been made, the complete integration of next-generation energy storage systems into our society will require the performance of many barriers to be solved. Large-scale deployment of these technologies means that issues such as efficiency, cost, scale, and environmental sustainability must be solved. Ongoing investment in R&D and supportive policies will be essential to address these barriers, and enabling energy storage technologies will pave the way to enable it to be a key component of the global transition to a renewable energy future. As the world strives to lower its carbon emissions and grow the proportion of renewable energy in the global energy mix, game-changing energy storage developments will be critical to realize a sustainable, resilient and low-carbon energy grid.

THEORETICAL FRAMEWORK

Several theoretical backgrounds support this research and the understanding of the challenges and opportunities in the energy storage sector:

Technology Adoption Model (TAM): A model that can help understand how the new technology gets adopted by the users, with various factors including perceived ease of use and perceived use. TAM has been used to explore what implements the acceptance of new -generation energy storage systems (Al-Gahtani et al., 2007; Venkatesh & Davis, 2000)

Innovation Diffusion Theory (IDT): This theory, developed by Everett Rogers, deals with how innovations diffuse out across a social system, specifically the stages of knowledge, persuasion, decision, implementation, and confirmation. This framework enables such analysis of the spread of advanced energy storage across various sectors (Mahajan et al., 1990; Stacks et al., 2019).

Systems Theory: This is a grand theory that aims to understand how energy storage systems function in the larger context of the energy grid. It shows how different technologies, infrastructures and stakeholders in renewable energy storage are interconnected and depend on one another (Aghmadi & Mohammed, 2024; Rufer, 2017).

The **Sustainability Theory** focuses on the principle that the economy should develop sustainably, addressing economic, environmental, and social objectives. This framework will inform the evaluation of both the financial and ecological sustainability of emerging energy storage technologies (Hunt et al., 2022).

CONCEPTUAL FRAMEWORK

The study's conceptual framework incorporates major factors defining the betterment and usage of next-gen energy storage methods. These are classified as advanced materials (solid-state and lithium-air batteries), hybrid storage technology (batteries with supercapacitors), decentralized energy storage (microgrids, residential storage) and policy/regulations (incentives, subsidies). It can be measured by metrics such as technological innovation (R&D), public-private partnerships, and the market adoption rate. Finally, these mediators impact the outcome measures of interest, such as the performance of energy storage systems (from economic and environmental perspectives), their contribution to efficient grid integration of renewable energy sources, and, ultimately, the sustainability of energy storage solutions. This framework demonstrates the convergence of contemporary energy technologies and solutions on energy storage and how technological, institutional, and market-driven inputs together determine the ability of energy storage to change the modern energy structure meaningfully.

RESEARCH PROBLEM

The technology behind renewable energy sources, such as solar and wind energy, has come a long way, and yet their intermittent nature still proves to be a substantial problem. To tackle this challenge, energy storage systems are crucial, but current technologies struggle with limitations regarding efficiency, scalability, costs, or the environmental impact of materials. Moreover, energy storage in decentralized energy grids remains a true-for-all complex problem. The increasing need for next-generation energy storage systems that break through these limitations is essential to enable the world to transition to a more renewable, energy-powered future.

RESEARCH QUESTIONS

1. What are the key innovations in energy storage technologies that will enhance the integration of renewable energy into the grid?
2. How can next-generation energy storage systems improve the efficiency, scalability, and sustainability of renewable energy storage?
3. What are emerging energy storage technologies' policy, economic, and environmental implications in supporting global energy transitions?

RESEARCH OBJECTIVES

The main objectives of this research are:

1. To analyze the innovations in energy storage materials and technologies, including advanced batteries, hybrid storage systems, and decentralized energy storage solutions.
2. To evaluate the scalability, efficiency, and sustainability of next-generation energy storage solutions in facilitating the transition to renewable energy.
3. To explore the policy, regulatory, and economic factors influencing the adoption and implementation of advanced energy storage technologies.
4. To assess the role of energy storage in enhancing grid stability, reducing carbon footprints, and achieving long-term energy security.

The chapter aims to provide insights into the cutting-edge developments in energy storage and their potential impact on the future energy landscape. It emphasizes the importance of continued research, investment, and policy support in realizing a sustainable energy future.

THE NEED FOR ADVANCED ENERGY STORAGE IN RENEWABLE ENERGY

The shift towards renewables at a global level is essential for the mitigation of climate change and eliminating dependence on finite fossil fuel reserves. The reason is, however, that renewable energy sources, such as solar and wind, are fundamentally intermittent, leading to new and significant problems for grid reliability and energy availability. To tackle these challenges, advanced energy storage systems are critical to maintaining a stable, efficient and sustainable energy supply.

The availability of renewable energy sources is often unpredictable and variable, depending on environmental conditions. Solar power generation, for example, can only occur in the daytime and is also impacted by weather conditions, while wind energy only happens when wind patterns are present and can be co-located at any time of the hour, day, or even season. These fluctuations lead to surges in energy creation, followed by shortages and a subsequent misalignment between energy availability and requirements.

Practical storage solutions will ensure renewable energy systems do not overload the grid during peak times of generation or fail to meet consumption requirements in lulls. This intermittency makes it difficult for renewables to be widely adopted, especially in areas with high energy consumption or poor complementary energy sources. As a result, advanced energy storage technologies are essential in smoothing out these fluctuations, guaranteeing uninterrupted energy delivery and facilitating seamless integration of renewables into the energy grid.

Energy storage systems play a critical role in stabilizing the grid with intermittent renewable energy inputs. These systems absorb energy when generation is high and release energy when generation is low, balancing supply and demand and avoiding blackouts while reducing reliance on fossil-fuel-based Peaker plants.

Energy storage does not just make reliability possible — it makes the energy grid more flexible. It enables load shifting, soaking up energy when demand and prices are low and delivering it when demand is high. In addition, storage systems enhance the grid's capacity to integrate and distribute distributed energy resources, like rooftop solar, and enable the establishment of microgrids in isolated or off-grid regions.

Energy storage is also critical for modernizing the grid, allowing a transition away from the traditional centralized grids to more resilient and decentralized energy systems. These smart grids are self-sufficient, quickly return to normal after disruptions and control energy flows through real-time data analysis and machine intelligence.

Although lithium-ion batteries, lead-acid batteries, and pumped hydro storage are some of the significant existing energy storage technologies that are adopted in various applications today, several limitations hinder their scalability, sustainability and, ultimately, a fully renewable energy future. One of the most significant issues stems from the resource limitations of traditional storage. As one example, lithium-ion batteries — among the most common types of storage technology — rely on materials like lithium, cobalt and nickel. Such materials are limited, costly, and frequently derived in environmentally and socially unsustainable circumstances. The extraction of these materials can result in ecological disaster, human rights abuses, and geopolitical conflict, thus ensuring a sustainable and ethical supply chain for the mass use of these technologies, which is becoming progressively challenging and must be prioritized. In addition, their extraction makes a significant impact on the environment, causing water pollution, habitat loss, and energy consumption during mining processes (Shu et al., 2023; Yao et al., 2016).

The other significant disadvantage of traditional energy storage systems is their environmental impact. Lead-acid and lithium-ion batteries produce waste and pose serious recycling issues. These batteries can also contribute to pollution at the end of their useful life. If they are not recycled, many of the materials in their construction are toxic and hard to recycle. In addition, the increasing number of batteries requiring disposal or recycling might put a burden on the current recycling infrastructure, further complicating waste management issues. Pumped hydro storage is cost-effective and utilizes existing infrastructure in some areas, but it is also environmentally controversial. The input contains a substantial amount of land and water, making it unviable for regions with insufficient geographic features, such as elevated lands with proximity to large water bodies. The construction of reservoirs and dams to enable pumped hydro storage also has catastrophic impacts on local ecosystems (e.g., Amusa et al., 2024; Li et al., 2024), causing harm to communities and wildlife, the degradation of water quality and habitat destruction.

Another important challenge is the efficiency and lifetime of existing energy storage solutions. Charges and discharges of many conventional systems (like lead-acid and lithium-ion batteries) are susceptible to losing energy in the process of charging and discharging. Such losses decrease the efficiency of the system as a whole and can contribute to increased operational costs over time. Moreover, the frequent need to replace these batteries, which have a relatively short operational life, adds to the expense and generates more waste. Lithium-ion batteries, for example, have a life of about 5-15 years before needing to be replaced, leading to an unsustainable process of unnecessary waste and resource consumption. Furthermore, the frequent replacements of these technologies lead to a financial burden for the consumers and industries that use them for large-scale energy storage (li et al., 2024; Nikolaos et al., 2023).

Another major challenge facing current storage technology is scalability. Most traditional systems are limited to particular geography, materials, or infrastructure conditions, which restricts their wide-scale deployment. For example, pumped hydro storage requires certain geographical features, like appropriate topography and access to water bodies, and it is only available in some places. Lithium-ion and lead-acid batteries are more flexible in deployment, but their scalability is hindered by the high cost of raw materials and the complexities of recycling or disposal. However, as energy storage demand increases, these systems could lack the global capacity for energy transition, particularly in regions with no infrastructure or resources to add (Saldarini et al., 2023; Yao et al., 2016).

Overcoming these limitations is key to advancing energy-storage solutions for a renewable-powered future. To create technology that is more efficient, cheaper, and less harmful, we will need to innovate in a range of areas, including materials science, battery chemistry and storage systems. Improvements like solid-state batteries that offer more energy density and longer life, as well as hybrid systems that combine multiple technologies for increased effectiveness, could yield considerable advancements. Moreover, distributed energy storage technologies, such as microgrids and community-scale storage, may overcome some of the scalability and geographical barriers of traditional systems. Developing and deploying next-generation energy storage solutions that target these challenges upfront is critical to facilitating the outsized spread of renewables and the global transition to a more permanent and sustainable energy future.

INNOVATIONS IN ENERGY STORAGE MATERIALS

Advances in innovative energy storage materials underpin the transformative potential of renewable energy systems, facilitating advances in the efficiency, capacity and sustainability of neurogenesis. Emerging technologies address substantive limitations associated with existing materials focused on improving performance, including energy density, life span, and environmental and economic costs. We discuss the most recent developments in battery technologies, the promise of novel materials, and the overall impact on society.

Solid-state batteries are a significant advancement in energy storage technology, providing many benefits over traditional liquid-based batteries. The main distinction is that they employ solid electrolytes instead of their liquid counterparts, paving the way for multiple potential doping for enhanced battery performance and safety. Benefits of solid-state batteries: One of the most impressive benefits of solid-state batteries is their much gassier energy density. Since they are physically smaller, these batteries are able to store a higher amount of power and are optimal for the many EV brands that require them, as well as small electronics, aero-techs, etc. As industries continue to thrive on power-dense applications, the increased energy density of solid-state batteries makes them a copious investment. For more information, please read Solid State Battery Market Research: Complementing Innovation and Investment.

The higher energy density of solid-state batteries means longer battery life for devices and vehicles, which is key to the adoption of electric vehicles. For instance, a solid-state battery-powered electric vehicle may be able to travel longer distances on a single charge than conventional lithium-ion batteries. This compactness also allows for more versatile design possibilities, empowering manufacturers to create slimmer, more efficient devices that are not weighed down by cumbersome power storage solutions. Portable electronics, where size, weight, and battery longevity are crucial to user experience and satisfaction (Machín, Morant, et al., 2024), necessitate this feature.

A key advantage of solid-state batteries over traditional lithium-ion batteries is the increased safety they provide. Conventional lithium-ion batteries contain liquid electrolytes, which can be highly flammable and dangerous when the battery is damaged or treated improperly and may lead to fire or thermal runaway. Conversely, solid-state batteries use a solid material in place of a liquid electrolyte, which minimizes the chances of combustion, resulting in these batteries being far safer to utilize in dangerous environments. This safety improvement is especially critical with EVs, as battery failure can be disastrous. Also, in addition to their safety and performance benefits over traditional batteries, they are less likely to catch fire or experience thermal runaway, yielding a safer energy storage system while boosting consumer confidence in electric vehicles and other battery-powered devices (Adeyinka et al., 2024; Machín et al., 2024).

For a start, SSBs have a much longer shelf life and are considered safer than traditional lithium-ion batteries. Traditional batteries undergo wear and tear during charge-discharge cycles, leading to capacity degradation and the need for replacements. However, solid-state batteries are more robust and do not degrade as much over time with every cycle, resulting in a longer usable lifespan. With its long life cycle, gensets are less likely to need replacing, which reduces the overall cost of ownership and the impact on the environment related to battery disposal and recycling. Moreover, solid-state batteries have a higher life expectancy which can be beneficial for large-scale energy storage systems, where their durability and long-life stability are crucial (Machín et al., 2024; Wang et al., 2023).

While the potential benefits of solid-state batteries are promising, there are still hurdles to overcome before they can be produced economically and at scale. Costs of manufacturing are still high, and ramping up production to meet global demand is still a work in progress. Moreover, there are still some problems that need to be solved, such as the conductivity of the solid electrolytes dealing with low temperatures so as to perform rotating machines regularly in different environments. Nonetheless, continued research and development in materials science and battery chemistry will ultimately overcome these obstacles in the near future. Solid-state battery tech: A game changer for energy storage Solid-state battery technology has the potential to change the way we store energy, as it provides a safer, more efficient and longer-lasting alternative to traditional battery technologies. You are going to want to be using those very soon with electric vehicles hitting the road, solar energy mitigation, and portable e-devices.

Lithium-air batteries are considered one of the most up-and-coming developments in energy storage, offering ultra-high energy density capabilities that can transform multiple sectors. Conventional lithium-ion batteries store energy via the electrochemical interaction between lithium and a liquid electrolyte. However, lithium-air batteries utilize oxygen from the air as a reactant in the charging and discharging cycles. This new methodology generates energy densities in lithium-air batteries that could rival gasoline, which is a tremendously disruptive technology for long-duration energy storage and high-energy uses. Collaboration and comprehensive analysis of data thus far have proved a highlight in lithium-air batteries, having the possible capacity to deliver up to 10 times the energy density of traditional lithium-ion batteries, which could greatly benefit electrified systems such as electric vehicles (EVs) and grid storage, where lightweight energy storage solutions with high energy density are urgently needed (Abraham & Jiang, 1996; Chen et al., 2022).

The extreme energy density of lithium-air batteries makes them highly desirable for applications with a high energy storage need. For electric vehicles, lithium-air batteries could be the solution to achieving the power required to drive long distances with one charge, alleviating the need for more frequent recharging and allowing EVs to be a viable option for long-distance driving. This capacity to store much more energy in the same physical space might help resolve one of the frictions around current-day lithium-ion batteries, which need large battery packs to deliver more extended ranges. In grid storage applications, lithium-air batteries may perhaps be even more competitive in both efficiency and cost when addressing the retention of the energy generated by renewable sources such as solar, wind, or any other intermittent energy source, in order to offer approximate permanent energy maintenance (Arévalo et al., 2024; Vedhanarayanan & Lakshmi, 2024).

The other important advantage of lithium-air batteries is their sustainability. Also, conventional lithium-ion batteries use limited materials like lithium, cobalt, and nickel, and their mining and processing have considerable environmental and social issues. Lithium-air batteries, on the other hand, use readily available materials such as oxygen found in the air, which lowers the possibility of environmental pollution and raw materials used in battery production. Coupling the use of oxygen as a reactant also helps reduce the dependence on high-cost rare materials, potentially reducing manufacturing costs over time, making this type of battery widely accessible worldwide (Llamas-Orozco et al., 2023).

Lithium-air batteries could also offer significant benefits over traditional lithium-ion batteries and have a broad application range — but they face a number of hurdles that will need to be overcome before they are commercially viable. Batteries have a limited cycle life, for example, which is one of the biggest challenges. The oxygen contained in the battery can react with lithium electrodes at the charge and discharge cycles, generating by-products that can lead to battery performance degradation over time. This degradation reduces the number of charge-discharge cycles another lithium-air battery can undergo until it loses its ability to hold a charge. This is a big issue for applications in which longevity is paramount, such as battery-powered vehicles and large-scale energy-storage systems. Scientists are constantly investigating a means to enhance the stability of the electrodes and reduce the growth of unwanted by-products. Still, this is currently a key barrier to the commercial application (Tan et al., 2017; Wang et al., 2019).

Another technical hurdle is controlling the battery's oxygen. Oxygen is highly reactive, so controlling its entry into and exit from the battery during operation is critical for efficiency and safety. Improper management of oxygen can lead to poor battery performance and safety hazards like thermal runaway or corrosion of battery components (Jung et al., 2016; Naqvi et al., 2022). Oxygen Endurance: A new generation of electrolysis systems and catalysts are being designed to handle the destructive effects of oxygen. However, this hurdle will have to be cleared before lithium-air batteries are mass-produced and deployed at scale.

However, the ability of lithium-air batteries to offer energy storage with ultra-high capacity, along with their sustainability benefits, makes them an exciting avenue of research and development. If the pursuits in the fields of materials science, oxygen management, and battery architecture persist, lithium-air batteries may form a foundation for the energy storage landscape in the future. Just as it was for the original version, overcoming the technical obstacles will be key to making these batteries commercially viable. However, their energy density, cost savings, and sustainability advantages are key to hastening the push toward a cleaner and renewable energy future.

EMERGING MATERIALS FOR ENHANCED STORAGE CAPACITY

When searching for new types of materials that will enhance energy storage capacity, several common discoveries have great potential to enhance the strength and stability of energy storage methods. Perhaps the most promising new development involves nanostructured materials — substances like graphene and carbon nanotubes, which are engineered to the nanoscale and can dramatically enhance capacity for energy storage. Graphene — a single layer of carbon atoms packed in a two-dimensional honeycomb lattice — possesses outstanding electrical conductivity, mechanical rigidity, and specific surface area. These properties make graphene an excellent candidate for improving energy storage devices, such as batteries and supercapacitors. In energy storage systems, graphene acts as a barrier, improving conductivity, charge-discharge rates, and overall energy density, providing faster charging times and more power delivered than traditional battery designs. Carbon nanotubes are a kind of carbon-coated structure and have similar advantages to graphene, including a large surface area and high electrical conductivity, thus providing themselves a good opportunity to increase performance in energy storage devices, notably in high-power applications (Chen et al., 2017; Farooq et al., 2024).

Metal-organic frameworks (MOFs) are another class of materials that researchers are increasingly looking at for energy storage. These materials are known as MOFs, a class of ultra-porous materials in which metal ions are coordinated to organic ligands to form an open framework structure that can accommodate large quantities of gas or liquid. Both materials have been garnering interest as potential candidates for supercapacitors and even next-generation batteries thanks to their extraordinary energy storage and release capabilities. MOFs, with their high surface area and tunable porosity , are promising candidates for energy storage applications, as they can not only boost energy density but also enhance charge storage capacity. Furthermore, MOFs are tenable materials in terms of their properties, such as conductivity and stability, making them suitable for use in a wide range of energy storage and conversions. The tunability of structures in MOFs allows for the design of high-performance energy storage systems with low-cost benefits (Ansari et al., 2024; Dutt et al., 2023).

A newer alternative, a growing class of materials for energy storage utilization, is called phase-change materials (PCMs). The PCMs can store energy in latent heat, the heat required for a substance to undergo a change of state, such as a solid or a liquid. Because of their ability to "store" energy in the form of heat, they are exceptionally well placed for thermal energy storage, which has huge use cases. In energy, the stake can take up excess heat during times of energy production and release it when demand or production is low, contributing to balancing energy supply and demand. This technology is of particular importance for balancing non-dispatchable renewables, such as solar and wind, which have variable generation. The potential use of PCMs for thermal storage applications can enhance the overall performance, reliability, efficiency and renewable energy integration into existing energy infrastructure (Kumar et al., 2023; Venkateswarlu & Ramakrishna, 2021).

Besides that, Researchers are also looking to see if they can make bio-inspired materials, materials built like natural systems that can increase the efficiency of lightweight energy storage. Nature is full of functional and efficient structures and many materials inspired by nature, such as biomimetic materials. For instance, energy is stored and transported most efficiently with natural nanostructures in plant cells. Scientists will try to mimic these structures in order to come up with ways to manufacture sustainable energy storage material to deal with the increasing demand for more energy storage. Bio-inspired materials can help enhance sustainable energy storage by serving as a greener alternative to conventional materials and offering new approaches to enable lightweight, cost-effective, and high-performance energy storage devices(Bertaglia et al., 2024; Kapoor et al., 2022).

This cutting-edge energy storage technology includes nanostructured materials, MOFs, PCMs, and bio-inspired designs. They provide outstanding opportunities for overcoming existing limitations of conventional storage technologies, e.g. low energy density, inefficiency and environmental consequences. These materials can help make renewable energy more viable and reliable on a global scale . Research in these areas is ongoing.

ENVIRONMENTAL AND ECONOMIC IMPACTS OF NEW MATERIALS

This Guide will deliver insight into evolving advanced materials for energy storage with more excellent performance coupled with mitigation of economic and environmental impacts based on traditional systems. One of the most important issues to address is reducing our dependency on resources, especially critical minerals, such as cobalt, which are used to produce traditional battery technologies, including lithium-ion batteries. Such practices cause considerable purity loss and ethical issues associated with cobalt extraction, including landscapes such as human rights violations in some mining areas. Consequently, increasing attention is being paid to identifying new materials that are more abundant, affordable, and eco-friendly. Sodium-ion and zinc-based batteries have also been attracting interest as potential alternatives to lithium-ion batteries. These alternatives employ commonly available and more affordable materials, providing a path toward a more sustainable future for energy storage technologies. These innovations can contribute to mitigating adverse environmental and social impacts associated with energy storage production (Hwang et al., 2017; Wu et al., 2024) by reducing dependence on variables in supply and politically problematic materials.

Recyclability is another important aspect of advanced materials development. Recycling of traditional energy storage systems, especially lithium-ion batteries, presents a variety of significant challenges. Their complex assembly and incorporation of rare materials also render their recycling process costly and inefficient, with toxic waste piling up." In response to this, researchers are developing energy storage systems that are made from materials that are easier to recycle or biodegrade. For instance, solid-state batteries that substitute the liquid electrolytes in traditional batteries with solid materials can be recycled more easily and may lessen the overall environmental impacts of battery disposal. Continued development of recyclability is fundamental to the sustainable adoption of energy storage technologies, particularly in a boundary-spanning circular economy in which materials are recycled and waste generated is minimized (Machín et al., 2024).

Cost efficiency is another crucial aspect of developing advanced energy materials. The performance of these new materials is of utmost importance, but it is equally important that these innovations be financially viable and competitive in a global market. These factors include the underlying domain for the manufacturing, processed raw materials, and system integration for deriving energy storage products. How ever, advances in manufacturing techniques and the discovery of cheaper materials have lowered the production costs of advanced batteries quite a bit. Units with such materials as sodium and zinc are also cheaper to extract and process than lithium and cobalt. These novel materials cost less to produce than traditional energy storage technologies, enabling the development of large-scale storage systems for renewable energy applications (Hwang et al., 2017; Lu et al., 2016).

Which is, in fact, a calming agent between the electrodes that helps in lithium-s batteries; LFP is an even better material in this case. By improving the performance and lifetime of storage technologies, these materials help reduce the carbon footprint associated with their manufacture, use and end-of-life. Materials with higher energy densities, longer lifespans, and better recyclability minimize the frequency of replacements, hence lessening the environmental impact throughout their lifecycle. Additionally, due to allowing better energy storage, these materials aid in the utilization of renewable energy and reduce reliance on fossil fuels, which will lead to a n increase in global emissions.

It will be important to develop and integrate advanced materials and materials processes to address these dual challenges of sustainability and scale as energy storage technologies continue to evolve. These innovations will unlock the full potential of renewable energy systems, making them far more reliable and efficient and marking a stepping stone toward a more resilient, sustainable energy future. Very sustainable and available, recyclable, affordable, and low-carbon materials make it possible for the energy storage industry to support the shift toward a more sustainable, clean energy future whilst tackling the economic issues of widespread renewable energy adoption.

HYBRID ENERGY STORAGE SYSTEMS

The hybridization of multiple energy storage technologies is a breakthrough in renewable terminals. Hybrid systems are a solution for varied requirements of renewable energy integration, delivering high performance, flexibility, and resilience by combining the advantages of various storage methods. It covers hybrid energy storage systems, their advantages, and successful implementations.

Hybrid energy storage systems (HESS) are emerging as a possible way to overcome the economic disadvantages of traditional storage technologies by coupling two or more energy storage technologies with complementary strengths, ultimately resulting in improved energy management and system performance. Full utilization of storage technologies One of the most common types of combinations is that of batteries and supercapacitors — batteries offer high energy density, which helps them store large amounts of energy for long periods. Supercapacitors, on the other hand, provide fast charging and discharging ability to meet peak power requirements. This coupling of the two allows for the effective handling of both short-term dips and long-term energy storage, contributing towards the utility of HESS in electric vehicles and grid-scale energy storage applications that require both speedy power output and long-term energy supply (Adeyinka et al., 2024; Dong et al., 2022).

Another exciting hybrid model includes thermal and electrical storage systems. Thermal storage technologies — molten salt or phase-change materials — store excess energy in heat, which can later be converted to electricity or used for heating. Thermal, along with electrical storage (batteries or supercapacitors), can result in a HESS, which allows for excess energy to be used as efficiently as possible for heating or power generation. Energy better integrated: You are utilizing energy in a more optimized manner, as your energy is stored in a suitable form and can be conversed and re-applied without wastage. For renewables — like solar energy that produces energy on a variable basis — the pairing of thermal and electrical storage systems has particular value. The surge in renewable energy production calls for efficient energy storage solutions to address intermittency and ensure a reliable energy supply; an emerging approach is to utilize electrolyzers powered by excess renewable energy to produce hydrogen, which can be stored and converted back into electricity when needed, creating a more stable energy grid (Arévalo et al., 2024).

Pumped hydro storage, compressed air energy storage (CAES), and other mechanical and chemical methods for storing energy may also complement battery technologies as part of hybrid systems. Pumped hydro storage is one of the oldest and most commonly used forms of energy storage, where excess electricity is used to pump water up a hill to be released later to generate power. CAES operates like that, compressing air and storing it in underground caverns or shafts and then releasing it to produce electricity when needed. When combined with batteries, these mechanical and chemical storage technologies offer high capacity at utility scale, energy delivery efficiencies, and long durations of energy supply. By combining two technology paths, HESS can address scalability discrepancies inherent in purely battery systems, enabling a low-temperature solution suitable for large items such as grid stabilization and/or backup power systems (Nikolaos et al., 2023; Rabi et al., 2023).

In such a hybrid energy storage system, the limitations of individual technologies, such as energy density, charge-discharge rates, scalability, and efficiency, can be overcome by taking advantage of the strengths of different components. HESS enable flexible storage options by integrating different energy storage technologies, allowing for a broader range of applications and use cases—from seconds (e.g., frequency regulation) to hours and days (e.g., load shifting) and even seasonal storage (e.g., thermal storage). Hybrid systems will be at the forefront of the way forward as demand for renewable energy continues to increase, allowing for more effective energy storage and integration into smart grids, making energy systems less fragile and more sustainable and reliable. They achieve peak performance as well as minimize energy waste while ensuring that stored energy is available when and where it is needed, accelerating the transition to a clean, sustainable energy future.

HESS have many admirable virtues and are a suitable solution for integrated renewable energy in power grids, as renewable energy sources such as wind and sun have the characteristic of being intermittent and unstable. Improved reliability is one of the most notable benefits of hybrid systems. HESS can alternatively be held in batteries, supercapacitors and thermal storage to offer an energy supply that is smoother and more continuous despite the intermittent nature of renewable energy generation. As such, hybrid systems can dynamically deploy different storage methods as needed depending on local or even seasonal conditions, which contributes towards the overall reliability of the distribution system (Adeyinka et al., 2024; Reilly et al., 2022).

Apart from reliability, HESS improves efficiency as well. This allows the individual technologies to perform better by assigning tasks to the best-suited storage mechanism. Batteries, for example, work well for storing significant energy for long periods. Conversely, supercapacitors will suit applications that require short-term, high-power usage. HESS minimizes energy losses and assures the use of the most efficient method for each scenario by distributing energy storage tasks according to the strengths of each of these technologies. This optimization level leads to minimal wastage and the best utilization of energy storage systems, which is essential for integrating renewable energy sources. (Ilyas et al., 2021; Zheng et al., 2021)

Flexibility is another key advantage of the hybrid system. Applications: HESS can be tailored to fulfil various energy requirements ranging from long-term storage for grid stabilization to short-duration high-power operations such as frequency regulation, unlike other energy systems. For example, grid stabilization, as well as other large-scale applications, benefit from batteries with long-duration energy storage and fast-charging and discharging capabilities, such as what is expected of supercapacitors used for frequency regulation and peak shaving applications. Hybrid systems exhibit flexibility, allowing them to be tailored for use in different sectors or end uses (Adeyinka et al., 2024; Simão & Ramos, 2020).

A hybrid also has the advantage of cost optimization as well. HESS combines different energy storage technologies and helps reduce dependency on high-cost energy storage technology, thus becoming more cost-efficient. Just in one example, although batteries can be relatively costly for long-duration storage, co-locating batteries with less expensive technologies, such as thermal storage, or mechanical storage technologies, such as compressed air or pumped hydro storage, can achieve cost reductions. HESS can balance the cost and performance trade-off by using complementary technologies, which can improve the economic feasibility of HESS technologies for a broad range of energy storage applications. This cost optimization becomes crucial as the demand for renewable energy storage increases and hybrid systems start to outperform other systems in global markets (Rafał et al., 2022).

Lastly, scalability and adaptability are some of the most important characteristics of hybrid energy storage systems. Since HESS can be tailored to suit the energy storage requirements of different industries, areas, and applications, they are highly scalable. From residential applications to commercial and utility-scale projects, such hybrid systems are capable of providing appropriate levels of storage to fulfil varied demands (Raza et al., 2024; Schubert et al., 2023). They can also be tailored to regional or industrial requirements depending upon area-specific things like climate, sources of energy production, and infrastructure. Hybrid systems are advantageous because they have high scalability and adaptability with regard to various approaches, making them highly suitable in a multitude of scenarios.

With advantages such as increased performance, higher efficiency, flexibility, economic effectiveness, and scalability, hybrid energy storage systems will be pivotal in the transition towards a renewable energy future. With renewable energy sources increasingly dominating the energy market, hybrid systems will be essential to tackle energy storage problems and contribute to efficient renewable energy implementation in a more sustainable and resilient energy infrastructure.

CASE STUDIES

Case Study 1: Tesla Hornsdale Power Reserve, Australia

One such example is the Tesla Hornsdale Power Reserve in South Australia. This project, which relies on a 150 MW/193.5 MWh lithium-ion battery system, is integrated into the regional energy grid, which is supported by renewable energy such as wind and solar. The reserve, designed to tackle South Australia's recurrent problems of grid instability, has established a standard for affordable, environmentally friendly energy storage options.

In addition to its core capability as a battery storage unit, the Hornsdale Power Reserve deployment also embraced a sophisticated grid management apparatus that has transformed the operations of energy distribution and the consumption patterns of energy. These tools allow for fine frequency control that maintains grid reliability despite changes in energy supply and demand. It has also reduced the chances of power failure and quickly responds to unprecedented energy demands.

Arguably, Hornsdale's most significant accomplishment has been displacing fossil-fuel Peaker use, as the plant has helped to reduce the need to fire up Peaker plants to address high energy demand during peak periods. The energy is delivered from the battery system instantaneously, thereby needing less expensive and harmful-to-the-environment backup power inputs. It has achieved significant reductions in energy costs, saving South Australian consumers millions of dollars each year and making energy more affordable.

The case of the Hornsdale Power Reserve shows how large-scale energy storage can facilitate the integration of renewables, improve grid stability, and increase reductions in carbon emissions. An ounce of this ambitious innovation and strategic implementation could go a long way in achieving a sustainable and reliable energy future in minuscule amounts across the world, making this project a success story. The Tesla Hornsdale Power Reserve in Australia integrates cutting-edge battery technology with innovative grid solutions—establishing a model for what energy systems could look like in a decarbonized future.

Case Study 2: Crescent Dunes Solar Energy Project, USA

One innovative approach to deliver 24/7 renewable energy availability is the Crescent Dunes Solar Energy Project in Nevada. For instance, the combination of a 110 MW concentrated solar power (CSP) plant with a molten salt thermal storage system for fulfilling continuous energy needs illustrates the effective use of solar energy generation and thermal energy storage.

At the heart of the project is a molten salt storage system that collects and holds heat from the sun's energy, which is focused on a central receiver by a field of hundreds of heliostats. Molten salt, capable of providing high-temperature thermal energy storage, also constitutes a powerful energy reservoir. This heat is then utilized to produce steam that turns turbines, generating electricity on demand during the night and/or on cloudy days. This ability fills the gaps associated with solar energy, providing constant energy availability day and night.

With its use of solar thermal technology that can be complemented with electrolyzers, the Crescent Dunes project is an example of how hybrid systems combining thermal and electrical storage technologies may offer flexible, sustainable energy solutions. The use of molten salts for long-duration storage also minimizes the need for fossil fuels for backup power and thus dramatically cuts greenhouse gas emissions. Additionally, grid-level solar energy storage and dispatch improves stability and maximizes the use of renewable energy sources.

Now, the Crescent Dunes Solar Energy Project is emerging as a model for how thermal storage can play a part in the systemic transition to renewable energy despite some difficulties in the early going. Such storage can help smooth out the intermittent supply of clean power from renewables by ensuring always-on electricity availability. As world energy needs continue to expand, projects such as Crescent Dunes act as a model for scaling sustainable energy solutions in support of our de-carbonization goals and energy independence.

Case Study 3: Renewable Energy Microgrid in Kodiak Island, Alaska

Kodiak Island, Alaska, has set a global benchmark for integrating renewable energy. The island is home to more than 14,000 residents and relies almost fully on renewable energy thanks to a combination of wind turbines, a lithium-ion battery system and flywheel flywheel energy storage. Moreover, it is composed of a hybrid methodology that guarantees energy consistency and reliability across the remote, eco-sensitive region by surmounting issues of geographic division and variable energy requirements.

The microgrid consists of wind turbines as the primary source of renewable energy, with the high-rate response ability of the lithium-ion battery system acting as a buffer. The batteries store excess energy during high-wind generation and are then discharged during low-wind periods. At the same time, flywheel flywheel energy storage offers demand power modulation for grid stabilization and the regulation of short-term imbalances. By coupling both long-term energy storage and fast-acting stabilization technologies, this approach delivers seamless energy without overreliance on diesel generators.

This has cut Kodiak Island's greenhouse gas emissions and slashed the local utility's operating costs as it transitioned to the hybrid renewable energy system. The island has become energy independent and has long-term economic upside by eliminating dirty and expensive diesel buses with clean energy.

A hybrid battery storage system that works: Kodiak Island, i.e., HESS for renewable energy integration; summary: measurable energy storage systems. These systems leverage complementary technologies to improve performance, increase reliability, and reduce cost. With global uptake of renewable energy on the rise, hybrid systems like the one offered by Kodiak will be critical to creating sustainable and resilient energy infrastructures, particularly in remote or off-grid areas. This case study underscores the importance of hybrid storage solutions that pave the way to carbon neutrality in the energy future while enhancing energy security and reliability for communities of all types.

DECENTRALIZED AND DISTRIBUTED ENERGY STORAGE

Energy decentralization and distribution represent a paradigm shift in the world energy landscape. Decentralized energy grids, unlike their centralized counterparts, harness localized energy generation and storage capabilities to improve flexibility, resilience, and accessibility. At the heart of these systems, energy storage contributes to the smooth integration of renewable energy sources and solves the key problems of intermittence and grid stability. Here, we explore the emergence of decentralized energy grids, the importance of storage technologies for decentralization, and the potential ramifications of this shift in access to and resilience from energy systems.

Decentralized energy grids are disrupting the way electricity is generated; they allow homes and businesses to produce (and store) their energy, such as via solar panels, wind turbines, and small-scale hydro systems. There are a number of factors driving this change. Renewable technologies have then significantly dropped their prices with solar panels and wind turbines. Concurrently, advancements in energy storage solutions have enabled decentralized energy generation to be both economically viable and efficient. Furthermore, there is an increasing need for resiliency as decentralized grids provide better protection against large-scale disruptions (natural disasters, cyberattacks, etc.) than traditional centralized power systems. A decentralized grid refers to the infrastructure and systems that enable the generation and distribution of energy at a local or community level rather than relying on centralized power plants and networks. The decentralized systems have received policy support from governments and organizations worldwide that provide incentives, subsidies, and regulations to promote sustainability and reduce greenhouse gas emissions. Thus, these energy models are increasingly deployed in urban settings, remote areas, and developing countries since they provide customized, adaptable energy services that meet the unique energy demands of various communities and contribute to a more robust and sustainable energy framework (Aoun et al., 2024; Pop et al., 2018).

Energy storage has become a game-changer in decentralized (or distributed) energy systems by helping to solve the intermittency of renewable energy sources. One of its key roles is the stabilization of supply and demand, while storage systems can capture excess energy in periods of high production and deliver it at times of high demand, providing a continuous and reliable energy supply. Energy storage further empowers grid independence, decentralizing energy production, and allowing communities to reach energy independence while minimizing transmission losses. It can also enable microgrids and nano grids, meaning even the smallest, self-contained grids can operate autonomously or connect to more extensive networks on demand (Adeyinka et al., 2024; Glowacki & Hanley, 2016). Additionally, energy storage helps integrate renewable energy by allowing more straightforward incorporation of variable sources like solar and wind into decentralized grids, increasing the percentage of clean energy in the total energy mix. Storage systems enhance the efficiency of decentralized grids by allowing load shifting, peak shaving and demand response through advanced intelligent energy management. Together, these capabilities make energy storage an integral part of strengthening energy systems for resilience and sustainability.

IMPLICATIONS FOR ENERGY ACCESSIBILITY AND SECURITY

Decentralized and distributed energy storage systems have far-reaching effects on energy accessibility and security, especially in regions that are underserved and vulnerable to energy disruptions. Distributed energy systems with storage technologies play a vital role in targeted power delivery to isolated and off-grid areas—remote communities in which the extension of conventional, centralized networks is often unfeasible or prohibitively expensive. By allowing communities to produce and store their energy, these systems grant them more power over their resources. Distributed grids complement traditional infrastructure in developing areas, providing clean and accessible power for economic growth and better living standards, education, and health. These systems also enable the electrification of previously unelectrified areas, conducive to economic development and social equity, through the availability of modern energy solutions even in the most hinterland places (Al-Saadi et al., 2021; Wu et al., 2021).

Decentralized grids can significantly improve the overall energy security of a region by expanding the number of potential energy sources and decreasing vulnerability to supply chain disruption on a large scale. 8†source These systems boost resilience, providing a continuous energy supply during grid failures, natural disasters, or geopolitical disruptions. Distributing energy assets throughout the grid reduces the chance of widespread blackouts and enhances recovery capabilities. Moreover, due to their proximity to end-users, they minimize the energy losses due to transmission and distribution, thereby improving overall system efficiency. Distributed and decentralized energy storage systems are the new way to generate, deliver, and consume energy. These systems provide solutions to vital challenges of renewable energy adoption while improving resilience and equity of access to energy through the integration of advanced storage technologies (Aoun et al., 2024). With the progress of the global energy transition, decentralized grids with robust storage solutions will play a crucial role in securing a sustainable and resilient energy future.

SUSTAINABILITY AND LIFE CYCLE ANALYSIS

Sustainability is fundamental to the development of next-generation energy storage technologies. #W: These systems have great potential to facilitate the adoption of renewable energy. However, the environmental impacts and life cycle performance must be evaluated to ensure a long-term advantage. This subsection presents the findings of research looking into the environmental impacts of energy storage technologies, the role that life cycle assessments (LCA) play in gaining an understanding of these impacts, and how the sustainability of energy storage solutions can be improved.

Energy storage systems play a critical role in the transition to clean energy, but the production, operation, and disposal of energy storage systems can have substantial environmental challenges. The extraction of rare earth elements and critical minerals needed to produce technologies like lithium-ion batteries, including lithium, cobalt, and nickel, can lead to habitat destruction, water pollution, and high energy consumption. Furthermore, mining often takes place in areas with weak environmental and labour regulations, creating significant issues for ethical sourcing and chaining sustainability. The manufacturing and production processes that these storage technologies undergo can also lead to adverse environmental consequences, as they tend to be energy-demanding and lead to significant greenhouse gas emissions, particularly when fossil fuels are involved. Furthermore, production inefficiencies waste even more crucial resources, contributing to an even higher environmental footprint. Energy storage as a technology has many chemistries, each with its complexities, and recycling systems have not kept pace, creating problems at the end of life. Improper disposal can release hazardous materials and contaminate soil and water, complicating attempts to ameliorate the environmental burden posed by these technologies (Llamas-Orozco et al., 2023; Machín, Cotto et al., 2024).

Life cycle assessment has a comprehensive system framework for evaluating the environmental performance of energy storage technologies, embracing all stages from raw material extraction to disposal or recycling. The LCA of energy storage technologies consists of key stages that assess the environmental impact. The first stage — raw material extraction — examines the environmental costs that come with mining and processing the metals such as lithium, cobalt and rare earth elements that will power the batteries of the future, which may result in habitat destruction, water contamination and a considerable energy footprint. Stage 2, manufacturing: considers what inputs of process energy and resources are needed during production, including the greenhouse gas emissions resulting from energy-intensive manufacturing processes. LCA assesses the performance of the storage system, including energy savings and greenhouse gas emissions reduction potential according to its operational life, as part of the operational phase. Finally, the end-of-life phase focuses on the disposal, recycling, and material recovery processes, emphasizing opportunities for circular economy practices to reduce waste, recover valuable materials, and decrease the overall environmental impact of the technology. The stages together give a complete overview of the environmental performance of energy storage systems along their entire lifecycle (Haase et al., 2022; Yamaki et al., 2024).

LCAs conducted on various energy storage technology types, including lithium-ion, solid-state and flow batteries, show some key differences in resource intensity and environmental impact. By assessing each technology at all stages of its life cycle, LCAs can highlight specific areas, or "hotspots," where appropriate changes may be especially effective in limiting environmental consequences. The specific details of energy-intensive manufacturing processes or unsustainable approaches to extracting raw materials, for instance, can give rise to the kind of targeted innovations that limit environmental damage. Moreover, the knowledge and understanding derived from the LCA outcomes allow for the prioritization of priorities in the development of next-generation energy storage components which utilize less materials, require less energy, and have lower cumulative environmental impacts, which would then lead to more sustainable options for energy storage systems (Le et al., 2023; Yamaki et al., 2024).

Such approaches are needed at all times in addressing environmental and sustainability challenges to add up within the energy storage systems life cycle and must integrate novel solutions and approaches. One of the key concepts is, however, material innovation: How about alternative, abundant, non-toxic materials such as sodium, zinc, or organic compounds to minimize dependence on scarce or harmful resources? Biopolymers or bio-based materials can be less damaging to the environment. Another necessary approach is designed for circularity, where modular storage systems make it easier to repair, upgrade, and recycle. At the same time, recyclability is prioritized to enable better material recovery at end-of-life. Green manufacturing practices, such as switching to renewable energy during production and using as little material as possible, also reduce emissions and waste. Advanced recycling technologies, such as chemical recovery techniques for recovering vital elements, including lithium and cobalt, facilitate the establishment of closed-loop systems in which collected materials may be reused in fresh storage systems. There are several areas where policy and regulation have an essential role to play in this: EPR (extended producer responsibility) programs that require manufacturers to take care of the end-of-life phase and sustainability standards that decide the target for environmental performance (Thakur et al., 2023; Vaghela et al., 2023). Ultimately, consumer awareness and engagement would be vital, as well as promoting responsible disposal practices and even participating in recycling programs. Holistic approaches toward developing energy storage systems that recognize the environmental impact, champion the circular economy, and prepare for an ultimately sustainable and clean source of energy can follow from the diverse pathways their strategies will inspire.

Table 1. Comparison table for various battery technologies

Feature	Solid-State Battery	Lithium-Air Battery	Lithium-Ion Battery	Sodium-Ion Battery	Flow Battery
Energy Density	High	Very High	Moderate to High	Moderate	Low to Moderate
Power Density	High	Moderate	High	Moderate	Low
Lifespan	Long	Short (due to degradation)	Moderate to Long	Moderate	Long
Safety	Very Safe (non-flammable)	Risk of degradation	Moderate (risk of thermal runaway)	Safe (but with some risks)	Very Safe (liquid-based)
Operating Temperature	Wide range	Limited	Moderate	Wide range	Moderate
Charge Time	Fast	Slow	Fast	Moderate	Slow
Cost	High	Very High	Moderate	Lower than lithium-ion	High (due to infrastructure)
Scalability	High (for small devices)	Low (still in R&D)	High	High	High

continued on following page

Table 1. Continued

Feature	Solid-State Battery	Lithium-Air Battery	Lithium-Ion Battery	Sodium-Ion Battery	Flow Battery
Environmental Impact	Low	Low	Moderate	Low	Low
Current Stage	Emerging, commercializing	Experimental, early-stage	Mature, widely used	Emerging	Used in grid storage

Table 2. Comparing the technological, financial, and legal obstacles

Technology	Technological Obstacles	Financial Obstacles	Legal & Regulatory Obstacles	Key Benefits
Solid-State Battery	Manufacturing complexity, material stability, scalability	High production costs, limited supply chain	Standardization and safety certifications needed	Higher energy density, enhanced safety, longer lifespan
Lithium-Air Battery	Short lifespan, efficiency losses, slow recharge rates	Expensive R&D, rare materials	Environmental concerns, lack of industry standards	Extremely high energy density, potential for ultra-light batteries
Lithium-Ion Battery	Thermal runaway risks, limited raw material availability	Cost fluctuations due to lithium and cobalt demand	Strict recycling regulations, safety compliance	Proven reliability, high efficiency, scalability
Sodium-Ion Battery	Lower energy density needs better anode materials	In emerging markets, the production scale is not optimized	There is less regulatory clarity compared to lithium-based	Lower cost, abundant materials, environmentally friendly
Flow Battery	Large footprint, complex maintenance	High initial costs require infrastructure investment	Grid integration regulations, environmental permits	Long lifespan, ideal for grid storage, easy scalability
Hydrogen Storage	Energy conversion losses, hydrogen leakage risks	High cost of electrolysis, storage, and transport	Stringent transport and safety regulations	Long-term storage potential, high energy capacity

RECOMMENDATIONS AND SOLUTIONS

Advanced energy storage technologies are crucial for the transition to a sustainable energy future, and their potential must be realized through collaborative efforts among governments, industries, research institutions, and consumers to promote innovation, adoption, and scalability. Actions are needed to accelerate the development and deployment of next-generation storage systems. Prepare some significant investment in research and development (R&D) to investigate new materials and processes first; a most notable area is in solid-state batteries, as well as hydrogen materials and storage types. Pushes for public-private sectors to build innovation centres to speed up ideas and breakthroughs. In addition, supportive policies and a regulatory framework, such as subsidies and tax benefits, and mandates establishing that power storage systems are a part of every renewable energy development to guarantee a proper equilibrium of the energy grid and minimize evictions and environmental impacts, must be enacted. This also means promoting hybrid storage systems and finding better ways to combine different kinds of technologies to achieve the most efficient processing of data. Governments should also present incentives for decentralized energy storage solutions, like residential units or microgrids, to supplement energy security.

Additionally, better collaboration with stakeholders through forums to share ideas and align research would also be a significant indication of progress. Sustainable storage technologies that use renewable, non-toxic materials must be developed, and life cycle assessments must be conducted to ensure minimal environmental impacts. Public awareness campaigns are essential to help consumers understand the advantages of energy storage. It is necessary to standardize the storage systems in order to ensure the systems remain safe and compatible and to identify that storage with grids ensures an intelligent approach towards providing the system with efficient and lower operational costs. Lastly, there should be a circular economy approach for energy storage systems so we can recycle and reuse them to reduce waste. Embracing these recommendations will be an essential step towards accelerating the transition to a cleaner and more sustainable energy future, ensuring the stability of our grids and provision of economic opportunities.

FUTURE TRENDS AND PROSPECTS

Energy storage technologies are set to evolve significantly in the future, spurred by rapid innovation, changing policy frameworks, and the need for sustainable energy solutions. Energy storage systems maintain a central position in the coming era of energy generation, storage, and consumption as the energy landscape changes profoundly. During the next 10 years, we expect energy storage technologies to evolve dramatically through key trends that include the following: advanced battery technologies such as solid-state batteries with higher energy density, faster charge time and improved safety; metal-air batteries such as lithium-air and zinc-air with ultra-high energy density characteristics well suited for grid-scale storage; and sodium-ion batteries as a low-cost alternative that provides adequate performances thanks to the abundance of raw materials. Distancing from traditional batteries, innovative technologies like flow batteries, supercapacitors, and hydrogen storage will further complement existing solutions, where the flow batteries offer backend scalable, long-duration storage opportunities; supercapacitors provide charge/discharge cycles in seconds to minutes for places that require high power; and hydrogen will play an essential, versatile role in decarbonizing heavy industries and transportation. Renewable energy integration will be more seamless, with energy-storage-coupled systems and AI-driven technologies enabling real-time optimization of storage performance for greater energy efficiency. Additionally, microgrids and peer-to-peer energy trading will continue to fuel decentralized storage systems, elevating local energy sovereignty and people-powered resilience.

Policy and regulation will be key to accelerating energy storage deployment. We anticipate governments scaling up financial incentives such as tax credits, feed-in tariffs and grants to support research, development and deployment of energy storage. The stabilizing role of energy storage integrated with renewable will indeed be guaranteed in new regulatory frameworks establishing this obligation for renewable projects — possibly whether it will be required to contribute to grid stability, and if capacity markets, the financial reward will be paid for energy storage contributing to grid reliability. Circular economy regulations (e.g. Extended Producer Responsibility, or EPR) will motivate recycling and sustainable manufacturing. Simultaneously, the implementation of policies designed to support community-owned storage systems and energy-sharing platforms will encourage the development of decentralized energy solutions. This growing trend will enable the balanced and responsible use of storage technologies while avoiding unnecessary environmental and social pressures.

Future advancements in energy storage will rely heavily on innovations in material science, manufacturing, digital transformation, and global collaboration. Innovative research on alternative materials, such as organic polymers, graphene, and materials inspired by biology, will enable storage systems that are more efficient and environmentally friendly than their predecessors. Additive manufacturing and automation will reduce production costs and increase scalability. Digital transformation, such as IoT-enabled innovative storage solutions, blockchain solutions, and integration with connected supply chains, will improve storage efficiency, predictive maintenance and energy transaction visibility. International collaboration on research and development will be bolstered by public-private partnership models that accelerate innovation, drive standardization of best practices and create pathways to market for next-generation technologies.

With these trends coming together, energy storage will be paramount in moving to a sustainable, resilient, decarbonized future, ensuring it integrates with renewable energy, a reliable grid and global energy equity.

DISCUSSION

Advanced energy storage technologies are at the heart of the transition to a sustainable energy future. However, while cloudy days and variable wind conditions can be rewarding for conventional, non-renewable sources of energy, the same is not true for solar or wind. Next-generation energy storage systems can help relieve these challenges by capturing excess energy during high production hours and releasing it during periods of demand exceeding generation. However, existing energy storage solutions have significant limitations that will need to be addressed to allow for widespread adoption and integration, including efficiency, scalability, environmental impact and cost. This discussion focuses on the significant innovations in energy storage technologies, their beneficial potential, and the challenges that must be resolved in order to unleash their potential in facilitating renewable energy integration.

Advanced battery technologies, such as solid-state and lithium-air batteries, are one of the most promising innovations in energy storage. Next-gen batteries are designed to push the boundaries beyond traditional lithium-ion batteries in parameters such as energy density, safety, and longevity. One such type of battery is solid-state batteries, which substitute the liquid electrolyte in traditional batteries for a solid electrolyte, improving performance and decreasing the risk of fires and other safety hazards. Lithium-air batteries work similarly, but they use oxygen from the air as one of the components of the energy storage reaction, and they have the theoretical potential to provide much higher specific energy than conventional batteries. This could lead to more efficient and compact energy storage systems, reducing the need for large-scale storage facilities and making the renewable energy system more viable in remote or off-grid locations.

Apart from these individual innovations, hybrid energy storage systems that incorporate different storage technologies, like batteries, supercapacitors, and pumped hydro storage, are becoming increasingly popular in optimizing energy storage solutions. Hybrid systems leverage several technologies and enable the decoupling of time scales that allow various renewable energy sources to use them and do so with different energy storage capabilities. Batteries, for example, are helpful for storing energy over long periods, while supercapacitors are designed for quickly delivering energy. Pumped hydro storage, by contrast, is designed for bulk, long-duration storage. The integration of hydrogen and battery technologies in hybrid systems could enhance energy storage efficiency and flexibility, enabling real-time balancing of supply and demand, especially in regions with high variability in renewable energy generation.

Furthermore, the integration of other technologies, such as residential energy storage and community-based microgrids, can strengthen grid resilience and contribute to energy security. Locally generated and stored energy provided by these systems allows for less dependency on centralized power grids and provides a more resilient infrastructure to withstand natural disasters or grid failures. Microgrids are energy systems that can be powered independently from the primary grid. They can include renewable energy sources and storage to maintain a continuous power supply in areas where access to electricity is limited. Decentralized energy storage offers customers energy independence by giving consumers control over their energy consumption and savings on energy.

However, multiple barriers still exist to the widespread adoption of next-generation energy storage technologies. Building these advanced storage systems is still very costly to develop and manufacture. Although the long-term benefits from increased efficiency and scalability will likely balance out the initial costs, low-cost-competitive pricing is essential for the technology to graduate to widespread deployment. Moreover, the environmentally hazardous production of new storage materials, such as the mining and end-of-life disposal of critical raw materials in batteries, among others, needs to be examined. A life cycle assessment (LCA) approach can also be used to identify potential environmental and economic trade-offs for different storage technologies to ensure long-term sustainability for these systems35.

Continued investment in research and development, public-private partnerships, and supportive policies will be critical to advancing these technologies further and accelerating the transition to a renewable energy future. Emerging technologies like advances in materials science and hybrid systems, combined with decentralized energy storage systems, are set to revolutionize the energy paradigm, rendering renewable energy resources more trustworthy, accessible, and sustainable for generations to come.

CONCLUSION

Renewable energy storage —where are we headed? As global energy systems transition to cleaner, more sustainable sources, the development and deployment of advanced energy storage technologies are critical to addressing the intrinsic challenges of renewables intermittency.

Discoveries in advanced materials like solid-state and metal-air batteries are leading to storage systems that are more cost-effective, safer and sustainable. Hybrid energy storage is an artificial type of energy storage that combines multiple different types of energy storage systems in order to achieve the needed performance and support the various energy requirements. Decentralized and distributed storage systems push boundaries in energy management, making it accessible and secure for communities across the globe.

Such breakthroughs are central not only to technology but also to a necessary global transition in energy. Next-gen storage solutions will revolutionize how grids and homes optimize their usage and reduce carbon footprints, making them a stepping stone for building a sustainable future. Additionally, the emphasis on life cycle s, sustainability, and circular economy principles ensures that these technologies are applied in ways that are consistent with broader environmental and societal goals.

With further developments to come, the world's least fortunate will surely benefit, as will the world as a whole. Next-generation energy storage technologies are not just tools; they are the foundation on which we can build a robust, efficient, and environmentally friendly global energy system. The path towards a sustainable energy future will certainly hinge on the adoption and implementation of such ground-breaking solutions.

REFERENCES

Abraham, K. M., & Jiang, Z. (1996). A polymer Electrolyte-Based rechargeable Lithium/Oxygen battery. *Journal of the Electrochemical Society*, *143*(1), 1–5. DOI: 10.1149/1.1836378

Adeyinka, A. M., Esan, O. C., Ijaola, A. O., & Farayibi, P. K. (2024). Advancements in hybrid energy storage systems for enhancing renewable energy-to-grid integration. *Sustainable Energy Research*, *11*(1), 26. Advance online publication. DOI: 10.1186/s40807-024-00120-4

Aghmadi, A., & Mohammed, O. A. (2024). Energy Storage Systems: Technologies and High-Power Applications. *Batteries*, *10*(4), 141. DOI: 10.3390/batteries10040141

Al-Gahtani, S. S., Hubona, G. S., & Wang, J. (2007). Information technology (IT) in Saudi Arabia: Culture and the acceptance and use of IT. *Information & Management*, *44*(8), 681–691. DOI: 10.1016/j.im.2007.09.002

Amusa, H. K., Sadiq, M., Alam, G., Alam, R., Siefan, A., Ibrahim, H., Raza, A., & Yildiz, B. (2024). Electric vehicle batteries waste management and recycling challenges: A comprehensive review of green technologies and prospects. *Journal of Material Cycles and Waste Management*, *26*(4), 1959–1978. DOI: 10.1007/s10163-024-01982-y

Ansari, M. Z., Habib, F., Gupta, J., Alsubaie, A. S., Barsoum, I., Alfantazi, A., Yatoo, M. A., Ahmad, Z., & Hussain, I. (2024). Frontiers in metal-organic frameworks: Innovative nanomaterials for next-generation supercapacitors. *Advanced Composites and Hybrid Materials*, *7*(6), 215. Advance online publication. DOI: 10.1007/s42114-024-00996-2

Aoun, A., Kashmar, N., Adda, M., & Ibrahim, H. (2024). From Bottom-Up towards a completely decentralized autonomous electric grid based on the concept of a decentralized autonomous substation. *Electronics (Basel)*, *13*(18), 3683. DOI: 10.3390/electronics13183683

Arévalo, P., Ochoa-Correa, D., & Villa-Ávila, E. (2024). Advances in thermal energy storage Systems for Renewable Energy: A review of recent developments. *Processes (Basel, Switzerland)*, *12*(9), 1844. DOI: 10.3390/pr12091844

Bertaglia, T., Costa, C. M., Lanceros-Méndez, S., & Crespilho, F. N. (2024). Eco-friendly, sustainable, and safe Energy Storage: A Nature-Inspired Materials Paradigm Shift. *Materials Advances*, *5*(19), 7534–7547. DOI: 10.1039/D4MA00363B

Chen, P., Bai, F., Deng, J. W., Liu, B., & Zhang, T. (2022). Recent progresses and challenges in aqueous lithium-air batteries relating to the solid electrolyte separator: A mini-review. *Frontiers in Chemistry*, *10*, 1035691. Advance online publication. DOI: 10.3389/fchem.2022.1035691 PMID: 36300027

Dong, Z., Zhang, Z., Li, Z., Li, X., Qin, J., Liang, C., Han, M., Yin, Y., Bai, J., Wang, C., & Wang, R. (2022). A survey of Battery–Supercapacitor Hybrid Energy Storage Systems: Concept, Topology, control and Application. *Symmetry*, *14*(6), 1085. DOI: 10.3390/sym14061085

Dutt, S., Kumar, A., & Singh, S. (2023). Synthesis of metal-organic frameworks (MOFs) and their derived materials for energy storage applications. *Cleanroom Technology*, *5*(1), 140–166. DOI: 10.3390/cleantechnol5010009

Farooq, N., Rehman, Z. U., Khan, M. I., Asghar, S., Saleem, M., Irshad, R., Sheikh, A., Shanableh, A., Manzoor, S., & Khan, Z. U. (2024). Nanomaterial-based energy conversion and energy storage devices: A comprehensive review. *New Journal of Chemistry*, *48*(19), 8933–8962. DOI: 10.1039/D3NJ04846B

Glowacki, B. A., & Hanley, E. S. (2016). Energy storage Technology for decentralized Energy Management: Prospects. In *InTech eBooks*. DOI: 10.5772/63415

Haase, M., Wulf, C., Baumann, M., Rösch, C., Weil, M., Zapp, P., & Naegler, T. (2022). Prospective assessment of energy technologies: A comprehensive approach for sustainability assessment. *Energy, Sustainability and Society*, *12*(1), 20. Advance online publication. DOI: 10.1186/s13705-022-00344-6

Hunt, J. D., Nascimento, A., Zakeri, B., Jurasz, J., Dąbek, P. B., Barbosa, P. S. F., Brandão, R., De Castro, N. J., Filho, W. L., & Riahi, K. (2022). Lift Energy Storage Technology: A solution for decentralized urban energy storage. *Energy*, *254*, 124102. DOI: 10.1016/j.energy.2022.124102

Hwang, J., Myung, S., & Sun, Y. (2017c). Sodium-ion batteries: Present and future. *Chemical Society Reviews*, *46*(12), 3529–3614. DOI: 10.1039/C6CS00776G PMID: 28349134

Ilyas, U., Khan, M. N. U., Ashfaq, U., & Qaiser, I. (2021). Energy optimization of hybrid energy storage system (HESS) for hybrid electric vehicle (HEV). *Engineering Proceedings, 12*(1), 75. Presented at the 1st International Conference on Energy, Power, and Environment, Gujrat, Pakistan, p. 11–12 November 2021. DOI: 10.3390/engproc2021012075

Jung, K., Kim, J., Yamauchi, Y., Park, M., Lee, J., & Kim, J. H. (2016). Rechargeable lithium-air batteries: A perspective on the development of oxygen electrodes. *Journal of Materials Chemistry. A, Materials for Energy and Sustainability*, *4*(37), 14050–14068. DOI: 10.1039/C6TA04510C

Kandari, R., Neeraj, N., & Micallef, A. (2022). Review recent strategies for integrating energy storage systems in microgrids. *Energies*, *16*(1), 317. DOI: 10.3390/en16010317

Kapoor, R. T., Rafatullah, M., Qamar, M., Qutob, M., Alosaimi, A. M., Alorfi, H. S., & Hussein, M. A. (2022). Review on recent developments in Bioinspired-Materials for sustainable energy and environmental applications. *Sustainability (Basel)*, *14*(24), 16931. DOI: 10.3390/su142416931

Kumar, R., Thakur, A. K., Gupta, L. R., Gehlot, A., & Sikarwar, V. S. (2023). Advances in phase change materials and nanomaterials for applications in thermal energy storage. *Environmental Science and Pollution Research International*, *31*(5), 6649–6677. DOI: 10.1007/s11356-023-31718-8 PMID: 38158531

Le, S. T., Nguyen, T. N., Bui, D., Teodosio, B., & Ngo, T. D. (2023). Comparative life cycle assessment of renewable energy storage systems for net-zero buildings with varying self-sufficient ratios. *Energy*, p. *290*, 130041. DOI: 10.1016/j.energy.2023.130041

Li, P., Luo, S., Lin, Y., Xiao, J., Xia, X., Liu, X., Wang, L., & He, X. (2024). Fundamentals of the recycling of spent lithium-ion batteries. *Chemical Society Reviews*, *53*(24), 11967–12013. Advance online publication. DOI: 10.1039/D4CS00362D PMID: 39471089

Llamas-Orozco, J. A., Meng, F., Walker, G. S., Abdul-Manan, A. F. N., MacLean, H. L., Posen, I. D., & McKechnie, J. (2023). Estimating the environmental impacts of global lithium-ion battery supply chain: A temporal, geographical, and technological perspective. *PNAS Nexus*, *2*(11), pgad361. Advance online publication. DOI: 10.1093/pnasnexus/pgad361 PMID: 38034093

Machín, A., Cotto, M. C., Díaz, F., Duconge, J., Morant, C., & Márquez, F. (2024). Environmental Aspects and Recycling of Solid-State Batteries: A Comprehensive Review. *Batteries*, *10*(7), 255. DOI: 10.3390/batteries10070255

Machín, A., Morant, C., & Márquez, F. (2024). Advancements and Challenges in Solid-State Battery Technology: An In-Depth Review of solid electrolytes and anode innovations. *Batteries*, *10*(1), 29. DOI: 10.3390/batteries10010029

Mahajan, V., Muller, E., & Bass, F. M. (1990). New product diffusion Models in Marketing: A review and directions for research. *Journal of Marketing*, *54*(1), 1–26. DOI: 10.1177/002224299005400101

Manthiram, A. (2020b). A reflection on lithium-ion battery cathode chemistry. *Nature Communications*, *11*(1), 1550. Advance online publication. DOI: 10.1038/s41467-020-15355-0 PMID: 32214093

Naqvi, A. A., Zahoor, A., Shaikh, A. A., Butt, F. A., Raza, F., & Ahad, I. U. (2022). Aprotic lithium-air batteries with oxygen-selective membranes. *Materials for Renewable and Sustainable Energy*, *11*(1), 33–46. DOI: 10.1007/s40243-021-00205-w

Nikolaos, P. C., Marios, F., & Dimitris, K. (2023). A review of pumped hydro storage systems. *Energies*, *16*(11), 4516. DOI: 10.3390/en16114516

Pop, C., Cioara, T., Antal, M., Anghel, I., Salomie, I., & Bertoncini, M. (2018). Blockchain-based decentralized management of demand response programs in smart energy grids. *Sensors (Basel)*, *18*(1), 162. DOI: 10.3390/s18010162 PMID: 29315250

Rabi, A., Radulovic, J., & Buick, J. (2023). Comprehensive Review of Compressed Air Energy Storage (CAES) technologies. *Thermo*, *3*(1), 104–126. DOI: 10.3390/thermo3010008

Rafał, K., Radziszewska, W., Grabowski, O., Biedka, H., & Verstraete, J. (2022). Energy Cost Minimization with Hybrid Energy Storage System Using Optimization Algorithm. *Applied Sciences (Basel, Switzerland)*, *13*(1), 518. DOI: 10.3390/app13010518

Raza, M. S., Abid, M. I., Akmal, M., Munir, H. M., Haider, Z. M., Khan, M. O., Alamri, B., & Alqarni, M. (2024). A comprehensive assessment of storage elements in hybrid energy systems to optimize energy reserves. *Sustainability (Basel)*, *16*(20), 8730. DOI: 10.3390/su16208730

Reilly, J., Poudel, R., Krishnan, V., Anderson, B., Rane, J., Baring-Gould, I., & Clark, C. (2022). *Hybrid distributed wind and battery energy storage systems* (NREL/TP-5000-77662). National Renewable Energy Laboratory. https://www.nrel.gov/docs/fy22osti/77662.pdf

Rufer, A. (2017). Energy storage. In *CRC Press eBooks*. DOI: 10.1201/b22265

Saldarini, A., Longo, M., Brenna, M., & Zaninelli, D. (2023). Battery Electric Storage Systems: Advances, challenges, and market trends. *Energies*, *16*(22), 7566. DOI: 10.3390/en16227566

Schubert, C., Hassen, W., Poisl, B., Seitz, S., Schubert, J., Usabiaga, E. O., Gaudo, P., & Pettinger, K. (2023). Hybrid energy storage systems based on Redox-Flow batteries: Recent developments, challenges, and future perspectives. *Batteries*, *9*(4), 211. DOI: 10.3390/batteries9040211

Shu, X., Kumar, R., Saha, R. K., Dev, N., Stević, Ž., Sharma, S., & Rafighi, M. (2023). Sustainability Assessment of Energy Storage Technologies Based on Commercialization Viability: MCDM Model. *Sustainability (Basel)*, *15*(6), 4707. DOI: 10.3390/su15064707

Simão, M., & Ramos, H. (2020). Hybrid Pumped Hydro Storage Energy Solutions towards Wind and PV Integration: Improvement on Flexibility, Reliability and Energy Costs. *Water (Basel)*, *12*(9), 2457. DOI: 10.3390/w12092457

Stacks, D. W., Salwen, M. B., & Eichhorn, K. C. (2019). An integrated approach to communication theory and research. In *Routledge eBooks*. DOI: 10.4324/9780203710753

Tan, P., Jiang, H., Zhu, X., An, L., Jung, C., Wu, M., Shi, L., Shyy, W., & Zhao, T. (2017). Advances and challenges in lithium-air batteries. *Applied Energy, pp. 204*, 780–806. DOI: 10.1016/j.apenergy.2017.07.054

Thakur, N., Samanta, P., & Mishra, S. (2023). Methods and technologies for recycling energy storage materials and devices. In *Green energy and technology* (pp. 489–507). DOI: 10.1007/978-981-99-4193-3_28

Vaghela, P., Pandey, V., Sircar, A., Yadav, K., Bist, N., & Kumari, R. (2023). Energy storage techniques, applications, and recent trends: A sustainable solution for power storage. *MRS Energy & Sustainability : a Review Journal*, *10*(2), 261–276. DOI: 10.1557/s43581-023-00069-9

Vedhanarayanan, B., & Lakshmi, K. C. S. (2024). Beyond lithium-ion: Emerging frontiers in next-generation battery technologies. *Frontiers in Batteries and Electrochemistry*, *3*, 1377192. Advance online publication. DOI: 10.3389/fbael.2024.1377192

Venkatesh, V., & Davis, F. D. (2000). A theoretical extension of the technology acceptance model: Four longitudinal field studies. *Management Science*, *46*(2), 186–204. DOI: 10.1287/mnsc.46.2.186.11926

Venkateswarlu, K., & Ramakrishna, K. (2021). Recent advances in phase change materials for thermal energy storage review. *Journal of the Brazilian Society of Mechanical Sciences and Engineering*, *44*(1), 6. Advance online publication. DOI: 10.1007/s40430-021-03308-7

Wang, C., Xie, Z., & Zhou, Z. (2019). Lithium-air batteries: Challenges coexist with opportunities. *APL Materials*, *7*(4), 040701. Advance online publication. DOI: 10.1063/1.5091444

Wu, T., Xu, D., & Yang, J. (2021). Decentralized energy and its performance assessment models. *Frontiers of Engineering Management*, *8*(2), 183–198. DOI: 10.1007/s42524-020-0148-7

Yamaki, A., Fujii, S., Kanematsu, Y., & Kikuchi, Y. (2024). Comparative Life cycle Assessment of energy storage systems for connecting Large-Scale Wind Energy to the grid. *Journal of Chemical Engineering of Japan*, *57*(1), 2406316. Advance online publication. DOI: 10.1080/00219592.2024.2406316

Yao, L., Yang, B., Cui, H., Zhuang, J., Ye, J., & Xue, J. (2016). Challenges and progress of energy storage technology and its application in power systems. *Journal of Modern Power Systems and Clean Energy*, *4*(4), 519–528. DOI: 10.1007/s40565-016-0248-x

Zheng, C., Wang, Y., Liu, Z., Sun, T., Kim, N., Jeong, J., & Cha, S. W. (2021). A hybrid energy storage system for an electric vehicle and its effectiveness validation. *International Journal of Precision Engineering and Manufacturing-Green Technology*, *8*(5), 1739–1754. DOI: 10.1007/s40684-020-00304-5

KEY TERMS AND DEFINITIONS

Advanced Materials: Cutting-edge materials designed to improve energy storage systems, such as solid-state batteries or lithium-air, offering higher efficiency, durability, and energy density.

Carbon Footprint Reduction: The process of minimizing the environmental impact of energy use by lowering the amount of carbon dioxide and other greenhouse gases emitted into the atmosphere.

Decentralized Grid: A power system where energy generation and storage are distributed across multiple locations rather than relying on a central, large-scale grid, improving reliability and resilience.

Energy Storage: The capture of energy for later use, typically in systems such as batteries, capacitors, or pumped hydro storage, to manage supply and demand, especially with renewable sources.

Hybrid Storage Systems: Energy storage systems that combine multiple technologies, such as batteries and supercapacitors, to optimize efficiency, cost-effectiveness, and performance for renewable energy integration.

Renewable Energy: Energy derived from natural, replenishable sources like solar, wind, and hydropower, offering a sustainable alternative to fossil fuels and reducing environmental impacts.

Sustainable Energy: Energy produced in ways that meet current needs without compromising the ability of future generations to meet their own needs, emphasizing renewable sources and minimal environmental harm.

Chapter 10
Innovative Materials for Solar Cells, Photocatalysis, and Hydrogen Storage in Renewable Energy Systems

Siddharthsingh K. Chauhan
https://orcid.org/0000-0003-2216-616X
Nirma University, India

Vineeta Chauhan
https://orcid.org/0000-0002-6147-3755
Indus University, India

ABSTRACT

The transition to renewable energy systems demands advanced materials capable of enhancing the efficiency and performance of solar cells, photocatalysis, and hydrogen storage technologies. This chapter explores innovative materials that are shaping the future of sustainable energy. In solar cells, new generations of materials such as perovskites and organic photovoltaics are achieving higher efficiencies with lower production costs. These materials exhibit superior light absorption and charge carrier mobility, making them promising for next-generation solar energy harvesting. In photocatalysis, nanostructured semiconductors, including titanium dioxide (TiO_2) and graphene-based composites, are emerging as catalysts for environmental purification and hydrogen production through water splitting. By focusing

DOI: 10.4018/979-8-3693-9316-1.ch010

on material innovations in these areas, this chapter highlights the pivotal role of cutting-edge research in advancing renewable energy technologies and paving the way for sustainable, high-performance energy systems

INTRODUCTION

The pressing need of transition to sustainable energy sources has led to a surge in research on innovative materials for renewable energy systems. Solar cells, photocatalysis, and hydrogen storage stand as three critical pillars in this quest to curb greenhouse gas emissions and meet the world's increasing energy demand. These technologies harness, convert, and store energy from the sun—an abundant and clean energy source—offering a promising alternative to fossil fuels. However, their effectiveness largely depends on the development of high-performance, efficient, and scalable materials capable of overcoming current limitations.

Solar cells, which convert sunlight into electricity, are at the forefront of renewable energy systems. Traditional silicon-based solar cells have dominated the market due to their well-established manufacturing processes and relatively high efficiency. However, their production can be energy-intensive, and performance under low-light conditions is limited. Thus, the development of next-generation materials such as perovskites, organic photovoltaics (OPVs), and quantum dots is transforming the field (Bai, Y., Zhang, J., & Liu, C. 2018). Perovskite solar cells (PSCs), in particular, have garnered substantial attention due to their remarkable light-absorption properties, tunable bandgap, and the potential for low-cost fabrication. These materials have achieved power conversion efficiencies comparable to silicon in just over a decade of development, sparking excitement for future scalability and commercialization.

Photocatalysis, another crucial aspect of renewable energy systems, leverages light to drive chemical reactions. The most common application is photocatalytic water splitting, where light energy splits water molecules into hydrogen and oxygen. This process produces hydrogen—a clean energy carrier that can be utilized in fuel cells or as an alternative fuel source. The challenge lies in creating photocatalysts that are both efficient and durable. Titanium dioxide (TiO_2) has been widely used due to its stability and strong oxidative power, but it requires ultraviolet light for activation, limiting its efficiency under natural sunlight.Consequently, scientists have been exploring innovative materials, including metal-organic frameworks (MOFs), carbon-based compounds, and doped semiconductors, to enhance the visible-light responsiveness and overall performance of photocatalytic systems (Batmunkh, M., Shapter, J. G., & Shearer, C. J. 2020).

Hydrogen storage, a key component of realizing a sustainable hydrogen economy, is critical to bridging the gap between hydrogen production and its utilization in energy applications. Current storage technologies face challenges related to the high pressures and cryogenic temperatures required to store hydrogen gas or liquid. This has spurred interest in innovative materials that can safely store hydrogen at ambient conditions while maintaining high energy densities. Metal hydrides, complex hydrides, and nanomaterials such as graphene and carbon nanotubes are emerging as promising one for hydrogen storage. These materials exhibit high hydrogen uptake, stability, and favorable release kinetics, but ongoing research aims to further optimize their properties for practical applications.

Solar Photo Voltaic System

Solar energy can be harnessed and utilized in two primary ways: through the solar thermal route and the solar electricity route.

Solar Thermal Route: This method involves the use of devices and technologies that capture and use the sun's heat. Examples include solar collectors, solar water heaters, solar dryers, and more. These systems are designed to absorb and transfer heat for applications such as heating water, air conditioning, drying crops, or even powering certain industrial processes (Chen S, Takata T, Domen K, 2017). The key idea is to directly convert sunlight into thermal energy for immediate use or storage.

Solar Electricity Route (Photovoltaics): This method generates electricity from sunlight using solar photovoltaic (SPV) systems. An SPV system typically comprises several interconnected components, including:

PV Panels (Arrays): These panels consist of solar cells that convert sunlight into direct current (DC) electricity.

Charge Controller: This device regulates the flow of electricity from the PV panels to batteries or the load, preventing overcharging or excessive discharge and ensuring system longevity.

DC-to-DC Converter: Used to optimize the voltage levels from the PV panels to suit various components or batteries.

Inverter (DC-to-AC): Converts the direct current produced by the PV panels into alternating current (AC), which is used by most household appliances and can be fed into the electrical grid.

Power Meter and Breaker: These components monitor and protect the electrical system, tracking energy production and usage, while preventing overloads.

Battery Storage System: Depending on the size and design of the SPV system, energy storage is typically managed through one or more batteries, enabling the use of solar electricity during cloudy periods or at night.Fig.1 shows schematic diagram of the solar photovoltaic system (Cho, K. T., Bae, H. S., & Kim, S. 2018).

Figure 1. Schematic diagram of the solar photovoltaic systems

Figure 1. Schematic diagram of the solar photovoltaic systems

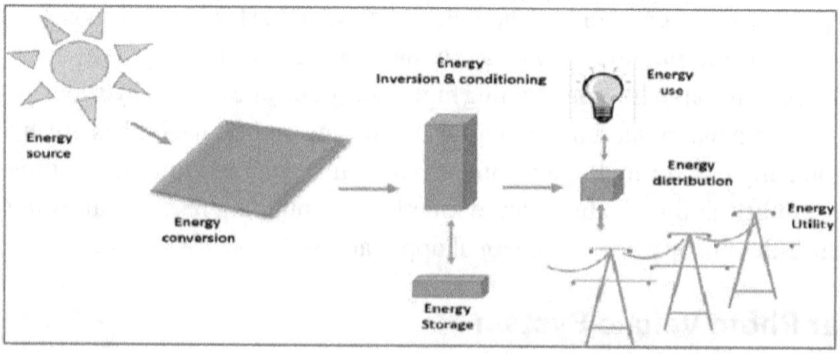

Solar photovoltaic technology directly converts sunlight into electricity without the need for intermediate conversion processes. This approach offers numerous benefits, including being pollutant-free during operation, robust in structure, and straightforward in design. PV systems also play a role in mitigating global warming issues by reducing carbon emissions. They are modular in nature, offer lower operational costs, require minimal maintenance, and can generate power across a wide range of scales, from microwatts to megawatts. Among renewable energy technologies, solar PV boasts the highest power density (. Choi, J., Kim, D., & Song, C. 2021). The increasing interest in PV technology is evident from significant growth trends. In 2013, for example, 100 MW of new capacity were installed daily. By 2015, the cumulative global PV capacity had reached a record 177 GW, reflecting rapid advancements and large-scale adoption. Based on the current rate of 100 MW of new capacity installed daily, the cumulative global photovoltaic capacity by 2030 could reach approximately 725 GW.

Despite its advantages, solar photovoltaic systems face notable challenges. As noted by Nadia et al., they remain relatively expensive compared to conventional fossil fuel energy sources, suffer from low conversion efficiencies, and experience intermittent power output due to the variability of sunlight. These limitations continue to drive research and innovation aimed at improving the cost-effectiveness, efficiency, and reliability of PV technology

The convergence of innovative materials across these three areas—solar cells, photocatalysis, and hydrogen storage—represents a powerful synergy within renewable energy systems. Advances in nanotechnology, material synthesis, and computational modeling have accelerated the discovery of new materials and the optimization of existing ones. Moving forward, multidisciplinary collaboration will be essential to address challenges related to cost, stability, and large-scale deployment. Emphasizing advancements in materials allows us to fully harness the capabilities of renewable energy technologies, paving the way for a greener and more sustainable future. This ongoing progress not only holds promise for reducing dependence on fossil fuels but also offers pathways to cleaner air, energy security, and a resilient economy.

Advanced Materials for Renewable Energy Systems

Advanced materials are revolutionizing renewable energy systems, enhancing the performance of solar cells, photocatalysis, and hydrogen storage technologies. Innovations such as perovskites and organic photovoltaics in solar cells offer high efficiency and flexibility at lower costs. Photocatalytic materials, like nanostructured semiconductors and graphene composites, enable environmental purification and efficient hydrogen production through water splitting. Hydrogen storage is advancing with metal-organic frameworks and complex hydrides that offer high density and faster absorption kinetics. These materials offer enhanced efficiency, scalability, and sustainability, driving the shift toward cleaner energy solutions at a faster pace (Geng T, Cai W, Wu L,2022)

The integration of renewable energy technologies like solar energy, photocatalysis, and hydrogen storage into a unified energy system offers promising pathways for sustainable energy solutions. Recent advancements focus on combining these technologies to optimize energy conversion, storage, and utilization. For example, Zhang et al. (2023) developed integrated systems combining solar-driven water splitting with efficient hydrogen storage, which significantly reduces energy loss and enhances overall system performance (Hou, Y., Zhuang, S., & Liu, Y., 2017). This integration addresses the intermittency of solar energy by coupling it with reliable hydrogen storage, creating a more consistent and scalable energy source.

OVERVIEW OF SOLAR CELLS, PHOTOCATALYSIS, AND HYDROGEN STORAGE TECHNOLOGIES

Recent research efforts are transforming renewable energy technologies with innovative materials designed to address efficiency and sustainability challenges. In solar cells, perovskite materials are a leading focus due to their remarkable ability to convert sunlight into electricity with high efficiency and relatively low manufacturing costs. A team led by researchers Hongguang Meng et al. developed a perovskite-based p–i–n structure that resists halide oxidation, enhancing its air stability and potential scalability for commercial applications (Meng et al., 2024). Similarly, integrated solar cell-energy storage systems, which combine solar cells with super capacitors or batteries, are gaining traction for their ability to manage intermittent energy production and improve grid reliability (Lu et al., 2024)

Photocatalysis is advancing through the development of nanostructured semiconductors. Materials like titanium dioxide and composite graphene are being tailored for applications such as environmental purification and water splitting to produce hydrogen. These nanomaterials demonstrate improved charge separation, increased surface area, and chemical resilience, making them effective for decomposing pollutants and generating clean fuel (Huang, Q., Liu, S., & Yu, X. 2016). Cutting-edge studies are exploring hybrid photocatalysts, including modifications that incorporate carbon-based materials, to achieve superior performance in light absorption and catalytic reactions.

Innovative Materials for Solar Cells

Silicon Photovoltaics

Silicon dominates as the primary semiconductor material in solar cells, making up approximately 95% of solar modules currently on the market. As the second most abundant element on Earth after oxygen, silicon is extensively used as a semiconductor, particularly in computer chips. Crystalline silicon cells consist of silicon atoms organized in a crystal lattice structure, a configuration that facilitates the efficient transformation of sunlight into electricity (Laufkötter C, Zscheischler J, Frölicher TL 2020)

Silicon-based solar cells offer a compelling combination of high efficiency, affordability, and long service life. Modules typically retain over 80% of their initial power output even after 25 years of operation, making them a durable and reliable option for solar energy generation.

Thin-Film Photovoltaics

Thin-film solar cells are created by layering one or more photovoltaic (PV) materials onto substrates like glass, metal, or plastic. The most commonly used thin-film PV semiconductors are cadmium telluride (CdTe) and copper indium gallium diselenide (CIGS). These materials can be applied to either the front or back surfaces of solar modules Lee, H., Kang, M., & Yoo, S. 2020).

CdTe is the second most utilized PV material after silicon and is known for its cost-efficient production processes, although its efficiency falls short of silicon cells. In contrast, CIGS cells achieve high efficiency in laboratory conditions due to their excellent PV properties, but the complexity of integrating four elements makes large-scale manufacturing more challenging. To ensure durability in outdoor environments, both CdTe and CIGS thin-film modules require additional protective measures (Qi Y, Liang X, 2021)

Perovskite Photovoltaics

Perovskite solar cells, a type of thin-film photovoltaic technology, derive their name from their unique crystal structure. These cells are formed by printing, coating, or vacuum-depositing layers onto a supporting substrate. Known for their relatively straightforward production process, they can achieve efficiencies comparable to crystalline silicon. Since their inception, perovskite solar cells have seen remarkable progress in efficiency, increasing from approximately 3% in 2009 to over 25% in laboratory settings by 2020 (& Wang, J., 2014). For commercial success, these cells need to be more durable, with the ability to endure 20 years of outdoor use. Current research focuses on improving their stability and creating scalable, cost-efficient production methods.

Figure 2 shows structure of Perovskite Solar Cells.Perovskite solar cells consist of several key layers that contribute to their high efficiency. The substrate, often glass, plastic, or metal, is typically coated with a transparent conductive oxide like indium tin oxide or fluorine-doped tin oxide (FTO) and acts as the front electrode, providing a stable base for the cell. The electron transport layer (ETL), made from materials such as titanium dioxide or zinc oxide, facilitates the flow of electrons from the perovskite layer to the electrode, enhancing charge extraction. The perovskite absorber layer is the heart of the cell and features compounds like methylammonium lead iodide ($CH_3NH_3PbI_3$) that absorb sunlight and generate charge carriers (& Wang, J.2014). Next, the hole transport layer (HTL) selectively moves holes to the back electrode while blocking electrons, with common materials including spiro-OMeTAD and PEDOT. Finally, the back electrode, often made from metals such as gold or aluminum, completes the electrical circuit by collecting holes and conducting them to an external load, ensuring efficient power output(Miele, A., & Kanoh, 2018)

Figure 2. Structure of perovskite solar cells

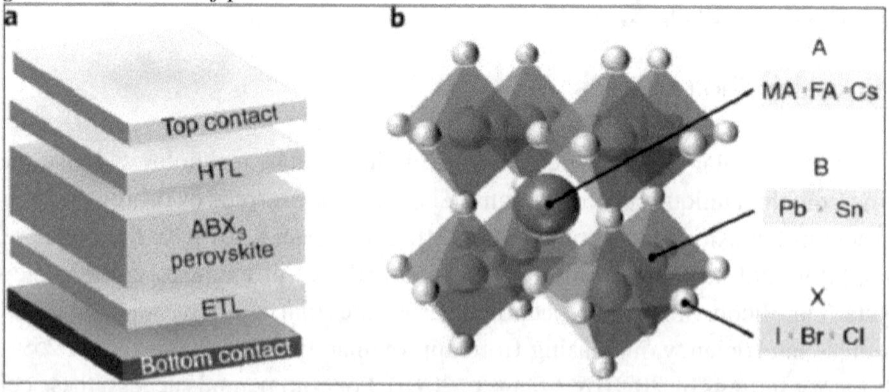

Organic Photovoltaics

Organic photovoltaic (OPV) cells are constructed from carbon-based compounds and can be designed to optimize characteristics like bandgap, transparency, or color. Although their efficiency is currently about half that of crystalline silicon cells and their operational lifespan is relatively shorter, they present the potential for significantly lower production costs at large scales. OPVs are compatible with various flexible substrates, including plastic, enabling versatile and innovative uses, such as in lightweight and portable solar devices.

Quantum Dot Solar Cells

Quantum dot solar cells generate electricity using tiny semiconductor particles known as quantum dots, which are just a few nanometers in size. These particles offer a novel approach to processing semiconductor materials. However, one of the primary challenges is establishing effective electrical connections between the quantum dots, which currently limits their overall efficiency. Despite this, quantum dot solar cells are relatively easy to produce and can be fabricated using techniques such as spin-coating, spraying, or roll-to-roll printing, similar to newspaper printing processes (Satyapal, S., & Aines, R. 2016)

Quantum dots are distinguished by their adjustable bandgaps, enabling them to absorb light wavelengths that are usually harder to capture. This adaptability makes them ideal for integration with other semiconductors, like perovskites, to boost the efficiency of multijunction solar cells. By utilizing this tunable property, quantum dot solar cells offer significant potential for enhancing light absorption and improving the overall performance of solar cells.

PHOTOCATALYSIS & IMPORTANCE OF MATERIALS IN RENEWABLE ENERGY

The importance of innovative materials in renewable energy systems cannot be overstated, as they directly influence system performance, cost efficiency, and scalability. Perovskite solar cells, for instance, are a major driver of next-generation photovoltaics due to their high power conversion efficiency and potential for flexible, lightweight applications. Researchers from the University of Colorado, including Michael D. McGehee and colleagues, emphasize that precise material engineering, such as controlling halide composition, can significantly improve perovskite durability, addressing a major hurdle to commercialization.

In photocatalysis, graphene-enhanced materials offer high electrical conductivity and large surface areas, leading to increased efficiency in hydrogen production through water splitting. This makes it possible to harness sunlight for clean hydrogen generation, a critical component for energy storage and fuel cell technology. Advancements in hydrogen storage materials, such as metal-organic frameworks (MOFs), offer higher storage densities and improved adsorption/desorption kinetics, effectively tackling key challenges related to safety and capacity.

Together, these advancements demonstrate that material innovation is at the core of building efficient, high-performance, and sustainable energy systems, making renewable energy more viable for widespread adoption. The combination of advanced material synthesis, design, and integration into devices holds the key to realizing the potential of clean energy technologies and reducing dependence on fossil fuels. Few of them are described below:

Nanostructured Semiconductors

Nanostructured semiconductors are at the forefront of photocatalytic innovation due to their superior surface-to-volume ratios and enhanced reactivity. Recent studies have focused on modifying surface morphology, doping, and heterojunction formation to optimize light absorption and charge separation. For example, Li et al. (2020) developed nanostructured titanium-based compounds combined with rare earth elements to improve visible light utilization. Such modifications reduce recombination rates, enhancing overall photocatalytic efficiency. Applications include environmental purification and hydrogen generation via water splitting.

Titanium Dioxide in Photocatalysis

Titanium dioxide remains a staple in photocatalytic research owing to its stability, affordability, and nontoxicity. However, its wide band-gap limits activity to UV light. Research by Zhao et al. (2019) explored doping TiO_2 with transition metals and coupling it with carbon-based materials to extend absorption into the visible spectrum. Modifications such as anatase-rutile heterostructures have also been employed to promote electron-hole separation, increasing efficiency for pollutant degradation and water splitting (Scott, L., & Hileman, E. 2015)

Graphene-Based Composites for Photocatalytic Applications

Graphene composites are increasingly used to boost the catalytic efficiency of semiconductors. Graphene's high conductivity and large surface area enhance charge transport and reduce recombination in photocatalytic systems. A recent study by Zhang et al. (2021) demonstrated that coupling TiO_2 with graphene oxide significantly improved photocatalytic degradation of organic pollutants under sunlight. Graphene-based hybrids with metal oxides and quantum dots are also being explored for improved light absorption and high reactivity, making them suitable for environmental remediation and hydrogen production.

Applications of Photocatalysis in Water and Air Purification

Photocatalysis is being widely applied in water and air purification systems to remove organic pollutants, heavy metals, and pathogens. TiO_2-based photocatalysts, often doped with metals or combined with carbon-based materials, enhance efficiency by improving light absorption. Recent studies have focused on scaling these technologies for real-world applications. For instance, Zhang et al. (2020) developed photocatalytic systems for the removal of pharmaceutical contaminants from wastewater. In air purification, photocatalysts degrade volatile organic compounds (VOCs), as demonstrated in the work by Wang et al. (2019), where TiO_2 nanostructures were employed to clean indoor air

Advances in Photocatalytic Efficiency

Enhancing photocatalytic efficiency involves optimizing charge separation and extending the light absorption spectrum. Research from Liu et al. (2021) demonstrated that coupling TiO_2 with graphene oxide significantly reduced recombination of electron-hole pairs, boosting the photocatalytic degradation of organic contaminants. Another study by Sun et al. (2022) introduced multi-component photocatalysts based on metal-organic frameworks (MOFs), which exhibited improved photocatalytic activity under visible light, opening pathways for more energy-efficient purification systems.These innovations are critical for advancing large-scale photocatalytic environmental purification technologies.

GREEN H_2 PRODUCTION METHODS

Renewable energy sources like solar, wind, and biomass can be utilized to generate sustainable biochemical, photonic, electrical, and thermal energy. These secondary energy forms are vital for enabling hydrogen production via water splitting processes. At present, thermal energy from fossil fuel-based steam reforming is the primary method of hydrogen generation. In contrast, water electrolysis, which offers environmental advantages and significant future potential, can use solar radiation as a major energy source. Furthermore, biological methods, such as dark fermentation and photo-fermentation by microorganisms, provide additional pathways for hydrogen production.

Hydrogen can be produced through various methods that use both renewable and non-renewable energy sources. Figure 3 showcases different water-splitting techniques powered by biochemical, photonic, electrical, and thermal energy, such as biolysis, photolysis, electrolysis, and thermolysis. Hybrid energy systems, which combine multiple energy types—like electrical with photonic, photonic with biochemical, or thermal with electrical—offer significant advantages in terms of thermodynamic efficiency, cost-effectiveness, and enhanced reaction kinetics compared to single-energy systems. Developments in catalyst technologies have sparked considerable interest in new water-splitting methods, including electrocatalytic, photocatalytic, photoelectrochemical, and photobiological processes, all focused on improving hydrogen production efficiency (Sun, G., Wang, W., & Liu, X. 2020).

Figure 3. Energy-driven methods for green hydrogen production

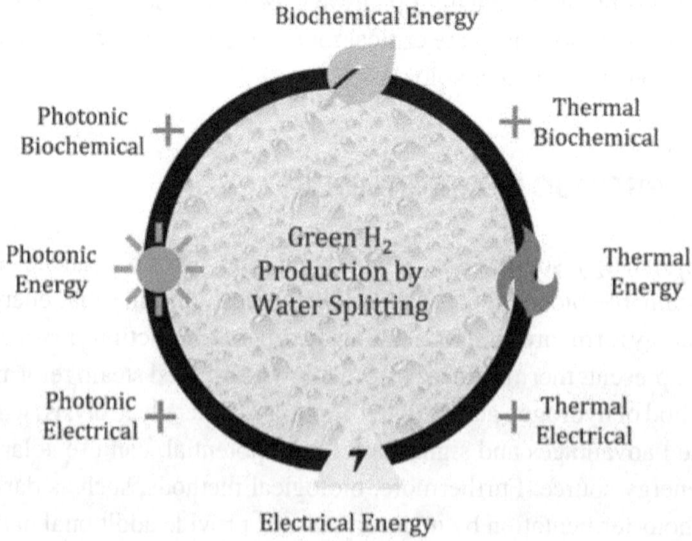

Green hydrogen production encompasses a variety of methods that utilize renewable energy sources to produce hydrogen with minimal environmental impact. Electrolysis of water is a key approach, where water (H_2O) is split into hydrogen (H_2) and oxygen (O_2) using an electric current. This method becomes "green" when powered by renewable energy sources like solar, wind, or hydropower. Among the types of electrolysis, alkaline electrolysis uses an alkaline electrolyte solution and is relatively mature and cost-effective, while proton exchange membrane (PEM) electrolysis employs a solid polymer membrane, offering high efficiency and flexibility in handling variable electricity inputs. Solid oxide electrolysis (SOEC) operates at high temperatures, enhancing efficiency through the use of both electricity and heat, but it remains a less mature technology. While electrolysis offers scalability and environmental benefits, it faces challenges such as electricity costs, sustainable electrode materials, and efficiency improvements.

Another promising method is photocatalytic water splitting, which utilizes sunlight and semiconductor photocatalysts to directly split water molecules into hydrogen and oxygen. This process is advantageous for its direct conversion of solar energy to chemical energy and potential for large-scale deployment at lower costs, though it currently suffers from low efficiency and the need for stable, effective photocatalysts. Relatedly, photoelectrochemical (PEC) water splitting combines elements of electrolysis and photocatalysis, using a photoelectrochemical cell to convert solar energy directly into hydrogen. This approach integrates light-absorbing semiconductors and electrolytes to generate current for water splitting, offering potential for high efficiencies, but faces challenges with material stability and large-scale implementation.

Biological hydrogen production leverages microorganisms such as bacteria and algae to generate hydrogen through various metabolic processes. Dark fermentation involves anaerobic bacteria breaking down biomass in the absence of light, yielding hydrogen and organic byproducts, while photo-fermentation uses photosynthetic bacteria that convert organic substrates into hydrogen with the aid of light. Another method, algal photolysis, employs algae to produce hydrogen through photosynthesis under specific conditions. While these methods utilize renewable organic materials and often operate under mild conditions, they face issues such as low yields and complex system control.

Thermolysis and thermochemical cycles present additional methods for green hydrogen production. Thermolysis uses high temperatures to directly split water molecules, requiring significant energy input and advanced reactor designs. Thermochemical cycles, on the other hand, rely on a series of chemical reactions driven by high temperatures—often supplied by concentrated solar power (CSP)—to generate hydrogen without directly splitting water molecules. These processes offer high potential efficiency, especially when coupled with solar heat, but face challenges related to high operational temperatures and the durability of reactor materials.

Hybrid energy systems that integrate two or more energy sources, such as electrical and photonic energy or thermal and electrical inputs, are designed to boost efficiency, lower costs, and improve reaction kinetics over single-energy systems. While these combinations create synergistic benefits, they also require intricate integration and sophisticated control mechanisms.

Finally, emerging electrocatalytic and photocatalytic methods seek to accelerate hydrogen production reactions by developing advanced catalysts, improving the efficiency and cost-effectiveness of processes such as electrolysis, photocatalysis, and photoelectrochemical methods. Collectively, these approaches represent the diverse and evolving strategies aimed at achieving scalable, efficient, and sustainable green hydrogen production for a clean energy future.

ROLE OF HYDROGEN IN RENEWABLE ENERGY SYSTEMS

Hydrogen is seen as a crucial energy source in the shift toward cleaner energy systems, with global efforts aimed at accelerating its use due to its transformative potential in multiple industries. As an energy carrier, hydrogen provides a route to significantly lower the carbon emissions from the power, transport, and industrial sectors. This promising capability positions hydrogen as a vital component in the strategies of nations striving for climate neutrality. The production of hydrogen from renewable sources such as solar and wind, known as green hydrogen, is particularly attractive due to its zero-emission potential and its role in energy storage, making it a key player in a sustainable energy future.

Hydrogen is crucial not only for energy storage and grid balancing but also for a variety of other applications. It plays a central role in hybrid renewable energy systems, as shown by Alzahrani et al., who studied the integration of hydrogen into hybrid microgrids to improve system flexibility and efficiency. Similarly, Bekele et al. investigated off-grid systems that combine hydrogen fuel cells with photovoltaic and wind power, demonstrating significant reductions in carbon emissions and excess electricity curtailment. In addition, Liponi et al. performed a techno-economic analysis, emphasizing hydrogen's role in optimizing energy generation and lowering costs in renewable hybrid systems.

Hydrogen production via water splitting, or electrolysis, presents a promising route to clean energy when powered by renewable sources like solar and wind. This method can produce green hydrogen with a significantly lower carbon footprint, leveraging the abundance and renewability of these energy sources while also helping to overcome energy storage challenges associated with renewable systems.

Hydrogen storage can be achieved through various physical methods, including gaseous hydrogen storage, liquid hydrogen storage, and storage using adsorbent materials. These methods are crucial for both stationary and transport applications. Among these, compressed hydrogen storage is the most advanced and commonly used, thanks to its simplicity and the availability of existing infrastructure.

Hydrogen storage is also essential for grid storage and balancing, helping integrate renewable energy systems (RES) such as wind and solar into power grid operations by managing the fluctuations in energy production. Advanced optimization techniques, such as the sparrow's search method, can enhance both local and global grid planning and stability. This approach typically involves electrolyzers, fuel cells, and hydrogen storage tanks, all working together to ensure efficient energy storage and retrieval.

In the transportation sector, hydrogen fuel cell vehicles (FCVs) offer a promising clean technology, providing long driving ranges, quick refueling, and zero greenhouse gas emissions. FCVs generate electricity by reacting hydrogen gas with oxygen, with water and heat as the only byproducts. According to Boichenko et al., hydrogen fuel cells can increase the efficiency of traditional energy systems by up to 85%, making them applicable in a variety of settings.

Recent Developments in Hydrogen Energy Storage

Reversible hydrogen storage can be accomplished using metal hydrides and adsorbent materials, with the hydrogen absorption capacity of these materials being a crucial factor for efficient storage. Important thermodynamic properties, such as the enthalpy of molecular hydrogen adsorption and the enthalpy related to the formation or decomposition of hydrides, are key to the storage process. Furthermore, factors like the kinetics of hydrogen adsorption, activation energy, hydrogen diffusion coefficient, and the rates of hydrogen absorption and desorption are essential considerations for effective storage.

Various laboratory techniques are available for measuring gas sorption, which help assess the hydrogen storage capacity of materials. In addition to capacity, optimizing solid hydrogen storage materials requires a thorough understanding of their physical and chemical properties, including microstructural features, which can be analyzed using X-ray diffraction and electron microscopy. Spectroscopic techniques provide further insight into the chemical and molecular characteristics of the materials being studied. However, these property measurements alone may not completely predict the sorption capacity of potential hydrogen storage materials. A recent comprehensive review examined the opportunities and challenges in hydrogen storage, with a particular emphasis on metal hydrides for sustainable energy applications.

Hydrogen Storage Using Metal Hydrides

Metal hydrides are compounds formed between metals and hydrogen, where the metal can exist as an element, alloy, or metal complex. The bonding type between the metal and hydrogen depends on the metal's electronegativity, which dictates how electrons are shared. For metals with low electronegativity, such as alkali metals (Group 1A) or alkaline earth metals (Group 2A), electrons are transferred from the metal to hydrogen, resulting in an ionic bond (Tao, J., Qiu, Z., & Pan, C. 2019). As the metal's electronegativity increases, the metal-hydrogen bond takes on a more metallic character. When the electronegativity is even higher, the bond becomes covalent. Metallic hydrides are commonly formed by transition metals, lanthanides, and actinides, where the overlap of electron energy levels between the metal and hydrogen leads to delocalized electrons and metallic bonding. In these compounds, hydrogen atoms typically occupy tetrahedral or octahedral voids within the metal lattice.

Hydrogen Storage for Mobility and Energy Applications

The rising demand for energy, combined with the need to reduce resource depletion and environmental impact, has accelerated the development of efficient, affordable, and environmentally friendly energy production and storage technologies. The heavy reliance on fossil fuels contributes significantly to greenhouse gas emissions, highlighting the urgent need for alternative, sustainable energy sources to meet the growing global electricity demand. In this regard, hydrogen is poised to play a key role in the future energy landscape, offering numerous benefits as a clean fuel, such as its light weight, high energy density, and the absence of harmful by-products during combustion.

Hydrogen is regarded as a green energy source due to its potential for production from renewable resources and its non-polluting nature. However, one of the main barriers to its widespread use, particularly in mobile transportation, is the challenge of providing safe and efficient storage solutions. To improve the performance of energy storage systems such as supercapacitors, lithium-ion batteries, fuel cells, and hydrogen storage technologies, researchers are increasingly concentrating on hydrogen as an energy carrier in the developing hydrogen economy (Tsai, Y. H., & Hwang, S.2021).

The adoption and commercialization of fuel cell electric vehicles (FCEVs) could be accelerated with the development of low-cost, lightweight, and compact hydrogen storage systems. If hydrogen is produced using low-carbon methods, these systems could play a significant role in reducing overall carbon emissions. To speed up this transition, it is crucial to deepen our understanding of the hydrogen storage technologies needed for transportation. An ideal storage solution should offer high volumetric and gravimetric energy densities, rapid fuel absorption and release, and efficient operation at room temperature and atmospheric pressure. Optimizing these characteristics is essential for creating a safe, effective, and affordable storage system. Currently, hydrogen storage technologies face challenges such as complex heat management, boil-off, low efficiency, expensive catalysts, stability concerns, slow response times, high operational pressures, low energy densities, and the potential for uncontrolled reactions. Despite these obstacles, compressed hydrogen storage remains the dominant solution in the industry and continues to provide a practical alternative for mobility compared to other technologies.

CHALLENGES AND CURRENT INNOVATIONS IN HYDROGEN PRODUCTION

Photocatalytic water splitting for hydrogen production faces several challenges, including the need for efficient, durable, and cost-effective materials. Key issues include improving the overall solar-to-hydrogen efficiency, overcoming the high recombination rate of charge carriers, and addressing the instability of photocatalysts under prolonged exposure to light and water. Recent innovations, such as doping photocatalysts with metals or coupling them with carbon-based materials, have improved charge separation and stability. For example, studies by Hisatomi et al.

(2021) focused on optimizing perovskite-based photocatalysts for higher efficiency in solar hydrogen production.

New Material Compositions for Improved Water Splitting Efficiency

Recent research has explored various material compositions to enhance the efficiency of photocatalytic water splitting. Heterojunctions, combining two semiconductors with complementary properties, have been successful in improving charge carrier separation. In 2022, Liu et al. reported on the use of metal-organic frameworks (MOFs) integrated with semiconductor materials, showing promising efficiency improvements under visible light. Additionally, recent studies on 2D materials like graphene and transition metal dichalcogenides have enhanced the photocatalytic performance of hydrogen production systems due to their high surface area and efficient electron mobility (Wei, Y., Liu, C., & Zhang, X. 2017)

Metal-Organic Frameworks: Design and Functionality

Metal-organic frameworks (MOFs) have emerged as promising materials for hydrogen storage due to their tunable porosity, high surface area, and the ability to accommodate large amounts of hydrogen. Recent research has focused on optimizing MOF structures by doping with metal centers or functionalizing organic linkers to enhance their hydrogen adsorption capacity. For example, in 2023, Zhang et al. developed a series of MOFs with tailored pore sizes, achieving significant improvements in hydrogen uptake at ambient conditions.

Complex Metal Hydrides: Capabilities and Limitations

Complex metal hydrides, such as lithium hydride (LiH) and magnesium hydride (MgH_2), offer high volumetric hydrogen storage densities. However, they face challenges including slow kinetics and the need for high temperatures to release hydrogen. Recent advancements, such as the work by Xie et al. (2022), have focused on improving the hydrogen release properties of metal hydrides by developing novel catalytic approaches and alloying techniques that allow faster hydrogen absorption/desorption at lower temperatures.

Nanostructured Materials for Enhanced Hydrogen Storage

Nanostructuring materials can enhance hydrogen storage by increasing surface area and optimizing adsorption characteristics. Recent studies, including those by Liu et al. (2021), demonstrated that nanostructured carbon-based materials and metal nanoparticles exhibit improved hydrogen storage capacities through increased interaction with hydrogen molecules. These materials have also been shown to have faster absorption kinetics, making them ideal candidates for efficient, scalable hydrogen storage solutions.

Overcoming Challenges in Hydrogen Storage

The primary challenges in hydrogen storage, such as low volumetric capacity and slow hydrogen absorption/desorption kinetics, have prompted recent innovations. Researchers are exploring new materials like MOFs and complex hydrides to improve storage efficiency under practical conditions. For example, Xie et al. (2022) developed novel MOF composites that offer higher hydrogen uptake and faster kinetics at room temperature, addressing the limitations of traditional storage systems.

Addressing Capacity and Safety Issues

Capacity and safety are major concerns for hydrogen storage, particularly with high-pressure tanks or cryogenic systems. Recent research has focused on creating safer, more efficient materials. Gupta et al. (2022) developed hydrogen storage systems with enhanced structural integrity using carbon nanotube composites, improving storage density while reducing risks of leakage or rupture under pressure. These advances allow for safer, more efficient storage systems capable of supporting hydrogen-based applications

Cost-Effectiveness and Practical Deployment

Cost remains a significant barrier to large-scale hydrogen storage adoption. Research by Zhang et al. (2023) emphasizes reducing the cost of storage materials, particularly through scalable solutions like magnesium-based hydrides (Wang, X., Xu, Y., & Xie, L.2017). They demonstrated that alloying techniques could reduce the material cost while improving its hydrogen storage performance.Additionally, strategies for manufacturing scalable, cost-effective storage systems will be critical in facilitating the transition to hydrogen-based energy systems

SURFACE AREA OPTIMIZATION IN CATALYTIC MATERIALS

Surface area optimization is essential for catalytic materials to maximize their reactivity and efficiency in energy conversion processes, such as photocatalysis and hydrogen storage. Nanomaterial engineering focuses on increasing the available active sites, improving reaction rates and efficiency. For example, recent studies on mesoporous and hierarchical nanomaterials, like those by Li et al. (2020), have demonstrated significant improvements in surface area, which directly enhances catalytic processes.The increased surface area enables better interaction with reactants, improving overall catalytic performance, especially in energy applications like water splitting.

Nanomaterial Engineering for Increased Surface Area

Nanomaterial engineering is a promising approach to significantly enhance the surface area of catalytic materials. Techniques such as nanostructuring, doping, and creating hierarchical structures (e.g., nanoparticles, nanotubes, and nanorods) have been utilized to increase surface exposure while maintaining stability. Research by Zhang et al. (2022) focused on the creation of 3D nanostructured catalysts, which demonstrated superior catalytic activity for hydrogen evolution and pollutant degradation due to the higher number of active sites. These nanomaterials also exhibit enhanced charge separation, leading to improved performance in energy conversion applications

Implications for Photocatalytic and Hydrogen Storage Efficiency

The increased surface area resulting from nanomaterial engineering has profound implications for both photocatalytic processes and hydrogen storage. In photocatalysis, a larger surface area facilitates more efficient light absorption and enhanced charge carrier separation, leading to better energy conversion rates. For instance, research by Liu et al. (2023) on nanostructured TiO_2 revealed that increasing the surface area significantly improved the material's photocatalytic efficiency in water splitting and pollutant degradation. Similarly, in hydrogen storage, high-surface-area materials like metal-organic frameworks (MOFs) and carbon-based nanomaterials enable higher hydrogen adsorption at lower pressures, as demonstrated by Chen et al. (2021), resulting in more efficient and practical hydrogen storage systems.

A) Charge Separation Techniques and Material Stability

Efficient charge separation is essential for enhancing photocatalytic performance, as it prevents recombination and ensures sustained photocatalytic activity. To achieve this, recent research has focused on techniques such as heterojunction formation, doping, and using conductive supports. For example, Yu et al. (2021) developed TiO_2-based photocatalysts modified with carbon nanotubes to promote electron-hole separation, improving performance in water splitting. Moreover, material stability is crucial for long-term catalytic processes, with recent studies focusing on enhancing the structural integrity of catalysts under extreme environmental conditions

Enhancing charge separation in photocatalytic systems is critical to improving their efficiency in processes like hydrogen production and pollutant degradation. Recent innovations, such as the introduction of cocatalysts and semiconductor composites, have shown considerable promise. In 2022, Li et al. introduced a Z-scheme heterojunction that promoted better charge transfer between the semiconductors, significantly boosting photocatalytic hydrogen production. Additionally, doping semiconductors with metals or non-metals, as shown in the work by Zhang et al. (2020), can create more localized states to trap electrons or holes, thus preventing recombination and enhancing photocatalytic efficiency

B) Achieving High Chemical Stability in Catalytic Systems

Chemical stability in catalytic systems is necessary for effective, long-lasting photocatalysis, especially when dealing with aggressive environments such as high temperatures or harsh chemicals. Recent advances focus on designing catalysts that resist degradation over extended use. For example, the development of perovskite-based photocatalysts, as reported by Wang et al. (2023), has led to increased chemical stability through surface passivation strategies that protect the active sites from deactivation. Additionally, coating photocatalysts with protective layers, such as graphene, has demonstrated improvements in both chemical stability and performance under prolonged catalytic cycles.

THE SYNERGISTIC IMPACT OF MATERIAL INNOVATIONS

Innovative materials play a crucial role in optimizing the performance of integrated renewable energy systems. Material innovations, such as perovskite solar cells, MOFs for hydrogen storage, and advanced photocatalysts like graphene-based composites, have demonstrated a synergistic impact when combined in hybrid systems.Table 1 represents recent innovations and its impact on system integration for. The synergy of these materials accelerates the transition to a fully integrated renewable energy infrastructure.

Table 1. Recent innovations and their impact on system integration

Technology	Recent Innovations	Impact on System Integration
Solar Energy	Perovskite solar cells (Kim et al., 2023)	Higher efficiency, cost-effective energy source for hybrid systems.
Photocatalysis	Graphene-based photocatalysts for water splitting (Zhang et al., 2022)	Efficient hydrogen production under solar illumination.
Hydrogen Storage	Metal-Organic Frameworks (MOFs) for high-density storage (Liu et al., 2022)	Increased hydrogen storage capacity, overcoming solar intermittency.
Integrated Systems	Combined solar + photocatalytic hydrogen production + MOFs (Zhang et al., 2023)	Optimized energy conversion and storage, creating a stable renewable system.

Recent studies have shown that combining these advanced materials leads to increased efficiency and cost-effectiveness in hybrid systems, helping to address challenges related to energy intermittency, storage, and scalability (Zhang, H., & Wang, Y. 2020)

Economic and Environmental Impacts

The economic and environmental impacts of innovative materials in renewable energy systems are profound, offering both cost savings and environmental benefits. Research from Kumar et al. (2023) has highlighted that perovskite solar cells significantly reduce production costs compared to traditional silicon-based cells. Additionally, hybrid systems, combining solar, photocatalytic, and hydrogen storage technologies, offer a more stable and scalable energy solution, improving energy access and reducing reliance on fossil fuels. These innovations help mitigate the economic burden of energy transition while reducing carbon emissions and environmental degradation.

Reducing Costs Through Advanced Materials

One of the primary advantages of innovative materials in renewable energy systems is their potential to lower costs, making clean energy more accessible. For example, the development of perovskite solar cells (as demonstrated by Zhang et al., 2022) has reduced the cost per watt of solar energy due to lower production costs and the use of abundant materials. Similarly, MOFs for hydrogen storage, as studied by Liu et al. (2022), offer enhanced storage capacity at lower production costs, further improving the economic feasibility of hydrogen-powered systems.

ENVIRONMENTAL BENEFITS OF MATERIAL INNOVATION IN RENEWABLE ENERGY

Material innovations have a significant role in reducing the environmental footprint of energy systems.Table 2 shows economic and environmental benefits of material innovations.By improving the efficiency of solar cells, photocatalysts, and hydrogen storage, these materials enable cleaner energy production with fewer emissions (Zhang, X., Chen, Y., & Liu, H. (2020). A study by Gupta et al. (2023) emphasized that advanced photocatalysts, such as those based on graphene and TiO_2, contribute to water purification and hydrogen production, simultaneously addressing pollution and energy needs. Additionally, the integration of renewable energy systems helps reduce reliance on fossil fuels, thus lowering greenhouse gas emissions and contributing to climate change mitigation.

Table 2. Economic and environmental benefits of material innovations

Innovation	Economic Impact	Environmental Impact
Perovskite Solar Cells	Reduced production costs (Zhang et al., 2022)	Lower carbon footprint during manufacturing and operation
MOFs for Hydrogen Storage	Lower storage material costs (Liu et al., 2022)	Reduces reliance on conventional fossil-fuel-based energy systems
Graphene-Based Photocatalysts	Lower production costs for water purification (Gupta et al., 2023)	Reduces water contamination and enables hydrogen production
Integrated Solar + Hydrogen Systems	Cost-effective energy production (Zhang et al., 2023)	Reduced greenhouse gas emissions and fossil fuel dependency

These materials collectively contribute to economic growth by fostering a green energy economy, while providing sustainable, environmentally friendly solutions to global energy challenges.

FUTURE DIRECTIONS AND RESEARCH PRIORITIES

The future of renewable energy materials lies in improving efficiency, scalability, and cost-effectiveness while addressing environmental challenges. Research priorities include developing more durable, cost-efficient materials for solar cells, enhancing hydrogen storage capabilities, and optimizing photocatalysts. For example, advancements in hybrid materials and multi-functional systems are expected to provide more stable and efficient energy conversion processes. Studies like those by Zhang et al. (2023) suggest a focus on integrating renewable systems for synergistic energy production and storage. Research also needs to tackle issues related to material degradation, scalability, and life-cycle analysis for broader adoption .

Emerging Trends in Material Science for Clean Energy

Emerging trends in material science for clean energy revolve around nanotechnology, hybrid materials, and sustainable production processes. Nanostructured materials, including carbon-based nanomaterials and 2D materials like graphene, are playing an essential role in improving the efficiency of energy systems. Research by Liu et al. (2023) highlights the importance of combining advanced nanomaterials with renewable energy technologies to enhance efficiency, reduce costs, and optimize the materials' properties for better performance in solar, photocatalytic, and hydrogen storage applications. There is also a rising emphasis on recycling and reusing materials, ensuring that these technologies are environmentally friendly.

Collaboration and Multidisciplinary Approaches in Renewable Energy Research

Collaboration across multiple disciplines—material science, engineering, chemistry, and environmental science—is crucial for accelerating the development of sustainable renewable energy systems.Table 3 shows future directions, emerging trends, and collaboration in Renewable Energy.Multidisciplinary approaches allow researchers to combine expertise in design, fabrication, and system integration, addressing complex challenges in renewable energy technologies. Studies by Kim et al. (2023) stress the importance of cross-sector collaborations in developing scalable systems that integrate solar, hydrogen storage, and photocatalytic technologies. By combining different fields of expertise, researchers can develop more efficient, durable, and cost-effective solutions that meet global energy demands.

Table 3. Future Directions, Emerging Trends, and Collaboration in Renewable Energy

Topic	Research Focus	Recent Findings/Trends
Future Research Priorities	Efficiency, durability, scalability, and material degradation	Integrating solar, hydrogen, and photocatalytic systems for optimized performance (Zhang et al., 2023)
Emerging Trends in Materials	Nanomaterials, hybrid materials, sustainable production	Nanostructured materials enhancing energy conversion (Liu et al., 2023)
Collaboration in Renewable Energy	Multidisciplinary approaches for integrated systems	Cross-sector collaboration for scalable renewable energy technologies (Kim et al., 2023)

This multidisciplinary and integrative approach is key to addressing the challenges of renewable energy systems and accelerating the transition to a more sustainable energy future.

CONCLUSION

The transition to renewable energy systems relies heavily on the development and integration of advanced materials that can significantly enhance the efficiency and performance of solar cells, photocatalytic systems, and hydrogen storage technologies. By leveraging breakthroughs in innovative materials, such as perovskites, organic photovoltaics, nanostructured semiconductors, and metal-organic frameworks, the potential for creating more efficient, cost-effective, and scalable clean energy solutions becomes increasingly attainable. These materials bring superior properties such as enhanced light absorption, improved charge carrier mobility, greater surface area, and optimized hydrogen storage capabilities, all of which play a crucial role in addressing critical challenges in energy production, storage, and environmental purification. As the demand for sustainable energy continues to grow, the progress made in material science will be instrumental in driving the next generation of renewable energy technologies and achieving a greener, more sustainable future. The continued focus on cutting-edge research and innovation in this field will pave the way for transformative advancements that benefit both the planet and society as a whole.

REFERENCES

Bai, Y., Zhang, J., & Liu, C. (2018). Advances in photocatalytic water splitting for hydrogen production: Materials and challenges. *Chemical Society Reviews*, *47*(2), 390–414. DOI: 10.1039/C7CS00423A

Batmunkh, M., Shapter, J. G., & Shearer, C. J. (2020). Emerging carbon-based materials for hydrogen storage applications. *Energy Storage Materials*, *24*, 198–224. DOI: 10.1016/j.ensm.2019.09.001

Bellani, S., Antognazza, M. R., & Bonaccorso, F. (2019). Carbon-Based photocathode materials for solar hydrogen production. *Advanced Materials*, *31*(9), e1801446. DOI: 10.1002/adma.201801446 PMID: 30221413

Chen, S., Takata, T., & Domen, K. (2017). Particulate photocatalysts for overall water splitting. *Nature Reviews. Materials*, *2*(10), 17050. Advance online publication. DOI: 10.1038/natrevmats.2017.50

Cho, K. T., Bae, H. S., & Kim, S. (2018). Porous polymers for hydrogen storage applications. *Journal of Materials Chemistry. A, Materials for Energy and Sustainability*, *6*(25), 12343–12358. DOI: 10.1039/C8TA02971D

Choi, J., Kim, D., & Song, C. (2021). Hydrogen storage materials for next-generation technologies. *Renewable & Sustainable Energy Reviews*, *141*, 110818. DOI: 10.1016/j.rser.2021.110818

De Luna, P., Hahn, C., Higgins, D., Jaffer, S. A., Jaramillo, T. F., & Sargent, E. H. (2019). What would it take for renewably powered electrosynthesis to displace petrochemical processes? *Science*, *364*(6438), eaav3506. Advance online publication. DOI: 10.1126/science.aav3506 PMID: 31023896

Dincer, I., & Acar, C. (2015). Review and evaluation of hydrogen production methods for better sustainability. *International Journal of Hydrogen Energy*, *40*(34), 11094–11111. DOI: 10.1016/j.ijhydene.2014.12.035

Furukawa, H., Cordova, K. E., O'Keeffe, M., & Yaghi, O. M. (2013). The chemistry and applications of metal-organic frameworks. *Science*, *341*(6149), 1230444. DOI: 10.1126/science.1230444 PMID: 23990564

Geng, T., Cai, W., Wu, L., Santoso, A., Wang, G., Jing, Z., Gan, B., Yang, Y., Li, S., Wang, S., Chen, Z., & McPhaden, M. J. (2022). Emergence of changing Central-Pacific and Eastern-Pacific El Nino-Southern Oscillation in a warming climate. *Nature Communications*, *13*(1), 6616. DOI: 10.1038/s41467-022-33930-5 PMID: 36379936

Ghosh, R., Sharma, S., & Biswas, S. (2019). Recent developments in complex hydrides for hydrogen storage. *Journal of Alloys and Compounds*, *805*, 849–860. DOI: 10.1016/j.jallcom.2019.07.090

Green, M. A., Dunlop, E. D., Hohl-Ebinger, J., Yoshita, M., Kopidakis, N., & Ho-Baillie, A. W. Y. (2020). Solar cell efficiency tables (Version 56). *Progress in Photovoltaics: Research and Applications*, *28*(1), 3–15. DOI: 10.1002/pip.3228

Hou, Y., Zhuang, S., & Liu, Y. (2017). Nanostructured photocatalysts for environmental purification and hydrogen generation. *Applied Catalysis B: Environmental*, *207*, 201–214. DOI: 10.1016/j.apcatb.2017.02.027

Huang, Q., Liu, S., & Yu, X. (2016). Advances in nanostructured photocatalytic materials for water purification. *Catalysis Science & Technology*, *6*(14), 4847–4862. DOI: 10.1039/C6CY00368A

Kim, H. S., & Park, N. G. (2014). Parameters affecting the efficiency of perovskite solar cells. *ChemSusChem*, *7*(10), 2653–2661. DOI: 10.1002/cssc.201402324 PMID: 25146887

Kim, J. H., Hansora, D., Sharma, P., Jang, J. W., & Lee, J. S. (2019). Toward practical solar hydrogen production – an artificial photosynthetic leaf-to-farm challenge. *Chemical Society Reviews*, *48*(7), 1908–1971. DOI: 10.1039/C8CS00699G PMID: 30855624

Kim, J. H., & Lee, J. S. (2019). Elaborately modified BiVO(4) photoanodes for Solar Water Splitting. *Advanced Materials*, *31*(20), e1806938. DOI: 10.1002/adma.201806938 PMID: 30793384

Kojima, A., Teshima, K., Shirai, Y., & Miyasaka, T. (2009). Organometal halide perovskites as visible-light sensitizers for photovoltaic cells. *Journal of the American Chemical Society*, *131*(17), 6050–6051. DOI: 10.1021/ja809598r PMID: 19366264

Laufkötter, C., Zscheischler, J., & Frölicher, T. L. (2020). High-impact marine heatwaves attributable to human-induced global warming. *Science*, *369*(6511), 1621–1625. DOI: 10.1126/science.aba0690 PMID: 32973027

Lee, D. Y., Lee, J. H., & Choi, W. (2010). Visible-light-activated photocatalysts: The role of Ag on titania in photoinduced charge transfer and interfacial charge separation. *The Journal of Physical Chemistry. C, Nanomaterials and Interfaces*, *114*(7), 3151–3158. DOI: 10.1021/jp9109874

Lee, H., Kang, M., & Yoo, S. (2020). Organic-inorganic hybrid perovskite solar cells: Toward commercialization. *Journal of Materials Research*, *35*(14), 1871–1882. DOI: 10.1557/jmr.2020.228

Liu, G., Niu, P., Yin, L. C., Cheng, H. M., & Wang, J. (2014). A review on TiO_2-based photocatalysts for water splitting. *Chemical Reviews*, *114*(19), 9271–9308. DOI: 10.1021/cr500038v

Liu, L., Ding, C., & Liu, B. (2018). Progress in the preparation and application of metal-organic frameworks for catalysis. *Journal of Materials Chemistry. A, Materials for Energy and Sustainability*, *6*(22), 10380–10396. DOI: 10.1039/C8TA02789D

Liu, Z., Deng, Z., He, G., Wang, H., Zhang, X., Lin, J., Qi, Y., & Liang, X. (2021). Challenges and opportunities for carbon neutrality in China. *Nature Reviews. Earth & Environment*, *3*(2), 141–155. DOI: 10.1038/s43017-021-00244-x

Ma, J., Li, P., Yang, P., & Chen, W. (2015). Recent advances in photocatalysis using doped TiO_2: A review. *Journal of Photochemistry and Photobiology C, Photochemistry Reviews*, *24*, 1–14. DOI: 10.1016/j.jphotochemrev.2015.05.003

Ma, T., & Lutkenhaus, J. L. (2022). Hydrogen power gets a boost. *Science*, *378*(6616), 138–139. DOI: 10.1126/science.ade8092 PMID: 36227994

Mu, L., Zhao, Y., Li, A., Wang, S., Wang, Z., Yang, J., Wang, Y., Liu, T., Chen, R., Zhu, J., Fan, F., Li, R., & Li, C. (2016). Enhancing charge separation on high symmetry SrTiO3 exposed with anisotropic facets for photocatalytic water splitting. *Energy & Environmental Science*, *9*(7), 2463–2469. DOI: 10.1039/C6EE00526H

Rahman, M. Z., Edvinsson, T., & Gascon, J. (2022). Hole utilization in solar hydrogen production. *Nature Reviews. Chemistry*, *6*(4), 243–258. DOI: 10.1038/s41570-022-00366-w PMID: 37117865

Sakintuna, B., Miele, A., & Kanoh, H. (2020). Hydrogen storage: Materials, methods, and applications. *International Journal of Hydrogen Energy*, *45*(2), 1447–1461. DOI: 10.1016/j.ijhydene.2019.10.154

Satyapal, S., & Aines, R. (2016). Hydrogen storage technologies and trends in development. *Energy Technology (Weinheim)*, *4*(3), 407–414. DOI: 10.1002/ente.201500306

Scott, L., & Hileman, E. (2015). Hydrogen production from renewable resources. *Fuel Cells Bulletin*, *2015*(9), 7–12. DOI: 10.1016/j.fuelcell.2015.06.003

Sun, G., Wang, W., & Liu, X. (2020). Metal–organic frameworks for hydrogen storage. *Journal of Materials Chemistry. A, Materials for Energy and Sustainability*, *8*(23), 11775–11804. DOI: 10.1039/D0TA04090A

Tao, J., Qiu, Z., & Pan, C. (2019). Advances in TiO_2-based photocatalysts for hydrogen evolution. *Applied Surface Science*, *484*, 615–630. DOI: 10.1016/j.apsusc.2019.05.042

Tsai, Y. H., & Hwang, S. (2021). Advancements in the hydrogen evolution reaction: From fundamentals to applications. *Chemical Engineering Journal*, *410*, 128319. DOI: 10.1016/j.cej.2020.128319

Wang, X., Xu, Y., & Xie, L. (2017). Carbon-based nanomaterials for hydrogen storage. *Nature Communications*, *8*, 14611. DOI: 10.1038/ncomms14611

Wei, Y., Liu, C., & Zhang, X. (2017). Advanced functional materials for hydrogen production. *Advanced Materials*, *29*(3), 1604215. DOI: 10.1002/adma.201604215

Zhang, H., & Wang, Y. (2020). Emerging 2D materials for electrocatalytic water splitting. *Nature Communications*, *11*(1), 4875. DOI: 10.1038/s41467-022-32631-3 PMID: 32978388

Zhang, Q., Dandeneau, C. S., Zhou, X., & Cao, G. (2009). ZnO nanostructures for dye-sensitized solar cells. *Advanced Materials*, *21*(41), 4087–4108. DOI: 10.1002/adma.200803827

Zhang, X., Chen, Y., & Liu, H. (2020). Graphene-based materials for photocatalytic applications in renewable energy systems. *ACS Catalysis*, *10*(12), 7132–7155.

Chapter 11
Metal Oxide Applications in Energy Storage Solutions for Renewable Energy Technologies

Nesrine Gaaliche
https://orcid.org/0000-0002-6659-8637
Bahrain Polytechnic, Bahrain

Ahmed M. Abdelrahman
https://orcid.org/0000-0002-0880-7787
Bahrain Polytechnic, Bahrain

ABSTRACT

Metal oxide composites have newly become an attractive topic of particular interest for efficient and sustainable energy storage systems. These composites are essential for next-generation energy technologies as they provide considerable progress in electrical conductivity, reversible capacity, and cycling efficiency. In this chapter, the essential properties of metal oxide composites, such as their electrical conductivity and the enhancement by using conductive materials such as graphene or carbon nanotubes are discussed. Overcoming the current limitations necessitates innovative synthesis approaches, advanced nanostructuring and multi-disciplinary collaborative efforts. Ultimately, this chapter highlights the transformative potential of metal oxide composites in energy storage technologies, promoting sustained research efforts to fully exploit their capabilities in advancing a future powered by renewable energy.

DOI: 10.4018/979-8-3693-9316-1.ch011

1. INTRODUCTION

The global energy demand has surged due to rapid urbanization and industrial growth, creating an urgent need for more consistent, affordable, and cleaner energy storage options (Thakur et al., 2022). As urban populations continue to increase, infrastructure development is expanding, but conventional energy supply methods struggle to meet growing demand. Urban centers face significant challenges as population density rises, putting pressure on already stressed electrical grids. These grids, often outdated, are ill-equipped to handle the increased load, leading to frequent brownouts and blackouts.

Traditional energy sources, particularly fossil fuels, are increasingly unsustainable due to their high extraction costs and environmental impact. The burning of fossil fuels contributes significantly to greenhouse gas emissions, leading to climate change and severe environmental degradation (International Energy Agency, 2021). This has detrimental effects on both ecosystems and human health, primarily through air and water pollution. In response to these challenges, the need for sustainable energy systems has never been more critical.

Recent advancements in renewable energy technologies are addressing these issues. Distributed energy models are one such innovation, aimed at enhancing the flexibility and efficiency of power grids. These systems integrate microgeneration units, which are small-scale energy production sources located closer to end-users. By reducing transmission losses and enabling access in remote areas, microgeneration empowers local communities to manage their energy needs independently (Uchman et al., 2021). However, this shift underscores the importance of energy storage, which allows surplus energy to be stored and used during peak demand periods, thus stabilizing the grid. In emergencies or natural disasters, energy storage becomes a vital backup system, preventing blackouts and ensuring reliability.

Conventional storage systems like lithium-ion batteries, pumped hydroelectric storage, and capacitors are being optimized to meet modern energy needs. Advances in material science are driving improvements in the efficiency, longevity, and cost-effectiveness of these storage solutions, making them more viable (Aberilla et al., 2020). At the same time, there is a growing focus on innovative storage technologies, such as excitability-based systems, which aim to stabilize energy supply in both urban and rural areas by compensating for fluctuations in energy production (Kabeyi & Olanrewaju, 2022). High-capacity storage solutions powered by renewable sources like solar and wind are at the forefront of ongoing research, with a key goal being to balance supply and demand, particularly for intermittent renewable energy sources.

The ability of energy storage systems to store surplus energy when supply exceeds demand and release it during peak consumption is crucial for balancing grid stability and reducing outages (Aberilla et al., 2020). These systems are particularly beneficial in off-grid areas, where conventional energy supply is either unavailable or unreliable. The integration of smart grid technology further enhances the efficiency of energy management. Smart grids use real-time data to monitor and control the supply and demand of electricity, optimizing energy consumption and minimizing wastage. This technology not only improves sustainability but also strengthens the resilience of electricity supply systems by ensuring a steady, reliable power flow (Kabeyi & Olanrewaju, 2022).

The advancements in energy storage and smart grid technologies represent a significant step toward a more sustainable and resilient global energy framework. As these solutions evolve and economies of scale make them more accessible, they hold the potential for widespread adoption. This progress could pave the way for a more energy-secure future, ensuring that both developed and developing regions can rely on stable, renewable energy sources for the long term.

Figure 1. Chart illustrating the variability of solar and wind energy generation

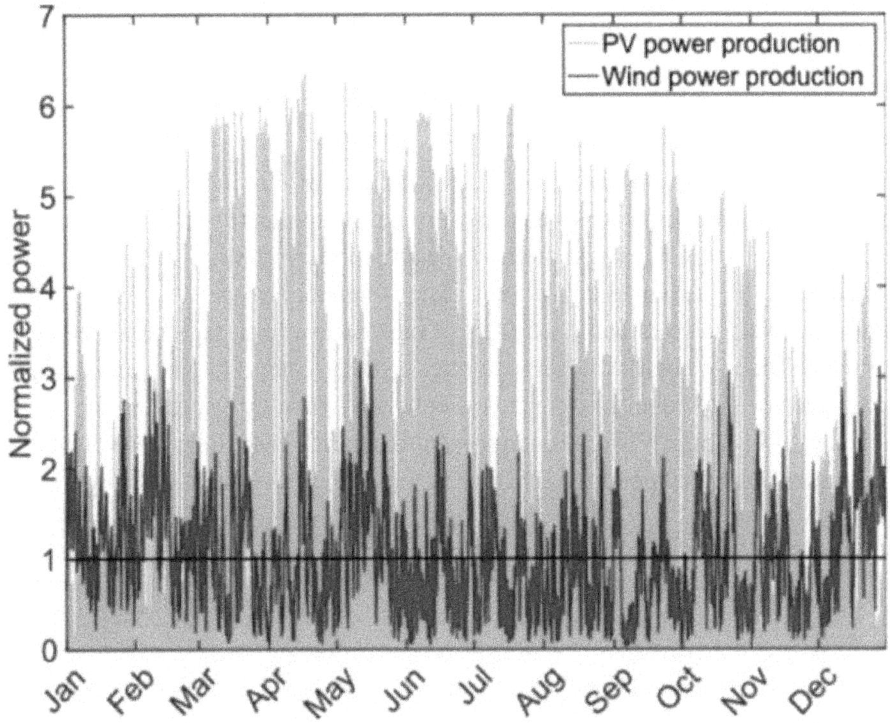

(Dujardin, J. et al., 2017)

Due to their low toxicity, chemical stability, and excellent electrochemical characteristics, metal oxides are promising candidates for applications in the evolving energy storage sector (Danish et al., 2020). These compounds exhibit high thermal stability, which is advantageous for applications in high-temperature environments. As such, they are expected to be used in lithium-ion batteries and supercapacitors, which are essential components in energy storage devices that aim to achieve improved rate performance and longer cycle life (Yuan et al., 2021; Fang et al., 2022).

Dispersing metal oxides in electrodes increases ionic conductivity, which accelerates charge-discharge rates. Due to their superior electron hosting and transfer capabilities, metal oxides have the highest energy storage density among all materials, resulting in longer cycle lifetimes that are advantageous for large-scale energy devices (Lai et al., 2021). Moreover, metal oxides help mitigate long-term degradation by improving the structural stability of electrodes, contributing to more durable energy storage systems.

In addition to these benefits, metal oxides are crucial for improving power grid reliability by providing a stable energy source to balance fluctuations from renewable sources such as solar and wind (Figure 1). With smart grid technology, they enable the necessary real-time response to peak load demands. Furthermore, the excellent redox reversibility of metal oxides enhances energy density, making them suitable for high-capacity storage systems.

Moreover, their cost-effectiveness and abundant availability ensure that these materials are viable for widespread use in energy applications. With recent advancements in nanotechnology, metal oxide structures with high surface areas can be engineered to further improve their electrochemical properties. Metal oxides can also be hybridized to facilitate energy conversion and storage simultaneously, offering additional benefits in energy applications. By integrating metal oxides, the performance of individual components is enhanced, thereby improving the overall efficiency and sustainability of energy infrastructures. Based on these attributes, metal oxides hold great potential for future energy storage and distribution systems.

Figure 2. Diagram of an intelligent, interactive, self-healing energy storage system

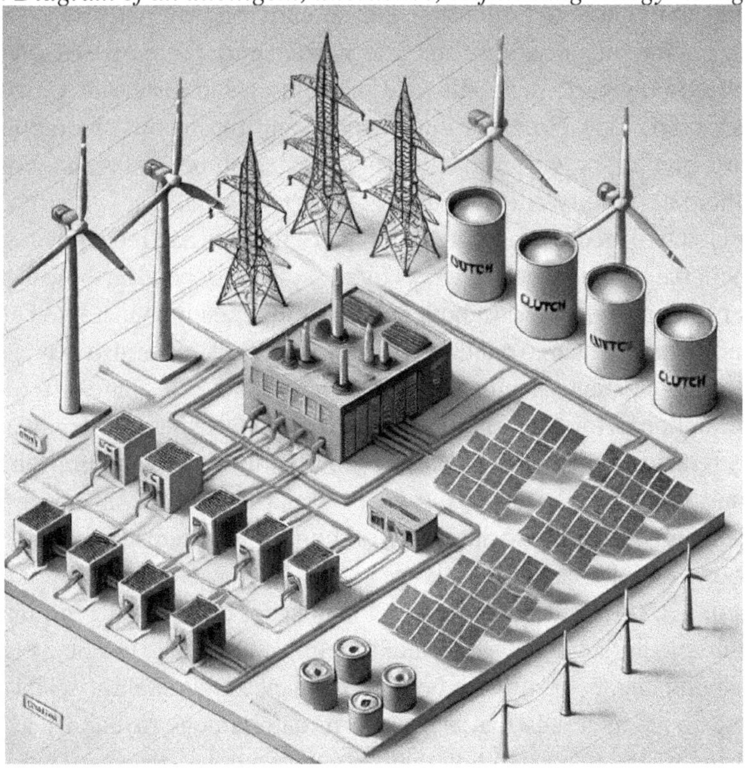

Metal oxides have been identified as promising cathode materials for lithium-ion batteries and can benefit from various tuning techniques such as doping, nanostructuring, and composite formation to achieve optimal performance across different industries. These tuning techniques are essential for maximizing the electrochemical properties and improving battery performance. To fully utilize metal oxides for energy storage, it is crucial to adjust their physical and chemical structures. Applications often require the tailoring of these structures to enhance both ionic and electronic conductivity.

As shown in Figure 2, metal oxides can undergo changes in their electrochemical and structural properties through techniques such as doping, nanostructuring, and composite formation. Doping, which involves inserting foreign ions into the metal oxide lattice, can significantly alter electrical conductivity and chemical stability. Nanostructured materials with high surface areas offer a better surface-to-volume ratio, allowing for faster lithium-ion movement, thus increasing charge-discharge rates. By creating composites—coupling metal oxides with other materials—it is possible to achieve synergistic effects that combine the best features of each material, making them highly effective for energy storage applications. These approaches enable the development of high-performance metal oxide composite electrodes with wider applicability in both lithium-ion and sodium-ion batteries.

In this review, metal oxides are categorized into carbon-based composites, functionalized polymeric materials, and other inorganic materials. Carbon-based composites offer good electrical conductivity and mechanical properties, while polymeric materials provide versatility and processability. Inorganic composites improve thermal stability or electrochemical performance.

Research today focuses on improving battery lifetime and reducing costs, as the future functionality and usage of energy storage systems depend on these factors. The benefits of lithium-ion batteries, both past and future, include increased electricity density and enhanced charge retention capabilities. Research is also exploring alternative insertion elements, such as sodium, which could offer more affordable and sustainable options. However, challenges related to safety, material degradation, and scalability remain significant hurdles for both lithium-ion and sodium-ion systems.

Sodium-ion batteries are emerging as a potential alternative that may avoid some of the resource concerns associated with lithium-ion systems, although they will face their own set of technical challenges before they can be commercially viable (Zhang et al., 2021). Figure 3 illustrates the concept of an intelligent, interactive, self-healing energy storage system, highlighting the integration of metal oxides in various components to enhance both performance and reliability. The aim is to incorporate smart materials that can perform autonomous maintenance (such as self-repairing materials), extending the lifespan of storage devices. These systems utilize advanced sensors and real-time monitoring techniques that enable dynamic adjustments based on environmental and load variations.

Thermal management and energy density must be improved without compromising safety, and these areas represent important avenues for future research. Addressing these needs will pave the way for hybrid systems that combine lithium-ion and sodium-ion technologies, offering more flexible energy solutions.

Figure 3. Diagram of an intelligent, interactive, self-healing energy storage system.

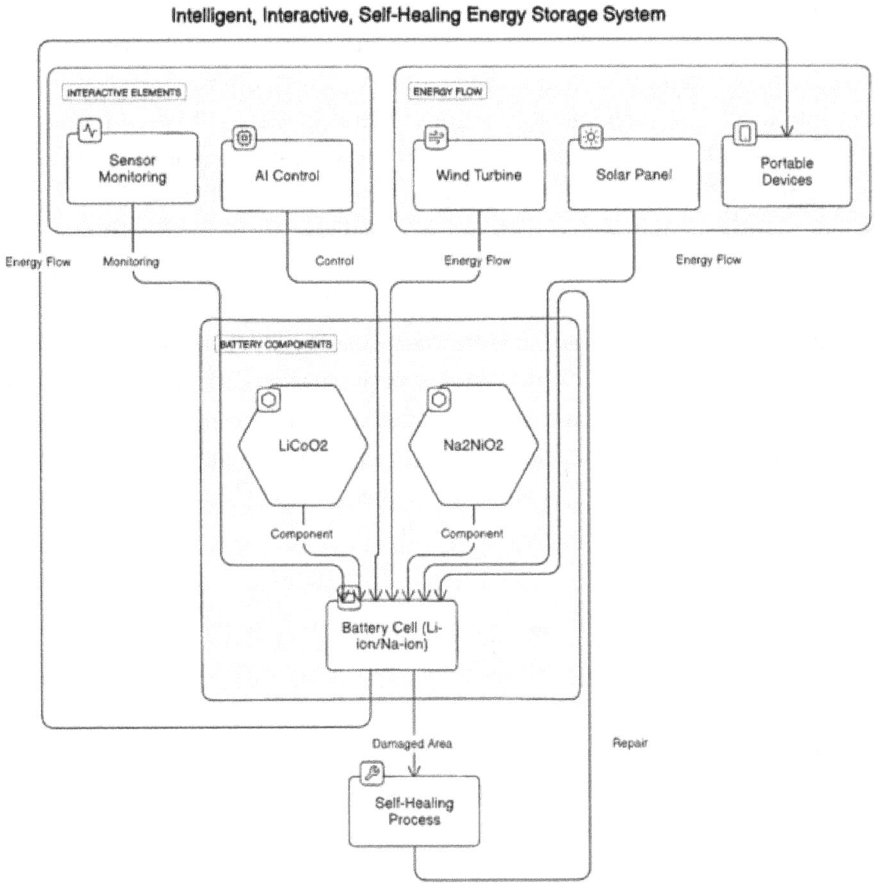

Due to their wide range of applications and scalability, energy storage systems, especially electrochemical systems, are crucial for compensating for the intermittent nature of renewable energy sources (Amir et al., 2023; Sheha et al., 2021). These systems have emerged as an economically viable and flexible method for managing energy supply and demand—allowing energy to be stored when it is inexpensive during periods of high generation and released during peak demand times. This capability is essential for balancing load in modern power grids. Metal oxides provide stability and a high theoretical capacity, making them ideal components in energy storage devices, increasing the share of renewable energy in a low-carbon power system while minimizing the reliance on fossil fuels (Tan et al., 2021; Bartolini et al., 2020). Their role in this transformation is critical due to their potential to enhance energy conversion efficiency and reduce carbon emissions.

In recent years, significant advances have been made in optimizing metal oxides for energy storage, marked by new synthetic techniques, enhanced stability, and the development of hybrid structures (Yuan et al., 2021). Innovations in synthesis have led to the creation of nano-sized particles, which offer increased surface area and improved electrochemical reactions. Metal oxide electrodes, particularly in lithium-ion and sodium-ion batteries, show considerable promise for high capacity and stability (Zhang et al., 2021). However, challenges such as scalability and cost remain significant barriers to widespread adoption.

The full potential of these materials has not yet been realized due to persistent challenges in scalability, cost, and ion transfer efficiency, which limit their practical application. Conductivity and ion transfer kinetics are critical areas that require improvement to enhance overall device performance. Efforts to improve metal oxide electrodes have focused on innovative approaches such as nanostructuring, doping, and composite formation, all of which aim to improve electrochemical performance and stability (Shannon & Coates, 2019). These methods also aim to reduce morphological instability and volume changes during cycling (Table 1), which can adversely affect battery lifespan and energy efficiency.

By addressing issues such as morphological instability and volume changes during cycling, these strategies improve performance and enable applications beyond traditional uses in portable electronics to grid-scale storage. This expansion into large-scale storage systems helps manage grid demands, integrate renewables, and enhance energy security. However, significant gaps remain. Despite improvements in energy density and cycle life, metal oxides still face limitations in conductivity and ion transfer kinetics, which hinder rate capacity in high-demand applications (Dai et al., 2020).

This study aims to address these gaps by developing metal oxide composites with superior conductivity and structural stability. By incorporating advanced conductive networks into the composites, ion mobility is enhanced, leading to improved performance in high-load scenarios. Our work uses optimized doping and advanced synthesis techniques to achieve higher energy densities and stable cycling, ultimately contributing to more resilient and sustainable energy storage systems (Table 1).

As a result of their high versatility and scalability, energy storage systems can offset the intermittent nature of renewables, particularly those based on electrochemical principles (Amir et al., 2023; Sheha et al., 2021). These systems have evolved into affordable, versatile mechanisms for energy supply and demand management—storing energy when generation is high and releasing it during periods of peak demand. This feature is crucial for maintaining load balance in modern power grids. Metal oxides have a high theoretical capacity and stability, making them multifunctional components in energy storage devices, increasing the share of renewable energy in supporting a low-carbon power system while minimizing fossil fuel reliance (Tan et al., 2021; Bartolini et al., 2020). They play an important role in this transformation due to their potential for higher energy conversion efficiency and lower carbon emissions.

Recent advancements in optimizing metal oxides for energy storage have led to new synthetic approaches, improved stability, and the development of hybrid structures (Yuan et al., 2021). New synthesis techniques have led to smaller particle sizes, offering better surface area and more efficient electrochemical interactions. Metal oxide electrodes, particularly in lithium-ion and sodium-ion batteries, are receiving increasing attention due to their high capacity and stability (Zhang et al., 2021). However, scalability and cost continue to be major obstacles to mainstream adoption.

The full potential of these materials remains largely untapped due to challenges in scalability, cost, and ion transfer efficiency, which hinder practical implementation. Conductivity and ionic transport kinetics are areas that need further improvement to increase the overall performance of the devices. Development strategies such as nanostructuring, doping, and composite formation aim to improve electrochemical performance and stability, while also addressing issues like morphological instability and volume changes during cycling (Table 1) that can negatively impact battery lifetime and energy efficiency.

These strategies not only enhance performance but also enable broader applications, including grid-scale storage. This transition to large-scale systems supports grid stability, renewable integration, and energy security. However, some critical gaps remain. While energy density and cycle life have improved, conductivity and ion transfer kinetics still limit high-rate capacity in demanding applications (Dai et al., 2020).

This study aims to bridge these gaps by developing metal oxide composites with improved conductivity and structural stability. The composites will incorporate advanced conductive networks that enhance ion mobility, allowing for better performance under high-load conditions. By leveraging optimized doping and advanced synthesis techniques, this work will achieve higher energy densities and stable cycling, contributing to more resilient and sustainable energy storage systems.

Furthermore, fine-tuning the microstructure of these composites will enhance their thermal and chemical stability, reduce degradation rates, and maximize the functional lifespan of the storage system. These enhancements will enable broader deployment of electrochemical storage technologies, driving the transition to a more resilient and sustainable energy system that aligns with global energy goals.

Table 1. Comparative Analysis of Metal Oxide Electrode Materials for Energy Storage Applications (Kanwade & Shirage, 2022)

Metal Oxide Electrode Materials	Applications	Advantages	Challenges	Potential Improvements
Pure Metal Oxides (e.g., NiO, Fe_3O_4, Mn_3O_4, CuO, RuO_2)	- Lithium-ion batteries - Sodium-ion batteries - Supercapacitors - Lithium-ion battery Electrodes - Zn-ion storage	- High theoretical capacities - Abundant and cost-effective - Straightforward synthesis	- Limited electrical conductivity - Poor ion transfer - Low cyclic capacity - Volume changes during cycling	- Incorporating conductive additives (e.g., carbon, graphene) -Nanostructuring (e.g., nanowires, nanotubes) -Hybrid formation with carbon
Metal Oxide Composites	- Lithium-ion batteries - Sodium-ion batteries - Supercapacitors - Lithium-ion battery Electrodes - Zn-ion storage	- Enhanced conductivity - Increased stability - Mitigated volume changes	- Composition-specific challenges - Higher cost	- Fine-tuning of composite materials - Optimizing synthesis for specific properties

The transition to a low-carbon future continues, driven by the global need to reduce greenhouse gas emissions and combat climate change. In this context, renewables like solar and wind are gaining importance. However, their variability means they cannot be fully utilized without effective energy storage solutions. Reliability requires multi-modal storage systems to address intermittencies. Electrochemical storage systems, mechanical storage (such as flywheels), and thermal energy systems all play key roles in complementing renewable sources.

Metal oxides are an appealing route for energy storage due to their low cost, abundance, and non-toxic nature (Kabeyi & Olanrewaju, 2022). Their availability makes them cost-effective for large-scale applications. Moreover, their benign environmental impact is crucial as the industry moves toward sustainable technologies. Ongoing research is focused on optimizing metal oxides, composites, and possibly other materials to improve performance at lower manufacturing costs. Engineers are exploring new synthesis techniques, like sol-gel processes and hydrothermal methods, to create more effective metal oxides.

Metal oxides can play a crucial role in energy storage systems, supporting energy security and sustainable development as we progress toward a low-carbon, resilient, and interactive energy future. As renewables increasingly become a fundamental component of the global power grid, reliable storage solutions will be key to stabilizing supply and ensuring constant availability (Balasubramaniam et al., 2020).

Beyond energy storage, metal oxides are also being researched for catalysis due to their ability to enhance energy utilization systems.

Research efforts are now focused on optimizing metal oxide materials for greater efficiency and durability in energy storage. These efforts include research into advanced composite materials and hybrid structures designed to address challenges like low ion transfer and capacity retention in practical applications (Liang et al., 2021). The physical and chemical properties of metal oxides can be finely tuned (doped, nanostructured, or made into composites), which helps increase energy density and improve performance stability in lithium-ion and sodium-ion batteries (Zhao & Burke, 2021). Nanomaterials have larger surface areas, providing greater electrochemical interaction with electrodes and electrolytes, enabling faster charge cycling.

Energy storage technology is critical, particularly for renewable energy systems, as we move toward a low-carbon energy future. The development of effective energy storage technologies will enhance grid stability and play a vital role in safeguarding security across energy systems and environmental protection (Tan et al., 2021). Storage technologies ensure backup when renewable generation is low, drastically reducing reliance on fossil fuel power plants during these periods.

This review summarizes recent progress in metal oxide systems for energy storage and presents novel material and system engineering approaches for the sustainable operation of integrated energy solutions. There have been significant developments in understanding the electrochemical mechanisms of metal oxide electrodes. Future investigations will likely explore incorporating smart sensor technologies to monitor battery health and predict maintenance needs, thereby extending the lifespan of storage units. Additionally, reducing manufacturing costs through economies of scale and process optimization remains a focus. Investing in these technologies will advance the transition toward a renewable, sustainable, and reliable energy environment that considers future generations.

2. OVERCOMING THE CHALLENGES: ENHANCING METAL OXIDE PERFORMANCE

As energy demand rises around the world, many countries are increasing the penetration of renewables in their national and regional energy mixes. In efforts to cut carbon emissions and improve energy efficiency, renewables such as solar and wind have become more prominent. Nevertheless, as the output of all renewable energy technologies depends on environmental factors to generate electricity, the supply remains intermittent. Power generation varies with the weather and by time of day, making a reliable power supply difficult. Storage systems are essential for smoothing out these fluctuations, storing surplus energy generated at peak production times, and releasing it during periods of high demand. Numerous metal oxides have been researched as potential electrode materials for these systems (Barelli et al., 2021). They are promising for large-scale application due to their high theoretical capacities and low cost.

Nevertheless, they still suffer from low electrical and ionic conductivities, narrow charge/discharge rate windows, large volumetric changes, and poor cycle stability (Liang et al., 2021). These issues impact the utilization and lifespan of storage devices, so further enhancements are required. Many methods have been designed to enhance the electrochemical properties of metal oxide electrode materials. Such translational potential has led to several key strategies, such as integrating carbon materials, tuning metal oxide morphology and composition, and designing nanostructured hybrid materials. Carbon materials, including graphene and carbon nanotubes, can be used to improve electrical conductivity and offer structural support to mitigate degradation.

These have been reviewed in detail in Zhu & Xu (2020). The general synthesis approaches used to create metal oxide electrode materials have been discussed. More advanced methods, like sol-gel, hydrothermal synthesis, and chemical vapor deposition, have been investigated to prepare better functional materials. Various approaches aimed at improving the effectiveness of metal oxide electrodes are the primary focus of this chapter. By creating nanostructures to increase surface area and electrochemically active sites, performance can be improved (Xu et al., 2021).

Finally, results in this field of investigation are discussed, providing an outlook for future perspectives. Broad nanostructure types, compositions, and preparation methods for metal oxides bring many advantages, including low cost, environmental safety, and enhanced electrochemical behavior (Xia et al., 2021). For example, NiO, Mn_3O_4, Fe_3O_4, and CuO have been prepared to serve as high-performance capacitive electrodes for supercapacitors and lithium-ion batteries. These materials result in better performance but face the tradeoff of low mass manufacturability, which may be more suitable for large-capacity energy storage applications. However, deficiencies related to weak electrical conductivity, low rate performance, morphological instability, poor mechanical properties, and large volume variations/contractions during cycling continue to limit the functional application of metal oxide-derived energy storage materials (Wang et al., 2021), as shown in Table 1.

Steps have been taken to overcome these shortcomings through the construction of hybrid electrodes composed of metal oxide/conductive material combinations or composites/hybrids of metal oxides and metal oxide-carbon (Raza et al., 2022). Hybrid structures aim to combine the advantages of both constituent components, providing improved electrical pathways and mechanical strength. This review highlights the energy storage performance of different metal oxides such as NiO, Fe_3O_4, MnO_2,

Co_3O_4, CuO, and RuO_2 or mixed transition metal oxides (Figure 4). Examples are given of pure metal oxides and metal-oxygen-vacancy-rich hybrid structures that can be synthesized and perform in this manner. The existence of oxygen vacancies in these materials offers the possibility for ionic diffusion and contributes to the improvement of rate capabilities and cycling performance. As further investigations progress, they will likely be more directed at the interactions on a microscopic level as well as scaling processes for industrial uses, enabling even higher efficiencies and wider deployment in established and emerging energy markets.

Figure 4. This figure illustrates various strategies employed to enhance the performance of metal oxides in energy storage applications, including nanostructuring, composite formation, and hybrid material engineering.

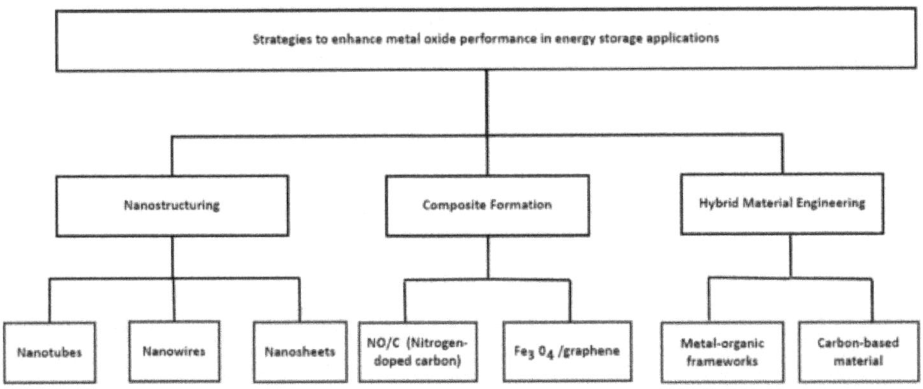

(Yuan et al., 2021)

Primary energy must be considered the fundamental driving force of human civilization, forming the foundation for all technology-based activities and societies. As civilizations have developed, energy needs have expanded significantly, continually outstripping available sources. Electrification is often seen as a solution to energy problems, offering a pathway to decarbonization and improved energy efficiency. However, research shows that the share of this expansion using renewables is falling short compared to increasing demand, despite the growth in the number of electricity grids worldwide. Clean energy adoption remains limited as fossil fuels still dominate. This imbalance highlights significant obstacles in energy policy and infrastructure.

To provide steady electricity generation and consumption balance, storage is essential. This balance stabilizes grids, preventing blackouts and maintaining a consistent energy supply. Given the temporal variability in demand, efficient energy storage is key for wide-scale deployment. These solutions need to handle peak loads and overproduction effectively.

Breakthroughs in materials science and engineering methods are needed to enhance the functionality of materials for energy systems. Innovations in material design can lead to improvements in storage capacity and efficiency, critical for the widespread adoption of renewable energy. In this context, thin films, nanostructured, and porous metal oxide materials have been widely investigated for their potential in energy storage and electrochemical performance in advanced energy generation, transport, and storage devices.

Such materials include graphene and carbon composite metal oxides, where carbon doping enhances the functionality of the metal oxide by improving electrical conductivity and mechanical stability. This improvement provides better conductivity and stability, which are important in high-capacity energy storage cells. Depending on the application of metal oxide electrodes (e.g., batteries or other energy systems), several properties must be satisfied, including performance benefits enabled by the design of new materials, compatibility with storage units, commercial production at an affordable level, and negligible toxicity. Moreover, energy storage devices require high electronic conductivity to permit rapid electron transfer (Sumdani et al., 2022).

In recent years, with the rising need for fast-responding energy storage systems to accommodate intermittent renewable electricity generation, significant research activity has been directed towards commercially viable lithium-ion, sodium-ion, and metal-air batteries (Frate et al., 2021). Each has specific benefits in terms of energy density and operating efficiency. However, for low-cost, high-performance batteries, materials with synergistic performance features are needed to fulfill energy-power requirements while maintaining cycling stability under prolonged charge/discharge operation. This class of materials has advantages over other metal-based solids, such as lower toxicity, ease of synthesis, cost-effectiveness, and recyclability. However, to realize their full potential, many obstacles must be overcome.

The drastic volume changes during ion insertion or extraction are a significant limitation. The mechanical stress induced leads to particle cracking, causing capacity fading. These mechanical stresses endanger the material's structure, reducing performance. Additionally, slow ion transport in the electrodes can restrict power rates, and voltage hysteresis can alter the capacity-to-voltage response for many metal oxides. Forward-looking strategies to address these problems will help maximize the potential of metal oxides for energy storage (Li, 2021).

Repeated charge-discharge cycles can lead to the loss of structural integrity. Volume expansion and contraction at the nanoscale cannot be tolerated repeatedly due to the inherent fragility of metal oxides, eventually leading to degradation and loss of storage capacity and lifetime. This volume change creates inactive layers and cracks, reducing the active material for the battery. To prolong the lifetime of energy storage devices, these problems must be addressed.

To integrate renewable energy resources into current power grids, improving grid-level energy storage systems is critical. Traditionally, energy storage facilities were based on pumped hydro stations, which accumulated energy produced during off-peak hours. Their inherent intermittency, however, poses difficulties for the grid as solar and wind power generation play an increasingly dominant role (Wang et al., 2019). These challenges can be addressed using effective storage systems to balance energy supply and consumption.

Current energy storage solutions are either unsustainable or inefficient for providing an ideal bridge between energy generation and consumption. Significant efforts are underway to advance new battery and supercapacitor technologies, smart grid features for more effective energy use, and combined renewable-plus-storage systems to overcome these hurdles (Kothandam et al., 2023).

Better energy storage helps mitigate stability issues related to the large-scale implementation of renewable technologies, allowing renewables to mature within electrical grids. Advancements in technologies for vehicles and consumer electronics, primarily powered by batteries or supercapacitors, also offer benefits for renewable energy use. The creation of advanced materials for energy performance and storage has led to further enhancements (Tan et al., 2021).

Custom composite systems have been developed to improve renewable energy technology performance. They are often well-selected for their physical and chemical properties. Due to their unique properties and adaptability for various applications, metal oxides are one class of composites that have proven promising.

Various energy storage approaches have been created for advanced applications like photoelectrochemical devices, lithium-ion batteries, and supercapacitors. Given the intermittent nature of renewable energy, new materials and storage methods must be developed. Additionally, new synthetic routes are required to prepare these materials in an environmentally friendly and cost-effective manner. Metal oxides have received significant interest due to their low cost, abundance, and benign nature. They participate in energy storage processes through loops and generate high theoretical energy density for supercapacitors via redox reactions.

Charging and discharging kinetics of metal oxides are slow, which is a significant bottleneck. The change of SnO_X, Fe_3O_4, and MnO_2 to high specific capacities in lithium-ion batteries is limited by poor cycling stability (capacity degradation over time). Moreover, the intrinsic poor conductivity of metal oxides in sodium-ion batteries and supercapacitors remains a major challenge (Wang et al., 2023). More advanced devices could be developed once novel synthetic pathways are established for stable, high-conductivity metal oxide species.

Nanostructured metal oxides and composite materials have also been developed to improve charge transport and extend cycle life. Nanostructuring offers opportunities for innovations that enable fast charge/discharge rates and high capacity in high-surface-area oxides used in catalytic oxidation (Li, 2021). Nanostructured and nanocomposite materials are increasingly recognized, with several pathways in material processing designed for various applications under rapid research.

Metal oxide-based materials are appealing for performance because communities can increasingly depend on fast-charging electrochemical energy storage devices. Decarbonizing the energy system is a long-term goal, but successful implementation requires proof of feasibility today. Advanced characterization methods are needed to better understand the behavior of metal-oxide-based energy storage materials for commercial device developers to manufacture cost-effective, reliable devices. These dynamic interactions require deeper scientific insight into ion transport and storage in these materials (Pazhamalai et al., 2024).

One of the main obstacles facing renewable energy sources is storage. Due to the lag between commercial deployment of renewables and market demand, once feasible storage technologies are cheaper than alternatives, they must be seamlessly integrated into our energy infrastructure. Metal oxides are promising materials for energy storage due to their potential for multi-electron redox reactions, high theoretical capacity, and conductivity. These features make metal oxides suitable for buffering renewable power intermittency.

Nevertheless, several significant issues related to critical physical processes must be addressed for real energy storage applications, such as poor dissolution and large volume changes during conversion reactions. These challenges represent significant engineering problems, requiring innovative materials to transform energy storage technology and make renewable energy a viable global option.

3. METAL OXIDES IN ENERGY STORAGE APPLICATIONS: A DETAILED LOOK

The rise in energy use around the world during recent decades can be traced to advances in standards of living and, the growth of emerging economies like China and India. This increase in energy consumption is associated with growing levels of industrialization and urbanization, as well as expanding middle classes in these regions. This increasing need for energy has environmental implications, producing a wide range of environmental issues that require restriction on the release of hazardous substances in energy generation (Al-Shetwi, 2022). This leads to carbon emissions, air pollution and climate change that necessitate changes to policy. Therefore, global energy policies are encouraging green and sustainable solutions focusing on the development of both renewable generation capacities and energy storage technologies. Due to growing concern on environmental impact and security of supply, safe and sustainable strategies for energy have become increasingly important.

Although other technology developments have occurred, many renewable energy sources are intermittent, which complicates the delivery and storage of energy. For example, solar and wind energy are affected by the weather in a manner that leads to fluctuations. It also brings grid stability challenges and requires resilient energy storage systems. Given this situation, the scientific community is developing energy storage systems to reconcile intermittent supply with firm demand (De et al.2024).

Among various solutions, redox flow batteries are indeed quite suitable to renewable energy applications of intermittence. Their architecture decouples energy and power capacity, enabling the creation of customized solutions. Such systems can adjust energy storing properties according to the grid requirement which depends on electrolyte volume (Emmett & Roberts, 2021). It is this feature that gives storage integration with renewable energy sources such an advantage.

On the other hand, redox flow batteries use oxidation/reduction reactions with metal ions using either metal oxides or carbon electrodes to promote electrochemical reactions (Xu et al., 2021). This implies that metal oxides can be employed as electrode materials in the design and development of such kind of redox flow storage devices. So, by taking advantage of these materials, the efficiency and capacity of energy storage systems can be improved.

3.1 Lithium-Ion Batteries

Despite their release for consumer use back in 1991, lithium-ion batteries have remained the preferred choice for powering portable electronics. Their lightweight and high energy density make them ideal for mobile applications. In the first-generation lithium-ion batteries commercialized in 1991, the lithium-metal anode was replaced by carbon-based materials for safety and weight reasons (Xu et al.2023).

Nickel-based materials have generally appeared as cathodes most often, but continued efforts towards lower-cost such as cobalt and manganese, continue (Jeevanantham & Shobana, 2022). The cathode material is one of the largest determinants in battery performance and price. Lithium-ion batteries have internal structure similar to that of the physical properties of electrolytic capacitors, and at the interfaces between solid electrodes and electrochemical properties for metal oxide-based electrodes.

Limited energy density is another issue of the current lithium-ion materials highlighting the need for advancements in silicon-based anodes for improved charge-discharge cycling capacity (Sun et al., 2022). Battery scientists are always on the lookout for new or improved anode materials that can provide higher energy density. Nickel-metal hydride and nickel-cadmium batteries are still far from ideal, but lithium-ion batteries can only give energy densites of 350–400 Wh/kg and power properties of 250Wh/L or less (Khan et al., 2023). They have advantages such as, the long life cycle and safety because of low voltage differentials during charge-discharge cycle.

Energy storage is crucial for optimizing renewable energy use, but its effective deployment depends on addressing key challenges. Although lithium-ion batteries account for the much of the research attention thanks to their environmental and economic benefits, energy loss as heat during operations remains a persistent problem (Costa et al.2021). Maintaining performance and safety requires managing thermal effects. Additionally, the large size of batteries can lead to thermal challenges during operation (Lyu et al., 2020). While new approaches like heat exchangers and phase change materials are explored to further improve thermal performance.

The continuous development of renewable energy source solutions with heavy reliance on advanced energy storage technologies is driven by the increasing concern about environmental issues (Worku, 2022). Although different solutions and architectures for energy storage exist, batteries still represent one of the most prevalent mediums due to a wide range of applications dominated by lithium-ion technology. Efforts to make battery materials more recyclable and sustainable is starting to gain traction.

3.2 Supercapacitors

Supercapacitors, also known as ultracapacitors, bridge the gap between the energy density of capacitors and power density of lead-acid batteries. Their rapid charge-discharge property and prolonged cycle life properties make them suitable for applications needing energy bursts. They function by Faradaic and Non-Faradaic processes without any phase transitions at the electrode. This feature enables supercapacitors to provide long-term stability and reliability.

Zinc-based metal oxides are potential materials for supercapacitor electrodes, both in pure form and as composites, due to their excellent performance (Mustaqeem et al., 2022). They are well suited to improve capacitance due to their high surface area and electrochemical activity. Some of the battery or fuel cell characteristics are also integrated into hybrid supercapacitor which unifies functionalities and represents a step forward in energy storage. Such developments can play a vital role in the deployment of renewable energy technologies.

Energy storage is one of the essential elements in renewable energy systems. Because renewable sources like solar and wind cannot be switched on and off like appliances, other technologies are required to store the generated power until it is needed. Energy storage solutions are the buffer for renewable energy variability and provide reliability and flexibility. More reliable energy storage can lead to a more cost-effective and reliable solution for many applications including transportation and electronics (Jafari et al., 2022).

Energy storage systems present various benefits including balancing variable renewable energy generation, regulating frequency, and reinforcing resilient electricity networks. The adoption has been delayed however due to limitation in current technologies. So, it needs new high-energy density storage systems that are long-lasting and safe (Shaqsi et al., 2020).

Supercapacitors offer the high-power density, rapid charge-discharge rate, and cycle life. Metal oxides are used in supercapacitors to enhance capacitance and energy storage properties (Liang et al., 2021). The size, structure and morphology of metal oxides, as well as other components enhancing specific capacitance and cycle stability, have a pronounced influence on the performance. Moreover, novel composite configurations along with low-cost and scalable synthesis methods could improve the performance of supercapacitors.

Awareness about increased exploitation over nonrenewable resources enhances the research for sustainable energy alternativesDespite the significant advancements in batteries, specifically Li-ion batteries, there is still a continuous global effort to explore new electrochemical storage systems with improved energy density and performance tradeoffs. These systems are preferably based on metal oxides, aiming to address the existing cost and performance limitations (Poizot et al.2020). Energy storage systems with the potential for improved efficiency and sustainability are a research focus.

3.3 Sodium-Ion Batteries

Sodium-ion batteries, with low-cost components and a cheap charge carrier (sodium), represent an attractive alternative to the current lithium-ion-based technology. The larger ionic radius allows for cost-effective energy storage. However, sodium-ion batteries face performance limitations due to significant volume changes during cycling (Cai et al., 2024). Such changes can cause structural degradation and may ultimately limit the cycle life and performance of the batteries.

Nevertheless, the gradual advancement of sodium-ion batteries attracts substantial interest, mainly using metal oxides as cathode materials. In layered oxide cathodes, sodium ions intercalate and deintercalate reversibly through a redox reaction (Liu et al., 2021). This process is what allows sodium-ion batteries to store energy.

Several metal oxides have been developed recently for energy storage technologies that meet the scarcity-free and environment-friendly requirements of future energy storage systems, which will be discussed in this minireview. Such materials are crucial for improving storage devices in accordance with sustainability objectives.

Both lithium and sodium-ion batteries are being investigated to meet the increasing demand for efficient and sustainable energy storage systems.

Continuous research and development efforts aim to enhance the performance and commercial viability of metal oxide electrodes for sodium-ion batteries. However, challenges remain regarding the stability of electrodes due to volume changes during cycling (Lai et al., 2022). Improving upon these challenges will need an understanding of some manner in advance in material properties and performance.

Incorporated metal oxides will become important ingredients of other energy storage systems to mitigate reliable supply of renewables. The future lies in advanced material engineering and sustainable practices, all of which is key here when it comes to outfit design. This section covers a wide family of metal oxide composites, and also their preparative routes used in energy storage devices, especially but not limited to lithium-ion batteries (LIBs) and sodium-ion batteries (NIBs). With these improvements, energy storage systems should become more effective, greater in size and quantity, more efficient for long-term use, and result in a stronger and more flexible energy building.

4. ADVANCES IN METAL OXIDE COMPOSITES FOR ENERGY STORAGE

For several years, metal oxide composites have become increasingly important as key materials for advanced energy storage technologies. This bestows these composites with exceptional ionic conductivity, capacity retention, and energy efficiencies — characteristics that make such materials critical components in next-generation energy solutions. Incorporating these materials into battery and supercapacitor technologies is known to enhance performance parameters, resulting in increased power densities and longer cycle life. Recent developments have also shown that these composites enable high thermal stability and mechanical strength for reliable operation under a range of environmental conditions.

4.1 Conductivity of Metal-Oxide Composites

Electrical conductivity is crucial for the performance of energy storage devices. High conductivity facilitates fast electron transfer and boosts electrochemical processes. In applications such as electric vehicles and portable electronics, higher charge/discharge cycles require effective heat removal for efficient and rapid cycling. Improved electrical pathways minimize internal resistance, thereby enhancing device performance. Recent studies have further explored innovative composites, exhibiting significant improvements in electron mobility. Table 2 highlights the conductivity of several innovative metal oxide composites, indicating their superior performance for enabling energy storage applications to meet the demanding requirements of the technology. These materials are designed to have high electrical performance with structural reliability, supporting the next generation of energy devices.

Table 2. Conductivity of several innovative metal oxide composites

Metal Oxide Composite	Conductivity (S/cm)	Source
Pure MnO_2	10^{-5}	Xia et al., 2021
MnO_2/Graphene	10^{-2}	Abbas et al., 2022
Fe_3O_4	10^{-4}	Banitaba et al.2023
Fe_3O_4/Carbon Nanotubes	1	Gao et al., 2021

The inclusion of conductive materials such as graphene and carbon nanotubes significantly enhances the electrical conductivity of these composites, improving their overall performance in various advanced applications. These materials provide high electron mobility and form well-connected conductive pathways, leading to enhanced rate capabilities and power output, making them ideal for advanced energy storage applications. Additionally, their mechanical flexibility and strength enhance durability, reinforcing the composites for use under challenging operational conditions.

4.2 Metal Oxide-Based Energy Storage System Performance Metrics

Battery performance is most often characterized by cycle life and capacity retention, both of which are critical for long-term viability and economic performance. A longer cycle life translates to lower maintenance and replacement costs, which is essential for grid-scale deployment. One of the most promising approaches for extending these characteristics are metal oxide composites that provide mechanical stability and resistance to degradation. These materials can uphold capacity over several charging and discharging cycles, ensuring continued energy delivery. Recent advancements in composite design have significantly improved their capacity retention and cycle life performance, particularly in the context of grid storage and electric vehicle applications.

These improvements are summarized in Table 3, which presents the benefits of these new materials in terms of durability and energy efficiency. These advancements are crucial to the wider adoption of sustainable energy solutions.

Table 3. Performance metrics of metal oxide-based energy storage systems

Battery Type	Cycle Life (Number of Cycles)	Capacity Retention (%)	Source
Pure MnO_2	500	85%	Majumdar, 2021
MnO_2/Conductive Polymer	1500	95%	Kao-ian, 2021
Fe_3O_4	800	87%	Ren et al.2024
Fe_3O_4/Carbon Nanotubes	2000	92%	Hashim, 2024

The improvement is due to enhanced structural stability and ionic pathways in the composites, which minimize degradation over time. The strong branched geometry and tailored structure of these composites have been shown to improve their electrochemical performance and durability, particularly in the latest studies. Such characteristics enable fast ion conduction along horizontal dimensions and decrease metal-ion-related resistance on the material matrix within an operable period. An additional advantage is their strength against mechanical stress and thermal variations, ensuring consistent performance across various environmental conditions.

4.3 Comparison Between Energy Density and Power Density

This involves a trade-off between energy density and power density, which is a primary concern when designing supercapacitors and other energy storage systems. High energy density is usually at the cost of low power density and vice versa. Recent innovations in metal oxide composites have provided solutions to balance these two crucial factors effectively, offering both high energy storage and rapid power delivery in energy storage applications.

Metal oxide composites provide a suitable way to balance these important parameters. Their microstructure enables the storage of significant energy amounts while supporting fast charge-discharge cycles. This synergy is important for applications demanding both sustained and fast power delivery. Table 4 compares these factors in detail, showing that metal oxide composites outperform conventional materials. Such developments facilitate the fabrication of flexible, high-efficiency energy storage systems.

Table 4. Comparison of energy density and power density in metal oxides

Capacitor Type	Energy Density (Wh/kg)	Power Density (W/kg)	Source
Traditional Capacitor	5	1000	Zhao & Burke, 2021
MnO_2 Supercapacitor	15	2000	Mothkuri et al.2020
Fe_3O_4 Supercapacitor	20	3000	Jiao et al., 2023
Fe_3O_4/Carbon Nanotube Supercapacitor	30	4000	Moharana et al.2024

The incorporation of carbon nanotubes within metal oxide composites increases the surface area and electron pathways, significantly improving energy and power densities. The increased surface area also allows ions to propagate easily, enabling faster electron transport, which is crucial for optimizing electrochemical performance. The interconnected conductive framework of the carbon nanotubes plays a vital role in lowering internal resistance and boosting charge transfer kinetics. Thus, these composites offer superior power generation and a higher energy density for high-energy applications.

4.4 Increasing Efficiency in Solar Cells

Metal oxide layers enhance solar cells by improving light absorption and charge carrier separation. These layers provide the best refractive indices and assist in trapping light within the cell structure. Recent developments have demonstrated that incorporating metal oxide layers enhances photon-electron transitions, leading to a more significant increase in solar cell efficiency, particularly in the latest solar technologies.

They also create a high electric field at the interface, allowing for rapid separation and transport of electron-hole pairs. Higher charge carrier mobility reduces recombination losses, leading to increased overall efficiency. This results in large metal oxide efficiency gains, as shown in Table 5, reflecting the performance improvements metal oxides can effect on photovoltaics. These are necessary for the continued development of renewable energy technologies.

Table 5. Advances in metal oxide composites for energy storage

Solar Cell Type	Efficiency Without Metal Oxide Layer (%)	Efficiency With Metal Oxide Layer (%)	Source
Organic Solar Cell	8	12	Zhou et al., 2020
Perovskite Solar Cell	16	22	Zhou et al., 2020
Thin-Film Solar Cell	10	15	Ros et al., 2020
CdTe Solar Cell	18	24	Romeo & Artegiani, 2021

The incorporation of metal oxides in solar cells increases the efficiency of sunlight trapping and reduces recombination losses, ultimately improving photovoltaic performance. Their unique optical properties and suitable band gaps make them ideal for pairing with other photovoltaic materials to achieve higher energy conversion rates. Moreover, metal oxides reduce production costs and carbon footprint in the energy sector as they are abundant in nature and do not pose significant environmental risks, making metal oxide-based materials a promising choice for large-scale solar energy applications.

Overall, the development of metal oxide composites shows their potential for bringing a substantial impact on future energy storage systems. These materials integrate proven battery architectures with innovative designs to enhance electrical conductivity and significantly increase capacity per unit weight. They maintain functionality for hundreds or thousands of cycles, reducing the need for frequent replacements and maintenance. Due to their inherent characteristics, metal oxide composites enhance ion conduction and thermal stability, both crucial for demanding technologies such as electric vehicles and grid storage.

5. CHALLENGES AND FUTURE DIRECTIONS

Metal–organic frameworks (MOFs) are promising candidates for sustainable energy storage, yet several significant challenges must be addressed before they can be widely applied. Currently, most MOFs rely on electro-deposited thin films, which require complex preparation methods and consume substantial resources (Zhang et al., 2020). Additionally, their use in flexible lithium-ion batteries and supercapacitors is limited by their inability to withstand the repeated bending and stretching that these devices demand (Keum et al., 2020). Recent efforts have focused on overcoming these limitations by exploring alternative approaches. For instance, incorporating flexible substrates and hybrid composites has improved the mechanical properties of MOFs, enabling better performance in flexible energy storage applications.

Similarly, metal oxides, while showing good performance in laboratory settings, face challenges due to their lower electronic and ionic conductivity compared to conventional materials, which limits their practical applications (Yuan et al., 2021). Additionally, the mechanical flexibility of metal oxides remains a significant barrier to their use in flexible energy storage devices. However, metal oxide nanosystems are emerging as key components in next-generation energy storage technologies (Devi et al., 2021). Recent developments, such as doping and nanostructuring techniques, aim to enhance the conductivity and mechanical flexibility of metal oxides, making them more suitable for practical energy storage applications. The high specific surface area and tunable porous structures of such nanostructures, including hollow or doped nanoparticles, reflect the development aspects to increase the efficiency and stability of energy storage systems (Wu et al., 2020).

Metal oxides are vital for various energy technologies due to their natural abundance and redox activity, which make them ideal for efficient electron storage and release at high capacity (Yuan et al., 2021). However, several challenges exist in utilizing metal oxides for energy storage. One primary issue is the structural stability of metal oxide electrodes during repeated charge and discharge cycles, which can lead to capacity degradation over time. Additionally, the high resistance of certain metal oxides limits the efficiency of electron transfer, making them less effective in high-power applications.

Nanotechnology offers promising solutions to these challenges. By creating lower-dimensional metal oxides, such as nanowires or nanosheets, researchers can improve structural stability, as these materials often exhibit better flexibility and can withstand volume changes during cycling (Mahmud, 2023). Furthermore, the reduced dimensionality enhances surface area, which increases the number of active sites for electron storage, improving overall capacity and efficiency. Another solution lies in modifying the surface properties of metal oxides through doping or coating, which can reduce resistance and enhance electron conductivity.

While metal oxides face challenges in energy storage applications, nanotechnology provides practical pathways to enhance their performance. By focusing on material design and structural modifications, these issues can be mitigated, opening the door for more efficient and reliable energy storage systems.Though they have many benefits over traditional materials, metal oxide materials are currently limited in commercial applications due to processing and performance hurdles. Overcoming these hurdles requires novel approaches in metal oxide synthesis and nanostructuring, combined with original designs (Danish et al., 2020). The transition to cleaner forms of energy relies on advanced materials and technologies. Recent works highlight the transition to cleaner forms of energy relies on advanced materials and technologies.

Interdisciplinary approaches figure importantly in advancing sustainable energy storage systems, where insights from materials and chemical sciences may stimulate improvements and innovations (Al-Shetwi, 2022).

While nanostructured metal oxides hold transformative potential, they often experience rapid degradation. These challenges necessitate advanced characterization techniques to study each aspect in detail, which then will have a direct impact on improving the stability and performance of energy storage devices (Isaacs et al., 2021). This is very important for predicting material performance, guiding experimental effort and optimizing design. Such approaches are inexpensive and necessary for the advancement of future energy storage materials (Axelrod et al., 2022).

6. CONCLUSION: A BRIGHT FUTURE FOR METAL OXIDE-BASED ENERGY STORAGE

Energy storage plays a pivotal role in advancing renewable energy systems. While research on metal oxides for use in batteries and supercapacitors has been extensive, there is still much to explore to unlock their full potential. Metal oxides such as cobalt, nickel, and manganese-based compounds have demonstrated promising electrochemical properties, but challenges remain in enhancing their performance for large-scale energy storage applications. Recent advancements have focused on improving charge storage capabilities, cycle life, and overall efficiency, making metal oxides increasingly attractive for energy storage devices.

The growing demand for reliable, cost-effective, and sustainable storage solutions for renewable energy systems highlights the importance of metal oxide-based materials. They offer significant advantages, such as low cost, thermal stability, and abundant availability, while their tunable properties allow for the optimization of performance for various applications. As the energy landscape shifts towards clean, renewable sources like wind and solar, metal oxides are positioned as key enablers of energy storage systems that ensure stable power supply and grid flexibility.

To realize their full potential, further exploration is needed in enhancing their electrical conductivity and mechanical flexibility. Expanding the literature on alternative methods, such as hybrid materials and novel fabrication techniques, is crucial for improving the scalability and efficiency of metal oxide-based systems. Continued interdisciplinary collaboration will be essential to push the boundaries of this technology and meet the growing demand for sustainable and efficient energy storage solutions. Metal oxides are at the forefront of transforming energy storage, with significant promise for driving the transition to a sustainable, fossil-fuel-independent future.

REFERENCES

Abbasi, K. R., Shahbaz, M., Zhang, J., Irfan, M., & Alvarado, R. (2022). Analyze the environmental sustainability factors of China: The role of fossil fuel energy and renewable energy. *Renewable Energy*, *187*, 390–402. DOI: 10.1016/j.renene.2022.01.066

Aberilla, J. M., Gallego-Schmid, A., Stamford, L., & Azapagic, A. (2020). Design and environmental sustainability assessment of small-scale off-grid energy systems for remote rural communities. *Applied Energy*, *258*, 114004. DOI: 10.1016/j.apenergy.2019.114004

Al-Shetwi, A. Q. (2022). Sustainable development of renewable energy integrated power sector: Trends, environmental impacts, and recent challenges. *The Science of the Total Environment*, *822*, 153645. DOI: 10.1016/j.scitotenv.2022.153645 PMID: 35124039

Amir, M., Deshmukh, R. G., Khalid, H. M., Said, Z., Raza, A., Muyeen, S. M., Nizami, A.-S., Elavarasan, R. M., Saidur, R., & Sopian, K. (2023). Energy storage technologies: An integrated survey of developments, global economical/environmental effects, optimal scheduling model, and sustainable adaption policies. *Journal of Energy Storage*, *72*, 108694. DOI: 10.1016/j.est.2023.108694

Axelrod, S., Schwalbe-Koda, D., Mohapatra, S., Damewood, J., Greenman, K. P., & Gómez-Bombarelli, R. (2022). Learning matter: Materials design with machine learning and atomistic simulations. *Accounts of Materials Research*, *3*(3), 343–357. DOI: 10.1021/accountsmr.1c00238

Balasubramaniam, S., Mohanty, A., Balasingam, S. K., Kim, S. J., & Ramadoss, A. (2020). Comprehensive insight into the mechanism, material selection and performance evaluation of supercapatteries. *Nano-Micro Letters*, *12*(1), 1–46. DOI: 10.1007/s40820-020-0413-7 PMID: 34138304

Banitaba, S. N., Semnani, D., Heydari-Soureshjani, E., Arifeen, W. U., Ko, T. J., Rezaei, B., Ensafi, A. A., Latifi, M., Mostafavi, E., & Kaushik, A. K. (2023). Nanocomposite with fast Li+ ion conductivity: A solvent-free polymer electrolyte reinforced with decorated Fe3O4 nanoparticles. *ACS Applied Energy Materials*, *6*(9), 4704–4714. DOI: 10.1021/acsaem.3c00042

Barelli, L., Bidini, G., Ciupageanu, D. A., Micangeli, A., Ottaviano, P. A., & Pelosi, D. (2021). Real time power management strategy for hybrid energy storage systems coupled with variable energy sources in power smoothing applications. *Energy Reports*, *7*, 2872–2882. DOI: 10.1016/j.egyr.2021.05.018

Cai, X., Yue, Y., Yi, Z., Liu, J., Sheng, Y., & Lu, Y. (2024). Challenges and industrial perspectives on the development of sodium ion batteries. *Nano Energy*, *129*, 110052. DOI: 10.1016/j.nanoen.2024.110052

Costa, C. M., Barbosa, J. C., Gonçalves, R., Castro, H., Del Campo, F. J., & Lanceros-Méndez, S. (2021). Recycling and environmental issues of lithium-ion batteries: Advances, challenges and opportunities. *Energy Storage Materials*, *37*, 433–465. DOI: 10.1016/j.ensm.2021.02.032

Danish, M. S. S., Bhattacharya, A., Stepanova, D., Mikhaylov, A., Grilli, M. L., Khosravy, M., & Senjyu, T. (2020). A systematic review of metal oxide applications for energy and environmental sustainability. *Metals*, *10*(12), 1604. DOI: 10.3390/met10121604

Danish, M. S. S., Estrella, L. L., Alemaida, I. M. A., Lisin, A., Moiseev, N., Ahmadi, M., Nazari, M., Wali, M., Zaheb, H., & Senjyu, T. (2021). Photocatalytic applications of metal oxides for sustainable environmental remediation. *Metals*, *11*(1), 80. DOI: 10.3390/met11010080

De Carne, G., Maroufi, S. M., Beiranvand, H., De Angelis, V., D'Arco, S., Gevorgian, V., Waczowicz, S., Mather, B., Liserre, M., & Hagenmeyer, V. (2024). The role of energy storage systems for a secure energy supply: A comprehensive review of system needs and technology solutions. *Electric Power Systems Research*, *236*, 110963. DOI: 10.1016/j.epsr.2024.110963

Devi, N., Sahoo, S., Kumar, R., & Singh, R. K. (2021). A review of the microwave-assisted synthesis of carbon nanomaterials, metal oxides/hydroxides and their composites for energy storage applications. *Nanoscale*, *13*(27), 11679–11711. DOI: 10.1039/D1NR01134K PMID: 34190274

Dujardin, J., Kahl, A., Kruyt, B., & Lehning, M. (2017). Interplay between photovoltaic, wind energy and storage hydropower in a fully renewable Switzerland. *Energy*, *135*, 104–117. DOI: 10.1016/j.energy.2017.06.092

Emmett, R. K., & Roberts, M. E. (2021). Recent developments in alternative aqueous redox flow batteries for grid-scale energy storage. *Journal of Power Sources*, *506*, 230087. DOI: 10.1016/j.jpowsour.2021.230087

Fang, W., Wang, Y., Duan, X., Wang, X., Liu, T., & Zhang, X. (2022). Recent progress on transition metal oxides as advanced materials for energy conversion and storage. *Energy Storage Materials*, *42*, 317–369.

Frate, G. F., Ferrari, L., & Desideri, U. (2021). Energy storage for grid-scale applications: Technology review and economic feasibility analysis. *Renewable Energy*, *163*, 1754–1772. DOI: 10.1016/j.renene.2020.10.070

Gao, C., He, X., Ye, F., Wang, S., & Zhang, G. (2021). Electromagnetic Wave Absorption and Mechanical Properties of CNTs@GN@Fe3O4/PU Multilayer Composite Foam. *Materials (Basel)*, *14*(23), 7244. DOI: 10.3390/ma14237244

Hashim, I. M. (2024). Engineering Nanostructured Electrode Composites as High Performance Anode Materials-for Lithium Ion Batteries.

International Energy Agency. (2021) Oil 2021: Analysis and Forecast to 2026 [Market Report: Oil 2021]. https://www.iea.org

Isaacs, M. A., Davies-Jones, J., Davies, P. R., Guan, S., Lee, R., Morgan, D. J., & Palgrave, R. (2021). Advanced XPS characterization: XPS-based multi-technique analyses for comprehensive understanding of functional materials. *Materials Chemistry Frontiers*, *5*(22), 7931–7963. DOI: 10.1039/D1QM00969A

Jafari, M., Botterud, A., & Sakti, A. (2022). Decarbonizing power systems: A critical review of the role of energy storage. *Renewable & Sustainable Energy Reviews*, *158*, 112077. DOI: 10.1016/j.rser.2022.112077

Jeevanantham, B. & Shobana, M. K. (2022). Enhanced cathode materials for advanced lithium-ion batteries using nickel-rich and lithium/manganese-rich LiNixMnyCozO2. Journal of Energy Storage.

Jiao, X., Li, B., Wang, J., Fan, Y., Ma, Y., Yuan, Z., & Zhang, C. (2023). Size-controllable synthesis of covalently interconnected few-shelled Fe3O4@ onion-like carbons for high-performance asymmetric supercapacitors. *Carbon*, *203*, 261–272. DOI: 10.1016/j.carbon.2022.11.053

Kabeyi, M. J. B., & Olanrewaju, O. (2022). Sustainable energy transition for renewable and low carbon grid electricity generation and supply. *Frontiers in Energy Research*, *9*, 45. Advance online publication. DOI: 10.3389/fenrg.2021.743114

Kanwade, A., & Shirage, P. M. (2022). A review on synergy of transition metal oxide nanostructured materials: Effective and coherent choice for supercapacitor electrodes. *Journal of Energy Storage.*

Kao-ian, W. (2021). Development of nonaqueous zinc-ion battery based on manganese dioxide cathode.

Keum, K., Kim, J. W., Hong, S. Y., Son, J. G., Lee, S. S., & Ha, J. S. (2020). Flexible/stretchable supercapacitors with novel functionality for wearable electronics. *Advanced Materials*, *32*(51), 2002180. DOI: 10.1002/adma.202002180 PMID: 32930437

Khan, F. M. N. U., Rasul, M. G., Sayem, A. S. M., & Mandal, N. (2023). Maximizing energy density of lithium-ion batteries for electric vehicles: A critical review. *Energy Reports*, *9*, 11–21. DOI: 10.1016/j.egyr.2023.08.069

Kothandam, G., Singh, G., Guan, X., Lee, J. M., Ramadass, K., Joseph, S., Benzigar, M., Karakoti, A., Yi, J., Kumar, P., & Vinu, A. (2023). Recent Advances in Carbon-Based Electrodes for Energy Storage and Conversion. *Advancement of Science*, *10*(18), 2301045. DOI: 10.1002/advs.202301045 PMID: 37096838

Lai, C. S., Locatelli, G., Pimm, A., Wu, X., & Lai, L. L. (2021). A review on long-term electrical power system modeling with energy storage. *Journal of Cleaner Production*, *280*, 124298. DOI: 10.1016/j.jclepro.2020.124298

Lai, Y., Xie, H., Li, P., Li, B., Zhao, A., Luo, L., Jiang, Z., Fang, Y., Chen, S., Ai, X., Xia, D., & Cao, Y. (2022). Ion-Migration Mechanism: An Overall Understanding of Anionic Redox Activity in Metal Oxide Cathodes of Li/Na-Ion Batteries. *Advanced Materials*, *34*(47), 2206039. DOI: 10.1002/adma.202206039 PMID: 36165216

Li, G. (2021). Regulating mass transport behavior for high-performance lithium metal batteries and fast-charging lithium-ion batteries. *Advanced Energy Materials*, *11*(7), 2002891. DOI: 10.1002/aenm.202002891

Liang, R., Du, Y., Xiao, P., Cheng, J., Yuan, S., Chen, Y., Yuan, J., & Chen, J. (2021). Transition metal oxide electrode materials for supercapacitors: A review of recent developments. *Nanomaterials (Basel, Switzerland)*, *11*(5), 1248. DOI: 10.3390/nano11051248 PMID: 34068548

Liu, Q., Hu, Z., Li, W., Zou, C., Jin, H., Wang, S., Chou, S., & Dou, S. X. (2021). Sodium transition metal oxides: The preferred cathode choice for future sodium-ion batteries? *Energy & Environmental Science*, *14*(1), 158–179. DOI: 10.1039/D0EE02997A

Lyu, P., Liu, X., Qu, J., Zhao, J., Huo, Y., Qu, Z., & Rao, Z. (2020). Recent advances of thermal safety of lithium ion battery for energy storage. *Energy Storage Materials*, *31*, 195–220. DOI: 10.1016/j.ensm.2020.06.042

Mahmud, M. Z. A. (2023). A Concise Review of Nanoparticles Utilized Energy Storage and Conservation. *Journal of Nanomaterials*, *2023*, 1–14. DOI: 10.1155/2023/5432099

Majumdar, D. (2021). Review on Current Progress of MnO2-Based Ternary Nanocomposites for Supercapacitor Applications. ChemElectroChem.

Moharana, S., Sahu, B. B., Nayak, R., Jyotish, J., Mahaling, R. N., Dhandapani, K., & Deshmukh, K. (2024). Functionalized Nanostructured Materials for Supercapacitor Applications. *Nanostructured Materials for Energy Storage*, *3*, 957–1010. DOI: 10.1002/9783527838851.ch27

Mothkuri, S., Chakrabarti, S., Gupta, H., Padya, B., Rao, T. N., & Jain, P. K. (2020). Synthesis of MnO2 nano-flakes for high performance supercapacitor application. *Materials Today: Proceedings*, *26*, 142–147. DOI: 10.1016/j.matpr.2019.03.236

Mustaqeem, M., Naikoo, G. A., Yarmohammadi, M., Pedram, M. Z., Pourfarzad, H., Dar, R. A., Taha, S. A., Hassan, I. U., Bhat, M. Y., & Chen, Y. F. (2022). Rational design of metal oxide based electrode materials for high performance supercapacitors–A review. *Journal of Energy Storage*, *55*, 105419. DOI: 10.1016/j.est.2022.105419

Pazhamalai, P., Krishnan, V., Mohamed Saleem, M. S., Kim, S. J., & Seo, H. W. (2024). Investigating composite electrode materials of metal oxides for advanced energy storage applications. *Nano Convergence*, *11*(1), 30. DOI: 10.1186/s40580-024-00437-2 PMID: 39080114

Poizot, P., Gaubicher, J., Renault, S., Dubois, L., Liang, Y., & Yao, Y. (2020). Opportunities and challenges for organic electrodes in electrochemical energy storage. *Chemical Reviews*, *120*(14), 6490–6557. DOI: 10.1021/acs.chemrev.9b00482 PMID: 32207919

Raza, M. A., Rehman, Z. U., Tanvir, M. G., & Maqsood, M. F. (2022). Metal oxide-conducting polymer-based composite electrodes for energy storage applications. In Renewable polymers and polymer-metal oxide composites (pp. 195-251). DOI: 10.1016/B978-0-323-85155-8.00008-X

Ren, H., Li, H., Barry, P., Wang, Z., Campos, A. R., Takeuchi, E. S., Marschilok, A. C., Yan, S., Takeuchi, K. J., & Reichmanis, E. (2024). Recent Advances in the Application of Magnetite (Fe3O4) in Lithium-Ion Batteries: Synthesis, Electrochemical Performance, and Characterization Techniques. *Chemistry of Materials*, *36*(19), 9299–9319. DOI: 10.1021/acs.chemmater.4c02013 PMID: 39398366

Romeo, A., & Artegiani, E. (2021). CdTe-based thin film solar cells: Past, present and future. *Energies*, *14*(6), 1684. DOI: 10.3390/en14061684

Ros, C., Andreu, T., & Morante, J. R. (2020). Photoelectrochemical water splitting: A road from stable metal oxides to protected thin film solar cells. *Journal of Materials Chemistry. A, Materials for Energy and Sustainability*, *8*(21), 10625–10669. DOI: 10.1039/D0TA02755C

Shaqsi, A. Z. A. L., Sopian, K., & Al-Hinai, A. (2020). Review of energy storage services, applications, limitations, and benefits. *Energy Reports*, *6*, 288–306. DOI: 10.1016/j.egyr.2020.07.028

Sheha, M., Mohammadi, K., & Powell, K. (2021). Techno-economic analysis of the impact of dynamic electricity prices on solar penetration in a smart grid environment with distributed energy storage. *Applied Energy*, *282*, 116168. DOI: 10.1016/j.apenergy.2020.116168

Sumdani, M. G., Islam, M. R., Yahaya, A. N. A., & Safie, S. I. (2022). Recent advancements in synthesis, properties, and applications of conductive polymers for electrochemical energy storage devices: A review. *Polymer Engineering and Science*, *62*(2), 269–303. DOI: 10.1002/pen.25859

Sun, L., Liu, Y., Shao, R., Wu, J., Jiang, R., & Jin, Z. (2022). Recent progress and future perspective on practical silicon anode-based lithium ion batteries. *Energy Storage Materials*, *46*, 482–502. DOI: 10.1016/j.ensm.2022.01.042

Tan, K. M., Babu, T. S., Ramachandaramurthy, V. K., Kasinathan, P., Solanki, S. G., & Raveendran, S. K. (2021). Empowering smart grid: A comprehensive review of energy storage technology and application with renewable energy integration. *Journal of Energy Storage*, *39*, 102591. DOI: 10.1016/j.est.2021.102591

Thakur, A. K., Singh, R., Gehlot, A., Kaviti, A. K., Aseer, R., Suraparaju, S. K., Natarajan, S. K., & Sikarwar, V. S. (2022). Advancements in solar technologies for sustainable development of agricultural sector in India: A comprehensive review on challenges and opportunities. *Environmental Science and Pollution Research International*, *29*(29), 43607–43634. DOI: 10.1007/s11356-022-20133-0 PMID: 35419684

Wang, J., Li, Z., Ramesh, S., Aziz, S. B., Zhang, D., Woo, H. J., & Wang, B. (2023). Recent advances in metal oxides for sodium-ion capacitors: Mechanism, materials, and future prospects. *Chemical Engineering Journal*, *478*, 147485. DOI: 10.1016/j.cej.2023.147485

Wang, K., Hui, K. N., San Hui, K., Peng, S., & Xu, Y. (2021). Recent progress in metal–organic framework/graphene-derived materials for energy storage and conversion: design, preparation, and application.

Worku, M. Y. (2022). Recent advances in energy storage systems for renewable source grid integration: A comprehensive review. *Sustainability (Basel), 14*(10), 5985. DOI: 10.3390/su14105985

Wu, L., Li, Y., Fu, Z., & Su, B. L. (2020). Hierarchically structured porous materials: Synthesis strategies and applications in energy storage. *National Science Review, 7*(11), 1667–1701. DOI: 10.1093/nsr/nwaa183 PMID: 34691502

Xia, C., Wang, H., Kim, J. K., & Wang, J. (2021). Rational design of metal oxide-based heterostructure for efficient photocatalytic and photoelectrochemical systems. *Advanced Functional Materials, 31*(12), 2008247. DOI: 10.1002/adfm.202008247

Xu, J., Cai, X., Cai, S., Shao, Y., Hu, C., Lu, S., & Ding, S. (2023). High-energy lithium-ion batteries: Recent progress and a promising future in applications. *Energy & Environmental Materials, 6*(5), e12450. DOI: 10.1002/eem2.12450

Xu, J., Zhang, Y., Huang, Z., Jia, C., & Wang, S. (2021). Surface modification of carbon-based electrodes for vanadium redox flow batteries. *Energy & Fuels, 35*(10), 8617–8633. DOI: 10.1021/acs.energyfuels.1c00722

Xu, R., Du, L., Adekoya, D., Zhang, G., Zhang, S., Sun, S., & Lei, Y. (2021). Well-defined nanostructures for electrochemical energy conversion and storage. *Advanced Energy Materials, 11*(15), 2001537. DOI: 10.1002/aenm.202001537

Yuan, S., Duan, X., Liu, J., Ye, Y., Lv, F., Liu, T., Wang, Q., & Zhang, X. (2021). Recent progress on transition metal oxides as advanced materials for energy conversion and storage. *Energy Storage Materials, 42*, 317–369. DOI: 10.1016/j.ensm.2021.07.007

Zhang, C., Liu, G., Geng, X., Wu, K., & Debliquy, M. (2020). Metal oxide semiconductors with highly concentrated oxygen vacancies for gas sensing materials: A review. *Sensors and Actuators. A, Physical, 309*, 112026. DOI: 10.1016/j.sna.2020.112026

Zhang, Z., Ding, T., Zhou, Q., Sun, Y., Qu, M., Zeng, Z., Ju, Y., Li, L., Wang, K., & Chi, F. (2021). A review of technologies and applications on versatile energy storage systems. *Renewable & Sustainable Energy Reviews, 148*, 111263. DOI: 10.1016/j.rser.2021.111263

Zhao, J., & Burke, A. F. (2021). Electrochemical capacitors: Materials, technologies and performance. *Energy Storage Materials.*

Zhou, Y., Li, X., & Lin, H. (2020). To be higher and stronger—Metal oxide electron transport materials for perovskite solar cells. *Small, 16*(15), 1902579. DOI: 10.1002/smll.201902579 PMID: 31389168

Zhu, Z., & Xu, Z. (2020). The rational design of biomass-derived carbon materials towards next-generation energy storage: A review. *Renewable & Sustainable Energy Reviews*, *134*, 110308. DOI: 10.1016/j.rser.2020.110308

Chapter 12
Next Generation Hybrid Energy Harvesters by Utilizing Renewable Technologies

Parkavi Valaguruvan
https://orcid.org/0009-0002-8921-2845
Indian Institute of Information Technology, Design, and Manufacturing, Kancheepuram, India

Pandiyarasan Veluswamy
https://orcid.org/0000-0002-2200-3929
Indian Institute of Information Technology, Design, and Manufacturing, Kancheepuram, India

Jayabal K.
Indian Institute of Information Technology, Design, and Manufacturing, Kancheepuram, India

ABSTRACT

This chapter delves into the critical role of renewable energy in addressing the sustainable global energy demands. It emphasizes the importance of renewable sources such as light, wind, and mechanical energy in reducing the carbon emissions and mitigating climate change. The chapter thoroughly explains the various renewable energy technologies and their transduction mechanisms, exploring for both renewable energy and hybrid energy harvesters. The hybrid system's ability to harness energy from multiple sources and convert it into electricity is highlighted as a promising solution to meet the power demands. Furthermore, the chapter discusses about the factors affecting energy conservation and the applications of these

DOI: 10.4018/979-8-3693-9316-1.ch012

technologies in domestic and industrial settings. Through a detailed comparison, the efficacy of hybrid mechanisms in enhancing energy conversion efficiency and their potential benefits for future generations are highlighted. The conclusion includes a comparative analysis of hybrid mechanisms, potential applications, and the market availability of these energy devices.

INTRODUCTION

The rising global energy demand is driven by population growth and advancements in industry. Notably, the population has been increased by 2 billion within a single generation. Hence, the global energy service demand is to be projected to rise by up to tenfold by 2050, while the primary energy consumption is anticipated to increase by 1.5 to 3 times (Dincer, 2001). Thus, the growing population drives a rapid increase in energy requirements worldwide. To address this, countries across the globe are implementing various strategies, plans, policies, and control measures to secure their energy needs and positioning them on the global stage. For achieving, financial stability energy sources are identified which are categorized as renewable and non-renewable source of energy. Among the above two energy sources, the finite sources cannot adequately meet the growing energy demand, as these resources are limited and consumable. It is well known that numerous controversial issues, often leading to severe conflicts, arise as dominant nations compete to control regions rich in fossil fuel reserves. Moreover, depending on non-renewable energy sources can influence the climate change, which in turn triggers devastating the natural disasters and cause significant harm to the planet's ecosystems. Thus, the over usage of these non-renewable waste leads to increase the content of $C0_2$ emissions, highlighting the urgent need to promote the clean environment (Kannan & Vakeesan, 2016).

Energy is a fundamental aspect of our daily lives, and it comes from a variety of sources. Fossil fuels such as coal, oil are formed from ancient organic materials were buried and then exposed to heat. (Youm et al., 2000) (Hiemstra-van der Horst & Hovorka, 2009). Nuclear energy comes from nuclear reactions, specifically fission of uranium or plutonium atoms. Renewable energy resources are crucial to the future of the world. They are called renewable because they replenish naturally on a human timescale. They can be reused to generate energy continuously that meet 14% of the world's total energy need. Renewable technologies are regarded as clean energy sources that, provide a sustainable solution that address the present and future economic and social requirements.(Dincer & Rosen, 1999), (Bilgen et al., 2004). Decentralized renewable energy harvesting is a viable option to satisfy both the rural areas and small scale communities with power consumption (Bergmann et al., 2008). However, the efficient capture and conversion of these abundant, yet intermittent energy sources faces unique challenges. This is where energy harvesters come into play.

Figure 1. Renewable energy sources

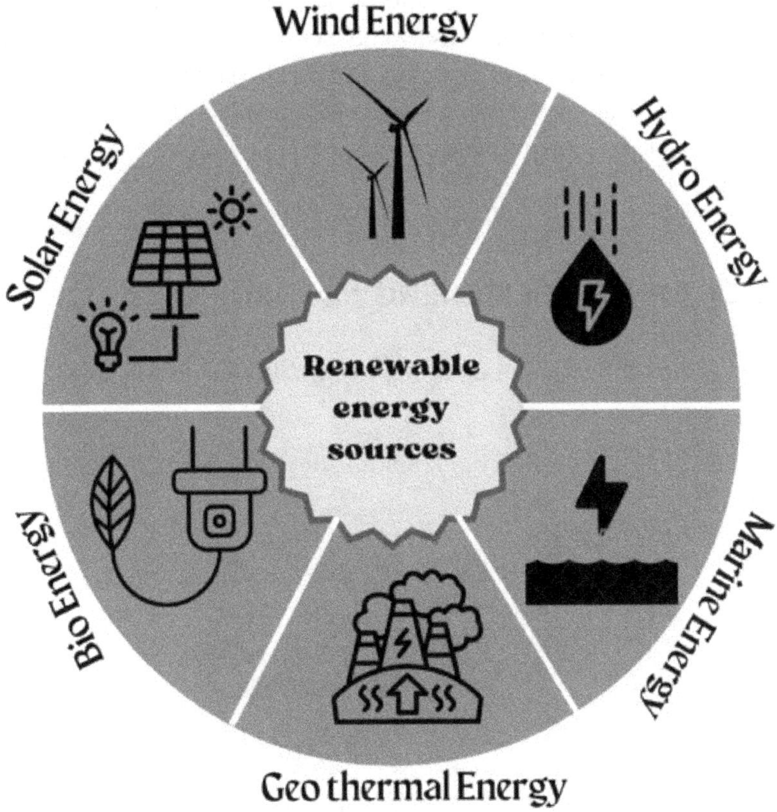

Energy harvesters or scavengers are the devices that are essential for maximizing the potential of renewable energy by capturing every bit of available energy and converting them it into a form that can be stored or used. They reduce reliance on batteries and minimizes the environmental impact. They ensures that the energy from renewable sources is harnessed efficiently, reducing the waste and improving the overall captured energy. These devices can be applied across various scales from powering small portable gadgets to supply electricity for entire communities showcasing their adaptability and crucial role in modern energy systems. They also facilitate innovative applications such as wearable technology, smart cities and remote monitoring systems. Figure 1 shows various forms of energy sources.

These renewable energy sources also posse's drawbacks due to the climatic conditions. Additionally, the availability of these sources may not utilized in continuous period due to the unpredictable nature. To address these challenges, combining the complementary characteristics of each resource could enhance system efficiency and reliability. Additionally, the combined use of renewable energy sources, known as hybrid systems, is becoming increasingly popular and is being widely adopted as an alternative to single energy harvesters. A hybrid system involves combing two or more different energy harvesting mechanisms to collect the energy from multiple sources leading to be more reliable and consistent power output compared to single energy sources.

NEED OF RENEWABLE ENERGY SOURCES

Figure 2 shows the need of choosing renewable sources of energy.

Figure 2. The growing necessity of renewable energy sources to ensure sustainable development

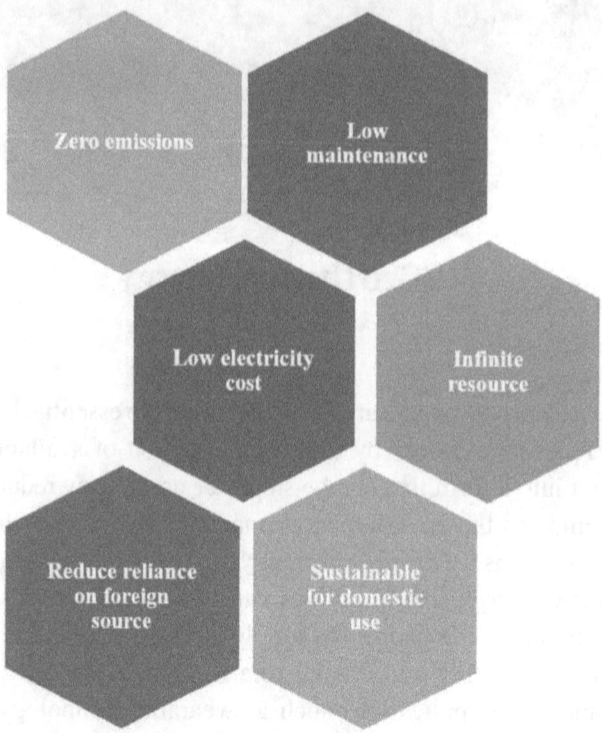

CLASSIFICATION OF ENERGY SOURCES BASED ON SCALE

The environment offers a diverse array of ambient energy source that includes mechanical, thermal, chemical, electromagnetic, and solar energy. Each has its own unique energy density. As electronic devices continue to shrink and consume less power, harnessing environmental energy to power these systems. The sources are categorized under macro and micro energy sources. Both macro-scale and micro-scale renewable energy sources play vital roles in transitioning to a sustainable energy future. Macro-scale systems are crucial for meeting substantial energy demands, whereas the micro-scale systems empower the individuals and communities to contribute to the overall energy supply.

Hydropower

Hydropower is categorized as one of the macro-scale energy source, which utilizes large dams to harness water flow, generating substantial amounts of electricity for widespread distribution. In 2005, hydropower produced approximately 2,950 TWh of electricity, (Darmawi et al., 2013). Small hydro plants, such as mini, micro and pico-hydropower systems avoids major environmental and social issues due to their smaller scale and minimal water storage, which generally does not obstruct the marine. They are considered as valuable technology for those who lack access to electricity. The potential of hydropower may not able fully meet the global demands and they are suitable to satisfy only half of their needs (S. Chen et al., 2015). But a well-planned hydropower development can significantly enhance the life style. The large volumes of water stored in reservoirs behind dams can be readily utilized, allowing the need to be satisfied. Moreover, large-scale hydropower generation primarily relies on substantial turbines that operate using high water flow, significant water head both typically necessitating dams of varying sizes (Pobering & Schwesinger, 2004),(Aryal et al., 2020).

Solar Power

Solar power stands out as one of the most dependable and eco-friendly energy sources. It involves harnessing energy for various purposes as it is a clean, abundant, and renewable source of energy, (Kannan & Vakeesan, 2016). However, its efficiency varies depending on geographic location and climate conditions. Regions like the Sahara Desert, the Arabian Desert, and the Mojave Desert have high solar insolation, while countries such as Spain, Italy, and Greece enjoy abundant sunshine. Solar energy is primarily used for electricity generation and serves as both macro and micro-scale applications. Solar is considered as common energy sources on

Earth, attracting global interest in areas such as photo catalysis, photovoltaics, artificial photosynthesis, and photothermal conversion. These technologies contribute to the advancement of green technologies (Hernandez et al., 2014). Solar-to-heat conversion in particular is promising due to its broadband absorption a straightforward operation and high efficiency. Applications of solar energy include solar water heaters, solar-driven water evaporation, and electricity generation. It can be employed in residential and industrial solar water heaters and provide electricity to remote areas without a grid access. Additionally, solar energy finds applications in transportation and agriculture too (Rabaia et al., 2021).

Geothermal Energy

Geothermal energy, sourced from the Earth's interior, is both sustainable and clean. Only a small portion of energy has been utilized which presents significant opportunities for expansion in both electricity generation and direct applications by leveraging stable underground temperatures and converting underground heat into electricity (Fridleifsson, 2001).Unlike fossil fuels, geothermal resources are renewable within technological and societal timeframes, making them a sustainable alternative. The heat, originating from the barysphere and the natural radioactive decay of minerals, can be utilized for various applications, including electricity generation and direct heating. The primary environmental impacts of geothermal development include surface disturbances, physical changes from fluid withdrawal, heat-related impacts, and releases the chemicals. It is essential to carefully evaluate both the positive and negative impacts of geothermal development, alongside potential mitigation measures, before proceeding with a project (Tester, 2007).Generally, the environmental effects of geothermal power generation as direct use are minor, controllable, or negligible, provided as there is full compliance with environmental act, which vary by countries (Rybach, 2003). Timely reclamation efforts in plant operations can substantially lower their long-term reclamation costs and aid in ecosystem recovery. (Anne Naeth et al., 2020).

Wind Energy

Wind energy is renowned as green technology for power generation due to its minimal environmental footprint. As environmental concerns escalate, generating clean energy becomes increasingly vital across all energy consumption sectors. Despite its cleanliness, wind energy's reliability over long periods can be inconsistent. Studies on wind energy potential indicate that fossil fuels must complement wind power (Sahin, 2004). Numerous scientific studies have explored this issue using different methodologies. The wind energy potential heavily depends various factors.

The energy used in manufacturing and transporting materials for building wind power plants is usually offset by the energy generated within a limited duration. Wind turbines does not emit harmful gases that contribute to acid rain or greenhouse gas emissions and capable to produce a zero direct air pollution. Although a small amount of CO2 is released during the construction and maintenance phases of wind energy projects, it is significantly less than that from fossil fuel-based power plants and can be easily absorbed by trees through photosynthesis. (Saidur et al., 2011). The expanding wind energy industry necessitates a comprehensive understanding of the environmental impacts caused by wind farms. It can be utilized in various applications, from operating windmills and pumping water to powering sailing boats (Dai et al., 2015).

Marine Energy

Marine energy, is indeed a renewable energy source harnessed from the ocean's natural movements and from their properties. It is essential for achieving long-term renewable energy goals and advancing a low-carbon economy. This is particularly significant for the United Kingdom. This approach aims to guide future research and promote environmentally conscious development. The ecological impacts of marine energy extraction are complex, they are not well understood, as the interactions between devices, and ecosystems are constantly changing, making it difficult them to establish clear cause and effect relationships. Marine energy includes tidal energy, which involves tidal barrages or underwater turbines to capture energy, as well as converting wave energy for power generation (Bonar et al., 2015).

Bioenergy

Biomass is a term that broadly includes both plant biomass (phytomass) and animal biomass (zoomass). It refers to the chemical energy stored by plants through the process photosynthesis by using sunlight to produce vegetation on land and in water. It consists of varying amounts of cellulose, hemicellulose, and lignin, along with small quantities of organic and inorganic substances. The specific composition of these organic materials is crucial for developing processes to produce alternative fuels and chemicals. Energy from-biomass programs typically focus on these more efficient species (Abbasi & Abbasi, 2010).In the European Union, bioenergy plays a crucial part in the renewable energy mix, contributing to a stable energy supply. The effects and efficiency of biomass production and its usage varies by region. Using biomass for energy is one of many land-use options, where the markets for bioenergy feedstock's are closely linked with those for agricultural commodities. Both biomass and biofuel markets operate globally but they face tariffs and nontariff trade barriers, resulting in a lower trade volumes compared to fossil fuels (Popp et al., 2014).

PORTABLE TECHNOLOGIES FOR ENERGY HARVESTING

As there are many energy sources available, Harvesters, or energy harvesting technologies, which is otherwise known as Scavengers are essential for effectively capturing and converting these natural energy sources into usable power. The electrical circuit is then amplified for regulating the signal to make it suitable for various applications. The captured energy is then stored in super capacitors or rechargeable batteries for later use (Hasan et al., 2021),(H. Liu et al., 2018). The energy output of an these various system varies greatly depending on the technology used in it and the efficiency of the conversion process involved. Factors such as the presence and strength of the environmental energy significantly influence the total energy output.

Advantages are

- Efficient Energy Conversion
- Maximizes Availability
- Utilizes wasted energy
- Environmentally friendly
- Economically beneficial by reducing cost

CLASSIFICATION

Based on the type of energy they harvest, the energy harvesters can be classified as

- Thermal Energy Harvesters which includes Thermoelectric Generators, Thermophotovoltaic systems, and pyroelectric generators
- Mechanical Energy Harvesters including Piezoelectric harvesters, Electromagnetic induction generators, Electrostatic generators
- Solar Energy Harvesters that includes Photovoltaic cells, Solar Thermal collectors, Concentrated Solar Power
- Wind energy harvesters that includes wind turbines
- Hydro and Marine Energy Harvesters which includes hydroelectric generators and tidal wave energy converters
- Chemical and Biological Energy Harvesters
- Radio Frequency Energy Harvesters

Figure 3 clearly depicts the classification of Nano generators / Energy harvesters

Figure 3. Classification of nano generators / energy harvesters

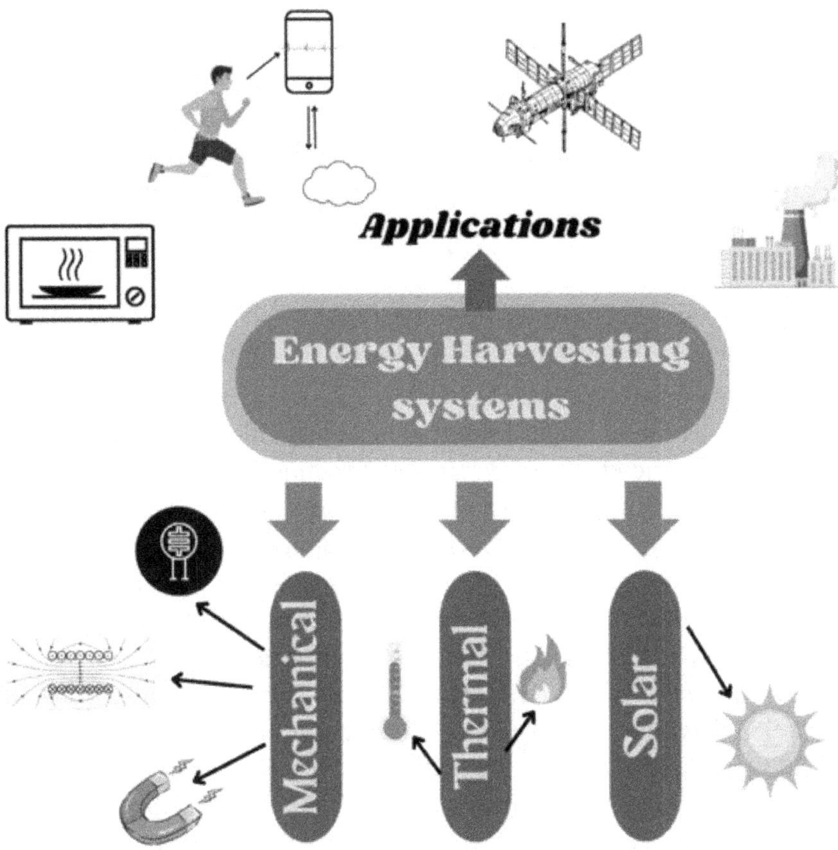

MECHANICAL ENERGY HARVESTERS

Mechanical energy harvesters are the devices, which are engineered to capture and convert the mechanical energy from different sources into electrical power. They can harness energy from various movements, vibrations, and mechanical stresses present in the environment. There are different types of mechanical energy harvesters namely piezoelectric harvesters, electromagnetic harvesters, triboelectric harvesters as shown In Figure 4.

Figure 4. Mechanisms of various mechanical energy harvesters (a) Piezo (b) Electromagnetic (c) Tribo

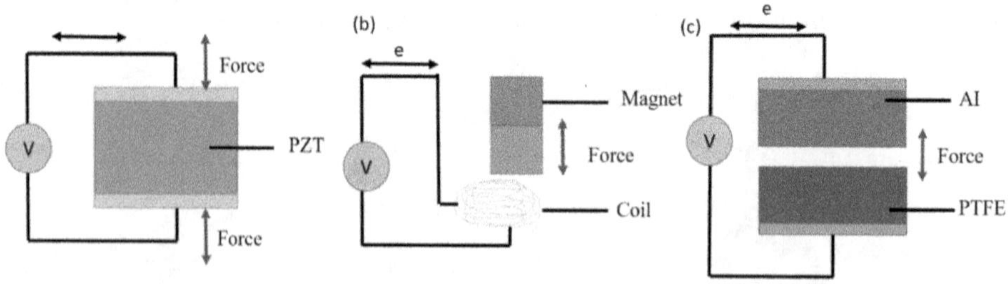

Piezoelectric Nano Generator (PENGs)

Piezoelectricity is the property of some materials that generate an electric charge when it is subjected to mechanical stress. Charge separation due to stress is called the direct piezoelectric effect, while stresses resulting from an electric field are known as the converse piezoelectric effect. Piezoelectric materials can function in different modes. In the 33-mode, stress is applied along the polar axis that produces voltage along the same axis. In the 31-mode, stress is applied perpendicular to the polar axis, that generates voltage at a right angle to the applied force. These materials can generate electrical charges in response to mechanical pressure through the reversible piezoelectric effect. (H. Liu et al., 2018). These materials are classified into natural and synthetic types namely Natural Piezoelectric Materials include quartz (SiO_2), topaz, tourmaline, Rochelle salt and some organic substances like silk, wood, rubber, bone, dentin, hair, and enamel. The other is the synthetic type which include polymers, ceramics, and composites. All ferroelectric materials are a subset of piezoelectric materials and have a molecular structure that allows them to exhibit an electric dipole. For energy, harvesting, piezoelectric materials typically have a defined polar axis, which influences performance. In ferroelectric ceramics like lead zirconate titanate (PZT) or polyvinylidene fluoride (PVDF), the polar axis is determined by the poling direction (Wu et al., 2013) whereas in non-ferroelectric materials like aluminium nitride (AlN) or zinc oxide (ZnO), the polar axis follows the crystal orientation (along the c-axis of the Wurtzite structure).Figure 5 shows the material evolution in the harvesters.

Figure 5. Evolution of materials used in generators across various years, highlighting changes in technology.

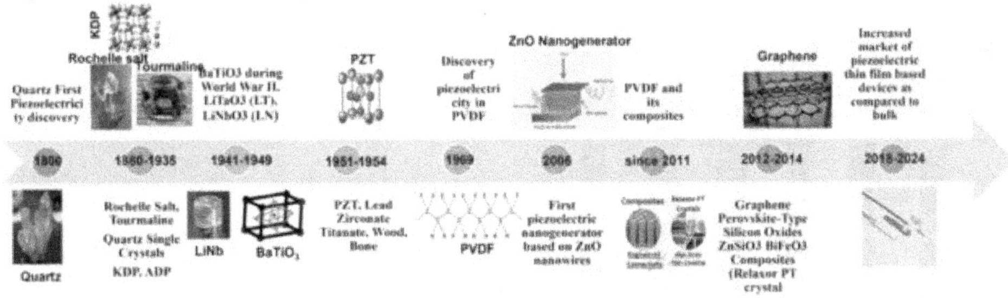

A piezoelectric energy harvester is a device designed to convert external forces acting on piezoelectric materials into electrical power. This technology operates based on the direct piezoelectric effect, where stress on the material generates electrical charges, creating current flow through an external circuit. Piezoelectric energy harvesters are into classified two types.

- Force Perpendicular to Nanowire Axis: When force is applied perpendicularly to a piezoelectric structure, deformation occurs, generating positive and negative electrical potentials on different parts of the structure as shown in Figure 6.
- Force Parallel to Nanowire Axis: When force is applied along the axis of the nanowire, uniaxial compression generates electrical potentials at the tips of the nanowire as shown in Figure 7.

These harvesters are efficient, low-profile power generators suitable for various applications, providing a sustainable way to harness ambient mechanical energy.

Figure 6. Structure of piezo harvester in which force is acting perpendicular to the axis

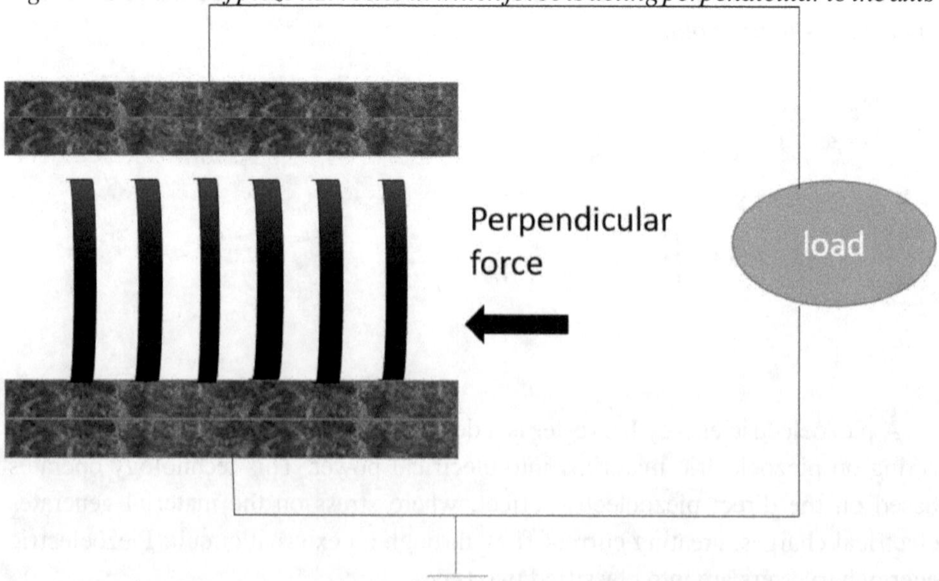

Figure 7. Structure of piezo harvester in which force is acting parallel to the axis.

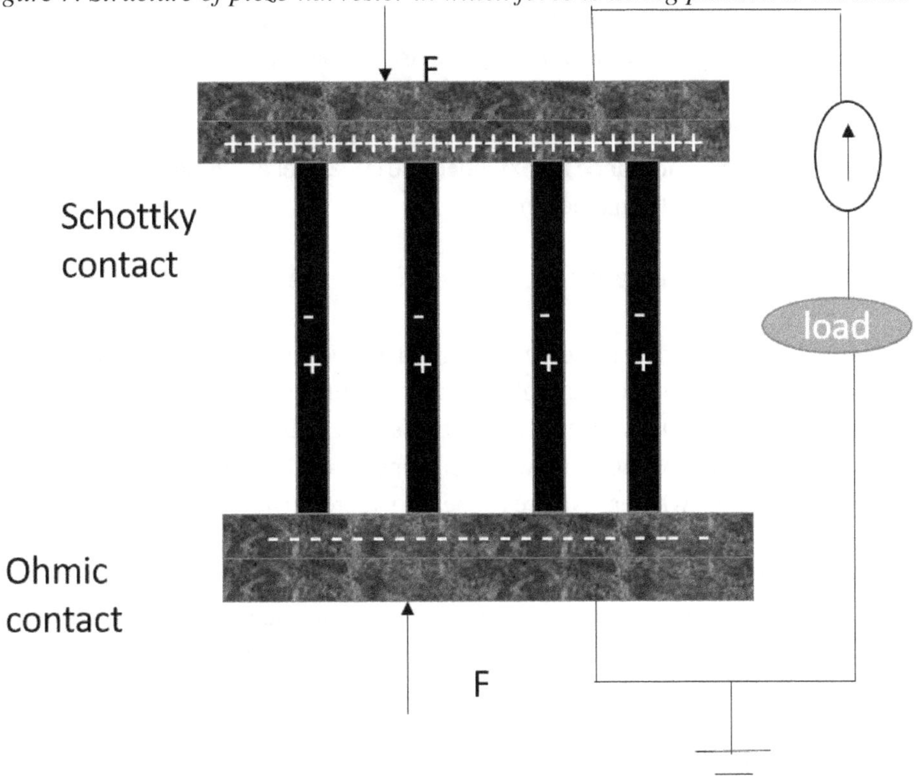

Electromagnetic Generator

The basic principle behind the electromagnetic generators is Faraday's law of electromagnetic induction. In 1831, Michael Faraday discovered that moving an electric conductor through a magnetic field induces a potential difference between the conductor's ends. Faraday's law states that the voltage, or electromotive force (emf), induced in a circuit is proportional to the rate of change of the magnetic flux linkage in that circuit. The electromagnetic generator (EMG) is an essential technology for transforming the mechanical energy into power. They operate based on converting the energy at high frequencies with excellent durability, generating and following the core principle. This technology is renowned for its optimal energy conversion at high frequencies and exceptional durability for long-term use.

The generators operate on rotational principles and they are employed in a wide range of applications, from large-scale power generation to smaller-scale uses like recharging car batteries. They can also harvest micro to milli watts levels of power using both rotational and linear mechanisms.. Rare earth magnets such as Neodymium Iron Boron (NdFeB) is known for their magnetic property and high energy density which could be operated at 120 °C. Similarly samarium cobalt with less power and less expensive than NdFeB is used to operated ate high temperatures of about 300°C. This technology provides a optimal energy conversion at high-frequency range and excellent endurance for extended use (Beeby et al., 2006). Figure 8 shows the milestones of electromagnetism.

Electrostatic Generator or Triboelectric Generator

The triboelectric effect is the electrification that occurs when materials come into contact and become charged through friction with different materials. The materials have both positive and negative effects. Materials such as Ethyl cellulose, Polyamide11, Polyamide 6-6, silk, Aluminium, woollen, paper, steel, wood, copper, and sulphur has a positive effect whereas Nature rubber, Polyacronitrile, Polystyrene, Polyester, Polyamide, Polyvinyl Chloride has an negative effects. Triboelectric Nanogenerators (TENGs) work by coupling tribo effect and static mechanism. TENG technology is a fascinating area of research in energy harvesting, nanoenergy, and nanosystems (NENS). It shows great potential across various field including mechanical, electrical, optical, acoustic and fluidic applications (Zhu et al., 2018). To enhance energy harvesting efficiency, extensive research has been focused on Nano and micro surface structuring to boost the charge generation. The sliding in generators can be either planar, cylindrical or disc in shape. In case of single electrode, one electrode is grounded and here the electron are exchanged between one electrode and ground to maintain the potential change. In a freestanding mode pair of symmetric electrode underneath a dielectric layer. TENGs have gained recognition as a promising method for energy harvesting across various applications. However, leveraging blue energy harvesting ocean, marine or wave energy through triboelectricity remains relatively novel. TENGs can effectively support marine activities, including underwater exploration, security surveillance, weather forecasting, oceanographic research, and maritime navigation (Radhakrishnan et al., 2022).

Figure 8. Timeline and key milestones in the discovery and development of electromagnetism.

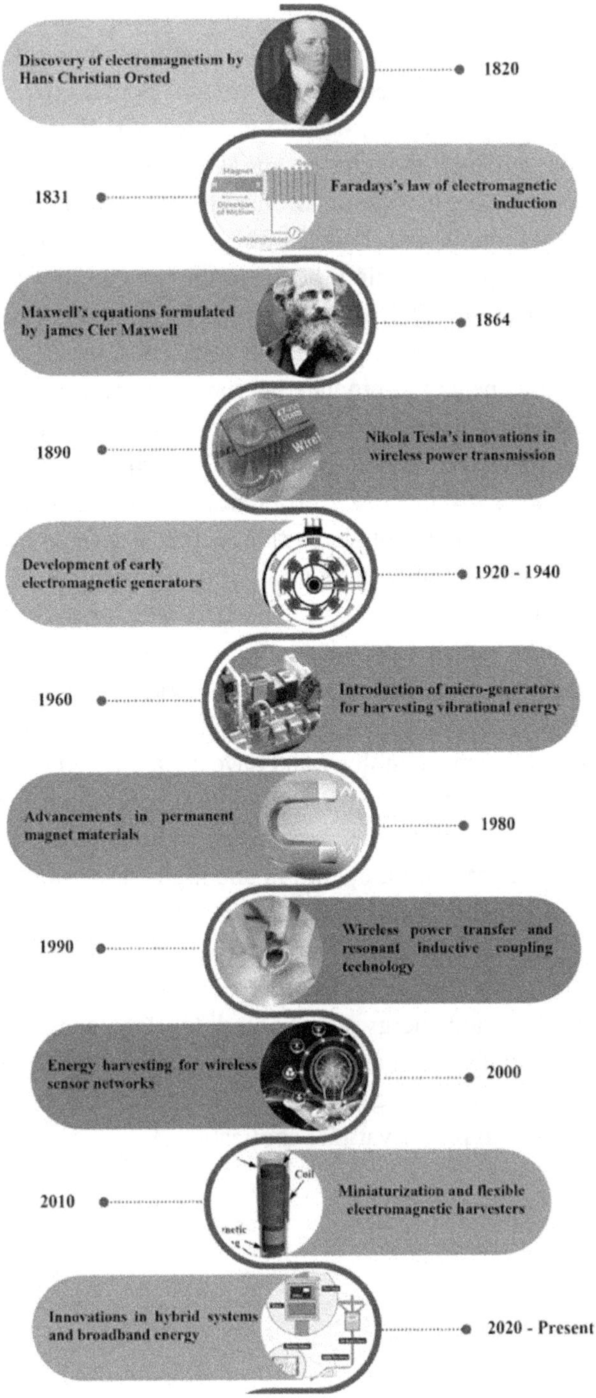

SOLAR ENERGY HARVESTER

Solar energy provides abundant radiation to the Earth's surface enabling self-powered photovoltaic applications. Photovoltaic technology is one of the most widely used methods for harnessing solar energy. A common approach to renewable energy estimation is the top-down method, used to analyse solar energy potential. Solar potential estimation generally includes potential in theoretical, technical and economic level. The process begins with theoretical potential analysis, which involves examining solar radiation. The technical level involves adjusting PV conversion efficiency. The final step is assessing the economic potential of the solar energy system. (Kannan & Vakeesan, 2016).Single crystalline semiconductor wafer are the first generation solar cells. Second-generation solar cells, which use an inorganic thin film structure, are cheaper to produce but have lower efficiency.

Amorphous thin film solar cells from the second generation have an efficiency of less than 14%, compared to the single-junction crystalline solar cells of the first generation, which can reach up to 27% efficiency. In theory, single-junction cells can attain a maximum efficiency of about 33%, which is defined by the Shockley–Queisser limit in thermodynamics. To surpass this efficiency while reducing production costs, new solar cell technology is needed. This breakthrough has led to the development of third-generation solar cells. Solar energy, being one of the most significant energy sources on Earth, has garnered global interest in various fields such as photocatalysis, photovoltaics, artificial photosynthesis, and photothermal conversion to advance green technology and have been utilized in various applications. Figure 9 depicts the solar harvesters evolution based on materials. Notably, solar-to-heat conversion is a particularly promising technology due to its broadband perfect absorption, straightforward operation and high conversion efficiency. This technology can be applied to solar water heaters, solar-driven water evaporation, and electricity generation. Advancements in PV self-powered applications are categorized with three key components namely the energy harvesting module, energy conversion module, and energy storage module (Rabaia et al., 2021). The PV panel captures solar radiation to generate electricity, which is then processed by the controller and inverter before being stored in an electricity storage device through a filtering circuit, ready to power various applications.

Figure 9. Overview of materials used in solar generation

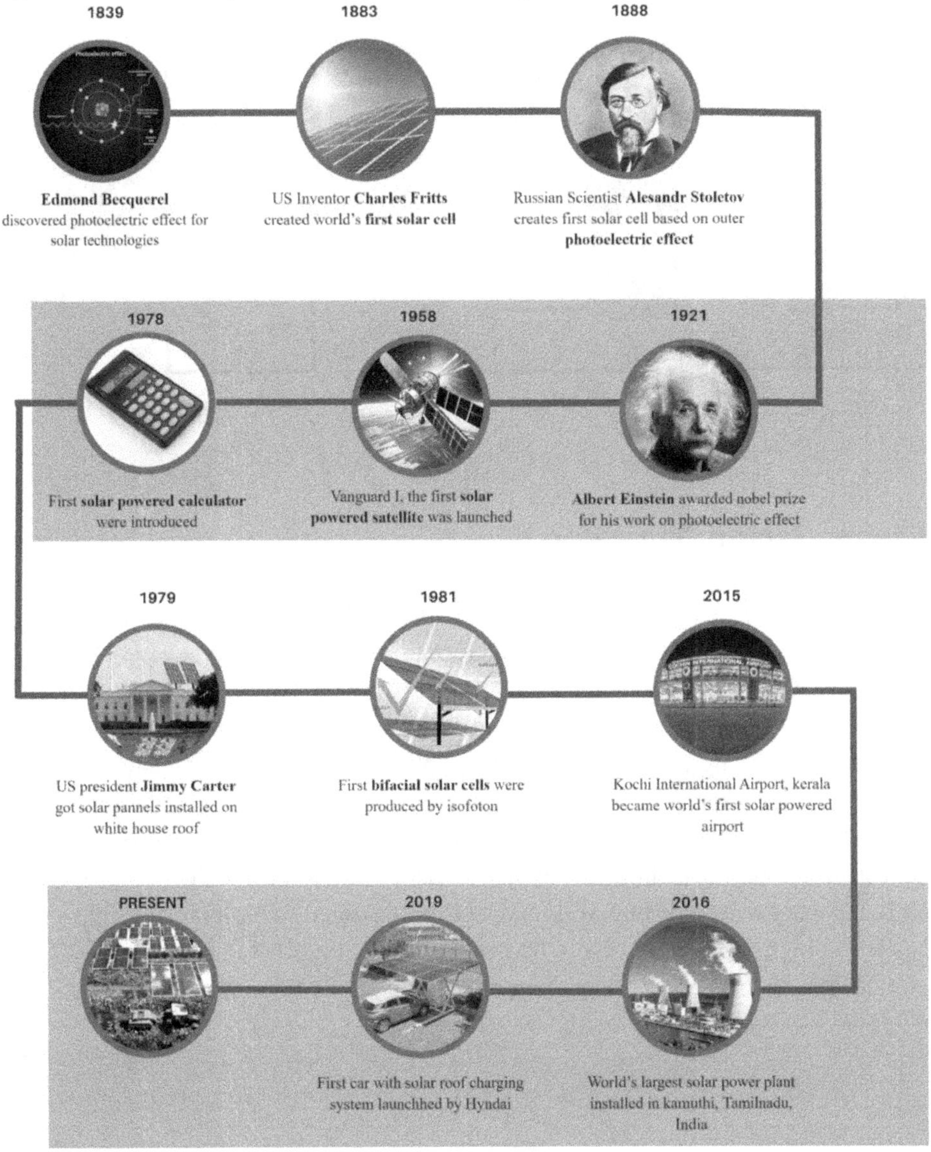

THERMAL ENERGY HARVESTERS

Figure 10 shows the different mechanism involved in thermal energy harvesters.

Figure 10. Mechanism involved in thermally based generators (a) Thermo (b) Pyro

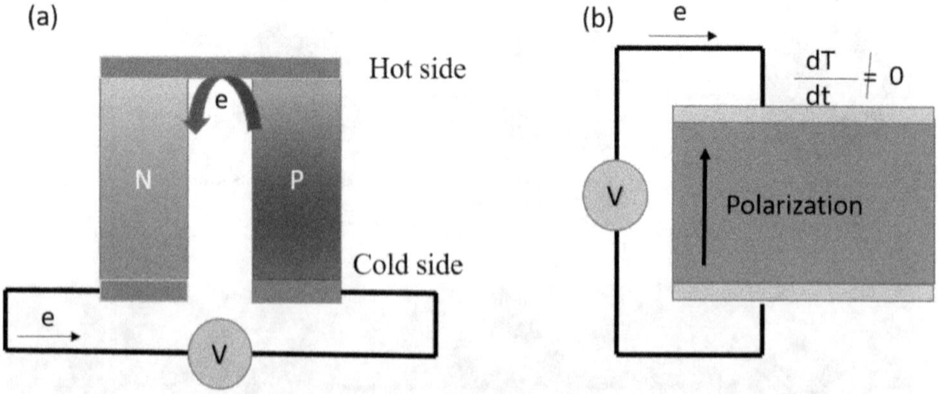

Thermoelectric Generator

Thermoelectricity involves converting thermal energy into electrical power through temperature gradient. Thermoelectric materials convert the heat directly into power when there is a temperature gradient between their ends. These materials are extremely versatile, as thermoelectric devices are scalable for energy conversion and have no moving parts. This design enables them to operate efficiently across various applications, from small-scale synthesis to large industrial facilities. The performance of thermoelectric devices can be significantly improved through nanostructure-based techniques, which enhance their thermoelectric properties. Utilizing nanostructured materials and low-dimensional super lattice morphologies has led to higher ZT values (X. Chen et al., 2018). Figure 11 shows the evolution of materials in thermoelectric generators.

Figure 11. Evolution history involved in thermoelectric generators.

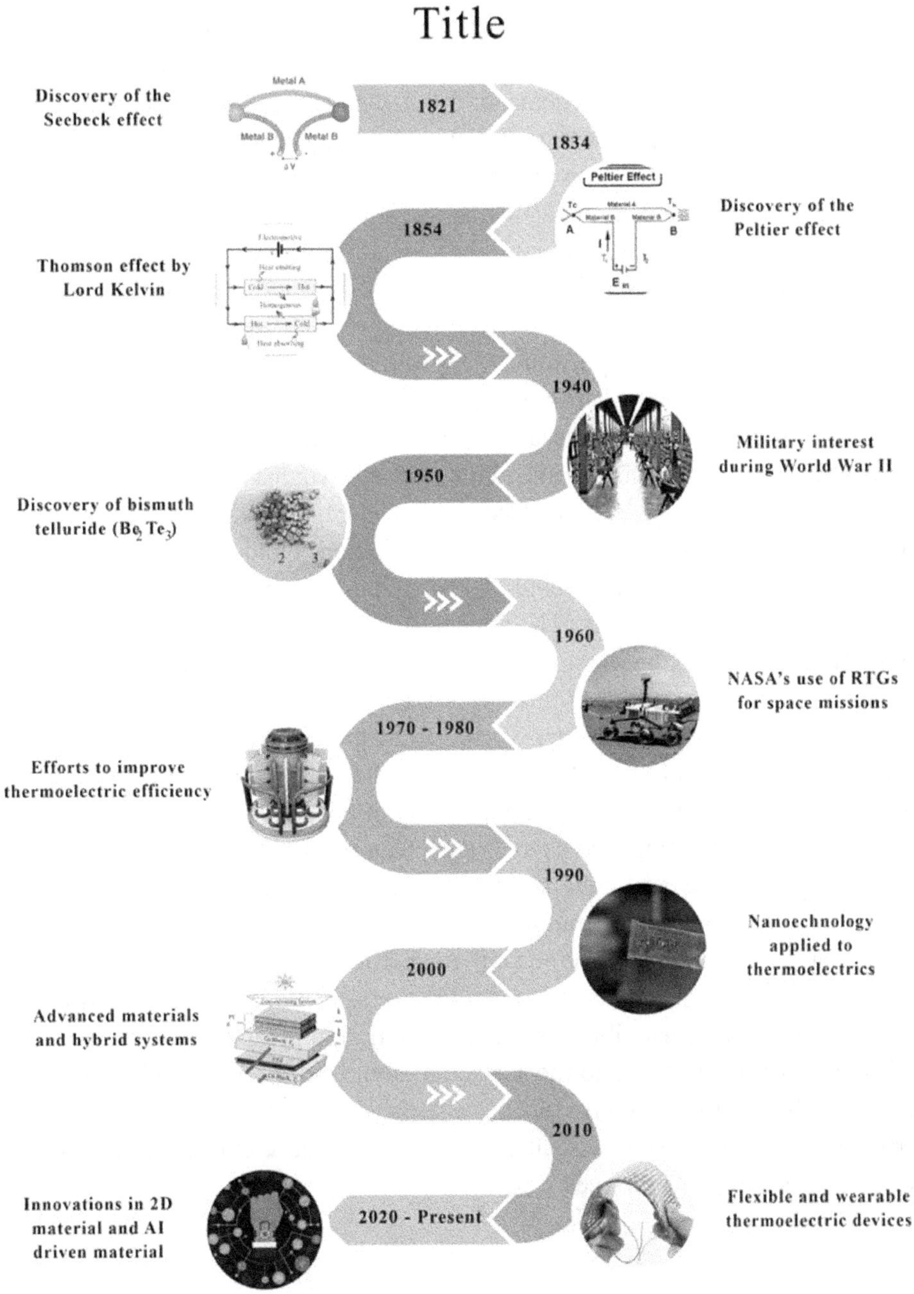

. When a temperature gradient exists in certain environments, thermal energy can be harnessed for power generation. The temperature difference acts as the driving force for energy conversion, while heat flow produces the energy. Although the efficiency of thermoelectric devices remains below the Carnot limit, they can used for applications with low power demands due to their capability of generating power under such conditions. They are used in energy harvesting applications for sensor data transmission in IOT, for cooling the lasers, microprocessors and in aerospace applications (Ando Junior et al., 2018).

Pyroelectric Generator

Pyroelectric effect is the phenomenon which is similar to thermoelectric effect. Here the intensity of pyroelectric materials changes with temperature which produces current. Pyroelectric materials generate power from the temperature fluctuations (dT/dt) and share similarities with piezoelectric harvesters, which convert mechanical oscillations (ds/dt) into electricity. These materials are particularly used due to their potential for high thermodynamic efficiency under suitable conditions. Unlike thermoelectric devices, materials used in this generator do not require any heat sinks for dissipating heat. They generally function at very low frequencies, typically below 1 Hz, due to slow temperature oscillations in systems with substantial thermal mass and heat transfer inertia. Cyclic pumping is used to address these slow oscillations. The energy consumed by the pumping process is generally minimal, accounting of about less than 2% of the harvested energy, making this approach viable. The spontaneous polarization in pyroelectric materials results in surface charges, attracting free charges such as ions or electrons to the material's surfaces. Pyroelectric behaviour arises from changes in surface charge in response to variations in ambient temperature, assuming that the polarization level depends on the material's temperature. When a pyroelectric material is heated (dT/dt > 0), its spontaneous polarization decreases as the dipoles within the material lose their alignment due to thermal vibrations. This reduction in polarization results in fewer free charges being bound to the surface region. Under open circuit conditions, these free charges remain on the electrode surface, generating an electric potential across the material. In short circuit conditions, an electric current flows between the material's two polar surfaces.

RF NANO GENERATOR

Radio Frequency Energy Harvesting is an advanced technology that captures electromagnetic waves to produce electricity for low-voltage electronic devices. To detect RF energy, specific antennas are used, such as ultra-wideband or narrowband types, depending on the targeted frequency band. For example, a narrowband antenna is suitable for detecting GSM-900 frequencies. Additionally, multiple-input-multiple-output (MIMO) systems are frequently employed to boost the efficiency and functionality of wireless communication systems. This innovative method facilitates wireless power delivery, making them ideal for portable devices (H. Wang et al., 2018).

BIO- ENERGY HARVESTERS

Bio energy harvesters generate power from biological sources, from the natural processes found in living organisms. These devices are promising for sustainable energy solutions, especially in regions where traditional power sources generation is not possible. Bioenergy harvesters encompass a diverse array of devices that produce energy through processes. At present, these advanced energy harvesters are capable of powering energy efficient electronic devices and they subject of extensive research for potential large-scale applications. (Cherubini & Strømman, 2011).One of the initial bio-energy harvesters is microbial fuel cells (MFCs), which have the potential to become significant future industrial energy solutions. MFCs utilize the catabolic metabolism of various microorganisms acting on diverse organic substrates, transforming them into micro bio-reactors to generate electrical energy. They are used for powering remote sensors, long-distance marine communication, wastewater processing with auxiliary power supply, and metal recovery processes. MFCs use bacteria to break down organic matter and produce electricity, while Enzyme-Based Fuel Cells utilize enzymes to catalyse reactions that generate electrical energy. Enzyme-based biofuel cells produces enzyme molecules which are immobilized on electrodes for power generation, are used in wearable electronics, medical devices, environmental monitoring, and industrial applications (Jeerapan et al., 2020). Advances in physical and chemical sciences have led to the creation of flexible and stretchable materials that conform to the body's contours. Functional polymer fibers have paved the way for the development of smart clothing that captures energy from the body's surface (Lv et al., 2021).

HYBRID ENERGY HARVESTING DEVICES

A single energy harvester often falls in lack of fulfilling the requirements of energy in electronic devices, which relies on availability of the energy source towards power generation. To address their limitations, hybrid energy harvesting is gaining popularity in recent years and will be advanced in upcoming years too. This approach harness energy from various origin sources and converts them it into power generation through transduction mechanisms. of hybrid energy harvesters (HEHs) are able to harness the energy from single source but they are lacking to harness sufficient energy due to climatic conditions. This has been led to the development of harnessing the energy from multi-source which is mainly based on energy transformation materials. Based on the availability of power source either single or both, the harnesses energy can be significantly boosted. Research on hybrid energy harvesters has been surged in recent years, leading to significant advancements in the field. Many studies have been focused on exploring various combinations of energy sources with their transformation mechanism to enhance power output and improve the energy efficiency. Integrating hybrid energy harvesters into infrastructure and vehicles can contribute to the development of eco-friendly transportation systems with smart mobility. Similarly, in healthcare applications, the creation of devices powered by hybrid energy harvesters can provide continuous and sustainable solutions (Selvan & Mohamed Ali, 2016),(Vullers et al., 2009). Moreover, integrating multiple energy harvesting technologies into single system can increase overall size and weight, posing challenges in applications that require portability and compactness. Additionally, the need for multiple energy conversions and storage components can make the deployment of them to be expensive. To understand the importance of hybrid energy harvesting and promoting towards a environment friendly solutions in the future, it would be crucial to address the challenges faced by them. Energy transformation mechanisms involved to have a unified design, with a focus on optimizing the interactions between the integrated components rather than just combining them. Designing and positioning the hybrid devices is very essentially to be noted as it avoids potential negative impacts and interference that could decrease the total efficiency. Owing to their combined efforts and their enhanced performance, hybrid energy harvesters acts as potential solutions for sustainable and greener environment (Bai et al., 2018), (Chandrasekaran et al., 2019).

Global Policies Promoting Hybrid Energy Technologies

Global incentives such as tax credits, grants, subsidies and Renewable Portfolio Standards (RPS) encourage the adoption of hybrid energy technologies. Tax deductions reduces the upfront costs for individuals and businesses, while feed-in tariffs provide a stable revenue for hybrid energy producers. RPS fosters are used in hybrid energy projects. Governments and international organizations fund R&D initiatives to advance hybrid energy technologies.

Piezoelectric –Electromagnetic Hybrid Energy Harvesters

Electromagnetism effects and piezoelectric conversion mechanisms from their harvesters harness insufficient energy. This insufficient synergy is not able to fully utilize the benefits of both various mechanical conversions in a single hybrid source design. To maximize their advantages, an efficient strategy to manage the two power sources is essential. The integration of these two mechanisms focussing on enhancing power. The aim is to optimize and match electrical damping with mechanical damping, thereby enhancing energy conversion efficiency in them. Both mechanical mechanism are frequently used for converting from kinetic to electric. In hybrid energy harvesters (HEHs), these techniques are combined to enhance the power density and increase energy generation capabilities. A primary objective of the harvesters is to optimize the system by aligning it electrically with mechanical damping and increases efficiency (Xia et al., 2015).Typically, piezoelectric and electromagnetic harvesters generate alternating current (AC) as output. To ensure a more stable energy supply, this AC needs to be rectified, stored, and the voltage has to be stabilized allowing the accumulated charges to be stored individually. These different output characteristics present significant challenges in designing an efficient interface for hybrid scavengers. One of the practical solution is to develop, separate conversion and storage systems for the piezoelectric and electromagnetic components.

Piezoelectric – Triboelectric Hybrid Energy Harvesters

This energy harvester's work under mechanical stress and friction. These effects share operational characteristics, and their responses making them ideal for integration to boost the overall energy output (Jung et al., 2015), (X. Wang et al., 2017). Both these effects exhibit similar output characteristics in various operating frequency ranges. Studies have demonstrated that combining both PENGs and TENGs can significantly enhance energy conversion efficiency from environmental sources. Hybrid Nano generator (NG) was designed with a wavy structure, where a piezo-electric film made of (P(VDF-TrFE)) nanofibers which was sandwiched between Kapton films. This combined design enabled the device to harvest the energy from both effects during both pressing and releasing cycle (Zhu et al., 2018).

Electromagnetic-Triboelectric Hybrid Systems

In the electromagnetic energy harvester, the movement between the magnet and coil can be achieved through various design. The structures are typically categorized into resonant, non-resonant, and rotational structures. Similarly, a triboelectric nanogenerators can be combined and performed based on its friction in different modes between the materials based on the applications. By integrating these two different energy harvesting methods, a single mechanical movement is capable to yield greater electricity generation. Furthermore, their distinct output characteristics impact the electric charging behaviour. If the structure is resonant-based then various mechanisms such as spring-mass, flexed-membrane, and magnetic-levitation designs can be utilized to create EM-TE hybrid system. Due to high voltage and low current output characteristics of the triboelectric generator, they are able to charge a capacitor to tens or hundreds of volts, albeit with a prolonged charging time. Conversely, if they produce low voltage of just a few volts but generates high current, allowing for rapid capacitor charging, though it quickly reaches saturation. To optimize the output power, each generator requires distinct matching impedances. Similarly, power density of an EMG and TENG has a relation with frequency. In EMG there is direct relation with frequency multiplies twice, whereas for a TENG, it has directly correlation. As a result, the output power of the triboelectric surpasses operates even at low frequencies to illuminate a light-emitting diode (LED) than that of EMG. Combining both TENG and EMG results proves as a promising strategy to leverage their auxiliary output features, thereby boosting the effectiveness (Zhou et al., 2018).A textile-based hybrid generator was designed and its performance exhibits excellent flexibility and air permeability, making them suitable for integration into clothing to harvest biomechanical energy. Moreover, a bendable and see-through triboelectric- piezoelectric composite generator with a single friction surface, was designed which utilizes a piezoelectric layer, along with zinc oxide as clear electrode. Consequently, some researchers have combined and analysed the performance of both electromagnetic generator and triboelectric generator in a unified device to harness the benefits of both systems.

Hybrid Solar and Triboelectric Nano Generators

Photovoltaic panels produce electricity by converting sunlight into energy through the photovoltaic effect and they are extensively used in harnessing solar power as a renewable resource. One significant drawback of solar cells is their reliance on daily light and weather conditions for generating energy. Hence, integrating solar cells with triboelectric generators allow the energy harvesters to function even without sunlight, thus enabling continuous and sustainable energy collection. In dual systems, it is crucial for exploring interactions between individual components of the generators and their impact on the device performance. The working mechanisms and designs in both TENGs and solar cells are distinct whereas the influence of TENGs on solar cells in HPTNGs is typically more significant than vice versa. There are three types of effects of TENGs on solar cell output namely no effects, negative effects and positive effects. In “no effects”, the TENGs do not affect the performance of solar cell whereas in the negative effects, they are noted with high-transparency TENGs and the other is the Positive effects, where the studies in hybrid system have observed performance enhancement mechanisms due to TENGs (Bai et al., 2018).

Mechanical and Photovoltaic Based Generator

In this hybrid generator, a piezoelectric potential is applied to photovoltaic panel to enhance the solar cell (SC) performance. The device's output voltage is increased and remains constant based on the applied pressure. This resulted in magnified voltage pulses influenced by the piezoelectric potential. Substantial negative charges gather at the ZnO/LDH interface, while positive charges accumulate at the ZnO/Au interface. Hence, ZnO nanosheets results in piezo resistance effect, which significantly enhances the electron flow from the solar cell through the PNG. Consequently, the pressure is applied in the hybrid device. These integrated devices aimed at eco-friendly power generators by utilizing textiles using cost-effective materials and scalable fabrication process. Fiber-based solar cells acts as the base for the PENG which allows the fibers to harvest both mechanical and light energy simultaneously. The components of the individual harvesters are combined together to enhance their output performance.

Solar and Thermal Energy Harvester

Hybridizing photovoltaic cells with other energy harvesters, especially thermal energy harvesters, is advantageous because sunlight includes thermal energy. Heat in the panel reduces the efficiency and life span of PV. Thermal energy harvesters absorb the heat from the panel and used it power source for TEG. Hybrid solar–thermal energy harvesters can capture both solar and thermal energy, thus improving the individual performance of either solar or thermoelectric energy harvester. This hybrid system has resulted in a 30% improvement in output power compared to using solar cells alone.Photothermal materials are determined with intrinsic properties such as light harvesting ability and light to heat conversion efficiency and possess a strong and wide absorbance. Materials have extensive light absorption range and high photothermal conversion efficiency. Similarly, PEDOT is used in multi-energy harvesters, due to its optimizing transparency and photothermal conversion properties. Moreover, using of non-ferric materials such as GaN, AIN, ZNO shows greater interest of using in photovoltaic and thermoelectric effects. A high open-circuit voltage in the solar cell was achieved which is crucial for improving energy conversion efficiency and driving high-voltage electronics. Moreover, a compact hybrid photovoltaic–photothermal / pyroelectric–thermoelectric energy cell functions utilizes the wider range of solar spectrum for energy harvesting applications (Mizoshiri et al., 2012).

Piezo-Pyro Energy Harvesters

In these hybrid energy harvesters, both the piezoelectric and pyroelectric effect takes place. The mechanical output of this hybrid device was produced through a process of stretching and releasing while the pyroelectric output voltage was produced by external temperature changes such as heating and cooling effect. To assess the power performance of individual harvesters, the piezoelectric and pyroelectric outputs were measured separately under various strains and thermal gradients. Since both effects rely on internal polarization changes, precisely controlling both energy harvesters can enhance the overall output performance. Materials used in this type of harvesters mainly includes ferromagnetic material such as ABO3 perovskite structure, polymers like PVDF and polycrystals like PZT, BTO.

Thermal, Kinetic and Solar

Both piezoelectric and pyroelectric effects can coexist in the same ferroelectric material for kinetic and thermal energy harvesting and they are even combined with a photovoltaic setup, allowing for the simultaneous harvesting of three energy sources. The main advantage is that the harvester remains functional and continuously generates electricity without interruptions, as long as one or more of the three abundant energy sources—thermal, kinetic, or solar—are available. (Bai et al., 2018).

Piezoelectric- Electromagnetic- Triboelectric Hybrid Energy Harvesters

Integrating all the mechanical energy harvesting systems into one offers a promising approach for enhancing the energy output performance. In a hybrid mechanical generator a magnet of buoyant structure was utilized with enhanced performance and it shows the results by its potential. They are capable of sensing acceleration along the distinct axes, specifically X, Y, and Z, to gauge angular velocity in multiple orientations. This feature underscores the harvester's potential use in healthcare monitoring, especially for observing human motion (Shaukat et al., 2023).

Thermo /Pyroelectric and Triboelectric Nanogenerators

This hybrid nanogenerator is capable of harvesting based on stress and strain and heat, which is ideal for applications where both mechanical vibrations and temperature differences are present. A design of a circular shaped. The rotating triboelectric nanogenerator equipped with thermoelectric nanogenerators was positioned between the terminals and the foundation. The high-speed relative rotation of the rotor and stator in the r-TENG induced electron flow.

INNOVATIONS IN MATERIALS SUPPORTING HYBRID SYSTEMS

The transition in renewable energy sources is driven by advancements by cutting-edge materials that are durable at high temperatures. Hybrid energy systems are beneficial particularly for high-performance battery technologies like lithium-ion batteries. In hydrogen storage systems, advanced materials in carbon is used and moreover the improvements in wind turbine blade materials that contribute to the efficiency of these systems. Materials like food waste and organic waste are also utilized to generate electricity. For example, lithium-ion batteries are widely used in energy storage while hydrogen storage materials that include metal hydrides like magnesium hydride and carbon fiber reinforced polymers. Geothermal energy systems employ high-temperature resistant materials and carbon-resistant coatings. Wind turbine materials includes carbon fiber composites bio-based composites hybrid composites and advanced polymer blends. In photovoltaics the materials used in bifacial solar panels, perovskite cells and solar balloons generate power even under cloudy conditions. Thermoelectric applications utilize materials like tin selenide (SnSe), lead telluride (PbTe), skutterudites ($CoSb3$), oxide thermoelectric $CaMnO3$, bismuth chalcogenides, MXenes and other metal carbides and nitrides which exhibit high thermal and electrical conductivity and high Seebeck coefficients.

Role of AI in Hybrid Systems

Hybrid systems needs a comprehensive analysis for location prediction that involves computer-based simulation and optimization. Numerous simulation tools such as HOMER, HYBRID 2, and HOGA plays a crucial role in this analysis. AI serves as a framework of intelligent factors for executing actions to maximize success, including branches like genetic algorithms, particle swarm optimization, simulated annealing, and artificial neural networks. Methods such as Tabu search, honey bee mating algorithm, bacterial foraging algorithm, game theory and combinations of metaheuristic algorithms enhance efficiency and reduce costs. Algorithms like

Adaboost, linear regression, logistic regression, decision trees, support vector machines, Naïve Bayes, KNN, and k-Means are employed to evaluate the performance of hybrid systems (S.M Zahree).

Advantages

Figure 12 mentions about the advantages of using hybrid energy harvesters.

Figure 12. Advantages of moving towards renewable sources of energy

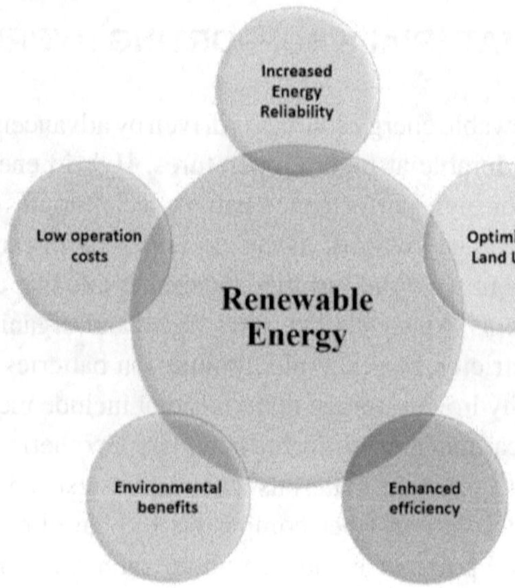

Limitations

Figure 13 provides the limitations of using hybrid energy harvesters.

Figure 13. Limitations using renewable sources of energy

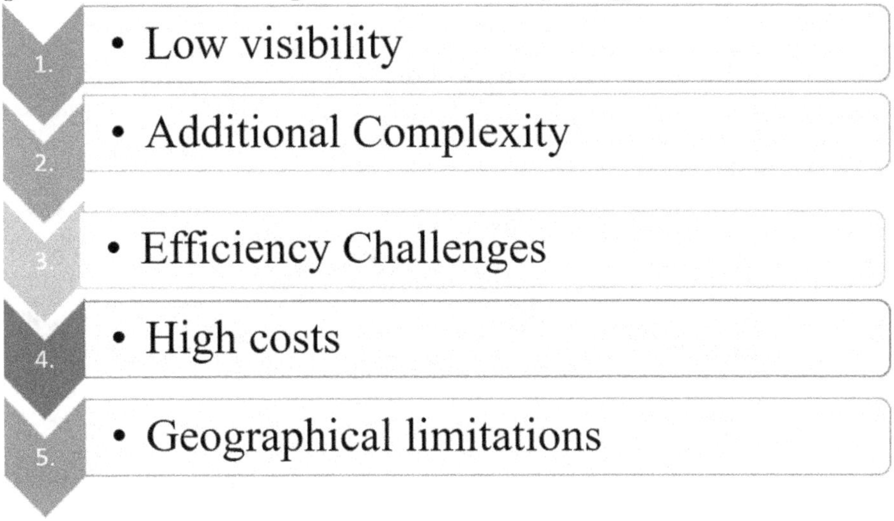

Applications

Figure 14 provides detailed applications of where hybrid energy harvesters are used.

Figure 14. Application of energy harvesters.

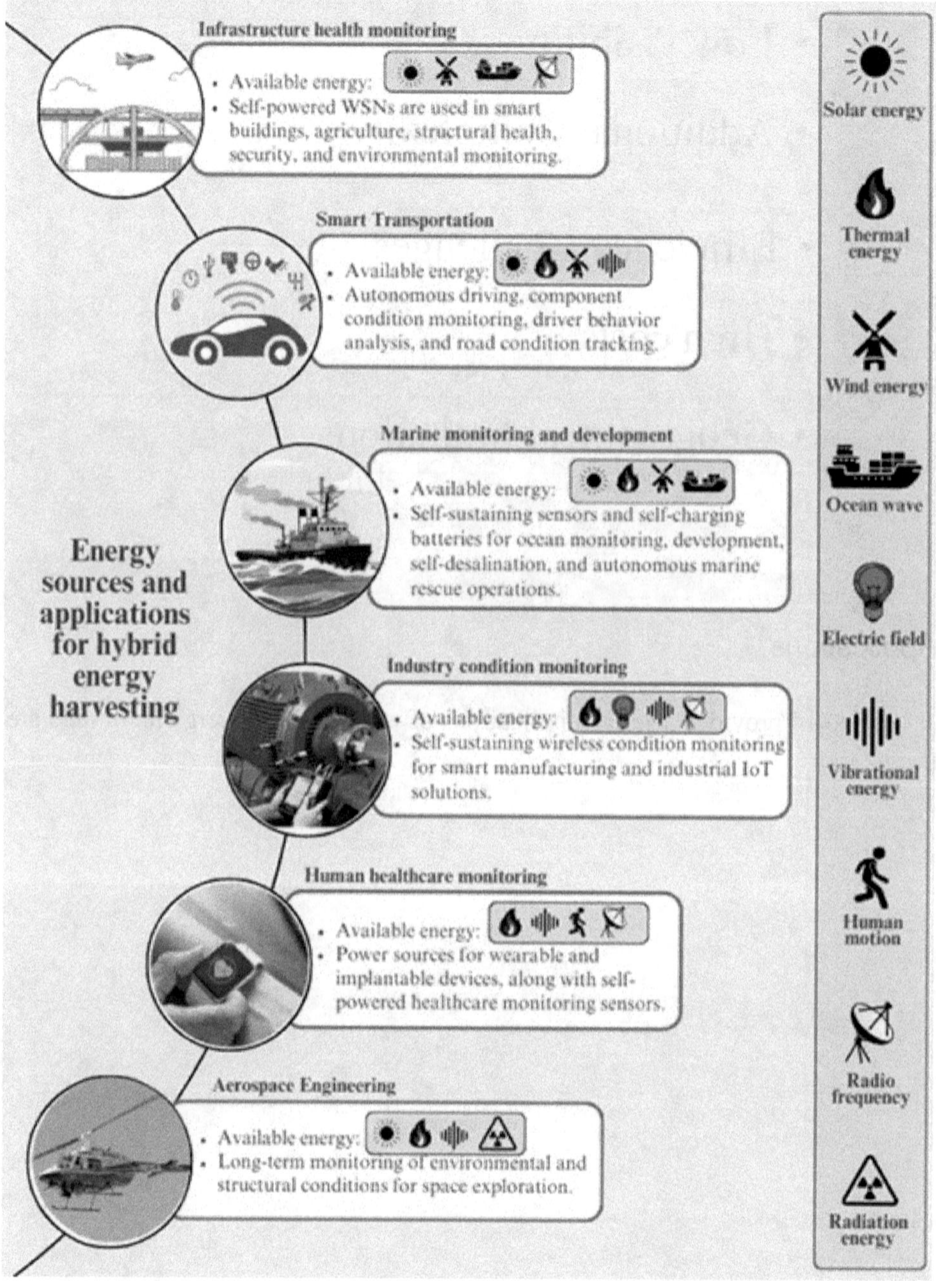

Smart Transportation

For addressing the increasing needs for both digitalization and automation, structures will be incorporated with electronics and sensors more extensively. By utilizing various energy sources, more dependable and efficient power supply can be achieved. Multiple energy sources are able to power different vehicles. By integrating versatile vibrational-based energy harvester that captures multi vibrational modes have been developed for self-powered monitoring systems in underground trains. Hybrid mechanical conversion methods are ideal for energy harvesting applications, particularly triboelectric methods can be used for applications like tire pressure monitoring. Nevertheless, material reliability remains a concern. Solar and thermal energy sources can also be employed, although they have some limitations. Electrification and autonomous driving are used in recent emerging trends, especially in the automotive industry. To satisfy the increasing demands for digital transformation and automation, sensors and electronics will be included into designs to make them more extensively in future (Morbiato et al., 2014).

Infrastructure Health Monitoring

To monitor infrastructure health, a huge network of wireless sensor nodes (WSNs) are often deployed, enabling a continuous detection. They are typically located outdoors; they can utilize conventional energy sources instead of relying on conventional batteries or wired connections. In recent years, the natural disasters or calamities creates significant risks to people and their property. To address this, autonomous hybrid mechanical energy harvester combined with a PV have been developed for monitoring natural disasters. Various display or sensor units, including flow meters and thermal sensors, can be integrated with renewable energy harvesters to power a smart agricultural irrigation system. (H. Liu et al., 2021).

Marine Monitoring

The monitoring of marine environments and development has received escalating attention from various countries and research groups. Harvesting in marine is considered as Blue-energy harvesting techniques which utilizes different conversion mechanisms and it has emerged as a promising solution. Mechanical harvesters are considered as potential way for capturing their energy known as blue energy, which offers opportunities for autonomous batteries and autonomous units in marine systems. This innovative approach does not enhance the efficiency alone in blue energy harvesting but also enables the creation of autonomous marine rescue devices and self-desalination systems. Combining water-wave power generators, solar panels, and wind turbines establishes a power generation network across the ocean's surface, making the vision of blue energy a reality (H. Liu et al., 2018).

Human Health Care Monitoring

The progress in implantable wearable electronic devices can help healthcare professional's early intervention in chronic diseases and enable real-time remote monitoring of patients' vital signs. One of the major challenge they faced while maintaining a sustainable power supply. The energy sources are generated from human body which includes muscle contractions, body movements, heat, cardiac and lung activities, and blood pulsation which typically exist in low-grade, low-frequency, and random forms. Numerous energy-capturing architectures has been developed to recharge power cells and ultimately create autonomous implantable or wearable devices capable of harnessing these energy sources. Developing efficient strategies to generate sufficient energy from them for monitoring and sensing in constrained environments presents an even greater challenge. To solve this, hybrid systems are integrated with multiple energy sources or conversion mechanisms which provide a promising solution. Recently, the impact of piezoelectric-thermoelectric hybrid nanogenerators (PE-TE HBNGs) has been studied for their ability to generate critical data for healthcare applications. The system monitors various parameters in real-time applications, including heartbeat, temperature, body acceleration, and blood oxygen saturation. It incorporates sensors in it for pulse oximetry, temperature, and acceleration, along with a microcontroller unit and a Bluetooth low-energy module. However, this sensor system typically relies on batteries for power, which have a limited lifespan (Chong et al., 2019).

Aerospace Engineering

Spacecraft and aircraft provides various energy sources for harvesting such as vibrations, temperature differences, and solar energy. Various analysis of technologies and energy sources are utilized for energy capturing in aerospace applications, mainly with thermal and vibration energy. Jet engine vibrations generally fall within the 20-500 Hz frequency range. Engines, gear trains, and hydraulic systems generate heat, which can serve as a thermal source. Another potential energy source for sensors on the fuselage where difference in temperature arises between the fuselage and the cabin. By integrating multiple transduction mechanisms into hybrid energy harvesters, the ability to monitor structural and environmental conditions over the long term is enhanced supporting both space exploration and continuous monitoring.

Self-Charging Power System

While self-charging, mechanical friction exists, where triboelectric generator is used to harvest these energies. However, these systems encounter obstacles due to their response such as unsteady pulsed output and minimal current generation, limiting their capacity to independently power most of the electronic devices. Due to their elevated energy concentration and stable performance, automatically charging devices mostly rely on accumulators for portable wearable electronics (X. Chen et al., 2018).

Self- Powered Biomedical Monitoring

The growing demand for personalized healthcare solution and real-time health monitoring systems in biomedical engineering possess a considerable challenge in ensuring long lasting power sources for electronic devices and wearables. Devices like neurostimulators and pacemakers are usually driven by external sources that are replaced through additional surgeries, which increases risk, cost, and discomfort. To tackle this issue, researchers developed a high-frequency vibrating implantable triboelectric generator for harvest ultrasound energy in vivo which reduces the risk in it (Hu et al., 2021).

Self- Powered Wearable Electronics

Wearable electronics offer great potential in various fields, including health monitoring, soft robotics, and in human-machine interactions. To provide a reliable power source for these devices, thermal energy and mechanical energy harvesters or innovative energy-harvesting methods are used whereas particularly hybrid TENG-based harvesters have been investigated in recent years. Considerable research has been focused on capturing energy from walking by integrating energy harvesters into shoes to power wearable sensors and electronics (F. Liu et al., 2019), (He et al., 2019)

Self-Powered Environment Monitoring

An autonomous wireless sensor network using triboelectric generator has been suggested as an effective monitoring method the environmental conditions and providing early warnings for natural disasters. This hybrid nanogenerators capture energy from the wind generated by moving vehicles, enabling the continuous powering of a wireless transmitter that can monitor traffic volume and it could be advantageous for intelligent traffic systems. Additionally, self-powered intelligent buoy system

(SIBS) that harvests wave energy for ocean monitoring was demonstrated and researchers on working on it (Shi et al., 2018), (Xi et al., 2019).

CONCLUSION

Energy is a vital component that supports human basic needs and significantly influences the country's economy growth. It is difficult to stable without an efficient energy supply to our day-to-day living. Hence, harvesters are required to harness the se energy to utilize them in efficient way. Hybrid energy harvesting holds promise for enhancing energy efficiency and contributing to a sustainable future. By converting multiple energy conversion mechanisms, hybrid energy scavengers have demonstrated improvements in energy output. They have great potential to meet the growing demand for eco-friendly energy alternatives in the upcoming years by addressing existing challenges and improving energy conversion efficiency and power density. Researchers have developed and implemented a variety of structural designs and tailored mechanisms to the specific characteristics of different energy sources and application scenarios, which has led them to increase energy production with improved power. While the field of hybrid energy harvesting is still evolving, it shows a significant potential for growth. The system combines different energy conversion processes, which allows them to generate power continuously under diverse environmental conditions acts as major advantage for them. This versatility opens up new possibilities and opportunities for sustainable energy generation across various applications. The main challenge of the hybrid energy harvester lies in their heightened structural complexity and the devices associated with various energy conversion materials from different energy sources. In current, research hybrid harvesters play a predominant role in hybrid energy harvesting by enhancing power generation. This prevalence underscores the need for further exploration of various conversion mechanisms and energy sources to fully unlock their potential of hybrid energy harvesting. This chapter covers a wide range of energy sources and conversion mechanisms, with a focus on various hybrid energy harvesters with qualitative assessments of application areas. They dominate the field, indicating that further exploration of diverse conversion mechanisms and energy sources is necessary. Despite the progress, hybrid energy harvesting technology is still in its early stages, and there is a lack of comprehensive comparison and discussion in the literature regarding hybrid versus single energy harvesting mechanisms. The potential improvements in enhanced power generation through hybrid energy harvesting technology require further research. We anticipate that hybrid energy harvesting technologies will continue to be a prominent area of research, with significant growth in their applications over the next decade.

ACKNOWLEDGEMENT

The authors would like to thank the innovation in Science Pursuit for Inspire Research through the Department of Science and Technology (DST) funded by the Ministry of Science and Technology by DST/INSPIRE/04/2017/ 002629 and Science and Engineering Research Board core research grant by CRG/2021/006188.

REFERENCES

Abbasi, T., & Abbasi, S. A. (2010). Biomass energy and the environmental impacts associated with its production and utilization. *Renewable & Sustainable Energy Reviews, 14*(3), 919–937. DOI: 10.1016/j.rser.2009.11.006

Ando, O. H.Junior, Maran, A. L. O., & Henao, N. C. (2018). A review of the development and applications of thermoelectric microgenerators for energy harvesting. *Renewable & Sustainable Energy Reviews, 91*(March), 376–393. DOI: 10.1016/j.rser.2018.03.052

Anne Naeth, M., Devereaux Jennings, P., Gamal El-Din, M., & Dhar, A. (2020). Geothermal Energy Resources: Potential Environmental Impact and 2 Land Reclamation 3 4 5 Amalesh. *Environmental Reviews, 28*(4), 415–427. www.nrcresearchpress.com. DOI: 10.1139/er-2019-0069

Aryal, R., Dokou, Z., Malla, R. B., & Bagtzoglou, A. C. (2020). Design optimization of a small-scale hydropower harvesting device. *Structural and Multidisciplinary Optimization, 61*(3), 1303–1318. DOI: 10.1007/s00158-019-02416-2

Bai, Y., Jantunen, H., & Juuti, J. (2018). Hybrid, multi-source, and integrated energy harvesters. *Frontiers in Materials, 5*, 65. Advance online publication. DOI: 10.3389/fmats.2018.00065

Beeby, S. P., Tudor, M. J., & White, N. M. (2006). Energy harvesting vibration sources for microsystems applications. *Measurement Science & Technology, 17*(12), R175–R195. Advance online publication. DOI: 10.1088/0957-0233/17/12/R01

Bergmann, A., Colombo, S., & Hanley, N. (2008). Rural versus urban preferences for renewable energy developments. *Ecological Economics, 65*(3), 616–625. DOI: 10.1016/j.ecolecon.2007.08.011

Bilgen, S., Kaygusuz, K., & Sari, A. (2004). Renewable energy for a clean and sustainable future. *Energy Sources, 26*(12), 1119–1129. DOI: 10.1080/00908310490441421

Bonar, P. A. J., Bryden, I. G., & Borthwick, A. G. L. (2015). Social and ecological impacts of marine energy development. *Renewable & Sustainable Energy Reviews*, *47*, 486–495. DOI: 10.1016/j.rser.2015.03.068

Chandrasekaran, S., Bowen, C., Roscow, J., Zhang, Y., Dang, D. K., Kim, E. J., Misra, R. D. K., Deng, L., Chung, J. S., & Hur, S. H. (2019). Micro-scale to nano-scale generators for energy harvesting: Self powered piezoelectric, triboelectric and hybrid devices. *Physics Reports*, *792*, 1–33. DOI: 10.1016/j.physrep.2018.11.001

Chen, S., Chen, B., & Fath, B. D. (2015). Assessing the cumulative environmental impact of hydropower construction on river systems based on energy network model. *Renewable & Sustainable Energy Reviews*, *42*(19), 78–92. DOI: 10.1016/j.rser.2014.10.017

Chen, X., Dai, W., Wu, T., Luo, W., Yang, J., Jiang, W., & Wang, L. (2018). Thin film thermoelectric materials: Classification, characterization, and potential for wearable applications. *Coatings*, *8*(7), 244. Advance online publication. DOI: 10.3390/coatings8070244

Cherubini, F., & Strømman, A. H. (2011). Life cycle assessment of bioenergy systems: State of the art and future challenges. *Bioresource Technology*, *102*(2), 437–451. DOI: 10.1016/j.biortech.2010.08.010 PMID: 20832298

Chong, Y. W., Ismail, W., Ko, K., & Lee, C. Y. (2019). Energy Harvesting for Wearable Devices: A Review. *IEEE Sensors Journal*, *19*(20), 9047–9062. DOI: 10.1109/JSEN.2019.2925638

Dai, K., Bergot, A., Liang, C., Xiang, W. N., & Huang, Z. (2015). Environmental issues associated with wind energy - A review. *Renewable Energy*, *75*, 911–921. DOI: 10.1016/j.renene.2014.10.074

Darmawi, S., Sipahutar, R., Bernas, S. M., & Imanuddin, M. S. (2013). Renewable energy and hydropower utilization tendency worldwide. *Renewable & Sustainable Energy Reviews*, *17*, 213–215. DOI: 10.1016/j.rser.2012.09.010

Dincer, I. (2001). Environmental issues: II-potential solutions. *Energy Sources*, *23*(1), 83–92. DOI: 10.1080/00908310151092218

Dincer, I., & Rosen, M. A. (1999). Energy, Environment and Sustainable Development Sub-programme Environment and Sustainable Development. *Sustainable Development, 64*(October), 0–41.

Fridleifsson, I. B. (2001). Geothermal energy for the benefit of the people. *Renewable & Sustainable Energy Reviews*, *5*(3), 299–312. DOI: 10.1016/S1364-0321(01)00002-8

Hasan, M. N., Sahlan, S., Osman, K., & Mohamed Ali, M. S. (2021). Energy Harvesters for Wearable Electronics and Biomedical Devices. *Advanced Materials Technologies*, *6*(3), 1–42. DOI: 10.1002/admt.202000771

He, T., Shi, Q., Wang, H., Wen, F., Chen, T., Ouyang, J., & Lee, C. (2019). Beyond energy harvesting - multi-functional triboelectric nanosensors on a textile. *Nano Energy*, *57*, 338–352. DOI: 10.1016/j.nanoen.2018.12.032

Hernandez, R. R., Easter, S. B., Murphy-Mariscal, M. L., Maestre, F. T., Tavassoli, M., Allen, E. B., Barrows, C. W., Belnap, J., Ochoa-Hueso, R., Ravi, S., & Allen, M. F. (2014). Environmental impacts of utility-scale solar energy. *Renewable & Sustainable Energy Reviews*, *29*, 766–779. DOI: 10.1016/j.rser.2013.08.041

Hiemstra-van der Horst, G., & Hovorka, A. J. (2009). Fuelwood: The "other" renewable energy source for Africa? *Biomass and Bioenergy*, *33*(11), 1605–1616. DOI: 10.1016/j.biombioe.2009.08.007

Hu, B., Wang, C., Jiang, K., Shen, Z., Yang, X., Yin, M., Liang, B., Xie, Q., Ye, Y., & Gao, Z. (2021). Development and validation of a novel diagnostic model for initially clinical diagnosed gastrointestinal stromal tumors using an extreme gradient-boosting machine. *BMC Gastroenterology*, *21*(1), 1–9. DOI: 10.1186/s12876-021-02048-1 PMID: 34922474

Jeerapan, I., Sempionatto, J. R., & Wang, J. (2020). On-Body Bioelectronics: Wearable Biofuel Cells for Bioenergy Harvesting and Self-Powered Biosensing. *Advanced Functional Materials*, *30*(29), 1–18. DOI: 10.1002/adfm.201906243

Jung, W. S., Kang, M. G., Moon, H. G., Baek, S. H., Yoon, S. J., Wang, Z. L., Kim, S. W., & Kang, C. Y. (2015). High Output Piezo/Triboelectric Hybrid Generator. *Scientific Reports*, *5*(1), 1–6. DOI: 10.1038/srep09309 PMID: 25791299

Kannan, N., & Vakeesan, D. (2016). Solar energy for future world: - A review. *Renewable & Sustainable Energy Reviews*, *62*, 1092–1105. DOI: 10.1016/j.rser.2016.05.022

Liu, F., Lai, Y., Zhao, B., Bradley, R., & Wu, W. (2019). Photothermal materials for efficient solar powered steam generation. *Frontiers of Chemical Science and Engineering*, *13*(4), 636–653. DOI: 10.1007/s11705-019-1824-1

Liu, H., Fu, H., Sun, L., Lee, C., & Yeatman, E. M. (2021). Hybrid energy harvesting technology: From materials, structural design, system integration to applications. In *Renewable and Sustainable Energy Reviews* (Vol. 137). Elsevier Ltd., DOI: 10.1016/j.rser.2020.110473

Liu, H., Zhang, J., Shi, Q., He, T., Chen, T., Sun, L., Dziuban, J. A., & Lee, C. (2018). Development of a thermoelectric and electromagnetic hybrid energy harvester from water flow in an irrigation system. *Micromachines*, *9*(8), 1–10. DOI: 10.3390/mi9080395 PMID: 30424328

Lv, J., Chen, J., & Lee, P. S. (2021). Sustainable wearable energy storage devices self-charged by human-body bioenergy. *SusMat*, *1*(2), 285–302. DOI: 10.1002/sus2.14

Mizoshiri, M., Mikami, M., & Ozaki, K. (2012). Thermal-photovoltaic hybrid solar generator using thin-film thermoelectric modules. *Japanese Journal of Applied Physics*, *51*(6 PART 2), 06FL07. Advance online publication. DOI: 10.1143/JJAP.51.06FL07

Morbiato, T., Borri, C., & Vitaliani, R. (2014). Wind energy harvesting from transport systems: A resource estimation assessment. *Applied Energy*, *133*, 152–168. DOI: 10.1016/j.apenergy.2014.07.055

Pobering, S., & Schwesinger, N. (2004). A novel hydropower harvesting device. *Proceedings - 2004 International Conference on MEMS, NANO and Smart Systems, ICMENS 2004*, 480–485. DOI: 10.1109/ICMENS.2004.1508997

Popp, J., Lakner, Z., Harangi-Rákos, M., & Fári, M. (2014). The effect of bioenergy expansion: Food, energy, and environment. *Renewable & Sustainable Energy Reviews*, *32*, 559–578. DOI: 10.1016/j.rser.2014.01.056

Rabaia, M. K. H., Abdelkareem, M. A., Sayed, E. T., Elsaid, K., Chae, K. J., Wilberforce, T., & Olabi, A. G. (2021). Environmental impacts of solar energy systems: A review. *The Science of the Total Environment*, *754*, 141989. DOI: 10.1016/j.scitotenv.2020.141989 PMID: 32920388

Radhakrishnan, S., Joseph, S., Jelmy, E. J., Saji, K. J., Sanathanakrishnan, T., & John, H. (2022). Triboelectric nanogenerators for marine energy harvesting and sensing applications. *Results in Engineering*, *15*(June), 100487. DOI: 10.1016/j.rineng.2022.100487

Rybach, L. (2003). Geothermal energy: Sustainability and the environment. *Geothermics*, *32*(4), 463–470. DOI: 10.1016/S0375-6505(03)00057-9

Sahin, A. D. (2004). Progress and recent trends in wind energy. *Progress in Energy and Combustion Science*, *30*(5), 501–543. DOI: 10.1016/j.pecs.2004.04.001

Saidur, R., Rahim, N. A., Islam, M. R., & Solangi, K. H. (2011). Environmental impact of wind energy. *Renewable & Sustainable Energy Reviews*, *15*(5), 2423–2430. DOI: 10.1016/j.rser.2011.02.024

Selvan, K. V., & Mohamed Ali, M. S. (2016). Micro-scale energy harvesting devices: Review of methodological performances in the last decade. *Renewable & Sustainable Energy Reviews, 54*, 1035–1047. DOI: 10.1016/j.rser.2015.10.046

Shaukat, H., Ali, A., Ali, S., Altabey, W. A., Noori, M., & Kouritem, S. A. (2023). Applications of Sustainable Hybrid Energy Harvesting: A Review. *Journal of Low Power Electronics and Applications, 13*(4), 62. Advance online publication. DOI: 10.3390/jlpea13040062

Shi, B., Li, Z., & Fan, Y. (2018). Implantable Energy-Harvesting Devices. *Advanced Materials, 30*(44), 1–18. DOI: 10.1002/adma.201801511 PMID: 30043422

Tester, J. (2007). *Congressional Testimony for*. 1–13.

Vullers, R. J. M., van Schaijk, R., Doms, I., Van Hoof, C., & Mertens, R. (2009). Micropower energy harvesting. *Solid-State Electronics, 53*(7), 684–693. DOI: 10.1016/j.sse.2008.12.011

Wang, H., Jasim, A., & Chen, X. (2018). Energy harvesting technologies in roadway and bridge for different applications – A comprehensive review. *Applied Energy, 212*(August 2017), 1083–1094. DOI: 10.1016/j.apenergy.2017.12.125

Wang, X., Yang, B., Liu, J., & Yang, C. (2017). A transparent and biocompatible single-friction-surface triboelectric and piezoelectric generator and body movement sensor. *Journal of Materials Chemistry. A, Materials for Energy and Sustainability, 5*(3), 1176–1183. DOI: 10.1039/C6TA09501A

Wu, N., Wang, Q., & Xie, X. (2013). Wind energy harvesting with a piezoelectric harvester. *Smart Materials and Structures, 22*(9), 095023. Advance online publication. DOI: 10.1088/0964-1726/22/9/095023

Xi, F., Pang, Y., Liu, G., Wang, S., Li, W., Zhang, C., & Wang, Z. L. (2019). Self-powered intelligent buoy system by water wave energy for sustainable and autonomous wireless sensing and data transmission. *Nano Energy, 61*(April), 1–9. DOI: 10.1016/j.nanoen.2019.04.026

Xia, H., Chen, R., & Ren, L. (2015). Analysis of piezoelectric-electromagnetic hybrid vibration energy harvester under different electrical boundary conditions. *Sensors and Actuators. A, Physical, 234*, 87–98. DOI: 10.1016/j.sna.2015.08.014

Youm, I., Sarr, J., Sall, M., & Kane, M. M. (2000). Renewable energy activities in Senegal: A review. *Renewable & Sustainable Energy Reviews, 4*(1), 75–89. DOI: 10.1016/S1364-0321(99)00009-X

Zhou, Z., Li, X., Wu, Y., Zhang, H., Lin, Z., Meng, K., Lin, Z., He, Q., Sun, C. C., Yang, J., & Wang, Z. L. (2018). Wireless self-powered sensor networks driven by triboelectric nanogenerator for in-situ real time survey of environmental monitoring. *Nano Energy*, *53*, 501–507. DOI: 10.1016/j.nanoen.2018.08.055

Zhu, J., Zhu, Y., & Wang, X. (2018). A Hybrid Piezoelectric and Triboelectric Nanogenerator with PVDF Nanoparticles and Leaf-Shaped Microstructure PTFE Film for Scavenging Mechanical Energy. *Advanced Materials Interfaces*, *5*(2), 1–6. DOI: 10.1002/admi.201700750

Chapter 13
Pioneering Advances in Energy Storage for Renewable Futures:
Harnessing Tomorrow

S. C. Vetrivel
https://orcid.org/0000-0003-3050-8211
Kongu Engineering College, India

P. Vidhyapriya
https://orcid.org/0000-0002-1421-8743
Kongu Engineering College, India

V. P. Arun
https://orcid.org/0009-0009-1211-2150
JKKN College of Engineering and Technology, India

V. Sabareeshwari
Amrita Vishwa Vidyapeetham, India

ABSTRACT

The transition to renewable energy sources is critical for achieving sustainability and combating climate change. However, the intermittent nature of solar and wind energy presents significant challenges for reliable energy supply. "Harnessing Tomorrow: Pioneering Advances in Energy Storage for Renewable Futures" explores innovative energy storage solutions that are essential for integrating renewable energy into the grid effectively. This chapter highlights the latest advancements in energy storage technologies, including lithium-ion batteries, flow batteries, supercapacitors, and emerging technologies such as solid-state batteries and hydrogen storage systems.

DOI: 10.4018/979-8-3693-9316-1.ch013

By examining case studies and pilot projects, the research emphasizes the role of energy storage in enhancing grid stability, enabling peak shaving, and providing backup power. Furthermore, the chapter discusses the economic implications of these technologies and their potential to reduce costs, improve efficiency, and support the widespread adoption of renewable energy.

1. INTRODUCTION

1.1 Overview of Renewable Energy

Renewable energy has become a cornerstone of modern energy policy, driven by the urgent need to mitigate climate change, reduce dependence on fossil fuels, and promote sustainable development. In 2023, renewables accounted for nearly 30% of global electricity generation, a significant increase from 24% in 2018 (Ahmad et al., 2023). This surge is attributed to substantial investments in solar, wind, hydro, and bioenergy technologies. Solar photovoltaic (PV) capacity alone grew by 22% in 2023, with China, the United States, and the European Union leading the expansion (Arya & Mahajan, 2023). Wind energy also saw remarkable growth, particularly offshore wind projects, which increased by 15% globally (Ahmad et al., 2022). Despite these advances, the intermittent nature of renewable sources remains a critical challenge, necessitating effective energy storage solutions to ensure a reliable and stable energy supply (Atawi et al., 2022).

1.2 Importance of Energy Storage in Renewable Integration

Energy storage is pivotal in addressing the variability and intermittency of renewable energy sources. By storing excess energy generated during peak production times and releasing it during periods of low production, storage technologies enhance grid stability, improve energy efficiency, and support the integration of higher shares of renewables (Arsad et al., 2022). For instance, in 2023, global energy storage installations surpassed 20 gigawatts (GW), a 60% increase from the previous year, underscoring the rapid adoption of storage solutions (Ahmad et al., 2023). Technologies such as lithium-ion batteries, pumped hydro storage, and emerging options like solid-state batteries and hydrogen storage are at the forefront of this transformation (Atawi et al., 2022). Effective energy storage not only mitigates the impact of fluctuations in energy supply but also contributes to the resilience of energy systems against natural disasters and other disruptions (Ahmad et al., 2022).

1.3 Objectives and Scope

This chapter aims to provide a comprehensive introduction to the critical role of energy storage in the renewable energy landscape. It will explore the current state of renewable energy adoption, highlighting the achievements and challenges faced in integrating these sources into the energy grid. The chapter will also delve into the importance of energy storage, examining how various technologies contribute to overcoming the inherent intermittency of renewables. By outlining the objectives and scope, this chapter sets the stage for a detailed discussion on the types of energy storage technologies, their integration with renewable systems, advancements, challenges, real-world applications, and future prospects. Through this exploration, readers will gain a deeper understanding of how pioneering advances in energy storage are driving the transition towards a sustainable and renewable energy future (Ahmad et al., 2023).

2. TYPES OF ENERGY STORAGE TECHNOLOGIES

2.1 Mechanical Energy Storage

Mechanical energy storage systems utilize physical processes such as gravity, inertia, or compressed air to store and release energy. These systems are known for their large-scale applications and ability to provide grid-level energy storage solutions. Among the most widely recognized mechanical energy storage technologies are pumped hydro storage, compressed air energy storage (CAES), and flywheel energy storage (Azarpour et al., 2022).

2.1.1 Pumped Hydro Storage

Pumped hydro storage (PHS) is the most widely deployed and mature form of mechanical energy storage, contributing significantly to global grid storage capacity. It works by using surplus electricity to pump water from a lower reservoir to an upper reservoir, typically located at a higher elevation. When energy is needed, the stored water is released back down through turbines, converting potential energy into electrical energy. Pumped hydro storage systems can provide large amounts of energy for several hours, making them an ideal solution for grid balancing and long-duration storage (Badi et al., 2023). They have high round-trip efficiencies (typically 70 to 90%) and are capable of rapid response times, which is critical for stabilizing power grids with high renewable energy penetration (Baria et al., 2023).

2.1.2 Compressed Air Energy Storage

Compressed air energy storage (CAES) is another significant mechanical energy storage technology that stores energy by compressing air in underground caverns or above-ground tanks. The compressed air is stored at high pressure and released when energy demand increases. Upon release, the compressed air is heated and expanded in a turbine generator, generating electricity (Bhuiyan et al., 2022).

There are two main types of CAES systems:

- **Adiabatic CAES:** This system stores energy by compressing air without losing heat in the process. The heat generated during compression is stored and later used to help expand the air, increasing efficiency.
- **Diabatic CAES:** In this system, the heat generated during compression is not stored but instead vented, resulting in lower efficiency (Bošnjaković et al., 2022).

CAES has several advantages, such as the ability to store large amounts of energy, long discharge durations, and high efficiency in certain configurations. The technology is particularly well-suited for applications where long-duration storage is required, such as balancing seasonal energy production from renewable sources like wind and solar. However, CAES faces challenges, including the need for specific geological formations for underground air storage, and energy losses due to heat management. The relatively high capital costs associated with CAES systems, especially for adiabatic configurations, also pose financial barriers. Despite these challenges, ongoing research into improving the efficiency and scalability of CAES continues to show promise for the technology's role in the future energy mix (Callegari et al., 2020).

2.1.3 Flywheel Energy Storage

Flywheel energy storage (FES) systems use the principle of rotational inertia to store energy. A flywheel consists of a rotating mass, usually made of steel or carbon fiber, that spins at high speeds to store kinetic energy. When energy is needed, the flywheel's rotation is slowed down, converting the kinetic energy back into electrical energy via a generator (Cantarero, 2020).

Flywheel energy storage systems offer high power density and rapid response times, making them ideal for applications that require quick bursts of energy, such as frequency regulation and grid stabilization. Flywheels have minimal degradation over time, offering long cycle lifetimes and a high number of charge/discharge cycles (typically several million). Additionally, they have a high round-trip efficiency of

85 to 95%, which makes them efficient in terms of both energy storage and retrieval (Chidolue & Iqbal, 2023).

2.2 Electrical Energy Storage

Electrical energy storage technologies store energy in the form of electrical charge, typically using electrostatic or electromagnetic mechanisms. These systems are crucial for applications that require rapid charge and discharge cycles, high power densities, and the ability to release stored energy quickly. Electrical energy storage technologies play a significant role in enhancing the flexibility and stability of electrical grids, enabling better integration of renewable energy sources, and improving power quality (Azarpour et al., 2022). Among the most common forms of electrical energy storage are capacitors, supercapacitors, and superconducting magnetic energy storage (SMES) systems. These technologies offer distinct advantages and limitations, depending on their application.

2.2.1 Capacitors and Supercapacitors

Capacitors and supercapacitors (also known as ultra-capacitors) are devices that store electrical energy electrostatically. A capacitor consists of two conductive plates separated by a dielectric material, while a supercapacitor has a much higher surface area, typically using activated carbon or similar materials for its electrodes, allowing it to store significantly more energy than conventional capacitors (Badi et al., 2023).

Capacitors store energy by accumulating charge on two conductive plates. When a voltage is applied, the plates accumulate an electric charge, creating an electric field across the dielectric material between them. The amount of energy stored in a capacitor is proportional to the voltage squared and the capacitance, with larger capacitors capable of storing more energy. However, capacitors have low energy density compared to other storage technologies, meaning they are suitable for applications requiring quick bursts of power rather than prolonged energy supply (Baria et al., 2023).

Supercapacitors, on the other hand, have much higher energy storage capacity and a significantly larger surface area for storing charge. Unlike traditional capacitors, supercapacitors store energy through electrostatic charge accumulation at the interface of the electrode material and the electrolyte (Bhuiyan et al., 2022).

2.2.2 Superconducting Magnetic Energy Storage

Superconducting Magnetic Energy Storage (SMES) is an advanced electrical energy storage technology that stores energy in the magnetic field created by the flow of direct current (DC) through a superconducting coil. Unlike traditional magnetic storage, which relies on conventional magnetic materials, SMES takes advantage of superconducting materials, which have zero electrical resistance when cooled below a certain temperature. This property allows SMES systems to store energy with minimal losses (Bošnjaković et al., 2022).

In SMES system, a superconducting coil is cooled to cryogenic temperatures using liquid helium or nitrogen, which enables it to carry a current indefinitely without resistance. When energy is stored, electrical energy is converted into a magnetic field within the coil, and when energy is needed, the current is extracted, and the magnetic field is converted back into electrical energy (Callegari et al., 2020).

SMES systems offer several advantages, including extremely fast charge and discharge cycles, high efficiency, and the ability to release large amounts of energy in short bursts (Chidolue & Iqbal, 2023). The primary application of SMES is in providing rapid response to grid fluctuations and stabilizing the power supply. Due to its high power density and quick discharge capabilities, SMES is ideal for applications where power quality is critical, such as voltage stabilization, frequency regulation, and load leveling on the electrical grid (Cantarero, 2020).

2.3 Electrochemical Energy Storage

Electrochemical energy storage is a critical class of energy storage technologies that converts and stores electrical energy in chemical form. This technology relies on electrochemical reactions to store and release energy, making it highly efficient for a variety of applications, from grid-scale storage to portable devices. The most common and well-known electrochemical energy storage systems are batteries, but they also include fuel cells and other forms of energy storage that depend on chemical reactions. Electrochemical systems are integral to the energy storage landscape due to their high efficiency, scalability, and ability to provide rapid response to changes in demand.

2.3.1 Batteries

Batteries are the most widely used form of electrochemical energy storage and come in various chemistries, each offering different advantages in terms of performance, cost, and application suitability. The three most common types of batteries used in energy storage systems are lithium-ion, lead-acid, and flow batteries, each with its own set of characteristics:

- **Lithium-ion Batteries:** Lithium-ion (Li-ion) batteries dominate the energy storage market due to their high energy density, long cycle life, and relatively light weight. They are widely used in electric vehicles, consumer electronics, and stationary energy storage systems. Li-ion batteries store energy through the movement of lithium ions between the cathode and anode, with the energy being released when the ions return to the anode. Over the past decade, significant advances have been made in improving the capacity, efficiency, and safety of lithium-ion batteries, making them the go-to solution for many energy storage applications. However, challenges such as resource availability (e.g., lithium, cobalt), recycling, and cost remain concerns that need to be addressed as demand for these batteries increases.
- **Lead-acid Batteries:** Lead-acid batteries, one of the oldest forms of rechargeable batteries, have been widely used for energy storage in applications such as backup power, off-grid energy storage, and automotive use (Sioshans 2025). These batteries store energy through a chemical reaction between lead and sulfuric acid. Despite their low energy density compared to lithium-ion batteries, lead-acid batteries are inexpensive, durable, and relatively safe. They have been a mainstay in large-scale energy storage for decades, but they are being gradually replaced by newer technologies like lithium-ion due to the latter's superior performance and longer lifespan.
- **Flow Batteries:** Flow batteries, including vanadium redox and all-vanadium flow batteries, are an emerging electrochemical technology that offers advantages in large-scale energy storage applications. Unlike conventional batteries, which store energy in solid electrodes, flow batteries store energy in liquid electrolytes. This allows for the separation of energy storage and power generation components, making it possible to scale each independently.

2.3.2 Hydrogen and Fuel Cells

Hydrogen and fuel cells represent another important form of electrochemical energy storage, providing a way to store energy in the form of hydrogen gas, which can be used in fuel cells to produce electricity when needed. Hydrogen has the potential to play a key role in the future of energy storage and transportation, especially when combined with renewable energy sources such as solar and wind. The electrochemical process in fuel cells is essentially the reverse of electrolysis: hydrogen gas reacts with oxygen in the fuel cell to produce electricity, water, and heat.

- **Hydrogen Storage:** Hydrogen is one of the most promising candidates for long-term energy storage, especially because it is energy-dense and can be stored in various forms, including compressed gas, liquid, or even solid-state storage. The main advantage of hydrogen as an energy carrier is its ability to store large amounts of energy for long durations, which is essential for decarbonizing sectors like heavy transportation, industry, and long-duration energy storage. However, hydrogen storage faces challenges related to cost, efficiency, and infrastructure, including the energy-intensive nature of hydrogen production (particularly when derived from non-renewable sources) and the need for advanced technologies to store and transport hydrogen safely and economically.
- **Fuel Cells:** Fuel cells are devices that convert stored hydrogen into electricity through an electrochemical process without combustion, making them highly efficient and environmentally friendly. The most common type of fuel cell used for energy storage and transportation is the proton exchange membrane (PEM) fuel cell, which is used in applications ranging from portable electronics to vehicles and stationary power systems. Fuel cells produce electricity as long as hydrogen and oxygen are supplied, offering a clean and efficient way to generate power. They have the advantage of being able to provide a continuous supply of energy without the limitations of battery cycle life. However, fuel cell technology is still expensive, and the lack of a widespread hydrogen infrastructure is a significant barrier to widespread adoption.

2.4 Thermal Energy Storage

Thermal energy storage (TES) involves storing energy in the form of heat, which can later be released for use in heating or power generation. This type of energy storage is particularly useful for integrating renewable energy sources, such as solar and wind, which are intermittent by nature. By capturing and storing heat when energy production exceeds demand, TES systems ensure a steady energy supply during periods of low generation or high demand. TES technologies are versatile, applicable in residential, commercial, and industrial applications, and are often used in conjunction with renewable energy systems to maximize their efficiency and reliability. Thermal energy storage systems can be broadly classified into three categories based on the physical mechanism of heat retention: sensible heat storage, latent heat storage, and thermochemical storage. Each of these methods offers distinct advantages and challenges in terms of energy density, cost, and complexity. The selection of a particular thermal energy storage technology depends on the application, scale, and requirements of the energy system, as well as environmental considerations.

2.4.1 Sensible Heat Storage

Sensible heat storage is the simplest and most commonly used form of thermal energy storage. It relies on the principle that the temperature of a material increases when heat is added and decreases when heat is removed. In sensible heat storage, energy is stored by raising the temperature of a medium—typically a solid or liquid material—such as water, sand, or molten salts. The amount of energy stored is directly proportional to the temperature change and the heat capacity of the material. One of the most commonly used materials for sensible heat storage is water, due to its high specific heat capacity and abundance. In large-scale applications, such as solar thermal power plants, molten salts are often used as the storage medium. These salts, which can remain in a liquid state at high temperatures, are ideal for storing thermal energy generated by concentrated solar power (CSP) systems. Sensible heat storage systems are relatively simple to design and operate, and they are typically more cost-effective than other forms of thermal storage. However, they require large volumes of storage material to achieve significant energy storage capacity, and the efficiency of these systems is limited by the temperature gradient between the stored energy and the surrounding environment.

2.4.2 Latent Heat Storage

Latent heat storage differs from sensible heat storage in that it stores energy during phase transitions, such as the melting or freezing of a material. The key advantage of latent heat storage is that it can store a large amount of energy without a significant change in temperature. When a material undergoes a phase change from solid to liquid (melting) or from liquid to gas (vaporization), it absorbs or releases a substantial amount of energy, known as the latent heat of fusion or vaporization. This process allows for the storage of large amounts of energy in a relatively small volume. Phase change materials (PCMs) are the most commonly used substances for latent heat storage. These materials are selected based on their ability to absorb and release energy at specific temperatures. PCMs can be organic (such as paraffin wax), inorganic (such as salts or salt hydrates), or eutectic mixtures (blends of two or more materials with a lower melting point than the individual components). PCMs are advantageous because they offer higher energy density compared to sensible heat storage materials, meaning that they require less space to store the same amount of energy.

2.5 Emerging Technologies

Energy storage technologies are continuously evolving to address the growing demand for more efficient, cost-effective, and sustainable solutions. While traditional energy storage systems such as lithium-ion batteries and pumped hydro storage have made significant impacts, emerging technologies are now being developed to offer higher performance, greater scalability, and enhanced efficiency. These next-generation solutions aim to overcome the limitations of current storage systems, such as energy density, lifespan, and cost, while also introducing new capabilities for large-scale and long-duration storage. These technologies are expected to transform the energy landscape, providing more flexible and resilient energy systems that can support the increasing penetration of renewable energy sources and contribute to global decarbonization efforts. Emerging energy storage technologies encompass a wide range of concepts, including advanced materials, hybrid storage solutions, and novel storage architectures. These innovations aim to optimize energy storage at various levels, from residential and commercial applications to utility-scale systems.

2.5.1 Advanced Materials

The development of advanced materials for energy storage is a cornerstone of the emerging technologies in this field. Researchers are focusing on discovering and optimizing materials that can improve the efficiency, capacity, and cost-effectiveness of energy storage systems. These materials are often designed to enhance the performance of batteries, capacitors, and other energy storage devices, and they include innovations in both electrodes and electrolytes, as well as new structural designs for energy storage systems.

2.5.1.1 New Battery Materials

One of the primary areas of research within advanced materials is the development of new and improved battery chemistries. While lithium-ion batteries dominate the current energy storage market, their performance limitations, such as relatively low energy density and reliance on finite resources, have spurred efforts to find alternatives. Materials such as sodium-ion, lithium-sulfur, and solid-state are at the forefront of these efforts.

- **Sodium-ion Batteries**: Sodium-ion batteries are being explored as a more cost-effective alternative to lithium-ion batteries, given the abundance and lower cost of sodium compared to lithium. Research is focused on improving the energy density and cycle life of sodium-ion batteries, which could make them particularly useful for large-scale storage applications such as grid energy storage.

- **Lithium-sulfur Batteries**: Lithium-sulfur (Li-S) batteries have the potential to offer significantly higher energy density than traditional lithium-ion batteries. Sulfur is abundant and cheap, making Li-S batteries a promising solution for both electric vehicle (EV) applications and grid storage. However, challenges remain related to their poor cycle life and efficiency, which researchers are working to address.
- **Solid-State Batteries**: Solid-state batteries replace the liquid electrolyte found in conventional batteries with a solid electrolyte, potentially improving energy density, safety, and longevity. Solid-state batteries are less prone to issues like leakage and flammability, which are common in liquid-based batteries. The improved safety profile and higher energy density make them a promising solution for EVs and portable electronics.

2.5.1.2 Advanced Supercapacitors

Supercapacitors, or ultracapacitors, are another area where advanced materials can make a significant impact. Supercapacitors store energy through electrostatic fields, offering faster charge and discharge rates compared to batteries, but traditionally with lower energy density. The development of advanced materials such as graphene and carbon nanotubes has the potential to significantly increase the energy density of supercapacitors while maintaining their fast charge-discharge capabilities.

- **Graphene-based Supercapacitors**: Graphene is a single layer of carbon atoms arranged in a two-dimensional lattice, known for its high conductivity, flexibility, and large surface area. Researchers are exploring graphene-based materials to enhance the performance of supercapacitors by increasing their energy density while retaining the high power output and rapid charging times.
- **Carbon Nanotube Supercapacitors**: Carbon nanotubes (CNTs) are another material that could revolutionize the supercapacitor market. CNTs have exceptional mechanical, electrical, and thermal properties, making them ideal for high-performance energy storage applications. By incorporating CNTs into the design of supercapacitors, it is possible to achieve both higher energy densities and faster charge/discharge cycles, which would expand the applicability of supercapacitors in various industries.

2.5.1.3 Advanced Electrolytes and Conductors

Innovations in electrolytes and conductors are also key to advancing energy storage materials. For example, solid electrolytes in solid-state batteries and liquid organic electrolytes in flow batteries are under investigation to improve performance and scalability. These advancements in electrolyte technology could enhance battery efficiency, safety, and overall energy storage capacity.

- **Organic Flow Batteries**: In organic flow batteries, organic molecules are used as the electrolyte instead of traditional metal-based electrolytes. Organic molecules are abundant, non-toxic, and potentially much cheaper than the materials used in conventional batteries. Organic flow batteries hold promise for long-duration storage, which is critical for balancing intermittent renewable energy generation, such as solar and wind.
- **High-Conductivity Solid Electrolytes**: Solid electrolytes, which are non-flammable and safer than liquid electrolytes, can be made using advanced materials like ceramic compounds or polymers (Pujari 2024). These solid electrolytes enable the development of batteries with improved safety profiles and better performance in terms of energy density and cycling stability.

3. INTEGRATION OF ENERGY STORAGE WITH RENEWABLE ENERGY SYSTEMS

3.1 Solar Energy Storage

Solar energy, one of the most abundant renewable resources, has been rapidly adopted worldwide. However, its intermittent nature necessitates effective energy storage solutions to ensure a consistent and reliable energy supply (Chu et al., 2023).

3.1.1 Photovoltaic Systems and Storage

Photovoltaic (PV) systems convert sunlight directly into electricity using solar cells. Integrating energy storage with PV systems is essential to manage the variability of solar power and provide a stable energy supply. Lithium-ion batteries are the most commonly used storage technology in residential and commercial PV installations due to their high energy density, efficiency, and declining costs (Hu et al., 2023). In 2023, the global market for solar-plus-storage systems grew by over 50%, driven by incentives and declining battery costs (Fan et al., 2023). Emerging technologies, such as solid-state batteries and flow batteries, are also being explored for their potential to offer higher safety and longer lifespans (Entezari et al., 2023).

3.1.2 Concentrated Solar Power and Storage

Concentrated Solar Power (CSP) systems use mirrors or lenses to concentrate sunlight onto a small area, converting it into heat to produce steam and drive turbines for electricity generation. Thermal energy storage (TES) is integral to CSP systems, allowing them to store excess heat during the day and release it during the night or cloudy periods (Hu et al., 2022). Molten salt is the most widely used TES medium in CSP plants due to its high heat capacity and stability at high temperatures (Islam et al., 2022). Notable CSP projects, such as the Noor Energy 1 plant in Dubai, integrate large-scale TES to provide continuous power generation, demonstrating the potential of CSP and TES in achieving grid stability (Gilmore et al., 2022).

3.2 Wind Energy Storage

Wind energy, another key renewable resource, also faces challenges due to its variability and intermittency. Effective energy storage solutions are essential to harness and stabilize wind power (Khan et al., 2021).

3.2.1 Onshore Wind and Storage Solutions

Onshore wind farms are widely deployed due to their lower installation costs compared to offshore wind farms. Integrating energy storage with onshore wind systems helps mitigate the impact of wind variability and enhances grid stability. Lithium-ion batteries are commonly used for short-term storage, while pumped hydro storage and compressed air energy storage (CAES) are explored for longer-duration applications (Kabeyi & Olanrewaju, 2022). In 2023, the deployment of battery storage in onshore wind farms increased by 35%, driven by technological advancements and cost reductions (Lee et al., 2019).

3.2.2 Offshore Wind and Storage Solutions

Offshore wind farms offer higher and more consistent wind speeds, making them an attractive option for large-scale renewable energy generation. However, the integration of energy storage with offshore wind presents unique challenges due to the harsh marine environment and higher installation costs (Ewim et al., 2021). Recent advancements in floating wind turbines and subsea energy storage systems, such as subsea batteries and compressed air storage, are being explored to address these challenges (Hu et al., 2022). These innovations aim to provide efficient and reliable storage solutions that can withstand the demanding conditions of offshore environments (Chu et al., 2023).

3.3 Other Renewable Sources and Storage

In addition to solar and wind, other renewable energy sources, such as biomass, bioenergy, ocean, and tidal energy, also benefit from effective storage solutions to enhance their viability and integration into the energy grid (Kumar et al., 2023).

3.3.1 Biomass and Bioenergy

Biomass and bioenergy systems convert organic materials into energy through processes like combustion, anaerobic digestion, and gasification (Lee et al., 2019). Energy storage can play a crucial role in optimizing these systems, particularly in balancing supply and demand and improving overall efficiency. Thermal energy storage and biogas storage are commonly used in bioenergy systems to store heat and biogas for later use, ensuring a consistent energy supply (Enebe et al., 2019).

3.3.2 Ocean and Tidal Energy

Ocean and tidal energy harness the power of ocean currents, waves, and tides to generate electricity. These sources are predictable but not constant, making energy storage essential for balancing supply and demand (Kabeyi & Olanrewaju, 2022). Emerging storage solutions for ocean and tidal energy include subsea batteries, which can be integrated with underwater turbines to store energy and release it during peak demand (Khan et al., 2021). Pumped hydro storage systems using seawater are also being explored to provide large-scale energy storage in coastal regions (Fan et al., 2023).

4. ADVANCEMENTS IN ENERGY STORAGE TECHNOLOGIES

The ongoing advancements in energy storage technologies are crucial for enhancing the efficiency, reliability, and affordability of renewable energy systems (Lu et al., 2020).

4.1 Innovations in Battery Technologies

Batteries are at the forefront of energy storage advancements, with significant research and development efforts aimed at improving their performance, safety, and cost-effectiveness (Panda & Das, 2021).

4.1.1 Solid-State Batteries

Solid-state batteries represent a major leap forward in battery technology. Unlike traditional lithium-ion batteries, which use liquid electrolytes, solid-state batteries employ solid electrolytes. This change enhances safety by eliminating the risk of leakage and thermal runaway (Maxmut O'g'li, 2023). Additionally, solid-state batteries offer higher energy densities and longer lifespans. Companies like Toyota and QuantumScape are leading the development of this technology, with commercial applications expected in the automotive sector by the mid-2020s (Rane, 2023a). The solid-state battery market is projected to reach $6 billion by 2030, driven by demand from electric vehicles and portable electronics (Oyekale et al., 2020).

4.1.2 Next-Generation Lithium-ion Batteries

Next-generation lithium-ion batteries are being developed to address the limitations of current lithium-ion technology. Innovations include the use of silicon anodes, which can store more lithium ions than traditional graphite anodes, thereby increasing energy density (Mohammad & Mahjabeen, 2023). Another advancement is the use of lithium-sulfur and lithium-air chemistries, which offer theoretical energy densities several times higher than conventional lithium-ion batteries. Researchers at MIT and Stanford University are making strides in these areas, aiming to bring these technologies to market within the next decade (Lu et al., 2020).

4.1.3 Alternative Battery Chemistries

Alternative battery chemistries, such as sodium-ion, zinc-air, and flow batteries, are gaining attention for their potential to provide cost-effective and scalable energy storage solutions (Manimegalai et al., 2023). Sodium-ion batteries, which use abundant and inexpensive materials, are being developed by companies like Faradion and Natron Energy (Oyekale et al., 2020). Zinc-air batteries offer high energy densities and are considered environmentally friendly due to their use of zinc and air (Panda & Das, 2021). Flow batteries, such as vanadium redox flow batteries, are particularly suited for large-scale energy storage due to their ability to decouple energy and power capacities (Rane, 2023b). These alternative chemistries are expected to complement lithium-ion batteries in various applications, from grid storage to electric vehicles (Maxmut O'g'li, 2023).

4.2 Breakthroughs in Hydrogen Storage and Fuel Cells

Hydrogen storage and fuel cell technologies are pivotal for achieving long-duration energy storage and decarbonizing various sectors, including transportation and industry (Mani & Goniewicz, 2023).

4.2.1 High-Density Hydrogen Storage

High-density hydrogen storage is critical for expanding the use of hydrogen as a clean energy carrier. Innovations in this area include the development of advanced materials, such as metal-organic frameworks (MOFs) and high-pressure tanks, which can store larger amounts of hydrogen at lower costs (Oyekale et al., 2020). Researchers at the National Renewable Energy Laboratory (NREL) and the Fraunhofer Institute are pioneering these technologies, aiming to improve storage capacity and reduce the overall system weight (Lu et al., 2020). Additionally, liquid organic hydrogen carriers (LOHCs) offer a promising approach to storing hydrogen in a liquid state, making it easier to transport and handle (Rane, 2023a).

4.2.2 Fuel Cell Efficiency Improvements

Fuel cells convert hydrogen into electricity through electrochemical reactions, offering high efficiency and zero emissions (Panda & Das, 2021). Recent advancements in fuel cell technology focus on improving catalyst performance and durability. Platinum-based catalysts, while effective, are expensive and scarce (Maxmut O'g'li, 2023). Researchers are exploring alternative catalysts, such as platinum-group metal-free (PGM-free) catalysts, which promise to reduce costs significantly (Mani & Goniewicz, 2023). Additionally, innovations in membrane technology are enhancing the efficiency and lifespan of fuel cells. The global fuel cell market is projected to grow at a compound annual growth rate (CAGR) of 15% over the next decade, driven by applications in transportation, stationary power, and portable devices (Rane, 2023b).

4.3 Advances in Thermal Storage

Thermal energy storage technologies are essential for balancing supply and demand in renewable energy systems, particularly in applications involving heat and power generation (Manimegalai et al., 2023).

4.3.1 Phase Change Materials

Phase change materials (PCMs) store and release thermal energy during the phase transition between solid and liquid states. PCMs offer high energy storage density and nearly constant temperature during the transition, making them ideal for applications in building heating and cooling, industrial processes, and concentrated solar power (CSP) plants (Lu et al., 2020). Recent developments in encapsulation techniques and composite materials are enhancing the thermal conductivity and stability of PCMs, thereby improving their efficiency and lifespan (Oyekale et al., 2020).

4.3.2 Thermochemical Energy Storage Systems

Thermochemical energy storage systems store energy through reversible chemical reactions. These systems have the potential to offer higher energy densities and longer storage durations compared to sensible and latent heat storage (Panda & Das, 2021). Current research focuses on identifying and optimizing suitable thermochemical materials, such as metal hydrides and salts, which can undergo reversible reactions with high efficiency (Maxmut O'g'li, 2023). Projects like the EU-funded CREATE and MERITS initiatives are advancing the development of thermochemical storage systems for residential and industrial applications (Rane, 2023b).

5. CHALLENGES AND SOLUTIONS IN ENERGY STORAGE

The integration of energy storage technologies into renewable energy systems presents several challenges that need to be addressed to maximize their potential. These challenges span technical, economic, and regulatory domains, and overcoming them is essential for the successful deployment and operation of energy storage systems (Rangel-Martinez et al., 2021).

5.1 Technical Challenges:

The technical challenges of energy storage technologies primarily concern energy density, efficiency, durability, and safety (Sayed et al., 2021).

5.1.1 Energy Density and Efficiency:

Energy density and efficiency are critical factors determining the viability of energy storage technologies. Higher energy density allows for more energy to be stored in a smaller volume, making storage solutions more compact and cost-effective. Current lithium-ion batteries offer a high energy density, but emerging technologies such as solid-state batteries and lithium-sulfur batteries promise even greater improvements (Tang & Li, 2023). Efficiency, the ratio of energy output to energy input, is another crucial aspect. Advancements in battery management systems (BMS) and materials science are focused on minimizing energy losses and improving overall system efficiency (Roga et al., 2022).

5.1.2 Durability and Lifespan:

The durability and lifespan of energy storage systems are essential for their economic viability. Frequent cycling of batteries can lead to degradation, reducing their capacity and efficiency over time. Researchers are exploring new materials and electrolytes that can withstand repeated charging and discharging cycles without significant degradation (Tester et al., 2021). Technologies such as flow batteries and solid-state batteries are being developed to offer longer lifespans compared to conventional lithium-ion batteries. Enhanced cooling systems and protective coatings are also being used to extend the lifespan of batteries and other storage technologies (Sanni et al., 2024).

5.2 Economic Challenges:

Economic challenges play a significant role in the widespread adoption of energy storage technologies. These challenges include the high cost of materials and manufacturing, as well as market adoption and scalability (Uddin et al., 2022).

5.2.1 Cost of Materials and Manufacturing:

The cost of materials and manufacturing remains one of the biggest barriers to the adoption of energy storage technologies. While the cost of lithium-ion batteries has decreased significantly over the past decade, it is still relatively high compared to traditional energy sources (Sayed et al., 2021). Innovations in battery chemistry, such as the use of abundant and inexpensive materials like sodium and zinc, are being pursued to reduce costs. Economies of scale, achieved through mass production and improved manufacturing processes, are also essential to bring down costs. Government incentives and subsidies can further support the reduction of upfront costs for energy storage systems (Rangel-Martinez et al., 2021).

5.2.2 Market Adoption and Scalability:

Market adoption and scalability are crucial for the successful deployment of energy storage technologies. The lack of standardization and the fragmented nature of the market can hinder large-scale adoption. Developing standardized protocols and interoperability frameworks can facilitate the integration of diverse energy storage systems (Tang & Li, 2023).

5.3 Regulatory and Policy Challenges:

Regulatory and policy challenges must be addressed to create a conducive environment for the growth of energy storage technologies. These challenges include the development of standards and certifications, as well as government incentives and support (Tester et al., 2021).

5.3.1 Standards and Certifications:

Standards and certifications are essential to ensure the safety, reliability, and interoperability of energy storage systems. The development of comprehensive standards can help guide the manufacturing, installation, and operation of storage technologies (Sayed et al., 2021). Organizations like the International Electrotechnical Commission (IEC) and the Institute of Electrical and Electronics Engineers (IEEE) are working on establishing global standards for energy storage systems. Certification programs can also provide assurance to consumers and investors about the quality and performance of storage technologies (Rangel-Martinez et al., 2021).

5.3.2 Government Incentives and Support:

Government incentives and support play a critical role in the development and deployment of energy storage technologies. Policies such as tax credits, subsidies, and grants can reduce the financial burden on developers and consumers, making storage solutions more affordable (Tang & Li, 2023). Regulatory frameworks that promote grid modernization and the integration of renewable energy can also drive the adoption of energy storage. Governments can support research and development initiatives to advance storage technologies and provide funding for pilot projects and demonstrations (Sanni et al., 2024).

The regulatory frameworks governing energy storage play a crucial role in shaping its adoption and integration into renewable energy systems. Expanding this discussion to include specific laws, incentives, and case studies from various regions can provide a deeper understanding of how policies influence market growth. Governments worldwide have implemented policies such as tax credits, feed-in tariffs, and capacity markets to promote energy storage deployment (Roga et al., 2022). For instance, the U.S. Investment Tax Credit (ITC) and Germany's KfW programs offer financial support for battery storage adoption, demonstrating how policy mechanisms can drive investment and innovation (Uddin et al., 2022).

Beyond regulatory considerations, economic models determine the financial viability of energy storage solutions. Incorporating insights on financing mechanisms such as power purchase agreements (PPAs), green bonds, and government subsidies can illustrate how businesses and utilities approach energy storage investments. Cost-benefit analyses of various storage technologies, including lithium-ion batteries, pumped hydro storage, and emerging alternatives like solid-state batteries, can highlight their economic feasibility across different scales. By evaluating factors such as installation costs, operational efficiency, and long-term returns, stakeholders can make informed decisions about energy storage deployment, ensuring sustainability and profitability in renewable energy transitions (Tester et al., 2021).

6. CASE STUDIES AND REAL-WORLD APPLICATIONS

This chapter explores various real-world applications and case studies that highlight the successful deployment of energy storage technologies across different sectors. These examples illustrate the practical benefits and challenges of integrating energy storage into renewable energy systems, offering valuable insights for future implementations.

6.1 Utility-Scale Energy Storage Projects

Utility-scale energy storage projects are crucial for balancing supply and demand on the grid, integrating renewable energy sources, and enhancing grid stability and reliability.

6.1.1 Hornsdale Power Reserve, South Australia

The Hornsdale Power Reserve in South Australia, powered by Tesla's lithium-ion battery technology, is one of the most well-known utility-scale energy storage projects. This 150 MW/193.5 MWh battery system has demonstrated the ability to provide rapid response to grid fluctuations, support frequency regulation, and reduce electricity costs (Ukoba & Jen, 2023). Since its commissioning in 2017, the Hornsdale Power Reserve has saved the grid millions of dollars in ancillary services costs and helped prevent blackouts during critical periods.

6.1.2 Moss Landing Energy Storage Facility, California

The Moss Landing Energy Storage Facility in California, operated by Vistra Energy, is currently the largest lithium-ion battery storage installation in the world, with a capacity of 400 MW/1,600 MWh (Adeyinka 2024). This project plays a significant role in stabilizing the California grid, providing peaking capacity, and supporting the integration of renewable energy. It demonstrates the scalability of lithium-ion battery technology and its potential to replace traditional peaking power plants (Ullah, Chen, & Choy, 2021).

6.2 Residential and Commercial Energy Storage Solutions

Energy storage solutions for residential and commercial applications are gaining traction as they offer numerous benefits, including energy cost savings, backup power, and enhanced energy independence.

6.2.1 Tesla Powerwall and Powerpack

Tesla's Powerwall and Powerpack are popular energy storage solutions for residential and commercial use, respectively. The Powerwall, with its compact design and integrated inverter, allows homeowners to store excess solar energy, reduce reliance on the grid, and provide backup power during outages (Ukoba, Fadare, & Jen, 2019). The Powerpack, designed for larger commercial installations, offers scalable energy storage solutions that can help businesses manage peak demand charges, enhance energy resilience, and participate in demand response programs.

6.2.2 Sonnen EcoLinx

The Sonnen EcoLinx is a smart energy storage system designed for residential use. It integrates with home automation systems to optimize energy usage, store solar energy, and provide backup power. The EcoLinx's advanced software enables intelligent energy management, allowing homeowners to maximize their solar investment and reduce electricity bills (Ullah et al., 2021). This system showcases the growing trend of integrating energy storage with smart home technologies.

6.3 Off-Grid and Rural Electrification

Energy storage technologies play a vital role in providing reliable and sustainable energy solutions for off-grid and rural areas, where access to the traditional power grid is limited or non-existent.

6.3.1 Solar Home Systems in Sub-Saharan Africa

Solar home systems (SHS) combined with battery storage are transforming energy access in Sub-Saharan Africa. Companies like M-KOPA and Off-Grid Electric are providing affordable SHS to households in remote areas, enabling them to generate and store their own electricity. These systems typically include a solar panel, battery, LED lights, and a few small appliances (Ukoba et al., 2019). By leveraging pay-as-you-go financing models, these companies are making clean energy accessible to low-income communities, improving quality of life and economic opportunities.

6.3.2 Microgrids in India

In India, microgrids powered by renewable energy sources and supported by battery storage are addressing the energy needs of rural villages. Organizations like SELCO India and Husk Power Systems are deploying solar and biomass microgrids with battery storage to provide reliable and affordable electricity to underserved communities (Ukoba & Jen, 2023). These microgrids not only improve energy access but also support local economic development by powering small businesses and enhancing education and healthcare services.

6.4 Industrial Applications of Energy Storage

Industrial applications of energy storage are critical for enhancing operational efficiency, reducing energy costs, and supporting sustainability goals in various sectors.

6.4.1 Renewable Integration at Manufacturing Facilities

Manufacturing facilities are increasingly integrating renewable energy sources with battery storage to reduce their carbon footprint and energy costs (Ullah et al., 2021)..

6.4.2 Energy Storage for Data Centers

Data centers, with their high and constant energy demands, are prime candidates for energy storage solutions. Google, for example, has implemented battery storage systems at its data centers to provide backup power, support grid stability, and enhance energy efficiency (Das et al.,2024). By storing excess renewable energy generated on-site, these systems help data centers reduce their reliance on fossil fuels and lower their overall carbon emissions (Ukoba et al., 2019). This case highlights the importance of energy storage in supporting the energy-intensive operations of data centers and contributing to broader sustainability efforts.

6.4.3 Environmental Impact of Energy Storage Solutions

The environmental impact of energy storage technologies is a critical consideration in their integration into renewable energy systems. While energy storage solutions are primarily aimed at enhancing the reliability and efficiency of renewable energy, their lifecycle, from production to disposal, can have significant environmental implications (Ukoba & Jen, 2023). The production of lithium-ion batteries requires the mining of raw materials such as lithium, cobalt, and nickel, which can lead to habitat disruption, water pollution, and high carbon emissions if not responsibly sourced. Additionally, the disposal of these materials at the end of their life cycle presents environmental challenges, with concerns about the accumulation of electronic waste and the need for efficient recycling methods.

However, advancements in sustainable storage technologies are mitigating these environmental concerns. New research is focused on developing more eco-friendly alternatives, such as sodium-ion batteries, which use more abundant and less environmentally harmful materials, and solid-state batteries, which offer higher efficiency with reduced material waste (Nikitina 2024). Moreover, integrating energy storage solutions with renewable energy generation, such as solar and wind, reduces the reliance on fossil fuels, helping to lower overall greenhouse gas emissions (Ullah et al., 2021).

Incorporating a lifecycle analysis of various energy storage technologies can provide a clearer picture of their true environmental impact. This includes assessing factors such as energy efficiency, material sourcing, and recyclability. Policymakers, manufacturers, and environmental advocates must work collaboratively to ensure that energy storage technologies are not only economically feasible but also environmentally sustainable, promoting a greener future for energy systems worldwide.

7. FUTURE TRENDS AND PROSPECTS

The landscape of energy storage technologies is evolving rapidly, driven by advances in materials science, engineering, and policy support. This chapter explores the predicted market growth for energy storage, emerging applications and innovations, and the role of energy storage in a sustainable future, culminating in a vision for the next decade in energy storage.

7.1 Predicted Market Growth for Energy Storage

The global energy storage market is poised for significant growth in the coming decade, driven by multiple factors including declining technology costs, supportive government policies, and advancements in energy storage solutions. According to projections, the market is expected to increase nearly 15-fold by 2030, reaching an estimated 1,095 gigawatt-hours (GWh), compared to just 74 GWh in 2021. A key driver of this growth is the sharp decline in the cost of lithium-ion batteries, the most widely used energy storage technology, which has fallen from around $1,100 per kilowatt-hour (kWh) in 2010 to approximately $130 per kWh in 2023. This cost reduction has made energy storage more accessible across various sectors, from residential to large-scale utility projects. Additionally, governments worldwide are implementing favorable policies such as tax incentives and renewable energy mandates to encourage the integration of energy storage systems, further fueling market expansion. Technological innovations, such as next-generation batteries and emerging alternatives like flow batteries and hydrogen storage, are diversifying the energy storage landscape, making it more adaptable to a range of applications.

7.2 Emerging Applications and Innovations

The versatility of energy storage technologies is expanding their applications across various sectors, driving innovation and new use cases.

7.2.1 Electric Vehicles

The electrification of transportation is a major trend, with electric vehicles (EVs) becoming more mainstream. Advances in battery technology are extending the range of EVs, reducing charging times, and lowering costs. Vehicle-to-grid (V2G) technology is an emerging innovation that allows EVs to act as mobile energy storage units, providing power back to the grid during peak demand periods.

7.2.2 Grid Modernization

Energy storage is playing a critical role in modernizing the grid, enabling better integration of renewable energy sources, enhancing grid stability, and providing ancillary services such as frequency regulation and voltage support. Advanced energy management systems and artificial intelligence (AI) are being integrated with storage solutions to optimize grid operations and efficiency.

7.2.3 Residential and Commercial Solutions

Energy storage solutions for homes and businesses are becoming more sophisticated, offering smart energy management, backup power, and cost savings. The integration of storage with renewable energy systems, smart home technologies, and demand response programs is creating new opportunities for consumers to participate in the energy market and achieve greater energy independence.

7.3 The Role of Energy Storage in a Sustainable Future

Energy storage plays a pivotal role in achieving a sustainable energy future by enabling the efficient integration of renewable energy sources, supporting decarbonization, and ensuring energy access in underserved regions. As renewable energy generation increases, the intermittency and variability of sources like solar and wind require effective storage solutions to stabilize the grid, maintain reliability, and ensure continuous energy supply. Energy storage systems (ESS) act as a buffer, capturing excess energy during periods of high generation and discharging it when demand exceeds supply or when generation is low. This ability to store and release energy as needed is key to balancing supply and demand, optimizing grid performance, and reducing reliance on fossil-fuel-based power plants, thus facilitating the transition to a low-carbon energy system.

7.3.1 Renewable Energy Integration

One of the most significant roles of energy storage in a sustainable future is its ability to facilitate the integration of renewable energy sources, such as solar and wind, into the grid. These energy sources, while abundant and clean, are intermittent by nature—solar power is only generated during daylight hours, and wind power fluctuates depending on wind conditions. Energy storage systems, including batteries, pumped hydro storage, and thermal storage, mitigate these challenges by storing surplus energy when renewable generation is high and releasing it when demand spikes or generation drops. This process ensures a stable and consistent power supply, making it easier to rely on renewable sources for the majority of the energy mix. For example, large-scale battery storage systems paired with solar and wind farms enable energy companies to store excess renewable power produced during sunny or windy periods and dispatch it to the grid during peak demand hours, reducing the need for fossil-fuel-based backup generation. This integration leads to lower emissions, increased grid reliability, and a smoother transition to a renewable-powered grid.

7.3.2 Decarbonization

Energy storage is critical for decarbonization efforts by reducing the reliance on fossil fuels and supporting the transition to cleaner, renewable energy sources. By enabling the integration of large-scale renewable energy generation, storage systems reduce the need for carbon-intensive power plants, which are traditionally used to meet peak demand. Energy storage systems, such as advanced batteries and hydrogen storage, allow for the utilization of excess renewable energy when it is abundant, and they provide backup power during periods when renewable generation is low. This reduces the need for natural gas or coal-fired plants, which are typically used to ensure grid stability.

7.3.3 Energy Access

Energy storage is also instrumental in expanding energy access to underserved and off-grid communities, which often lack reliable and affordable electricity. In remote areas, where extending the central grid infrastructure is often economically unfeasible, decentralized energy storage solutions can provide a cost-effective and sustainable alternative. Microgrids equipped with energy storage systems can operate independently from the main grid, using locally generated renewable energy, such as solar or wind, to power homes, schools, and businesses. These systems store energy produced during the day and provide power after dark or during periods of low generation (Horzela-Miś 2025).This ensures that communities have a constant, reliable source of energy, enabling better living conditions, economic development, and improved access to essential services such as healthcare and education.

7.4 Vision for the Next Decade in Energy Storage

The next decade promises transformative advancements and expanded applications for energy storage technologies, driven by continued innovation, policy support, and market growth.

8. CONCLUSION

Energy storage is a cornerstone of the transition to a sustainable, low-carbon energy future. By enabling the efficient integration of renewable energy, supporting decarbonization efforts, and enhancing energy access in underserved regions, energy storage technologies are transforming the global energy landscape. As renewable energy sources like solar and wind continue to grow, storage systems provide the flexibility needed to overcome their intermittency, ensuring a reliable and consistent energy supply. Additionally, energy storage is essential for reducing reliance on fossil fuels, accelerating the decarbonization of multiple sectors, and promoting energy equity in both developed and developing regions. The ongoing advancements in energy storage technologies promise to unlock new opportunities, fostering a cleaner, more resilient, and sustainable energy future for generations to come.

REFERENCES

Adebukola, A. A., Navya, A. N., Jordan, F. J., Jenifer, N. J., & Begley, R. D. (2022). Cybersecurity as a threat to healthcare. *Journal of Technology and Systems*, *4*(1), 32–64. DOI: 10.47941/jts.1149

Adegoke, A. (2023). Patients' reaction to online access to their electronic medical records: The case of diabetic patients in the US. *International Journal of Applied Sciences. Current and Future Research Trends*, *19*(1), 105–115.

Adeyinka, A. M., Esan, O. C., Ijaola, A. O., & Farayibi, P. K. (2024). Advancements in hybrid energy storage systems for enhancing renewable energy-to-grid integration. *Sustainable Energy Research*, *11*(1), 26. DOI: 10.1186/s40807-024-00120-4

Ahmad, M., Peng, T., Awan, A., & Ahmed, Z. (2023). Policy framework considering resource curse, renewable energy transition, and institutional issues: Fostering sustainable development and sustainable natural resource consumption practices. *Resources Policy*, *86*, 104173. DOI: 10.1016/j.resourpol.2023.104173

Ahmad, T., Madonski, R., Zhang, D., Huang, C., & Mujeeb, A. (2022). Data-driven probabilistic machine learning in sustainable smart energy/smart energy systems: Key developments, challenges, and future research opportunities in the context of smart grid paradigm. *Renewable & Sustainable Energy Reviews*, *160*, 112128. DOI: 10.1016/j.rser.2022.112128

Arsad, A. Z., Hannan, M. A., Al-Shetwi, A. Q., Mansur, M., Muttaqi, K. M., Dong, Z. Y., & Blaabjerg, F. (2022). Hydrogen energy storage integrated hybrid renewable energy systems: A review analysis for future research directions. *International Journal of Hydrogen Energy*, *47*(39), 17285–17312. DOI: 10.1016/j.ijhydene.2022.03.208

Arya, S., & Mahajan, P. (2023). Future in solar cell technology. In *Solar Cells: Types and Applications* (pp. 237-256). Springer Nature Singapore. DOI: 10.1007/978-981-99-7333-0_9

Atawi, I. E., Al-Shetwi, A. Q., Magableh, A. M., & Albalawi, O. H. (2022). Recent advances in hybrid energy storage system integrated renewable power generation: Configuration, control, applications, and future directions. *Batteries*, *9*(1), 29. DOI: 10.3390/batteries9010029

Azarpour, A., Mohammadzadeh, O., Rezaei, N., & Zendehboudi, S. (2022). Current status and future prospects of renewable and sustainable energy in North America: Progress and challenges. *Energy Conversion and Management*, *269*, 115945. DOI: 10.1016/j.enconman.2022.115945

Badi, I., Stević, Ž., & Bouraima, M. B. (2023). Overcoming obstacles to renewable energy development in Libya: An MCDM approach towards effective strategy formulation. *Decision Making Advances, 1*(1), 17–24. DOI: 10.31181/v120234

Baria, R., Baumgaertner, J., Glass, H., Jupe, A., Robertson-Tait, A., Beadsmore, G., . . . Kocis, I. (2023). International cooperation to address and mitigate the climate change issue using unconventional geothermal technology (EGS). In *Enhanced Geothermal Systems (EGS)* (pp. 175-191). CRC Press.

Bhuiyan, M. A., Hu, P., Khare, V., Hamaguchi, Y., Thakur, B. K., & Rahman, M. K. (2022). Economic feasibility of marine renewable energy. *Frontiers in Marine Science, 9*, 988513. DOI: 10.3389/fmars.2022.988513

Bošnjaković, M., Katinić, M., Santa, R., & Marić, D. (2022). Wind turbine technology trends. *Applied Sciences (Basel, Switzerland), 12*(17), 8653. DOI: 10.3390/app12178653

Callegari, A., Bolognesi, S., Cecconet, D., & Capodaglio, A. G. (2020). Production technologies, current role, and future prospects of biofuels feedstocks: A state-of-the-art review. *Critical Reviews in Environmental Science and Technology, 50*(4), 384–436. DOI: 10.1080/10643389.2019.1629801

Cantarero, M. M. V. (2020). Of renewable energy, energy democracy, and sustainable development: A roadmap to accelerate the energy transition in developing countries. *Energy Research & Social Science, 70*, 101716. DOI: 10.1016/j.erss.2020.101716

Chidolue, O., & Iqbal, T. (2023, March). System monitoring and data logging using PLX-DAQ for solar-powered oil well pumping. In *2023 IEEE 13th Annual Computing and Communication Workshop and Conference (CCWC)* (pp. 0690-0694). IEEE. DOI: 10.1109/CCWC57344.2023.10099099

Chu, W., Vicidomini, M., Calise, F., Duić, N., Østergaard, P. A., Wang, Q., & da Graça Carvalho, M. (2023). Review of hot topics in the sustainable development of energy, water, and environment systems conference in 2022. *Energies, 16*(23), 7897. DOI: 10.3390/en16237897

Das, D., Chakraborty, I., Bohre, A. K., Kumar, P., & Agarwala, R. (2024). Sustainable integration of green hydrogen in renewable energy systems for residential and EV applications. *International Journal of Energy Research, 2024*(1), 8258624. DOI: 10.1155/2024/8258624

. Enebe, G. C., Ukoba, K., & Jen, T. C. (2019). Numerical modeling of the effect of annealing on nanostructured CuO/TiO2 pn heterojunction solar cells using SCAPS.

Entezari, A., Aslani, A., Zahedi, R., & Noorollahi, Y. (2023). Artificial intelligence and machine learning in energy systems: A bibliographic perspective. *Energy Strategy Reviews*, *45*, 101017. DOI: 10.1016/j.esr.2022.101017

. Ewim, D. R. E., Okwu, M. O., Onyiriuka, E. J., Abiodun, A. S., Abolarin, S. M., & Kaood, A. (2021). A quick review of the applications of artificial neural networks (ANN) in the modeling of thermal systems.

Fan, Z., Yan, Z., & Wen, S. (2023). Deep learning and artificial intelligence in sustainability: A review of SDGs, renewable energy, and environmental health. *Sustainability (Basel)*, *15*(18), 13493. DOI: 10.3390/su151813493

Gilmore, N., Koskinen, I., van Gennip, D., Paget, G., Burr, P. A., Obbard, E. G., Daiyan, R., Sproul, A., Kay, M., Lennon, A., Konstantinou, G., Hemer, M., Gui, E. M., & Gurieff, N. (2022). Clean energy futures: An Australian-based foresight study. *Energy*, *260*, 125089. DOI: 10.1016/j.energy.2022.125089

Horzela-Miś, A., & Semrau, J. (2025). The role of renewable energy and storage technologies in sustainable development: Simulation in the construction industry. *Frontiers in Energy Research*, *13*, 1540423. DOI: 10.3389/fenrg.2025.1540423

Hu, X., Banks, J., Guo, Y., & Liu, W. V. (2022). Utilizing geothermal energy from enhanced geothermal systems as a heat source for oil sands separation: A numerical evaluation. *Energy*, *238*, 121676. DOI: 10.1016/j.energy.2021.121676

Hu, Z., Ran, C., Zhang, H., Chao, L., Chen, Y., & Huang, W. (2023). The current status and development trend of perovskite solar cells. *Engineering (Beijing)*, *21*, 15–19. DOI: 10.1016/j.eng.2022.10.012

Islam, M. R., Wu, Y., Liu, K., Wang, Z., Qu, S., & Wang, Z. (2022). Recent progress and future prospects for light management of all-perovskite tandem solar cells. *Advanced Materials Interfaces*, *9*(4), 2101144. DOI: 10.1002/admi.202101144

Kabeyi, M. J. B., & Olanrewaju, O. A. (2022). Geothermal wellhead technology power plants in grid electricity generation: A review. *Energy Strategy Reviews*, *39*, 100735. DOI: 10.1016/j.esr.2021.100735

Khan, A., Chenggang, Y., Hussain, J., & Kui, Z. (2021). Impact of technological innovation, financial development, and foreign direct investment on renewable energy, nonrenewable energy, and the environment in Belt & Road Initiative countries. *Renewable Energy*, *171*, 479–491. DOI: 10.1016/j.renene.2021.02.075

Kumar, S., Lalitha Kameswari, Y., & Koteswara Rao, S. (2023). Smart grid management for smart city infrastructure using wearable sensors. In *Data Analytics for Smart Grids Applications—A Key to Smart City Development* (pp. 39–63). Springer Nature Switzerland. DOI: 10.1007/978-3-031-46092-0_4

Lee, S. Y., Sankaran, R., Chew, K. W., Tan, C. H., Krishnamoorthy, R., Chu, D. T., & Show, P. L. (2019). Waste to bioenergy: A review on the recent conversion technologies. *BMC Energy*, *1*(1), 1–22. DOI: 10.1186/s42500-019-0004-7

Lu, Y., Khan, Z. A., Alvarez-Alvarado, M. S., Zhang, Y., Huang, Z., & Imran, M. (2020). A critical review of sustainable energy policies for the promotion of renewable energy sources. *Sustainability (Basel)*, *12*(12), 5078. DOI: 10.3390/su12125078

Mani, Z. A., & Goniewicz, K. (2023). Adapting disaster preparedness strategies to changing climate patterns in Saudi Arabia: A rapid review. *Sustainability (Basel)*, *15*(19), 14279. DOI: 10.3390/su151914279

. Manimegalai, V., Rukkumani, V., Gayathri, A., Pandiyan, P., & Mohanapriya, V. (2023). An overview of global renewable energy resources. *Renewable Energy and AI for Sustainable Development, 2*(2.4), 2-5.

Maxmut O'g'li, X. F. (2023). Renewable energy sources: Advancements, challenges, and prospects. *International Journal of Advances in Scientific Research*, *3*(08), 14–25.

. Mohammad, A., & Mahjabeen, F. (2023). From silicon to sunlight: Exploring the evolution of solar cell materials. *JURIHUM: Jurnal Inovasi dan Humaniora, 1*(2), 316-330.

Nikitina, A. A., Lavrentev, F. V., Yurova, V. Y., Piarnits, D. Y., Volkova, O. O., Skorb, E. V., & Shchukin, D. G. (2024). Layered nanomaterials for renewable energy generation and storage. *Materials Advances*, *5*(2), 394–408. DOI: 10.1039/D3MA00924F

Oyekale, J., Petrollese, M., Tola, V., & Cau, G. (2020). Impacts of renewable energy resources on effectiveness of grid-integrated systems: Succinct review of current challenges and potential solution strategies. *Energies*, *13*(18), 4856. DOI: 10.3390/en13184856

Panda, D. K., & Das, S. (2021). Smart grid architecture model for control, optimization and data analytics of future power networks with more renewable energy. *Journal of Cleaner Production*, *301*, 126877. DOI: 10.1016/j.jclepro.2021.126877

Pujari, H. K., Rudramoorthy, M., Gopi, R. R., Mishra, S., Alluraiah, N. C., & Vaishali, N. B. (2024). *Optimal reconfiguration, renewable DGs, and energy storage units' integration in distribution systems considering power generation uncertainty using hybrid GWO-SCA algorithms*. International Journal of Modelling and Simulation., DOI: 10.1080/02286203.2024.2363605

Rane, N. (2023). Contribution of ChatGPT and other generative artificial intelligence (AI) in renewable and sustainable energy. Available at *SSRN* 4597674. DOI: 10.2139/ssrn.4597674

Rane, N. (2023). Integrating leading-edge artificial intelligence (AI), Internet of Things (IoT), and big data technologies for smart and sustainable architecture, engineering and construction (AEC) industry: Challenges and future directions. (September 24, 2023).

Rangel-Martinez, D., Nigam, K. D. P., & Ricardez-Sandoval, L. A. (2021). Machine learning on sustainable energy: A review and outlook on renewable energy systems, catalysis, smart grid and energy storage. *Chemical Engineering Research & Design*, *174*, 414–441. DOI: 10.1016/j.cherd.2021.08.013

Roga, S., Bardhan, S., Kumar, Y., & Dubey, S. K. (2022). Recent technology and challenges of wind energy generation: A review. *Sustainable Energy Technologies and Assessments*, *52*, 102239. DOI: 10.1016/j.seta.2022.102239

Sanni, O., Adeleke, O., Ukoba, K., Ren, J., & Jen, T. C. (2024). Prediction of inhibition performance of agro-waste extract in simulated acidizing media via machine learning. *Fuel*, *356*, 129527. DOI: 10.1016/j.fuel.2023.129527

Sayed, E. T., Wilberforce, T., Elsaid, K., Rabaia, M. K. H., Abdelkareem, M. A., Chae, K. J., & Olabi, A. G. (2021). A critical review on environmental impacts of renewable energy systems and mitigation strategies: Wind, hydro, biomass and geothermal. *The Science of the Total Environment*, *766*, 144505. DOI: 10.1016/j.scitotenv.2020.144505 PMID: 33421793

Sioshansi, R., & Denholm, P. (2025). Long-duration energy storage in transmission-constrained variable renewable energy systems. *Cell Reports Sustainability*, *2*(1), 100285. DOI: 10.1016/j.crsus.2024.100285

Tang, X., & Li, X. (2023). Exploring carbon neutrality potential in the petroleum industry: The role of green monetary policies as motivators. *Resources Policy*, *86*, 104230. DOI: 10.1016/j.resourpol.2023.104230

Tester, J. W., Beckers, K. F., Hawkins, A. J., & Lukawski, M. Z. (2021). The evolving role of geothermal energy for decarbonizing the United States. *Energy & Environmental Science*, *14*(12), 6211–6241. DOI: 10.1039/D1EE02309H

Uddin, S. U., Chidolue, O., Azeez, A., & Iqbal, T. (2022, June). Design and analysis of a solar-powered water filtration system for a community in Black Tickle-Domino. In *2022 IEEE International IOT,Electronics and Mechatronics Conference (IEMTRONICS)* (pp. 1-6). IEEE. DOI: 10.1109/IEMTRONICS55184.2022.9795758

Ukoba, K., Fadare, O., & Jen, T. C. (2019, December). Powering Africa using an off-grid, standalone, solar photovoltaic model. []. IOP Publishing.]. *Journal of Physics: Conference Series*, *1378*(2), 022031. DOI: 10.1088/1742-6596/1378/2/022031

Ukoba, K., & Jen, T. C. (2023). *Thin films, atomic layer deposition, and 3D printing: Demystifying the concepts and their relevance in Industry 4.0*. CRC Press. DOI: 10.1201/9781003364481

Ullah, F., Chen, C. C., & Choy, W. C. (2021). Recent developments in organic tandem solar cells toward high efficiency. *Advanced Energy and Sustainability Research*, *2*(4), 2000050. DOI: 10.1002/aesr.202000050

Wang, Z., Guo, Y., & Wang, H. (2022). Review on monitoring and operation-maintenance technology of far-reaching sea smart wind farms. *Journal of Marine Science and Engineering*, *10*(6), 820. DOI: 10.3390/jmse10060820

Yang, X., He, L., Xia, Y., & Chen, Y. (2019). Effect of government subsidies on renewable energy investments: The threshold effect. *Energy Policy*, *132*, 156–166. DOI: 10.1016/j.enpol.2019.05.039

Chapter 14
Recent Advancement of Graphene–TiO_2 Composite Nanomaterials and Their Applications in Energy Storage and Conversion: Novel Functions and Future Prospects Explored

Hari Shankar Biswas
https://orcid.org/0000-0003-1883-5244
Surendranath College, India

Amit Kr. Kundu
https://orcid.org/0000-0002-7649-8154
Sripat Singh College, India

Shib Shankar Biswas
https://orcid.org/0000-0001-7767-2110
Surendranath College, India

ABSTRACT

Graphene-TiO2 composites are promising due to their synergistic properties, combining graphene's exceptional electronic and optical characteristics with TiO2's high photocatalytic efficiency. This review covers synthesis methods like chemical vapor

DOI: 10.4018/979-8-3693-9316-1.ch014

deposition (CVD), sol-gel, and hydrothermal techniques, highlighting their impact on composite structure and properties. Key factors such as graphene loading, morphology control, and interface engineering are discussed for their role in enhancing photocatalytic activity and charge transfer efficiency. Environmental applications include water purification through pollutant degradation and air filtration for VOCs removal. In energy, these composites improve solar cell efficiency and lithium-ion battery electrode performance. Challenges like scalability, stability, and cost are addressed, with future research directions focusing on innovative synthesis and interface engineering for commercialization.

1. INTRODUCTION

Graphene, a single layer of carbon atoms organized in a hexagonal lattice, has captivated the scientific community since its groundbreaking isolation in 2004 (Biswas et al., 2022; Santra et al., 2024). This discovery marked the beginning of a new chapter in nanomaterials research, unveiling an exceptional set of properties that position graphene as a promising candidate for diverse technological applications (Ahmad et al., 2024; Kumar et al., 2024).The structure of graphene, where each carbon atom is bonded in an sp^2 hybridization, forms a two-dimensional honeycomb lattice (Sahu et al., 2024). This atomic configuration is responsible for its exceptional mechanical properties, including a tensile strength exceeding 130 GPa and remarkable flexibility. These attributes make graphene one of the strongest known materials, capable of withstanding extreme mechanical stresses without losing its structural integrity. Beyond its mechanical prowess, graphene also exhibits unparalleled electrical conductivity, reaching values as high as 200,000 cm^2/Vs (Zheng et al., 2024). This extraordinary level of electrical performance has made it a highly promising candidate for use in electronic devices, sensors, and energy storage systems. Furthermore, graphene's thermal conductivity, which can reach up to 5000 W/mK, surpasses that of most materials, making it an ideal candidate for thermal management applications (Vairagade et al., 2024). In addition to these properties, graphene is also optically transparent, absorbing only 2.3% of visible light, which further enhances its appeal for use in optoelectronics and transparent conductive films (Ahmad et al., 2024). These unique properties, combined with its biocompatibility and impermeability to gases, have spurred rapid developments in a range of advanced applications (Figure 1), including drug delivery systems and biosensors (Singh et al., 2024).

Figure 1. Schematic drawing of various applications of Graphene-TiO_2 nanocomposite

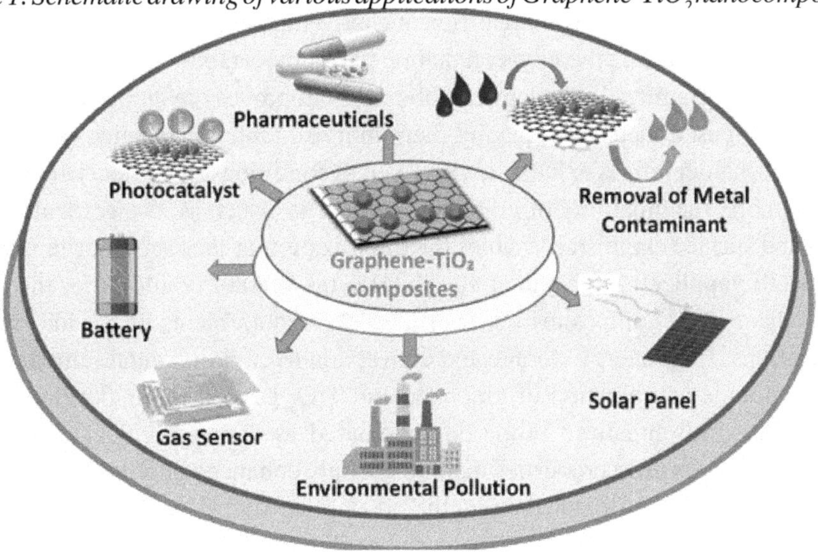

Despite graphene's remarkable properties, its integration with other materials has been a major focus of research, particularly for enhancing its functionality in composite materials. Among the various materials explored for combination with graphene, titanium dioxide (TiO_2) stands out as a particularly promising candidate (Farooq et al., 2024). TiO_2 is a well-established semiconductor known for its excellent photocatalytic activity, which is primarily driven by its wide bandgap. TiO_2 exists in several crystalline forms, with anatase and rutile being the most commonly studied phases (Thakur et al., 2024). The anatase phase, with a bandgap of approximately 3.2 eV, is particularly effective in absorbing ultraviolet (UV) light, making it a valuable material for photochemical applications (Seiß et al., 2022). This photocatalytic activity allows TiO_2 to catalyze a wide range of reactions under UV light, including the degradation of pollutants, water splitting, and the breakdown of harmful organic compounds. TiO_2's chemical stability, non-toxic nature, and abundance contribute to its widespread use in a variety of industrial applications, ranging from water purification and air detoxification to self-cleaning surfaces and environmental remediation (Dagar et al., 2024). Additionally, TiO_2 is commonly used in the production of paints, coatings, and cosmetics, where its inertness and UV-absorbing properties are particularly valuable (Egerton, et al., 2014).

The integration of graphene with TiO_2 in composite materials presents an exciting opportunity to combine the unique properties of both materials, resulting in synergistic effects that enhance their overall performance (Kaveh et al., 2024). Graphene's high surface area and excellent charge carrier mobility make it an ideal candidate for improving the charge transport properties within TiO_2 matrices (Jain et al., 2024). By facilitating more efficient charge transfer, graphene helps to reduce recombination losses, thereby enhancing the photocatalytic efficiency of TiO_2. Additionally, graphene's mechanical robustness improves the durability and stability of TiO_2-based materials, making them more suitable for practical applications where mechanical stress or harsh environmental conditions may affect performance. Furthermore, the tunability of graphene's properties, such as its electrical conductivity and surface chemistry, enables the optimization of these composite materials for specific applications (Shah et al., 2024). This versatility allows for the design of graphene-TiO_2 composites that can meet the requirements of a wide range of applications, from energy storage and conversion to environmental remediation.

The unique combination of graphene and TiO_2 has led to the development of composites with enhanced properties compared to their individual components alone. These improved properties include not only enhanced photocatalytic activity but also increased mechanical strength and stability under harsh conditions. Such advancements have expanded the potential applications of graphene-TiO_2 composites across a diverse array of fields. In environmental applications, for instance, these composites have shown significant promise in the photocatalytic degradation of pollutants, water splitting for hydrogen production, and air purification by decomposing volatile organic compounds (VOCs). In energy storage and conversion technologies, graphene-TiO_2 composites have been used to enhance the performance of lithium-ion batteries, supercapacitors, and dye-sensitized solar cells (DSSCs) (Davu et al., 2024). The improved charge storage and conversion efficiency offered by these composites have made them attractive candidates for use in next-generation energy technologies, which are critical for addressing global energy challenges.

The ongoing research into graphene-TiO_2 composites highlights their potential to drive technological advancements across a wide range of fields. This review explores the various synthesis strategies that have been developed for producing graphene-TiO_2 composites, as well as the characterization techniques used to analyze their properties. By focusing on the synergistic effects that arise from the combination of graphene and TiO_2, this review aims to provide insights into the ways in which these composites can be further optimized for specific applications. The versatility of graphene-TiO_2 composites positions them at the forefront of innovation in fields ranging from environmental sustainability to renewable energy technologies. As research continues to push the boundaries of what is possible with these materials, graphene-TiO_2 composites are likely to play an increasingly important role in addressing some of the most pressing challenges of the 21st century, from pollution control and resource conservation to the development of cleaner and more efficient energy systems.

Again, graphene-TiO_2 composite nanomaterials have emerged as promising candidates for energy storage and conversion due to their synergistic properties. Graphene offers high electrical conductivity, large surface area, and mechanical strength, while TiO_2 provides excellent chemical stability, photocatalytic activity, and tunable bandgap. These composites enhance charge transport, improve electrochemical performance, and enable efficient electron-hole separation, making them suitable for batteries, supercapacitors, and photocatalytic fuel cells. Recent advancements focus on optimizing synthesis methods and surface modifications to enhance energy efficiency and sustainability.

2. SYNTHESIS METHODS

Graphene-TiO_2 composites are synthesized using various methods tailored to achieve desired composite structures and properties. Among these, chemical vapor deposition (CVD) stands out for its ability to produce high-quality graphene films directly on substrates(Li et al., 2024). In CVD, hydrocarbon gases (e.g., methane) are decomposed at high temperatures (typically 800-1000°C) in the presence of a metal catalyst (often copper foil). The carbon atoms assemble into a graphene layer on the catalyst surface, which can then be transferred onto TiO_2 substrates through methods like chemical etching or polymer-assisted transfer. This approach offers precise control over graphene thickness and uniformity, yielding high-quality graphene-TiO_2 composites ideal for applications requiring large-area coverage and high performance. However, CVD requires specialized equipment and meticulous control of synthesis parameters, making it relatively complex and costly compared to other methods. Hydrothermal and solvothermal methods provide alternative routes for synthesizing graphene-TiO_2 composites under milder conditions, typically at lower temperatures and in aqueous or organic solvents (Ruan et al., 2024). In hydrothermal synthesis, graphene oxide (GO) or reduced graphene oxide (rGO) is combined with TiO_2 precursor solutions in a sealed reactor and heated under pressure at temperatures ranging from 100°C to 200°C. This process promotes the nucleation and growth of TiO2 nanoparticles onto the graphene surface, allowing for control over composite morphology and interface interactions. Solvothermal methods follow a similar principle but use organic solvents instead of water, offering different solvent-based dispersion properties and potentially modifying graphene-TiO2 interactions. These methods are advantageous for their simplicity, scalability, and ability to control the size and distribution of TiO_2 nanoparticles on graphene sheets, thereby influencing composite properties such as photocatalytic activity and mechanical strength.

Other synthesis techniques include mechanical mixing and electrophoretic deposition, each offering unique advantages and challenges. Mechanical mixing involves physically blending graphene and TiO_2 nanoparticles using techniques such as ball milling or ultrasonication. While straightforward and cost-effective, mechanical mixing may result in uneven dispersion and aggregation of nanoparticles, affecting composite homogeneity and performance consistency. Electrophoretic deposition, on the other hand, utilizes an electric field to deposit graphene and TiO_2 nanoparticles onto conductive substrates. This method allows for precise control over film thickness and composition, yielding dense and uniform coatings with enhanced adhesion (Escorcia-Díaz et al., 2023). However, electrophoretic deposition requires careful optimization of deposition parameters and may be limited by the complexity of electrode preparation and handling. Comparative analysis of these synthesis techniques reveals trade-offs between complexity, cost, scalability, and control over composite properties. CVD excels in producing high-quality graphene films but requires specialized equipment and precise control over synthesis conditions. Hydrothermal and solvothermal methods offer versatility and scalability while allowing for control over composite morphology and interface interactions. Mechanical mixing and electrophoretic deposition provide simpler and cost-effective alternatives but may sacrifice some level of control over composite uniformity and performance. The selection of synthesis method for graphene-TiO2 composites depends on specific application requirements, emphasizing the need for continued research to optimize synthesis techniques and advance the practical implementation of these materials in various technological fields.

3. CHARACTERIZATION TECHNIQUES

Graphene-TiO_2 composite nanomaterials are gaining significant attention due to their unique properties and potential for applications across diverse fields, including environmental remediation, energy storage, and catalysis. To harness their full potential, it is essential to perform comprehensive characterization of these composites, which provides insights into their structure, surface properties, and performance. The characterization of graphene-TiO_2 composites involves several techniques that address different aspects of the material, including structural, surface, electrical, and optical properties (Figure 2).

3.1. Structural Characterization

Structural characterization is fundamental for understanding the crystalline nature, composition, and arrangement of atoms in graphene-TiO_2 composites. X-ray diffraction (XRD) provides detailed information about the crystalline phases present in the composite by analyzing the diffraction patterns of X-rays scattered by the atoms in the material. For graphene-TiO_2 composites, XRD helps identify the crystalline phases of both graphene and TiO_2, distinguishing between different polymorphs of TiO_2, such as anatase and rutile. This information is critical because the photocatalytic and electronic properties of TiO_2 depend on its crystalline phase (Ahmed et al., 2024). Additionally, XRD can be used to estimate the crystallite size and degree of crystallinity of the composite, parameters that directly influence the material's performance in various applications.

Figure 2. Schematic diagram of three main characterization techniques of Graphene-TiO_2 composite

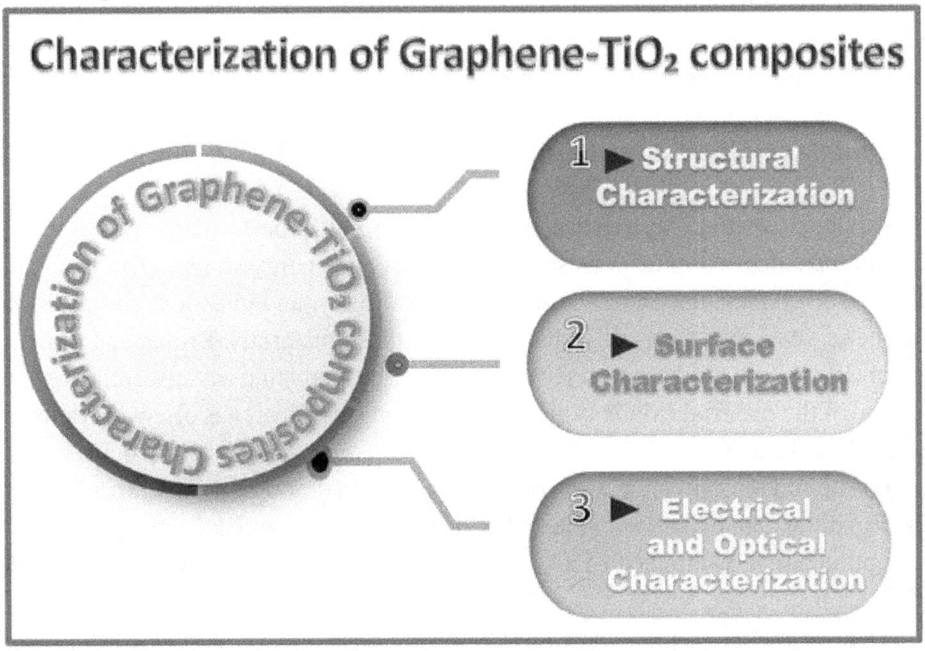

Complementing XRD, Raman spectroscopy serves as a powerful tool for probing the vibrational modes of atoms in the composite (Witte et al, 2024). Raman spectroscopy is particularly useful for detecting the presence of graphene and TiO_2 phases, as well as analyzing defects and disorder within the material. In the case of graphene-TiO_2 composites, characteristic peaks in the Raman spectra can confirm the presence of graphene through the G and 2D bands (Wang et al., 2022), while the Raman-active modes of TiO_2 provide information on its phase and crystallinity. Any shifts in the positions of these peaks may indicate interactions between graphene and TiO_2, such as chemical bonding or charge transfer, which can affect the composite's overall properties. Raman spectroscopy, therefore, offers a deeper understanding of the material's microstructural integrity and interactions at the atomic level.

For visualizing the morphology and microstructure of graphene-TiO_2 composites, transmission electron microscopy (TEM) and scanning electron microscopy (SEM) are invaluable tools (Kaveh et al., 2024; Song et al., 2024). TEM offers high-resolution images at the atomic scale, allowing researchers to observe the distribution, size, and orientation of graphene layers and TiO_2 nanoparticles within the composite. TEM is particularly useful for revealing the intimate contact between graphene and TiO_2, which is crucial for optimizing charge transfer processes in photocatalytic and electronic applications (Yang et al., 2024). SEM provides detailed surface topography and particle size distribution, helping researchers understand how the morphology of the composite impacts its performance. The combination of TEM and SEM offers a comprehensive view of the structural arrangement and surface characteristics of graphene-TiO_2 composites.

3.2. Surface Characterization

Surface properties play a critical role in determining the performance of graphene-TiO_2 composites in applications such as catalysis and energy storage (Kumar et al., 2024). Techniques that analyze surface area, porosity, and chemical composition are therefore essential for fully characterizing these materials. Brunauer-Emmett-Teller (BET) analysis determines the surface area by analyzing gas adsorption on the surface of the material (Mane et al., 2024). A high surface area is often desirable in catalysis and energy storage applications, as it increases the interaction between the composite and reactant molecules, thereby improving the composite's performance. By providing data on surface area and pore size distribution, BET analysis enables the optimization of graphene-TiO_2 composites for enhanced catalytic activity and adsorption capacity.

Another crucial surface characterization technique is X-ray photoelectron spectroscopy (XPS), which provides detailed information about the elemental composition and chemical states of the elements present on the surface of the composite (Simon et al., 2022). XPS is highly sensitive to surface chemistry and is used to determine the oxidation states of titanium and carbon, as well as the nature of any functional groups or chemical bonds between graphene and TiO_2. Understanding these interactions is key to optimizing the material for specific applications, such as improving photocatalytic efficiency or enhancing the composite's stability under harsh conditions. XPS can reveal the presence of defects, oxygen vacancies, or chemical modifications, all of which can significantly impact the material's performance.

Fourier-transform infrared (FTIR) spectroscopy is another valuable tool for surface characterization, providing insights into the chemical bonds and functional groups within the composite (Shui et al., 2024). FTIR spectra can identify specific functional groups, such as hydroxyl, carbonyl, or carboxyl groups, which can influence the composite's interaction with other materials and affect its stability and performance. For example, in graphene-TiO_2 composites, functional groups on the graphene surface can enhance its dispersion in the TiO_2 matrix and improve its compatibility with other components in composite materials. FTIR analysis is also useful for confirming the presence of chemical modifications or surface treatments that have been applied to tailor the properties of the composite for specific applications.

3.3. Electrical and Optical Characterization

Graphene-TiO_2 composites have garnered significant attention in recent years due to their exceptional electrical and optical properties, which make them promising candidates for a wide range of applications, including electronics, photodetectors, and solar cells. The unique synergy between graphene and TiO_2 enables these composites to exhibit enhanced electrical conductivity, charge carrier mobility, and optical characteristics, which are critical for improving the efficiency of various devices (Biswas et al., 2022; Pandey et al., 2024).One of the most crucial aspects of evaluating the electrical performance of graphene-TiO_2 composites is through conductivity measurements. Graphene is known for its exceptional electrical conductivity, stemming from its high carrier mobility and low resistivity (Janavika et al., 2023). When integrated into TiO_2, graphene can significantly boost the overall conductivity of the composite. This enhancement is due to the formation of conductive pathways within the composite, where graphene acts as a bridge for efficient charge transport(Ma et al., 2024; Ruan et al., 2024). Consequently, the incorporation of graphene into TiO_2 leads to an increase in charge carrier mobility and a reduction in recombination rates. These factors are pivotal in enhancing the performance of devices like solar cells and electronic components, where efficient charge transport and minimal energy loss are desired (Znidi et al., 2024).Moreover, the improved electrical properties of graphene-TiO_2 composites extend their utility in energy storage applications, such as supercapacitors and batteries. The high conductivity of graphene facilitates rapid electron transfer, which is essential for achieving high power densities and fast charge-discharge cycles in these devices (Ates et al., 2024). Therefore, by conducting detailed conductivity measurements, researchers can directly assess the suitability of graphene-TiO_2 composites for various electronic and energy storage applications, paving the way for the development of more efficient and high-performance devices.

In addition to electrical properties, the optical characteristics of graphene-TiO_2 composites play a crucial role in their functionality, particularly in light-driven processes (Migdadi et al., 2024). UV-visible (UV-Vis) method involves measuring the absorption and transmission of light by the material, thereby providing valuable information about its bandgap and electronic transitions. The incorporation of graphene into TiO_2 can significantly modify its optical properties (Kaur et al., 2024). Notably, it can extend the absorption range of TiO_2 into the visible spectrum, which is vital for applications like photocatalysis and photovoltaics. This extension into the visible range is attributed to the interaction between graphene and TiO_2, which alters the electronic structure of the composite, leading to enhanced light absorption and improved photocatalytic efficiency (Vembuli et al., 2024).UV-vis spectroscopy is instrumental in analyzing these modifications, enabling researchers to fine-tune the optical properties of the composite for specific applications. For instance, in photovoltaic devices, optimizing light absorption is crucial for maximizing the conversion of sunlight into electrical energy. By analyzing the UV-vis spectra, researchers can identify the optimal composition and structure of the graphene-TiO_2 composite (Basyach et al., 2022) to achieve the desired optical performance. Similarly, in photocatalytic applications, enhanced light absorption contributes to the effective degradation of pollutants, making graphene-TiO_2 composites highly efficient for environmental remediation.

4. PROPERTIES OF GRAPHENE-TIO_2 COMPOSITES

Graphene-TiO_2 composites exhibit a range of properties (Figure 3) that make them highly suitable for various advanced applications, particularly due to the unique combination of their individual components. The electrical conductivity of these composites is significantly enhanced compared to pure TiO_2, owing to the excellent conductive nature of graphene. Graphene, a two-dimensional sheet of carbon atoms arranged in a honeycomb lattice, is known for its exceptional electrical conductivity, which surpasses that of many other materials. When combined with TiO_2, graphene acts as a conductive network, facilitating efficient electron transport (Pervaiz et al., 2024). This enhancement in electrical conductivity is critical for applications in electronics and optoelectronics, where fast charge transfer and low resistance are essential (Krishnapriya et al., 2024). In photocatalytic applications, the improved conductivity reduces the recombination rate of electron-hole pairs generated in TiO_2 upon light irradiation, thereby increasing the photocatalytic efficiency (Pei et al., 2024). The synergistic effect of graphene's conductivity and TiO_2's photocatalytic properties creates a composite material that is highly effective for solar cells, photodetectors, and other electronic devices.The mechanical properties of graphene-TiO_2 composites also show significant improvement due to the incorporation of graphene. Graphene is renowned for its extraordinary mechanical strength, being one of the strongest materials known, with a tensile strength approximately 200 times greater than steel (Debnath et al., 2024). When graphene is integrated into TiO_2 matrices, it enhances the overall mechanical robustness of the composite (Acikgoz et al., 2024). This enhancement is particularly beneficial for applications requiring high durability and mechanical stability, such as in structural materials and coatings. The flexibility and strength of graphene help in distributing stress more evenly across the composite, preventing the propagation of cracks and improving the fracture toughness. Additionally, the mechanical reinforcement provided by graphene can lead to better wear resistance and longevity of the composite material in various applications (Maurya et al., 2024). The improved mechanical properties ensure that the composite can withstand harsh operational conditions without significant degradation, making it ideal for use in demanding environments.

Figure 3. Schematic diagram of various properties of graphene-TiO_2 Composites

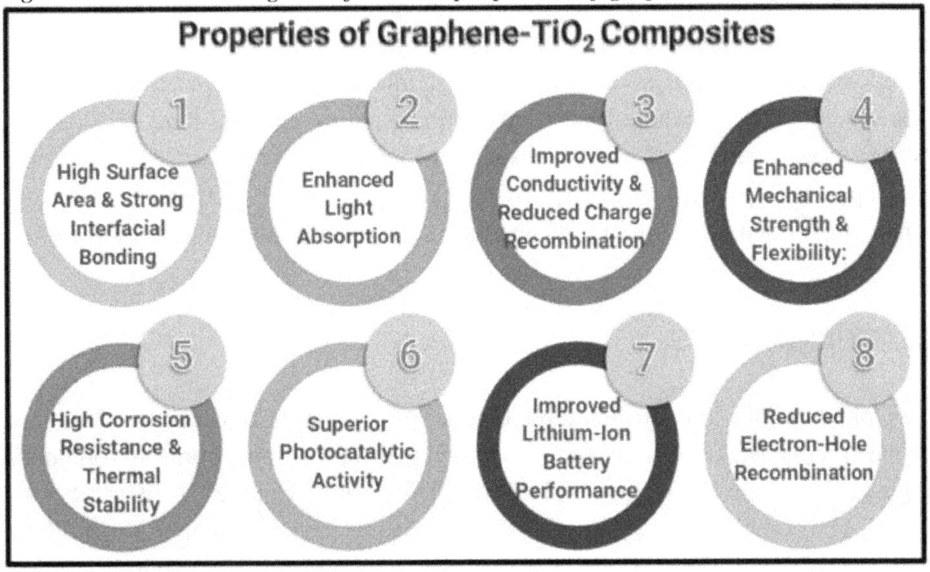

Optical properties of graphene-TiO_2 composites are another area where the combination of these materials yields significant advantages (Thakur et al., 2024). TiO_2 is well-known for its photocatalytic activity and its ability to absorb ultraviolet (UV) light. However, its efficiency is often limited to the UV region, which constitutes a small fraction of the solar spectrum. The incorporation of graphene can extend the light absorption range of TiO_2 into the visible spectrum, thereby enhancing its photocatalytic performance under natural sunlight. This is due to the ability of graphene to serve as a photosensitizer and facilitate electron transfer processes. UV-vis spectroscopy studies have shown that graphene-TiO_2 composites exhibit enhanced optical absorption, which is crucial for applications in solar energy conversion and environmental remediation. The interaction between graphene and TiO_2 not only broadens the absorption spectrum but also improves the separation and migration of photogenerated charge carriers, leading to higher photocatalytic efficiency. This makes the composite highly effective for degrading organic pollutants, producing hydrogen via water splitting, and other photochemical applications.

Chemical stability is another critical property of graphene-TiO_2 composites that makes them suitable for a wide range of applications (Acikgoz et al., 2024). Both graphene and TiO_2 are chemically stable materials, and their combination results in a composite that inherits this stability, often enhancing it further. This chemical stability is particularly important for applications in harsh chemical environments, such as in catalysis, environmental cleanup, and energy storage. The presence of graphene can protect TiO_2 from photocorrosion, a common issue that limits the long-term stability and efficiency of photocatalysts (Yang et al., 2024). Graphene's inert nature and strong covalent bonding can act as a protective layer, preventing the leaching and degradation of TiO_2 under reactive conditions. Additionally, the composite's stability is beneficial for electrochemical applications, such as in batteries and supercapacitors, where long-term cycling stability and resistance to chemical degradation are crucial for performance and longevity. The chemical robustness of graphene-TiO_2 composites ensures their reliability and effectiveness in various applications, making them highly desirable for industrial and environmental uses.

5. APPLICATIONS

Graphene-TiO_2 composites have emerged as versatile materials with unique properties that lend themselves to a broad range of applications, from environmental remediation to energy storage and advanced materials engineering (Kumar et al., 2024). The combination of graphene's exceptional electrical conductivity, mechanical strength, and high surface area with TiO_2's photocatalytic activity, chemical stability, and biocompatibility creates composites with enhanced performance and new functionalities. These composites are increasingly being utilized in various domains, including photocatalysis, energy storage and conversion, sensors, and material reinforcement, among others. Below, the applications of graphene-TiO_2 composites are explored in greater detail.

5.1. Photocatalysis for Environmental Remediation

One of the most promising and widely researched applications of graphene-TiO_2 composites is in photocatalysis, particularly for water and air purification (Wang et al., 2022). TiO_2 is a well-established photocatalyst that generates reactive oxygen species (ROS) under ultraviolet (UV) light, which can degrade organic pollutants and disinfect microorganisms (Chauke et al., 2024). However, TiO_2 alone is limited by its wide bandgap, which restricts its activity to UV light, accounting for only about 5% of the solar spectrum. By integrating graphene with TiO_2, researchers have successfully enhanced the photocatalytic efficiency of TiO_2, extending its light absorption range into the visible spectrum. This is due to graphene's ability to act as an electron acceptor, facilitating charge separation and reducing the recombination of electron-hole pairs generated during the photocatalytic process.

Figure 4. Schematic representation for organic pollutants photodegradation by Graphene-TiO2 nanocomposite.

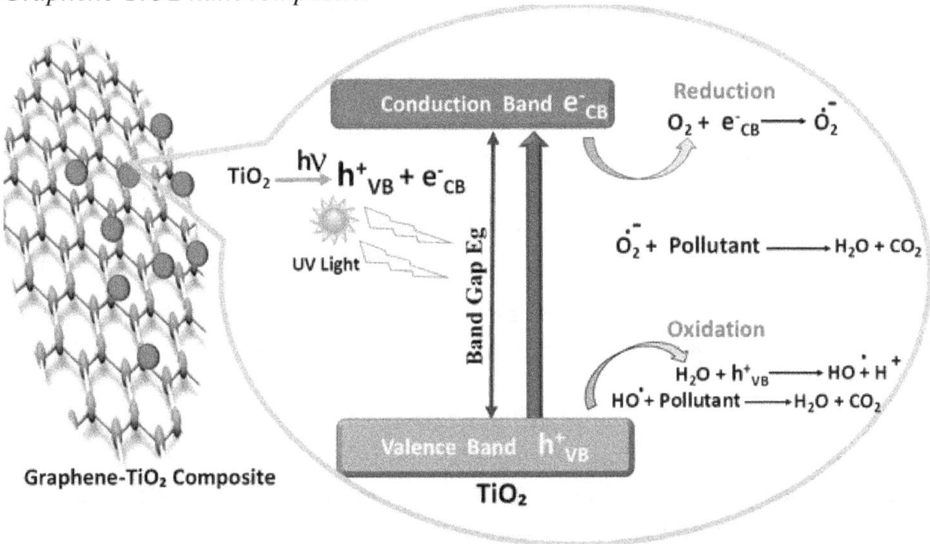

In water purification, graphene-TiO_2 composites have demonstrated high efficiency in the degradation of organic pollutants (Figure 4), such as dyes, pharmaceuticals, and pesticides, which are common contaminants in wastewater (Wang et al., 2022). The improved photocatalytic performance results in faster and more complete degradation of these harmful compounds, making the composites ideal for wastewater treatment plants and drinking water purification systems. Additionally, the composites exhibit excellent antibacterial properties, effectively inactivating pathogens, including bacteria and viruses, which are critical for ensuring water safety.Similarly, in air purification, graphene-TiO_2 composites are employed to remove volatile organic compounds (VOCs) and other harmful pollutants from indoor and outdoor air. VOCs, such as formaldehyde and benzene, are commonly found in household products and industrial emissions and pose significant health risks. The photocatalytic action of the composites breaks down these VOCs into harmless byproducts, improving indoor air quality and reducing the risk of respiratory illnesses. The enhanced photocatalytic activity of graphene-TiO_2 composites ensures the rapid and efficient removal of air pollutants, making them highly suitable for air purification devices, ventilation systems, and pollution control technologies.

5.2. Energy Storage and Conversion Technologies

Graphene-TiO_2 composites have also gained significant attention in the field of energy storage and conversion(Kumar et al., 2024), particularly in applications such as lithium-ion batteries, supercapacitors, and solar cells. In energy storage devices, the composites are utilized as electrode materials due to their excellent electrical conductivity, high surface area, and mechanical stability. Graphene's conductive network facilitates efficient charge transport, while TiO_2 provides structural stability and electrochemical activity. This synergy results in improved performance, including higher capacity, faster charge-discharge rates, and better cycling stability, making graphene-TiO_2 composites ideal for next-generation lithium-ion batteries and supercapacitors. In lithium-ion batteries, for example, graphene-TiO_2 composites are used to enhance the performance of both the anode and cathode materials. The incorporation of graphene helps mitigate the issue of poor electrical conductivity in TiO_2-based electrodes, resulting in faster electron transport and higher charge storage capacity. Additionally, graphene's flexibility and strength improve the mechanical stability of the electrodes, reducing the risk of degradation during repeated charge-discharge cycles (Sultanov et al., 2024). This leads to longer-lasting batteries with higher energy density, which are crucial for applications in electric vehicles, portable electronics, and renewable energy storage systems.

Supercapacitors, which require materials with high power density and rapid charge-discharge capabilities, also benefit from the use of graphene-TiO_2 composites. The high surface area of graphene enables the storage of a large amount of charge, while TiO_2 provides additional pseudocapacitance, resulting in enhanced overall performance. The composites' ability to deliver high energy and power densities makes them suitable for use in applications requiring rapid energy delivery, such as in regenerative braking systems, backup power supplies, and wearable electronics. In solar energy conversion, graphene-TiO_2 composites have shown promise in improving the efficiency of dye-sensitized solar cells (DSSCs) and perovskite solar cells. In DSSCs, TiO_2 is commonly used as a photoanode material due to its high light absorption and electron transport properties. However, the incorporation of graphene further enhances the conductivity and charge separation efficiency of the TiO_2-based photoanode, resulting in higher power conversion efficiency. Similarly, in perovskite solar cells, graphene-TiO_2 composites improve the stability and performance of the devices, making them more efficient and durable for large-scale solar energy harvesting.

5.3. Advanced Sensing Applications

Graphene-TiO_2 composites have gained significant attention in advanced sensing applications, particularly in the development of gas sensors and biosensors. Their high sensitivity, selectivity, and versatility make them ideal candidates for detecting a range of analytes with precision (Saxena et al., 2023). In gas sensing, graphene-TiO_2 composites demonstrate exceptional capabilities in detecting gases such as nitrogen dioxide (NO_2), ammonia (NH_3), and volatile organic compounds (VOCs). Graphene's large surface area offers abundant active sites for gas adsorption, enhancing the sensor's overall sensitivity. Concurrently, the semiconductor properties of TiO_2 contribute to the sensor's responsiveness by altering its electrical properties in the presence of target gases. This synergistic interaction enables graphene-TiO_2 based sensors to detect even low concentrations of gases, making them highly effective for environmental monitoring, industrial safety, and air quality control. For example, in environmental monitoring, these sensors can detect NO_2 emissions from vehicles and industrial activities, which are known to contribute significantly to air pollution and pose health risks. By providing real-time data on NO_2 levels, these sensors facilitate early intervention and help mitigate exposure to harmful pollutants. In industrial settings, graphene-TiO_2 sensors play a crucial role in detecting hazardous gases such as ammonia, thereby ensuring worker safety and compliance with environmental regulations. Their rapid response time and high accuracy make them indispensable tools for maintaining safe and healthy environments.

In biosensing applications, graphene-TiO_2 composites are utilized to detect various biomolecules, including glucose, proteins, and DNA, which are essential for medical diagnostics and disease monitoring. The unique properties of graphene, such as its high surface area and ability to be functionalized, allow for the immobilization of specific biomolecules on its surface. This enables the development of highly sensitive and specific biosensors (Saxena et al. 2024). TiO_2 further enhances these biosensors by providing biocompatibility and stability, which are crucial for reliable and consistent performance. The integration of graphene and TiO_2 in biosensors leads to the creation of devices capable of detecting a wide range of biological analytes with high precision, making them suitable for applications in medical diagnostics, point-of-care testing, and personalized medicine. For instance, graphene-TiO_2 based glucose sensors are being developed for continuous glucose monitoring in diabetic patients. These sensors offer real-time data, allowing for more effective management of blood sugar levels and reducing the risk of complications associated with diabetes. Additionally, the sensitivity and specificity of these biosensors extend to the detection of other biomolecules, enabling advancements in areas such as cancer diagnostics, infectious disease monitoring, and biomarker detection. The versatility and effectiveness of graphene-TiO_2 composites in sensing applications underscore their potential to revolutionize the field of sensor technology, offering innovative solutions for healthcare, environmental monitoring, and industrial safety.

5.4. Reinforcement of Composite Materials

Beyond environmental and energy applications, graphene-TiO_2 composites are also used to enhance the properties of polymer and ceramic composite materials (Dutt et al., 2020). In polymer composites, the incorporation of graphene-TiO_2 can significantly improve mechanical properties, such as tensile strength, Young's modulus, and fracture toughness. The strong interfacial interaction between the polymer matrix and the composite material leads to better load transfer and stress distribution, resulting in more durable and robust materials. These enhanced mechanical properties are particularly valuable in industries where high strength and lightweight materials are essential, such as aerospace, automotive, and construction.

For example, in the aerospace industry, graphene-TiO_2 reinforced polymer composites can be used to manufacture lightweight components with superior mechanical strength, reducing the overall weight of aircraft and improving fuel efficiency. Similarly, in the automotive industry, these composites can be used to produce lighter and stronger vehicle parts, contributing to increased fuel efficiency and safety (Sundar et al., 2024). In ceramic composites, graphene-TiO_2 enhances thermal conductivity and durability. The excellent thermal properties of graphene, combined with the stability of TiO_2, result in materials that can withstand high temperatures and thermal cycling. This makes graphene-TiO_2 ceramic composites ideal for use in thermal management systems, heat exchangers, and other high-temperature applications, where efficient heat dissipation and resistance to thermal degradation are critical.

6. CHALLENGES

6.1 Challenges in Previous Studies

Despite significant progress in graphene-TiO_2 composite nanomaterials, several challenges remain in their synthesis and application. One major issue is aggregation, where graphene sheets tend to restack, reducing available surface area and limiting charge transfer efficiency. Surface functionalization and optimized dispersion techniques are needed to address this. Chemical instability is another concern, particularly under harsh operational conditions, leading to material degradation and reduced photocatalytic activity. Stabilization strategies, such as doping or hybridization with other nanomaterials, have been explored to improve durability. Additionally, morphology control remains challenging, as precise regulation of the TiO_2 nanoparticle size and uniform distribution on graphene sheets significantly affects performance. Variations in synthesis methods, such as sol-gel, hydrothermal, or chemical vapor deposition, impact composite structure and reproducibility. Addressing these challenges requires innovative approaches in material engineering, improved synthesis techniques, and a deeper understanding of structure-property relationships to enhance practical applications in energy storage and conversion.

6.2 Current Challenge

Graphene-TiO_2 composites hold immense potential, but several challenges hinder their large-scale application. Scalability remains a major issue, as synthesis techniques like chemical vapor deposition (CVD) and liquid-phase exfoliation, though effective in the lab, struggle to maintain uniformity and performance in industrial-scale production. Additionally, high production costs pose a barrier, as the fabrication of high-quality graphene and its integration with TiO_2 significantly increase expenses. Alternative methods, such as using reduced graphene oxide (rGO), offer a more cost-effective solution while preserving key properties. Another critical concern is long-term stability, as TiO_2 undergoes photocorrosion and graphene degrades over time, affecting overall performance. Addressing this requires improved surface modifications and protective coatings. Lastly, environmental concerns arise due to the use of hazardous chemicals in synthesis and potential ecological risks during disposal. The development of greener synthesis methods and thorough toxicity assessments is essential for ensuring the safe and sustainable application of these composites.

7. FUTURE DIRECTION

To advance the practical application of graphene-TiO_2 composites, researchers are developing scalable and cost-effective synthesis techniques, including green chemistry approaches, to enhance commercial viability. Improving long-term stability is another key focus, with efforts directed toward surface modifications, protective coatings, and hybrid composite structures to prevent degradation. The adoption of sustainable and eco-friendly synthesis methods, such as using environmentally friendly precursors and energy-efficient processes, aims to minimize ecological impact. Additionally, material innovation is progressing with the development of doped graphene-TiO_2 composites to enhance functional properties. The integration of artificial intelligence (AI) and machine learning in material design is further accelerating composite optimization, enabling precise property control. These advancements will help graphene-TiO_2 composites achieve widespread adoption in energy storage and conversion applications.

8. CONCLUSION

Graphene-TiO_2 composites represent a promising class of materials with unique properties that combine the exceptional electrical conductivity of graphene with the photocatalytic and structural benefits of TiO_2. This synergy has led to significant advancements in various applications, including photocatalysis for water and air purification, energy storage and conversion in batteries, supercapacitors, and solar cells, as well as in sensors and advanced composite materials. Key findings highlight their enhanced photocatalytic efficiency, superior electrical and mechanical properties, and robust chemical stability, making them suitable for a broad range of industrial and environmental applications. Looking ahead, graphene-TiO_2 composites hold great promise for future applications, but their success hinges on overcoming key challenges. Among these, issues related to large-scale production and high manufacturing costs continue to pose significant obstacles to their broad commercial implementation. Innovations in synthesis techniques, such as green and cost-effective methods, will be essential. Long-term stability and environmental impact also require continued research to ensure these materials can perform reliably over extended periods and are safe for widespread use. Potential areas for further research include the development of new hybrid composites that incorporate additional functionalities, such as enhanced thermal conductivity or specific catalytic properties tailored for particular reactions. Advances in material design through the use of dopants or novel composite structures can further enhance the performance of graphene-TiO_2 composites. Additionally, integrating artificial intelligence and machine learning into the research process could expedite the discovery and optimization of these materials. Technological advancements in the synthesis and application of graphene-TiO_2 composites hold the potential to revolutionize fields ranging from renewable energy to environmental remediation. By overcoming current challenges and continuing to innovate, researchers can unlock the full potential of these versatile materials, paving the way for a more sustainable and technologically advanced future.

ACKNOWLEDGEMENT

The author (HSB) gratefully acknowledges the financial support from the UGC Minor Project (No. F.PSW-140/15-16). Sincere appreciation is extended to Dr. Indranil Kar, Principal of Surendranath College, for his invaluable guidance and cooperation. The author also expresses gratitude to Dr. Suchandra Chatterjee, Dr. Lalita Das, Dr. Harisadhan Ghosh, Dr. Apurba Biswas, Dr. Noor Salam, and Mr. Sourav Mishra for their insightful discussions and valuable contributions.

Conflict of interest: There is no conflict of interest.

REFERENCES

Acikgoz, A., Demircan, G., Aktas, B., Aktas, H., Yalcin, S., Kisa, M., & Das, R. (2024). Synthesis, Mechanical Characterization, and Biocompatibility Evaluation of Graphene-TiO2 Reinforced Alumina for Biomedical Applications. Ceramics International.

Ahmad, F., Ghazal, H., Rasheed, F., Shahid, M., Vasantham, S. K., Rafiq, W., Abbas, Z., Sarwar, S., Ain, Q. U., Waqar, A., Awais, M., Asim, M., & Atiq, S. (2024). Graphene and its derivatives in medical applications: A comprehensive review. *Synthetic Metals*, *304*, 117594. DOI: 10.1016/j.synthmet.2024.117594

Ahmad, T., & Ansari, M. Z. (2024). Optical transparency combined with electrical conductivity: Challenges and prospects. Metal Oxides for Next-generation Optoelectronic Photonic and Photovoltaic Applications, 3-43.

Ahmed, T. Y., Aziz, S. B., & Dannoun, E. M. (2024). Role of outer shell electron-nuclear distant of transition metal atoms (TMA) on band gap reduction and optical properties of TiO2 semiconductor. *Results in Engineering*, *23*, 102479. DOI: 10.1016/j.rineng.2024.102479

Ates, M., Yoruk, O., Bayrak, Y., Turkyilmaz, M., & Karabulut, H. (2024). An active carbon/graphene oxide/acid nanocomposite electropolymerized via a thiophene monomer: Supercapacitor device performances in a 2032 coin cell. *Ionics*, *30*(9), 1–20. DOI: 10.1007/s11581-024-05684-1

Basyach, P., Sonowal, K., Chetia, P., Kalita, L., Borthakur, S., & Saikia, L. (2022). Photocatalytic Conversion of CO 2 into Value Added and Renewable Fuels over Heterogeneous Nanocatalysts: A Green and Environmental Benign Approach. *Heterogeneous Nanocatalysis for Energy and Environmental Sustainability*, *2*, 221–274. DOI: 10.1002/9781119772057.ch23

Biswas, H. S., Bala, S., Kundu, A. K., Saha, I., Poddar, S., Sarkar, S., & Mandal, P. (2022). Tuned synthesis and designed characterization of graphene oxide thin film. *Inorganic Chemistry Communications*, *139*, 109356. DOI: 10.1016/j.inoche.2022.109356

Biswas, H. S., Datta, J., Mandal, P., Poddar, S., Kundu, A. K., & Saha, I. (2022). Optimized study of the annealing effect on the electrical and structural properties of HDLC thin-films. *RSC Advances*, *12*(46), 29805–29812. DOI: 10.1039/D2RA06255K PMID: 36321089

Chauke, N. M., Mohlala, R. L., Ngqoloda, S., & Raphulu, M. C. (2024). Harnessing visible light: Enhancing TiO2 photocatalysis with photosensitizers for sustainable and efficient environmental solutions. *Frontiers in Chemical Engineering*, *6*, 1356021. DOI: 10.3389/fceng.2024.1356021

Dagar, T., Sarkar, S., Choudhury, B. P., & De, S. (2024). Passageway into a Clean World with the Magical Nanomaterial TiO2. Nanotechnology for Environmental Management, 285.

Davu, S. R., Tejavathu, R., & Tummala, S. K. (2024). Characterization of graphene–TiO2-deposited semi-organic solar cell. Proceedings of the Institution of Mechanical Engineers, Part E: Journal of Process Mechanical Engineering, 09544089241258839. DOI: 10.1177/09544089241258839

Debnath, M., & Adhikary, A. (2024). *Use of graphene in rubber nanocomposite, its processability, and commercial advantages. Rubber Products: Technology and Cost Optimisation*. Walter de Gruyter GmbH & Co. KG.

Dutt, M. A., Hanif, M. A., Nadeem, F., & Bhatti, H. N. (2020). A review of advances in engineered composite materials popular for wastewater treatment. *Journal of Environmental Chemical Engineering*, *8*(5), 104073. DOI: 10.1016/j.jece.2020.104073

Egerton, T. A. (2014). UV-absorption—The primary process in photocatalysis and some practical consequences. *Molecules (Basel, Switzerland)*, *19*(11), 18192–18214. DOI: 10.3390/molecules191118192 PMID: 25383755

Escorcia-Díaz, D., García-Mora, S., Rendón-Castrillón, L., Ramírez-Carmona, M., & Ocampo-López, C. (2023). Advancements in nanoparticle deposition techniques for diverse substrates: A review. *Nanomaterials (Basel, Switzerland)*, *13*(18), 2586. DOI: 10.3390/nano13182586 PMID: 37764615

Farooq, N., Kallem, P., ur Rehman, Z., Imran Khan, M., Kumar Gupta, R., Tahseen, T., Mushtaq, Z., Ejaz, N., & Shanableh, A. (2024). Recent trends of titania (TiO2) based materials: A review on synthetic approaches and potential applications. *Journal of King Saud University. Science*, *36*(6), 103210. DOI: 10.1016/j.jksus.2024.103210

Jain, P., Rajput, R. S., Kumar, S., Sharma, A., Jain, A., Bora, B. J., Sharma, P., Kumar, R., Shahid, M., Rajhi, A. A., Alsubih, M., Shah, M. A., & Bhowmik, A. (2024). Recent Advances in Graphene-Enabled Materials for Photovoltaic Applications: A Comprehensive Review. *ACS Omega*, *9*(11), 12403–12425. DOI: 10.1021/acsomega.3c07994 PMID: 38524428

Janavika, K. M., & Thangaraj, R. P. (2023). Graphene and its application: A review. *Materials Today: Proceedings*. Advance online publication. DOI: 10.1016/j.matpr.2023.05.446

Kaur, D., & Pal, B. (2024). Highly enhanced photocatalytic hydrogen production via GO-modified Cu-TiO2 system for dehydrogenation of alcohols under UV & sunlight irradiation. *Solar Energy*, *282*, 112905. DOI: 10.1016/j.solener.2024.112905

Kaveh, R., Alijani, H., Falletta, E., Bianchi, C. L., Mokhtarifar, M., & Boffito, D. C. (2024). Advancements in superhydrophilic titanium dioxide/graphene oxide composites coatings for self-cleaning applications on glass substrates: A comprehensive review. *Progress in Organic Coatings*, *190*, 108347. DOI: 10.1016/j.porgcoat.2024.108347

Kumar, A., & Rai, R. S. (2024). Electrical, Mechanical, and Thermal Properties of Two-Dimensional Nanomaterials. Two-Dimensional Nanomaterials-Based Polymer Nanocomposites: Processing, Properties and Applications, 195-230.

Kumar, B. A., Ran, F., Maram, P. S., & Sangaraju, S. (2024). Progressive horizons of energy generation and storage: Nook and cranny of photo-supercapacitors. *Journal of Energy Storage*, *97*, 112876. DOI: 10.1016/j.est.2024.112876

Li, W., Liang, F., Sun, X., Zheng, K., Liu, R., Yuan, H., Cheng, S., Wang, J., Cheng, Y., Huang, K., Wang, K., Yang, Y., Yang, F., Tu, C., Mao, X., Yin, W., Cai, A., Wang, X., Qi, Y., & Liu, Z. (2024). Graphene-skinned alumina fiber fabricated through metalloid-catalytic graphene CVD growth on nonmetallic substrate and its mass production. *Nature Communications*, *15*(1), 6825. DOI: 10.1038/s41467-024-51118-x PMID: 39122739

Ma, Z., Tao, E., Yang, S., Liu, L., Li, Y., & Qian, J. (2023). Improved charge transfer by Ca2+ modified TiO2/graphene conductive material for enhancing conductivity. *Surfaces and Interfaces*, *38*, 102779. DOI: 10.1016/j.surfin.2023.102779

Mane, P. V., Rego, R. M., Yap, P. L., Losic, D., & Kurkuri, M. D. (2024). Unveiling cutting-edge advances in high surface area porous materials for the efficient removal of toxic metal ions from water. *Progress in Materials Science*, *146*, 101314. DOI: 10.1016/j.pmatsci.2024.101314

Maurya, S. S., Grain, A. K., Pandey, K. K., & Keshri, A. K. (2024). Graphene Nanoplatelets for Reinforced Cermet Coatings with Improved Mechanical and Tribological Properties. *ACS Applied Nano Materials*, *7*(16), 18811–18823. DOI: 10.1021/acsanm.4c02324

Migdadi, A. B., Al-Bataineh, Q. M., Ahmad, A. A., Al-Khateeb, H. M., & Telfah, A. (2024). Titanium dioxide/reduced graphene oxide nanocomposites as effective photocatalytic for hazardous 4-nitrophenol. *Journal of Alloys and Compounds*, *971*, 172794. DOI: 10.1016/j.jallcom.2023.172794

Pandey, M., Deshmukh, K., Raman, A., Asok, A., Appukuttan, S., & Suman, G. R. (2024). Prospects of MXene and graphene for energy storage and conversion. *Renewable & Sustainable Energy Reviews*, *189*, 114030. DOI: 10.1016/j.rser.2023.114030

Pei, C., Zhu, J. H., & Lv, L. (2024). Enhanced visible-light induced photocatalytic activity in core-shell structured graphene/titanium dioxide nanofiber mats. *Journal of Environmental Chemical Engineering*, *12*(4), 113112. DOI: 10.1016/j.jece.2024.113112

Pervaiz, H., Shahzad, N., & Jamil, Q. (2024). The impact of a TiO 2/r-GO composite material on the performance of electron transport electrodes of dye sensitized solar cells. *RSC Advances*, *14*(23), 15907–15914. DOI: 10.1039/D4RA00829D PMID: 38756851

Ruan, X., Li, S., Huang, C., Zheng, W., Cui, X., & Ravi, S. K. (2024). Catalyzing Artificial Photosynthesis with TiO2 Heterostructures and Hybrids: Emerging Trends in a Classical yet Contemporary Photocatalyst. *Advanced Materials*, *36*(17), 2305285. DOI: 10.1002/adma.202305285 PMID: 37818725

Sahu, S., & Tiwari, S. (2024). Carbon Allotropes: Fundamental, Synthesis, Characterization, and Properties Functionalization. In Carbon Allotropes (pp. 41-70). CRC Press.

Santra, S., Bose, A., Mitra, K., & Adalder, A. (2024). Exploring two decades of graphene: The jack of all trades. *Applied Materials Today*, *36*, 102066. DOI: 10.1016/j.apmt.2024.102066

Saxena, J., Singh, A., & Jyoti, A. (Eds.). (2023). *Nanobiotechnology: principles and applications*. Bentham Science Publishers. DOI: 10.2174/97898151235551230101

Saxena, S., Sen, P., Soleymani, L., & Hoare, T. (2024). Anti-Fouling Polymer or Peptide-Modified Electrochemical Biosensors for Improved Biosensing in Complex Media. *Advanced Sensor Research, 3*(5), 2300170. DOI: 10.1002/adsr.202300170

Seiß, V., Thiel, S., & Eichelbaum, M. (2022). Preparation and real world applications of titania composite materials for photocatalytic surface, air, and water purification: State of the art. *Inorganics, 10*(9), 139. DOI: 10.3390/inorganics10090139

Shah, S. S., Niaz, F., Ehsan, M. A., Das, H. T., Younas, M., Khan, A. S., Rahman, H. U., Nayem, S. M. A., Oyama, M., & Aziz, M. A. (2024). Advanced strategies in electrode engineering and nanomaterial modifications for supercapacitor performance enhancement: A comprehensive review. *Journal of Energy Storage, 79*, 110152. DOI: 10.1016/j.est.2023.110152

Shui, Y. J., Yao, W. H., Lin, J. H., Zhang, Y., Yu, Y., Wu, C. S., Zhang, X., & Tsou, C. H. (2024). Enhancing Polyvinyl Alcohol Nanocomposites with Carboxy-Functionalized Graphene: An In-Depth Analysis of Mechanical, Barrier, Electrical, Antibacterial, and Chemical Properties. *Polymers, 16*(8), 1070. DOI: 10.3390/polym16081070 PMID: 38674991

Simon, P., Baldovino-Medrano, V. G., & Wojcieszak, R. (2022). X-ray photoelectron spectroscopy (XPS): principles and application for the analysis of photoactive materials. In *Springer Handbook of Inorganic Photochemistry* (pp. 249–271). Springer International Publishing. DOI: 10.1007/978-3-030-63713-2_10

Singh, S., Rathi, R., Senger, A., Rawat, V., Srivastava, C. M., & Pathak, S. R. (2024). Recent advancements in fabrication strategies for functionalized nanomaterial-based biosensors. In *Functionalized Nanomaterials for Biosensing and Bioelectronics Applications* (pp. 143–171). Woodhead Publishing. DOI: 10.1016/B978-0-12-823829-5.00011-7

Song, C., Xiao, L., Chen, Y., Yang, F., Meng, H., Zhang, W., Zhang, Y., & Wu, Y. (2024). TiO2-Based Catalysts with Various Structures for Photocatalytic Application: A Review. *Catalysts, 14*(6), 366. DOI: 10.3390/catal14060366

Sultanov, F., Tatykayev, B., Bakenov, Z., & Mentbayeva, A. (2024). The role of graphene aerogels in rechargeable batteries. *Advances in Colloid and Interface Science, 331*, 103249. DOI: 10.1016/j.cis.2024.103249 PMID: 39032342

Sundar, L. S., Mir, M. A., & Ashraf, M. W. (2024). Graphene and Graphene-Based Composite Nanomaterials for Rechargeable Lithium-Ion Batteries. Nanomaterials for Energy Applications, 186-225.

Thakur, D., Thakur, V., Singh, N., Kumar, M., & Thakur, M. S. (2024). Graphene-Based Efficient Photocatalytic Materials for Hydrogen Generation. In Towards Sustainable and Green Hydrogen Production by Photocatalysis: Insights into Design and Development of Efficient Materials (Volume 2) (pp. 465-495). American Chemical Society. DOI: 10.1021/bk-2024-1468.ch018

Thakur, N., Thakur, N., Kumar, A., Thakur, V. K., Kalia, S., Arya, V., Kumar, A., Kumar, S., & Kyzas, G. Z. (2024). A critical review on the recent trends of photocatalytic, antibacterial, antioxidant and nanohybrid applications of anatase and rutile TiO2 nanoparticles. *The Science of the Total Environment*, *914*, 169815. DOI: 10.1016/j.scitotenv.2023.169815 PMID: 38184262

Vairagade, S., Kumar, N., & Singh, R. P. (2024). Recent advancements and applications in thermally conductive polymer nanocomposites. Polymer-Plastics Technology and Materials, 1-50.

Vembuli, T., Thiripuranthagan, S., & Kumaravel, S. (2024). Enhanced removal of hazardous organic contaminants with advanced visible light-active F-doped TiO2/rGO/PVDF photocatalytic membranes. *Journal of Alloys and Compounds*, *1005*, 175997. DOI: 10.1016/j.jallcom.2024.175997

Wang, J., Tian, J., Wang, W., Zhao, Z., Li, L., & Zhang, Z. (2022). Enhanced photocatalytic activity of graphene/TiO2 nanotubes composites prepared by wet transfer method. *Fullerenes, Nanotubes, and Carbon Nanostructures*, *30*(4), 495–502. DOI: 10.1080/1536383X.2021.1960510

Witte, A., & Garg, N. (2024). Particle shape, crystallinity, and degree of polymerization of fly ash via combined SEM-EDS and Raman spectroscopy. *Cement and Concrete Research*, *184*, 107612. DOI: 10.1016/j.cemconres.2024.107612

Yang, M., Jiang, R., Zhu, J., Zhang, X., Li, G., Li, W., Ma, F., Jiang, X., & Li, H. (2024). Review on the Solar-Driven Photocathodic Protection of Metals in the Marine Environment. *Coatings*, *14*(3), 276. DOI: 10.3390/coatings14030276

Yang, R., Wang, J., Zhou, K., Cai, D., Yang, S., Li, N., ... & Tao, E. (2024). Preparation of Self-Assembled Graphene Materials with Assisted Multi-structured TiO2: Internal Three-phase Heterojunction Built by Doping with Al3+. Ceramics International.

Zheng, D., Chen, P., Liu, Y., Li, X., Liu, K., Yin, Z., Frisenda, R., Zhao, Q., & Wang, T. (2024). Research progress on two-dimensional indium selenide crystals and optoelectronic devices. *Journal of Materials Chemistry. A, Materials for Energy and Sustainability*, *12*(28), 16952–16986. DOI: 10.1039/D4TA01584C

Znidi, F., Morsy, M., & Uddin, M. N. (2024). Recent advances of graphene-based materials in planar perovskite solar cells. *Next Nanotechnology*, *5*, 100061. DOI: 10.1016/j.nxnano.2024.100061

Chapter 15
Renewable Energy Integration:
Storage Solutions for a Sustainable Future

Ranjit Singha
https://orcid.org/0000-0002-3541-8752
Christ University, India

R. Sreethi Rebeka
https://orcid.org/0009-0008-9011-7048
Mount Carmel College, India

Surjit Singha
https://orcid.org/0000-0002-5730-8677
Kristu Jayanti College, India

Babitha Nadampalli
Christ University, India

Shruti Jose
https://orcid.org/0000-0003-3139-2505
Christ University, India

Alphonsa Diana Haokip
https://orcid.org/0000-0003-2578-0114
Don Bosco College, Maram, India

ABSTRACT

This chapter describes how energy storage technologies provide critical services for the integration of renewable energy sources, as those sources are variable and intermittent (e.g. solar and wind). It reviews different types of storage solutions — electrochemical, mechanical and thermal — including their relative advantages and drawbacks, as well as their appropriateness for various applications. It highlights the importance of adopting different energy storage types that are tailored to different energy needs. Global Sustainability and Energy Storage: Discussing the broader implications of energy storage systems for global sustainability, such as reducing reliance on fossil fuels, lowering carbon emissions, and improving access to energy in underserved regions. Energy storage technologies are crucial to a reliable, sustainable, low-carbon energy future.

DOI: 10.4018/979-8-3693-9316-1.ch015

INTRODUCTION

Transitioning to renewable energy resources has been central to global efforts to minimize climate change and attain sustainable development goals. Countries worldwide are adopting renewable energy systems, including solar, wind, and hydropower, to boost greenhouse gas emissions and transition away from finite fossil fuel resources. Despite their tremendous manifest potential for generating clean and nearly abundant energy, renewables have an inherent intermittent and site-dependent nature that poses major hurdles to their seamless integration into existing energy grid infrastructures. The must in the renewable variation can imbalance the energy supply/demand, which hinders the reliability and stability of power systems (Fekete et al., 2023; Mlilo et al., 2021).

These challenges are addressed by energy storage solutions, which, in most cases, are the critical enablers for continuous grid stability and operation, optimized energy usage, and constant power supply. Storage systems can be used to store surplus energy produced during times of high energy generation and release that energy when demand surpasses supply, helping to create a balance between generation and consumption. Not only do these solutions enhance the reliability of renewable energy systems, but they also play an essential role in boosting the scalability of such systems, enabling them to be adopted widely in different geographical and economic contexts (Adeyinka et al., 2024; Hunt et al., 2022).

This chapter discusses the pivotal role of energy storage in voicing its importance in the integration of renewable energy, thus enabling the advancement of sustainable and resilient energy systems. This chapter will focus on some exciting developments in key energy storage technologies, such as batteries, supercapacitors, and hydrogen storage, as well as how these technologies can be deployed to solve the unique challenges of renewable energy. It will also present the theoretical and conceptual foundations driving advances in next-generation energy storage systems.

This chapter seeks to explore in detail how innovative storage solutions are revolutionizing the energy sector and what role they will play in shaping the future of sustainable energy. The article will finish by providing some feasible recommendations for future work and development that adapts the progression of technology to the global sustainability goals.

BACKGROUND

The quest for renewable energy , a sustainable replacement for fossil fuels, spans several centuries and has seen dramatic evolution throughout that period. It was the first tap of renewable energy, but it was narrowly focused on hydropower, which was used to mill grain and produce electricity in the late 1800s. While industrialism expanded, resulting in the increased dependence on non-renewable and discarded substitutes, renewable alternatives were greatly overlooked. However, the environmental impacts of fossil fuel consumption, including climate change and resource depletion, sparked renewed interest in sustainable energy sources in the mid-20th century (Jacobson et al., 2015).

The energy crises of the 1970s were a turning point, driving global investment in research and development in renewable energy. It began with the first generation of renewable energy systems, including solar panels, wind turbines, and geothermal systems. However, the variable nature of solar and wind energy soon revealed the fantasy that we would need efficient technologies capable of storing the energy to match supply and demand (Hunt et al., 2022; Jacobson et al., 2015).

Since the introduction of lead-acid batteries back in the ' 90s of the 19th century, energy storage technologies have come a long way. These initial batteries were revolutionary but were not efficient or built to store large amounts of energy. The development of lithium-ion batteries in the 1990s brought a transformation in energy storage technology with a higher energy density, longer cycle life, and better performance. Lithium-ion technology emerged as a pillar of this change, fuelling climate change mitigations as widely used in residential, commercial and grid-scale applications (Abraham & Jiang, 1996; Fekete et al., 2023).

However, outside lithium-ion, the 21st century has already seen fast-paced innovation in energy storage — from supercapacitors to flow batteries to hydrogen storage systems. These technologies have opened new avenues for integrating renewables, tackling challenges such as energy intermittency, grid stability, and long-duration storage. One such promising option which has become a ballpark system thanks to its certified potential in storing energy as a chemical entity and facilitating a wide range of utilization in transport and industrial sectors (Jacobson et al., 2015; Rampai et al., 2024) is hydrogen storage.

Today, ambitious sustainability aspirations and rapidly motivated technologies have turned renewables into a growing component of the global energy mix. The 2023 global electricity generation share, according to the International Energy Agency (IEA), indicates around 30% from renewables and exceptional growth of wind and solar energy. Denmark, Germany, and China are some of the early adopters of renewables and have successfully integrated large shares of the energy demand from renewables (Jiang & Raza, 2024; Oyedotun et al., 2022).

Moreover, these developments do not come without challenges. The intermittency of renewable energy generation has been a fundamental challenge, especially in areas where solar and wind are heavily relied upon. All of these concerns can be addressed through energy storage technologies, making them a crucial factor for adopting renewables into power grids, improving curtailment, and increasing energy access in remote regions (Lew et al., 2013; Wei et al., 2023).

The continued development of energy storage technologies will be crucial to meet the ever-evolving energy needs as the global energy transition speeds up. This chapter lays the groundwork for introducing the transformative by exploring the historical context, key milestones, and current trends surrounding energy storage and their potential to create a more sustainable energy future.

THEORETICAL FRAMEWORK

Energy storage combines the disciplines of thermodynamics, electrochemistry and materials science and consists of converting energy from one form to another, storing it and releasing it when energy is needed. The field of thermodynamics describes energy conversions and efficiencies, and electrochemistry offers a molecular and atomic view of a class of devices such as batteries and supercapacitors. Energy storage techniques can be divided into three theoretical categories: electrochemical, mechanical, and thermal. Electrochemical storage devices such as batteries and supercapacitors depend on redox reactions and ionic migration, with Li-ion-based batteries utilizing the reversible diffusion of Li-I from beds to anodes and vice-versa as governed by the Ernst equation and Faraday's laws. Mechanical storage devices, like pumped hydro systems and flywheels, use gravitational potential and rotational kinetic energy according to the law of energy conservation and hydraulic dynamics. Thermal storage is an advanced technology that can capture and store heat energy for later use and is based on established principles of heat transfer, as in the case of sensible heat storage that uses materials like molten salts and latent heat storage that uses phase-change materials for heat storage. Together, these approaches allow for effective energy retention and release, facilitating various applications in energy systems (Ding, 2021; Detka & Górecki, 2023).

Round trip efficiency, energy density , and cycle life are key factors that are selective for energy storage systems that convert energy between stored and usable forms. Energy density, or the quantity of energy stored per unit of mass or volume, is vital for applications with size and weight restrictions, such as mobile and aerospace devices. The ratio of energy retrieved to the energy invested is the round-trip efficiency, which indicates system losses during conversion and storage that arise, leading to more comprehensive performance. Cycle life (determined by material properties and operational conditions) defines the total number of charge-discharge cycles a system can experience before significant degradation occurs. They are further observed in the principles guiding design and optimizing storage technologies, which were enabled to conform with energy storage techniques adopted/used due to importance (Adeyinka et al., 2024; Hossain et al., 2020).

The development of energy storage technologies is closely related to advances in materials science and engineering, with materials playing a key role in the performance, efficiency and scalability of these systems. For example, in batteries, advances in electrode materials (such as lithium iron phosphate (LFP) and nickel-rich cathodes) have advanced energy density as well as safety, and the development of solid electrolytes is leading to next-gen solid-state batteries. While with super-capacitors, the evolution of nanostructured materials, such as graphene and carbon nanotubes, have shown significant improvements in capacitance and charge-discharge rates, which makes them thereby practical in enhanced high-power applications. Relating to hydrogen storage, different kinds of materials are being researched to solve the challenges of energy density and system integration; metal hydrides and complex porous structures, namely metal-organic frameworks (MOFs), are high-storage capacity materials. These novel materials are powering the evolution of energy storage technologies to adapt to the increasing requirements of renewable energy systems (Detka & Górecki, 2023; Oyedotun et al., 2022).

Further energy loss, degradation, and scale-up hurdles can be resolved by using engineering methods (e.g., 3D printing and molecular modelling), which allow for accurate synthesis and fine-tuning of the material properties (Cheng & Jiang, 2021; Tang et al., 2019).

Energy storage technologies fundamentally rely on the principles of different branches of science and engineering fields, embedded within a theoretical framework to solve the challenges of integrating renewable energy into the electrical grid. Sustainable sector Energy continues to evolve toward efficient, reliable energy storage, utilizing new chemistry and materials science to address energy challenges by integrating advances in electrochemical, mechanical, and thermal storage. This framework not only informs continuing research but also forms the basis for the design of new storage technologies that are essential for the future of renewable energy.

CONCEPTUAL FRAMEWORK

Renewables are becoming interconnected with storage systems in a vast and dynamic interaction mode, whose technical, economic and operational challenges require a solid conceptual framework. It is based on the analysis of the role of storage technologies in the optimal use of renewables, the factors affecting their performance, and the implications for energy systems.

Renewable energy sources, such as solar and wind, are naturally variable, generating energy at times when environmental conditions allow. Storage systems play a critical role in balancing supply with demand by bridging the gap between generation and consumption. The balance energy transfer process draws on three distinct functions: capture —when energy storage systems use surplus energy during times of high generation; holding —when the captured energy is stored to smooth out the inconsistency of renewable energy sources; and dispatch —when stored energy is released to satisfy demand, balance the grid, or supply backup power. The interaction between these three components is highly dependent on the type of storage technology, the nature of the renewable energy source, and the conditions in which the energy system operates, making energy storage a key element for reliable and efficient renewable energy integration (Farghali et al., 2023; Worku, 2022).

There are a few key determinants that affect how effective and practical energy storage systems can be in integration with renewable energy. Its energy density, or the amount of energy stored per unit weight or volume, is critical for a range of applications, including grid-scale storage and electric vehicles – the higher it is, the smaller and more efficient the system can become. In other words, faster charges/discharges in energy storage mean it can address renewable energy fluctuations or real grid demands in real-time, with technologies such as supercapacitors being the fastest cycle and batteries somewhere between speed and capacity. Cost, which includes capital, operational, and maintenance costs, also continues to be a significant obstacle to overcome, which will require improved materials, manufacturing, and economies of scale in order to make hydrogen more economical. Furthermore, the cycle life and degradation of storage systems have a tremendous impact on their long-term performance and economic feasibility since durable materials and low degradation rates are crucial for sustainable and reliable operation (Farghali et al., 2023; Yao et al., 2016).

There are multiple challenges to integrating energy storage systems into larger energy systems. As the renewable share increases, maintaining grid stability becomes more challenging, and storage systems must not only store energy but also provide ancillary services, including frequency regulation, voltage support, and black start capabilities. A third critical challenge is managing the energy itself . Storage systems must interact with renewable energy sources, which requires advanced management systems that optimize energy matches and ensure that generation, storage, and consumption balance through time. Scalability and interoperability also play a part in meeting increasing energy demand while maintaining seamless integration with the current grid infrastructure. The widespread adoption of storage technologies requires speeds and interoperability of technologies that work well in networks of different types (Adeyinka et al., 2024; Prakash et al., 2022).

The interactive conceptual framework serves to synthesize these components and to illustrate the interaction of renewable energy sources, storage systems, and energy systems from a broader perspective. It demonstrates the dependencies between storage technology performance, operational efficiency, and system-level requirements. A guide for researchers, policymakers, and industry stakeholders, the framework pinpoints important factors and challenges to address in order to streamline the incorporation of renewables with the help of innovative storage solutions.

The benefits of bi-directional collaboration can be let loose along the entire value chain through this conceptual framework, establishing a compass to solve both technical challenges, such as efficiency and scalability and system-level challenges that disrupt the integration of renewable energy and storage systems. It underlines further types of renewable energy integration ways using key factors, such as energy density, charge/discharge rates, and grid stability.

RESEARCH PROBLEM

The downloaded data is until October 2023. However, the transition towards renewable energy policymakers faces a series of challenges when integrating such sources into existing energy production systems, many of which are related to current energy storage technologies. While there have been advancements in energy storage, there are still specific key barriers that need to be overcome for renewable energy to be deployed and used at scale. Based on the main challenges outlined in this section, the need for innovative solutions to overcome these challenges is discussed.

Energy storage continues to face multiple gaps and challenges that must be addressed to realize its full potential. We remain grounded in the technosphere, where lithium-ion carriers dominate, yet limited real estate for energy density, not to mention safety hazards and reliance on expensive and scarce materials such as cobalt and lithium are ubiquitous. These new alternatives—hydrogen storage, supercapacitors, flow batteries—are still in development, facing challenges too when it comes to cost, efficiency, and how large they can be made before they become impractical. These issues are exacerbated by high costs, especially in developing areas where there is a growing need for renewable energy options, but the capital required for storage infrastructure remains to be reduced. Further, many storage systems have low cycle life and performance degradation, which limit reliability and cost-effectiveness over time. Moreover, much higher energy density and efficiency in the existing technologies are required to make them suitable for applications that need long-duration storage or active grid stabilization. This highlights the necessity for innovation to fix these flaws.

The use of renewable energy is, however, not without its challenges, especially when converting them to the existing infrastructure where grid stability must be maintained; they must be compatible with legacy systems and get the energy to reach the farthest ends of the globe with little or no energy access. The unreliability of renewable energy sources such as solar and wind can create grid instability, blackout, or inappropriate resource use without adequate storage solutions since existing grids often must be prepared for real-time management of variable inputs. Moreover, countless energy systems, which were conceived for centralized fossil-fuel-based generation, struggle to adopt decentralized and variable renewables and require sophisticated storage technologies to close the technological loop and provide a higher service level. Renewable energy coupled with storage will grow in value, especially in remote and underdeveloped areas with low grid access. Nonetheless, their use is limited because many of the other storage methods are very expensive and not very efficient.

The gaps and limitations of existing energy storage technologies highlight the pressing challenge of finding new solutions to meet the needs of renewable energy integration. 4. Emerging storage tech needs superior price performance, scalability, efficiency and sustainability.

Furthermore, storage systems should be multifunctional, providing grid balancing, long-duration storage, and high-energy density solutions that allow a systematic transition to renewables. Meeting these needs will demand a multidisciplinary approach, integrating advances in materials science, engineering, and energy policy, as well as partnerships across industry, academia, and governments.

The specific research problem relates to technological, economic, and system-level challenges that complicate the effective integration of renewables into existing infrastructures over a wide range of overreaching terms. This chapter discusses the challenges and gaps of the energy storage concepts mentioned above. It lays the foundation for the subsequent analysis of state-of-the-art energy storage solutions that are critical to facilitating the global transition to a sustainable energy future.

RESEARCH QUESTIONS

- What are the key challenges in integrating renewable energy sources with existing energy storage technologies?
- How can next-generation energy storage solutions enhance the efficiency and reliability of renewable energy systems?

- What are the emerging trends and innovations in energy storage that could facilitate a sustainable energy future?

RESEARCH OBJECTIVES

- To analyze the current challenges in renewable energy integration and identify potential solutions.
- To explore the latest advancements in energy storage technologies and their applications in renewable energy systems.
- To provide practical recommendations for improving the scalability and efficiency of energy storage solutions.

Analyzing the Current Challenges in Renewable Energy Integration and Identifying Potential Solutions

Due to the growing need for sustainable and environmentally friendly power generation, transitioning to renewable energy sources has become a global imperative. However, there are multiple challenges to integrating renewable energy into existing power grids, many of which have to do with intermittency, grid stability, infrastructure limitations, and regulatory issues. One key problem that we face with renewable sources of energy, such as solar and wind, is that they are highly variable and often depend on weather patterns and time of day. Unlike traditional power generation from fossil-fuel sources, which can produce a steady and predictable rate of electricity, renewables are inherently intermittent, causing power supply variations that result in net energy deficits that may challenge grid operation.

Another major worry is grid stability, as conventional power grids were built on the assumption of centralized generation and predictable power flows. Meeting such focused challenges requires a renewed sophisticated electricity infrastructure, smart grids, and intelligence about renewables when used and in terms of power consumption. In many cases, this is coupled with a lack of appropriate transmission infrastructure, further compounding the challenges and resulting in losses/reduced efficiency of this energy. Moreover, outdated grid regulations and policies can create obstacles to the widespread deployment of renewables, and in some energy markets, fossil fuel-based generation is still preferred over cleaner alternatives (Akinsooto et al., 2024).

Financial and economic issues are key to renewable energy integration , too. Sky-high upfront investments, lack of financial incentives and volatile market prices may dissuade stakeholders from fully adopting renewables. Moreover, some countries have a shortage of technical knowledge and public awareness to facilitate the effective implementation and upkeep of renewable energy systems (Bensalah, 2024; Nassani et al., 2025).

It requires multi-faceted effort to overcome these challenges. Energy storage technologies, including batteries and pumped hydro storage, can help address intermittency problems by storing surplus energy and supplying it when necessary. Demand response programs, virtual power plants, and decentralized energy generation also contribute to increased overall grid flexibility and resilience. Evidence includes government policies and incentives, such as feed-in tariffs, tax credits and subsidies that can spur investment in renewable energy infrastructure. New technologies, such as artificial intelligence and blockchain-based energy trading, can better facilitate energy distribution and consumption. When these challenges are addressed holistically, it becomes possible to integrate renewable energy in a reliable, efficient and economically viable manner.

Exploring the Latest Advancements in Energy Storage Technologies and Their Applications in Renewable Energy Systems

Energy storage technologies have been developed as pivotal solutions to address the intermittency of renewable energy sources. Other fields, such as battery technology, thermal storage, and even mechanical storage, have developed faster, more efficient, scalable and cheaper energy storage solutions. Out of these, lithium-ion batteries are the most common storage technology deployed due to their high energy density, high efficiency, and decreasing costs. However, the issues of resource availability, safety, and environmental impact have prompted the consideration of alternative battery chemistries like solid-state batteries, sodium-ion batteries, and flow batteries.

For instance, solid-state batteries provide greater energy density, superior safety, and longer lifespan than traditional lithium-ion batteries, making them an attractive option for grid-scale energy storage. Just like their lithium counterparts, sodium-ion batteries are getting much interest, mainly because of the economy and availability of raw materials, which has the potential to make them a sustainable solution for your typical application (not just EVs) where you would have big packs. Flow batteries such as vanadium redox flow batteries are scalable and have a long cycle life, making them suitable for large-scale storage of excess energy from solar and wind farms (Qazi et al., 2025).

In addition to chemical energy storage, mechanical and thermal storage options are being developed. Pumped hydro storage is still one of the most stable large-scale energy storage methods, using gravitational potential energy to store and release energy. Compressed air energy storage (CAES) and flywheel energy storage systems offer supplementary solutions for grid stability and load balancing. Thermal energy storage is especially useful in concentrated solar power (CSP) plants, where its phase change material and molten salt can capture excess heat, which subsequently can be converted to electricity (Nikolaos et al., 2023; Rabi et al., 2023).

Grid stability is the use of these energy storage technologies in renewable energy systems. They provide fossil energy savings, energy arbitrage in terms of peaking, and decentralized energy management, which can reduce fossil fuel dependence and provide better energy security. Intelligent energy storage systems can also optimize charge and discharge cycles with the help of artificial intelligence technology and internet of things (IoT) technology, thus becoming more efficient and cost-effective—continued research and development of advanced energy storage.

Providing Practical Recommendations for Improving the Scalability and Efficiency of Energy Storage Solutions

To further optimize energy storage alternatives, there is a need for strategic guidance to accelerate the scalability, effectiveness, and implementation of renewable energy systems. Among the multiple factors impacting scalability, the cost of energy storage technologies is one of the key ones. Advancements in materials, manufacturing processes, and economies of scale will all contribute to reducing costs. Still, continued investment in research and development (R&D) is critical to achieving each. To further boost energy storage adoption, governments can implement policies and subsidies that provide financial incentives to consumers and industries.

The standardization of regulations and grid codes that allow for entry into systems is another essential scalability path. Many energy markets have not yet established clear policies that would guide storage deployment, creating uncertainty for investors and project developers. Transparent regulatory frameworks and market mechanisms that ensure energy storage operators are compensated for the services they provide, including frequency regulation and peak demand management, can promote further participation (Akinsooto et al., 2024; Zheng et al., 2022).

The energy storage solutions especially need to be not only technologically advanced but also intelligently managed. Unlike a doughnut refrigerator model, advanced battery management systems (BMS) can improve performance and longevity (when charge-discharge cycles are well optimized) and also allow the prevention of degradation. By leveraging predictive analytics and artificial intelligence integration, real-time monitoring and control can be achieved, significantly reducing energy losses and leading to improved system efficacy overall. In addition, hybrid energy storage systems offering multiple storage technologies (e.g., lithium-ion batteries with flywheels or supercapacitors) can offer more responsive and less expensive solutions for a variety of energy storage applications (Seh et al., 2023; Yang et al., 2022).

Technological improvements, along with the deployment of distributed energy storage systems, can improve grid resilience and flexibility. Localized storage for home, commercial and industrial use allows excess renewable energy to be stored nearer to the point of use, mitigating transmission losses and bolstering energy security. These platforms enable the exchange of energy between peers, enabling a more localized and democratized energy distribution network that uses blockchain technology to provide decentralized energy trading.

However, one way to speed up the deployment of scalable energy storage solutions is through public-private partnerships (PPP) and international collaboration. The collaboration here focuses on knowledge sharing, joint investments and global regulatory policies that can facilitate overcoming common challenges and innovation in the energy storage industry (Qazi et al., 2025). Overall, technological progress, regulatory support, financial incentives, and strategic collaborations will all play a vital role in achieving the integration of energy storage in renewable energy systems at the desired scale and efficiency.

RENEWABLE ENERGY AND STORAGE TECHNOLOGIES

Renewable sources will feature energy storage systems to mitigate the intermittency of renewables itself, which is one of their main limitations. Solar, wind, hydropower, and emerging technologies like geothermal and biomass are the cornerstone of the world's transition to sustainable energy. However, the output of these energy resources tends to be very variable, making their integration into the grid problematic. Energy storage systems play an essential role in this regard by storing excess energy produced at times of generation and consuming it at times of high demand or at times when renewable energy sources are not making (Farghali et al., 2023; Worku, 2022; Yao et al., 2016). In this section, the key renewable energy sources are explored before examining current energy storage technologies and case studies that illustrate the successful integration of these systems on modern energy grids.

Solar energy is one of the most widespread renewable resources available, and it is obtained through photovoltaic (PV) cells or solar thermal systems. Solar PV cells direct sunlight into direct current electricity while concentrating solar thermal systems. Mirrors or lenses direct sunlight to produce heat used to generate electricity, such as through a steam generator. Solar energy suffers from one of the biggest challenges of any renewable energy source: intermittency. Due to this variability, solar energy is less predictable, so robust energy storage solutions are required to ensure solar power is dispatchable and reliable. Batteries and thermal storage systems are energy storage systems that contribute to the solution to this problem and allow the storage of additional energy produced in sunlight so they can take advantage at night and on cloudy days. Improved or new storage technologies, such as lithium-ion batteries and solid-state batteries, are making these systems more efficient and less expensive to install, thus making solar energy more prominent on the grid (Vaghela et al., 2023; Yao et al., 2016).

The other primary renewable energy source is wind energy, which is kinetic wind energy converted to electricity by turbines. That said, like solar energy generation, which depends on the intensity of naturally occurring sunlight for energy production by solar panels, energy generation from wind is highly variable, depending on wind speed, which can vary over a day from season to season. This variability can result in overproduction periods when wind speeds are high and energy deficit periods when wind speeds are low. Energy storage systems are necessary to balance supply and demand, which helps mitigate these challenges. These days, wind energy generation can be stabilized with storage solutions like batteries, pumped hydro storage, and compressed air energy storage systems. Hybrid systems, therefore, such as wind plus solar, can help smooth out fluctuations, as the two will tend to generate at different times. Looking at it from another perspective, solar power may be accessible when the wind is quiet and vice versa, thus providing a more stable and continuous generation of electricity (Rekioua, 2023; Wei et al., 2023).

Hydropower, which utilizes the strength of moving water to produce electricity, is one of the most mature and reliable sources of renewable energy. Although hydropower is more stable relative to solar and wind power, seasonal variations and water levels can affect its reliability. Where rain patterns are not consistent or in times of drought, hydropower generation can vary widely. The role of the flow storage system is to optimize the capacity of the hydropower plants by collecting and TLC of excess energy from water-flow rates when they are higher and then providing the energy during the dry season of low flow or when more power is needed. A type of grid-scale energy storage, pumped hydro storage, has been used extensively to increase the flexibility and reliability of hydropower plants. This involves moving the excess energy to pump water from a lower reservoir to an upper one, which can then come down through turbines when demand peaks in the case of demand peaks or peaks in renewable generation (Alnaqbi et al., 2022; Nikolaos et al., 2023).

Emergent energy types like geothermal and biomass could also provide more stable output than solar and wind energy. Geothermal energy taps into the heat beneath the Earth's surface to produce electricity, offering a stable and dependable energy source. Continuous Energy Generation Biomass energy is generated primarily from organic materials like wood, agricultural residue, and, in principle, almost any plant life. However, both geothermal and biomass are struggling to scale up and integrate with modern grids. Again, geothermal power plants are highly location-dependent and expensive to build infrastructure for, and biomass energy depends on resource availability and how cumbersome its material transport plans can be. Energy storage systems can help integrate these renewables by buffering when supply does not match demand in both cases. Biomass energy, for instance, can be used as a fuel source to provide energy at times of peak demand, whereas geothermal sources can utilize thermal energy storage that retains excess heat until needed (Rezaei et al., 2021; Yadav et al., 2024).

Many case studies have successfully integrated renewable energy with storage systems around the world. For instance, "In California, the use of large-scale battery storage has allowed for more renewables, particularly solar and wind, to be integrated into the grid. Another example is the 100-megawatt Hornsdale Power Reserve in South Australia. The facility, which relies on lithium-ion batteries, serves an overarching purpose of keeping the grid stable by storing surplus renewable energy when supply runs higher than demand and sending it back to the grid when that balance shifts. It has been instrumental in reducing dependence on fossil fuel-based peaking facilities while showcasing the ability of energy storage systems to stabilize and fortify the grid (Aurecon, 2024; Kumar & Shrimali, 2024).

However, in Europe, the use of wind energy with energy storage systems has been an essential aspect of the region's renewable energy strategy. Denmark and Germany have been pioneers in the integration of wind energy with grid-connected storage technologies, including pumped hydro and grid-scale battery systems. These storage technologies enable the capturing of wind energy during periods of high generation to feed a grid that may be extremely high in demand or where the wind generation has begun to taper off. For example, wind energy comprised 47% of Denmark's electricity consumption in 2019, illustrating the necessity for storage systems to address variability (Hunt et al., 2022; Nikolaos et al., 2023).

Energy storage systems are necessary for the integration of renewable energy sources and the transition towards a reliable and sustainable energy future. Solar, wind, hydropower, and newer forms like geothermal and biomass all face their issues with intermittency and variability. Energy storage technologies are, however, increasingly demonstrating themselves as a panacea to such problems. Storage systems allow for the intermediated storage and dispatch of renewable energy, thereby improving grid stability and facilitating a higher share of renewables in the energy generation mix. Used in live news 4: As storage technologies improve - battery, thermal storage and other innovative solutions - the ability to integrate renewable energy into the energy grids of today will continue to reach new levels, driving a global transition toward clean and sustainable energy economies.

IN-DEPTH ANALYSIS OF CURRENT ENERGY STORAGE TECHNOLOGIES

An in-depth analysis of current energy storage technologies reveals a diverse landscape of solutions, each with its unique strengths and limitations. Among the most widely used technologies are lithium-ion batteries, which dominate the energy storage market due to their high energy density, fast response times, and scalability. These batteries are particularly suited for applications in consumer electronics, electric vehicles, and grid-scale storage. However, lithium-ion batteries face several challenges, including a limited cycle life, high costs, and safety concerns, especially when subjected to overcharging, overheating, or physical damage. Furthermore, lithium-ion batteries rely heavily on critical materials such as lithium and cobalt, which are scarce and expensive. The extraction of these materials also raises environmental and ethical concerns, as mining operations often have significant environmental impacts and are associated with labour issues in some regions (Hossain et al., 2020; Wei et al., 2023).

To overcome these limitations, research and development are focused on alternative battery chemistries. Solid-state batteries, for instance, have garnered significant attention due to their potential to offer higher energy densities and improved safety profiles. Unlike traditional lithium-ion batteries, solid-state batteries use a solid electrolyte instead of a liquid one, reducing the risks of leakage and flammability. Moreover, solid-state batteries are expected to last longer and operate more efficiently, with the promise of revolutionizing sectors like electric vehicles and grid storage. Additionally, sodium-ion batteries, which use sodium instead of lithium, are being explored as a more sustainable and cost-effective alternative. These batteries benefit from the abundance and lower cost of sodium, though they still face challenges in terms of energy density and efficiency compared to lithium-ion technology. Another promising alternative is redox flow batteries, which store energy in liquid electrolytes and offer the advantage of scalability, longer cycle life, and the ability to discharge energy over extended periods (Machín et al., 2024; Olabi et al., 2023).

Supercapacitors, another form of energy storage, operate differently from batteries by storing energy electrostatically. This enables supercapacitors to charge and discharge much more rapidly than batteries, with exceptionally high power density and long cycle life. They perform well even in extreme temperature conditions, making them ideal for applications that demand quick bursts of energy, such as grid stabilization, peak load management, and regenerative braking in electric vehicles. However, supercapacitors have lower energy density than batteries, which limits their ability to store large amounts of energy over long periods. While this makes them unsuitable for applications like long-duration energy storage, their ability to release energy in short intervals rapidly makes them valuable for complementing battery systems in hybrid configurations that balance power and energy needs (Libber et al., 2024; Nigam et al., 2021).

Hydrogen storage offers another promising solution for long-term energy storage, particularly in applications where high energy density is required. The process involves using electricity to convert water into hydrogen through electrolysis, which can then be stored and converted back into electricity or used as a fuel. Hydrogen is particularly attractive because it can store large amounts of energy and be used in a variety of sectors, including transportation, industry, and even power generation. One of the key advantages of hydrogen storage is its versatility—hydrogen can be stored for extended periods, making it ideal for seasonal storage or long-duration energy needs. Moreover, hydrogen can be used in fuel cells to produce electricity, with water as the only byproduct, which aligns with global goals for clean energy and carbon neutrality (Xie et al., 2024; Yang et al., 2023).

However, hydrogen storage faces several significant challenges that hinder its widespread adoption. One of the main issues is the high conversion losses associated with electrolyzers, which reduce the overall efficiency of the process. Additionally, the infrastructure required to produce, store, and transport hydrogen is expensive and yet to be widespread. High-pressure tanks and pipelines are necessary for storing and distributing hydrogen safely, adding complexity and cost to the system. Despite these challenges, recent advances are making hydrogen storage more viable. Improvements in electrolyzer efficiency are reducing the energy losses during hydrogen production, and innovations in solid hydrogen storage materials, such as metal hydrides, are making storage safer and more compact. These advances, coupled with growing investment in hydrogen infrastructure, are driving the technology toward broader adoption (Kalibek et al., 2024; Osman et al., 2024).

While each energy storage technology has its strengths, it also faces significant challenges that must be addressed to realize its full potential. Lithium-ion batteries, despite their popularity, need improvements in cost, safety, and sustainability. Alternative chemistries like solid-state, sodium-ion, and redox flow batteries are showing promise, offering solutions to some of these issues. Supercapacitors, while excellent for short-duration applications, need to be integrated with other storage technologies to address long-duration needs. Hydrogen storage holds considerable potential for long-term energy storage but requires further advancements in efficiency, infrastructure, and cost reduction. The ongoing research in these areas is critical for unlocking the full potential of energy storage systems, which will play a pivotal role in enabling the global transition to a renewable and sustainable energy future.

CASE STUDIES ILLUSTRATING SUCCESSFUL RENEWABLE ENERGY INTEGRATION PROJECTS

Hornsdale Power Reserve

The Hornsdale Power Reserve in South Australia is one of the world's largest and most powerful grid-scale battery systems, created to make renewable energy more beneficial and durable. Developed by Neoen, the project uses Tesla's lithium-ion battery technology and is connected to the Hornsdale Wind Farm. Initially launched in 2017, the Tesla megapack was expanded in 2020 to increase its potential, with a capacity of 150 MW as well as 193.5 MWh worth of storage. The Hornsdale Power Reserve's primary purpose is to store excess production during high-generation periods and release it during high-demand or renewable-generation low periods to balance the grid.

One of the main challenges for renewable sources like wind is their intermittency — wind generation can differ during the day and by season. This can make it challenging to maintain a consistent supply of energy because, during peak use times, fossil-fuel-based peaking plants are typically dispatched to supplement the existing grid. The Hornsdale Power Reserve meets this challenge by functioning as an energy buffer on a large scale. When there is too much wind energy, the battery will store that energy. When the wind generation drops off, or demand spikes, that energy stored is released back into the grid, providing electricity when you need it."

The Hornsdale Power Reserve has had a transformational impact on Australia's energy market. It has done wonders for grid stability with rapid-response capabilities that balance supply and demand. The rapid response time of the system — which can go from idle to activation in milliseconds — makes it critical for emergencies or generation swings. Moreover, by cutting down on the use of fossil fuels for peaking power, the battery system has contributed to lowering energy costs, which has benefited consumers and decreased carbon emissions. As a result, the system has reduced the need for gas-fired power plants, creating a cleaner, sustainable grid.

The Hornsdale Power Reserve has exposed the incredible potential of large-scale battery storage for renewable energy integration since its installation. In 2021, it was revealed that in its first four years of operation, the facility has saved South Australia more than $100 million by stabilizing the grid and avoiding costly fossil fuel imports at peak times. The project is seen as a role model for other nations seeking to incorporate renewables on a large scale. It demonstrates how energy storage technology can be influential in achieving a low-carbon, sustainable energy future.

Hywind Scotland, UK

Hywind Scotland is the world's first floating offshore wind farm, and it has taken a giant leap in renewable energy and storage technology integration in Aberdeenshire, UK. The project integrates floating wind turbines with battery storage systems to provide more steadiness and efficiency to offshore wind energy generation, which Equinor is developing. The plant is made up of five 6-megawatt turbines, each one capable of producing 30 MWs of electricity. To counter the intermittency of wind energy and improve the stability of the electricity supply, the wind farm is combined with a state-of-the-art battery storage system.

The primary challenge with offshore wind energy is the variability of wind speeds and the fact that it is not always possible to match generation with demand. This is especially true for offshore farms, where generation could take place at times of low demand or when other sources of power are widely available, creating the potential for wasted energy. The Hywind Scotland project mitigates the synchronization issue by attaching a battery storage system to the wind farm itself. The system stores excess power generated when the wind is plentiful, and power generation exceeds demand, then dispenses it when generation dips below what is needed to balance supply in order to keep the grid stable. The storage pad also holds a more uniform and predictable energy flow, similar to the integration of renewable sources into the grid.

Hywind Scotland has had a significant impact on you. It shows how hybridizing floating offshore wind technology with energy storage systems makes it possible to realize the full potential of renewable energy generation in offshore environments. Compared with fixed-bottom turbines, floating wind farms can be located in deeper water, where the wind blows better, which broadens the geographical range of offshore wind energy. This added to the overall potential of offshore wind energy as a source of clean energy. Alongside it, however, the storage system adds reliability to energy generation so that electricity can be generated even if the wind is not blowing.

Hywind Scotland provides a successful model upon which future offshore wind projects may be based. The project achieved the milestone of being the first to demonstrate energy storage-integrated floating wind technology, which will pave the way for deploying more advanced offshore wind farms featuring outstanding energy reliability and considerable use potential. It is also an example of how hybrid renewable energy systems can contribute to decreasing dependence on fossil fuels, increasing grid stability, and advancing the global transition to a more sustainable energy future. Hywind Scotland is likely to be emulated across the globe, offering lessons in pairing next-generation wind technology with energy storage solutions to learn to leverage the combined approach for improved renewable energy systems performance and scalability.

Noor Energy Complex, Morocco

Noor Energy Complex in Ouarzazate, Morocco The Noor Energy Complex is an industrial complex seeded by several of the world's largest solar thermal energy plants combined with a thermal energy storage system. The plant fits Morocco's ambitious renewable energy scheme, which is designed to reduce the North African nation's use of fossil fuels and help meet the world's sustainability objectives. Phase 1 of Noor is focused on photovoltaic (PV) technology, while phases 2 and 3 are concentrated mainly on solar thermal (CST) systems.

The signature innovation of Noor Energy is its use of molten salt for thermal energy storage. This enables the plant to produce electricity at times when the sun is not shining. CSP systems use mirrors or lenses to focus sunlight in order to heat a fluid that produces steam to drive a turbine to generate electricity. In the Noor plant, excess heat is stored in huge tanks full of molten salt, which can hold heat for long periods. The thermal storage capability enables the plant to deliver continuous electricity production, even after the sun has gone down or on cloudy days when solar generation stops. The largest of the three plants is Noor III, which has a 150 MW capacity and can store energy for up to 6 hours, allowing it to generate electricity even at peak demand hours in the evening — making it a dependable and dispatchable energy source.

Noor Energy has made a significant impact not just in terms of energy generation but also because it proves that thermal storage can be the answer to one of the biggest challenges of solar energy — intermittency. Standard thermal storage is added to the solar generation process to augment the reliability and stability of solar power and reduce reliance on fossil fuels during day and evening hours, providing a powerful renewable energy source. In addition, the plant can deliver stable, on-demand electricity to the national grid, thus helping Morocco move away from imported fossil fuels and helping the country meet its goal of generating 52% of its electricity from renewable sources by 2030.

In addition to providing electrical energy, the Noor Energy Complex has cemented its position as a leading site for renewable energy development in Morocco. It has also opted to showcase the possibility of innovative energy storage technologies to overcome the intermittency problem of sources, such as renewable energy. The project has established Morocco as a pioneer in renewable energy, proving that solar energy could be available at a later period through thermal storage, which is a renewable energy technology with similar effectiveness and usability as fossil plants for regions with great solar potential like North Africa. The Noor Energy

Complex demonstrates how renewable energy and state-of-the-art storage systems can help meet increasing energy needs while enabling a shift to a cleaner, more sustainable energy future.

California Energy Storage Projects, USA

California has also emerged as a global leader in embedding renewable energy onto its power grid thanks to its bold targets to slash carbon emissions and move to 100% clean energy. A key piece of this transition has been deploying hundreds of grid-scale battery storage projects that help integrate solar and wind, with the goal of mitigating the challenge renewables introduce through their intermittency and variability. The most high-profile example has been the pairing of large-scale battery storage systems with the state's growing generation of solar and wind power.

One such project is the "Moss Landing Energy Storage Facility," which is located on the Monterey Peninsula. It is among the largest of its kind globally, with a 300 MW capacity and the capacity to store 1,200 MWh of energy. It functions by capturing excess solar and wind energy produced throughout the day or when the wind is at or near power and when supply is outpacing demand. During peak hours, when renewable generation is lower, and demand is higher, this energy is released back into the grid. This helps California use its renewable resources more efficiently and reduce its use of natural gas and other fossil fuels to balance supply with demand.

In fact, these grid-scale battery storage projects have had an incredible effect. To begin with, it greatly enhanced grid flexibility and reliability. The grid in California can stay balanced and operable because these projects allow energy to be stored for when it is most needed and sent out to transmission lines when these conditions exist. That has been essential in staving off blackouts in moments of peak demand, especially on hot summer days when demand for electricity surges. More than this, they have helped California with its renewable energy ambitions, enabling the state to deliver on its aim of increasing the share of its electricity from renewables to 60% by 2030.

The projects will also help the state meet its broader clean energy goals, avoiding the need for fossil-fuel peaking plants that are used to deliver electricity during times of peak demand. California has made great strides toward its ambition of a carbon-free grid by 2045, with large-scale storage systems helping out. Additionally, these successful projects provide models for implementing battery storage to help integrate other renewables into the grid, pointing the way for other regions seeking to ramp up renewable energy systems.

California's battery storage projects are generating economic opportunities — on top of providing technical and environmental benefits. Each of these projects creates jobs in the fields of energy storage, renewable energy, and infrastructure development and contributes to driving the costs of energy storage technologies down. With our data availability up to October 2023, the ongoing development of battery storage in the state, driven and informed by these early projects, will be pivotal in setting the course for energy storage and, ultimately, renewable energy integration in California in particular and around the world.

Energy diversity from renewables, coupled with energy storage technology, is therefore fundamental to addressing issues of variability, reliability, and scalability in clean energy systems. Batteries, supercapacitors, and hydrogen storage offer essential solutions, each appropriate for particular use cases and energy needs. This chapter presents such case studies from across the globe, showcasing how renewable energy generation paired with energy storage can completely change the energy landscape for a given region.

TECHNOLOGICAL INNOVATIONS IN ENERGY STORAGE

California has also emerged as a global leader in embedding renewable energy onto its power grid thanks to its bold targets to slash carbon emissions and move to 100% clean energy. A key piece of this transition has been deploying hundreds of grid-scale battery storage projects that help integrate solar and wind, with the goal of mitigating the challenge renewables introduce through their intermittency and variability. The most high-profile example has been the pairing of large-scale battery storage systems with the state's growing generation of solar and wind power.

One such project is the "Moss Landing Energy Storage Facility," which is located on the Monterey Peninsula. It is among the largest of its kind globally, with a 300 MW capacity and the capacity to store 1,200 MWh of energy. It functions by capturing excess solar and wind energy produced throughout the day or when the wind is at or near power and when supply is outpacing demand. During peak hours, when renewable generation is lower, and demand is higher, this energy is released back into the grid. This helps California use its renewable resources more efficiently and reduce its use of natural gas and other fossil fuels to balance supply with demand.

In fact, these grid-scale battery storage projects have had an incredible effect. To begin with, it greatly enhanced grid flexibility and reliability. The grid in California can stay balanced and operable because these projects allow energy to be stored for when it is most needed and sent out to transmission lines when these conditions exist. That has been essential in staving off blackouts in moments of peak demand, especially on hot summer days when demand for electricity surges. More than this, they have helped California with its renewable energy ambitions, enabling the state to deliver on its aim of increasing the share of its electricity from renewables to 60% by 2030.

The projects will also help the state meet its broader clean energy goals, avoiding the need for fossil-fuel peaking plants that are used to deliver electricity during times of peak demand. California has made great strides toward its ambition of a carbon-free grid by 2045, with large-scale storage systems helping out. Additionally, these successful projects provide models for implementing battery storage to help integrate other renewables into the grid, pointing the way for other regions seeking to ramp up renewable energy systems.

California's battery storage projects are generating economic opportunities — on top of providing technical and environmental benefits. Each of these projects creates jobs in the fields of energy storage, renewable energy, and infrastructure development and contributes to driving the costs of energy storage technologies down. With our data availability up to October 2023, the ongoing development of battery storage in the state, driven and informed by these early projects, will be pivotal in setting the course for energy storage and, ultimately, renewable energy integration in California in particular and around the world.

Energy diversity from renewables, coupled with energy storage technology, is therefore fundamental to addressing issues of variability, reliability, and scalability in clean energy systems. Batteries, supercapacitors, and hydrogen storage offer essential solutions, each appropriate for particular use cases and energy needs. This article presents such case studies from across the globe, showcasing how renewable energy generation paired with energy storage can completely change the energy landscape for a given region.

GRID INTEGRATION AND STABILITY

Adding energy storage systems to existing grid system infrastructure is a vital challenge for the 21st-century modernization of energy systems and their integration with renewable energy sources such as solar and wind. Unlike conventional fossil fuel-based power plants, power generation from renewable sources is intermittent, with energy production varying depending on weather conditions, time of day, and season. This variability leads to a gap between demand and supply, something that has always been a concern for traditional grid systems that were designed to handle stationary, centralized power generation. Introducing energy storage technologies to the grid can alleviate these difficulties. They allow surplus renewable energy to be stored when supply surpasses demand and for the surplus to be released when demand exceeds generation (Adeyinka et al., 2024; Worku, 2022).

A significant issue in grid integration thus becomes the handling of this intermittent supply and the provision of continuous energy to fulfil demand. Traditional grid infrastructure was not developed to accommodate decentralized and variable renewable energy sources. Thus, to integrate these sources, a complete restructuring of grid operations is needed. Advanced energy storage systems — from lithium-ion batteries and pumped hydro storage to more novel solutions such as solid-state batteries — are crucial in buffering these oscillations in order to provide backup energy in times called weak renewable generation. These storage systems need to be synchronized with the grid operations in order to deliver power reliably and efficiently (Adeyinka et al., 2024).

Storage systems act in the management of ancillary services, which are essential for grid stabilization, such as frequency response, reactive power support, load shifting, and power flow control. These services are necessary to balance supply and demand in renewable energy-rich grids. The grid is kept in balance by frequency regularisation, which adjusts power generation to match instantaneous supply and demand. Voltage support must help keep electrical voltage within safe operating limits to protect the equipment and maintain a stable system. For instance, load shifting in energy storage systems allows power to be saved for use during low-demand periods and released during high-demand periods, eliminating the need to install costly peaker plants. These services make energy storage systems functional elements in modern grids due to their capability to manage the variability induced by the generation of renewable energy (Kumar & Palanisamy, 2020; Prakash et al., 2022).

Scalable approaches to increased grid resiliency arise from hybrid solutions, which comprise various types of storage technologies, demand-side management, and smart grid technologies. Battery storage can also be paired with demand response programs to offer a more dynamic solution to the intermittent nature of renewable energy generation. Batteries can store excess energy during peak renewable generation periods, and consumer demand can be curtailed and moved to periods of lower energy availability. This facilitates the supply-demand equilibrium for a particular commodity type and lowers the required supply of additional infrastructure. Innovative grid technologies, such as advanced metering infrastructure, demand response systems, and energy management software, complement these solutions and allow for real-time monitoring, predictive analytics, and automated control of energy flows. Digital tools such as artificial intelligence (AI) and the internet of things (IoT) play an essential role in optimized energy management through predicting demand patterns, grid condition monitoring, and real-time energy flow adjustment (Raza et al., 2024).

However, the integration of energy storage systems is likely only to be successful with the development of supportive policy and regulatory frameworks that mobilize investment into storage technologies. Governments provide subsidies, tax incentives, grants, and other R&D funding that play a key role in bringing down integration barriers. These incentives assist in decreasing the initial investment required to implement energy storage systems and promote private sector participation in energy storage advancement. There must be clear standards set out for interconnection to the grid, as well as a clear pathway for market participation so that energy storage systems can be incorporated safely and efficiently into the grid system. Regulators need to know how these storage systems would be compensated for providing grid services like energy storage, frequency and voltage support (Kumar & Palanisamy, 2020).

Another major regulatory challenge is the requirement for standardized energy storage policies. Grid interconnection standards are a requirement to enable efficient and safe operation of energy storage systems in the grid. Direct access to MAS must be modified to reflect energy storage requirements, including the unique capacity to charge or discharge energy when required. Moreover, it is also essential for policymakers to offer long-term planning frameworks that correspond with the objectives of integrating renewable energy." This includes the establishment of renewable energy targets and the identification of clear pathways through which energy storage can support the achievement of these goals. When the implications of energy storage are known, without uncertainty about future policy frameworks, the investment will not be deterred, and the pace of grid modernization will not slow (Lew et al., 2013).

Enabling the private sector to partner with the public sector to address key barriers to the integration of energy storage is also critical. Storage sites need to be sited and built; therefore, governments should work with industry, from utilities to energy storage manufacturers and technology enablers, to enable the successful deployment of storage systems. These stakeholders can work together to research and develop projects, tests, and pilots to identify best practices and optimize the deployment of energy storage technologies. Public-private partnerships can also be crucial for financing large-scale energy storage projects, which require high upfront investments. When governments and private sector actors collaborate closely, they can establish a conducive ecosystem for energy storage, which will catalyze innovations that will hasten the achievement of renewable energy in the future (Arshi et al., 2024).

Significantly, the successful integration of energy storage systems into the grid relies on a systems approach that takes into consideration not only technical solutions but also social, economic, and environmental factors. The potential benefits of energy storage in terms of energy equity are significant, including for low-income and vulnerable consumers. It can save you money on your energy bill by giving you the ability to use your renewable energy more efficiently, but energy storage's benefits must be shared equitably. In addition, the ecological consequences of energy storage technologies should be thoroughly examined, with a specific focus on resource extraction, recycling, and system disposal of the storage systems (Farghali et al., 2023; Sibilla & Kurul, 2023).

The device enables peak load shifting and load levelling in existing grid systems that are already served by renewable energy resources. Energy storage systems can serve as balancing solutions, stabilizing the grid while accommodating the increasing share of renewables in the energy mix and enabling societies to adapt to a renewable-powered world while dealing with renewable energy's variability challenges. Innovation technologies and regulatory changes with the help of government, industry, and consumers will be crucial in addressing integration challenges and achieving a reliable, affordable, low-carbon energy future for all.

RECOMMENDATIONS AND SOLUTIONS

The globe also needs innovative ideas to boost energy storage technologies that will facilitate the integration of renewable energy into the energy grid. That will require a multi-pronged approach, from investing more in R&D to alternative storage technology like solid-state batteries, flow batteries, and hydrogen storage. These innovations improve renewable energy reliability by increasing efficiency, scalability, and cost-effectiveness. Improvements in materials science can also help improve energy density, cycle life, and safety components so they will last in the long term.

Upscaling existing technologies, particularly lithium-ion batteries and supercapacitors, is also necessary. These systems have much more potential for grid-scale and transportation applications, with performance bolstered by advanced electrode materials and manufacturing methods. A diversified approach to energy storage involving mechanical (e.g., pumped hydro), thermal (e.g., molten salt) and electrochemical systems should provide tailored solutions depending on the needs of energy supply (from the hour today), long-duration storage and grid stabilization.

A set of strategic measures needs to be adopted to address technical, economic, and regulatory barriers. One major challenge is cost-effectiveness: This can be mitigated by saving on capital by increasing economies of scale, mass production, and more affordable materials. Large-scale deployment can be further accelerated through financial incentives, such as subsidies and tax breaks. The improved robustness of Storage Systems with optimized BMS and advanced materials will further increase the reliability and efficiency of storage systems.

It is also crucial to the widespread adoption of regulatory reforms. Governments must harmonize interconnection standards and regulatory environments, allowing for quicker and cheaper deployment. Robust policies should be established that enable energy storage to participate in competitive grid markets and facilitate participation in decentralized energy systems. Long-term supply commitments can promote the development of large-scale storage projects and provide additional confidence among participants.

Utilities, technology providers, grid operators , and policymakers must work together to develop integrated energy storage solutions. More importantly, policymakers are trained to ensure that an appropriate regulatory environment is conducive to innovation and sustainability. Researchers must collaborate with industry leaders to ensure nascent technologies are translated into deployable, scalable solutions that complement the broader objectives of the energy transition.

Building a skilled workforce is another key piece of making the transition to energy storage at scale. Tailored education and training for researchers, engineers, and technicians will help enable innovation and breakthroughs in solutions. Additionally, public awareness campaigns can educate consumers about the benefits of energy storage, building acceptance and engagement in the transition to a sustainable energy system.

These recommendations help the energy storage sector propel past current limitations and drive more widespread integration of renewables. The development of a sustainable and resilient global energy system will be faster through a synchronous, multi-use approach across technological development and implementation, economic sensing, regulatory facilitation, and the evolution of human resources.

FUTURE DIRECTIONS

Energy storage and renewables integration shall redefine the global energy domain, riding on the winds of next-generation cutting-edge technologies. One of the focus areas is next-gen storage technologies. Solid-state batteries, which substitute the traditional electrolytes in liquid form with solid-state materials, offer much higher energy densities, short charge times, and enhanced safety, eliminating the risk of flammability and degradation. Similarly, there has been growing interest in other forms of long-duration energy storage, including advanced flow batteries and hydrogen storage systems. Flow batteries, for instance, can separate power generation from energy storage, enabling them to be used at a much larger scale for grid stabilization and backup power. On the other hand, despite the challenges of hydrogen storage and delivery methods, metal hydrides and metal-organic frameworks (MOFs) are anticipated to provide high-capacity, efficient solutions in a variety of scenarios, such as grid storage and transportation. Such technologies, still in the early stages, could pave the way for energy systems capable of integrating renewable sources like solar and wind power much more efficiently and reliably.

Energy storage has also been the focus of research for innovations in materials science that could be game-changing. Energy storage systems have many emerging nanomaterials like carbon nanotubes, graphene, and other advanced composites, which can potentially enhance their conductivity, stability, and efficiency. These materials can help create smaller, higher-capacity storage devices that last longer before replacements are needed and charge/discharge cycles are faster. Moreover, research on new electrolytes aims to improve energy concentration and stability metrics in solid-state and lithium-ion batteries, among others. Nanotechnology, and in particular quantum dot-based storage systems, will open the door for higher energy amounts and more efficient ways to harvest them. In addition, nature-based processes inspired by energy storage systems are expected to serve us for a more sustainable and eco-friendly future. In conclusion, such breakthroughs will not only improve the storage technologies in the future if designed into these innovations. Still, they will also lead to more robust and sustainable energy systems that are always available.

Therefore, future development should focus on the viability of AI and ML to optimize energy storage and renewable energy systems. AI-enhanced algorithms can help match energy production/storage to consumption by forecasting energy needs, optimizing the grid and refining the efficiency of storage systems. Using AI and ML for Predictive Maintenance of Storage Systemsdatabricks.comAI and ML can be leveraged to optimize predictive maintenance for storage systems, increasing uptime and improving longevity. As storage systems develop, these technologies will also enable the smooth functioning of decentralized, renewable energy-based grids — a trend that will become more and more common. The dream is to sustain high penetration levels of renewable energy through new storage solutions and AI. These developments will serve a critical role in fighting climate change, promoting energy access, and enabling the transition to a clean, low-carbon energy system. When these technological developments reach a mature stage of progress in the future, they will play a prominent role in achieving global energy security, protecting the environment, and promoting equity in the distribution of energy.

DISCUSSION

As we discussed in this chapter, energy storage technologies are an integral element of the system in overcoming the unique challenges posed by the variability and intermittency of converting renewable resources to power. As renewables like solar and wind power become more prevalent, so does the need to capture excess energy when it is available and to free it up when generation falls short. As we work towards more stable, reliable, and sustainable energy delivery systems, energy storage has become a key piece of the energy puzzle, with rapid advancements driving progress towards low-carbon economies.

Key takeaway points and recommendations The take-home messages of this chapter identify the variety of energy storage technologies, all with unique benefits and shortcomings. - Electrochemical storage systems, most notably lithium-ion batteries, have emerged as the leading technology used for grid-scale storage as they offer high energy density, scalability, and existing infrastructure. These features render lithium-ion batteries open the doorway to applications, including renewable energy assimilation to electric vehicles (EVs). An additional challenge is the cost, cycle life, and scarcity of materials. The escalating demand for lithium-ion batteries has environmental and ethical issues surrounding the sourcing of materials such as lithium, cobalt and nickel. In addition, with battery costs decreasing, there is an ongoing need for innovations that will prolong and improve such batteries in terms of environmental impact and efficiency.

However, in addition to these electrochemical storage technologies at the forefront of the energy storage revolution, mechanical storage solutions, including pumped hydro and flywheels, present alternatives. Pumped hydro storage, which leverages the gravitational potential energy of water to produce electricity, offers large-scale storage with long discharge times. Nonetheless, its geographical limitations have made it a rather location-specific solution since it is predicated on topographical conditions that allow reservoirs to be built. Additionally, pumped hydro systems are often inefficient at converting energy, with energy loss through friction, evaporation, and the mechanics of the system. Similarly, flywheel storage provides high-speed discharge and can sustain high power requirements for very short periods. Its energy capacity is, however, most limited compared to other storage solutions; it is often only theoretically applicable in niches requiring fast response times (e.g. frequency regulation).

Thermal energy storage, which is helpful in capturing and storing heat for long-duration storage, is an excellent opportunity for long-duration storage, particularly for those requiring continuous, high-volume energy supply for weeks or months. Molten salt storage, as is used in some solar thermal power plants, enables heat to be stored and then converted back into electricity. While thermal storage has relatively modest losses over time, the overall system's efficiency from power to energy is lower than that of other technologies. Finally, the elevated temperatures associated with such systems can present operational and material challenges. That said, the long duration of energy storage potential and the ability to store renewable energy when it is generated in abundance are some of the reasons why thermal energy storage needs to be part of our future energy conversation.

For one, when exploring these energy storage technologies, it is evident that only specific solutions will suffice to address energy storage requirements in their entirety. The best storage technology is highly application-, energetics-, and location-dependent. Short-duration applications (e.g., to provide peak power support or frequency control, which requires a fast-response system, such as a supercapacitor or flywheel). These are great for delivering instant bursts of energy but are unable to store ample amounts over time. In contrast, long-duration applications that require energy storage over days or weeks will use more thermal storage or flow batteries , which gives them more capacity but less speed and efficiency. Combination approaches, which use multiple storage technologies to target different components of energy needs, will likely offer the most effective and reliable solutions.

Energy storage systems, by necessity, must fit into contexts broader than their individual technical and operational characteristics; integrating energy storage into the grid infrastructure itself will require comprehensive energy management systems that account for the range of evolving storage options. Real-time information, predictive models, and advanced control systems are necessary to create enough dispatchable energy from a mix of diverse renewables and storage systems, ensuring that energy flows to where it is most needed. Innovative grid technologies, along with AI and IoT, are becoming more instrumental in facilitating energy storage and generation coordination. Based on Hickey's vision, these digital tools not only optimize the flow of energy by predicting demand patterns, assessing grid conditions and enabling real-time decision-making but also ensure the stability of the grid, even as the penetration of renewable energy increases.

Beyond technological solutions, the fast-emerging landscape of energy storage has giant systemic ramifications with regard to sustainability. The more they mature with battery technologies, the more we start seeing with usage. These storage technologies are endless and can help reduce fossil fuel dependence and the carbon footprint closer to the toolbox. Energy storage lessens the need for backup power plants that burn fossil fuels — the backbone of electric grids — which are often turned on when large amounts of renewable energy are not on the grid, filling in the gaps with dirty energy sources. This change not only reduces the emissions of greenhouse gases but also decreases the impact of energy production on the environment. The extended storage capacity also allows the better use of renewable resources so that they can become a more stable and dependable source of energy, making renewable energy more viable.

Moreover, energy storage technologies can play a crucial role in enhancing global energy access in remote and marginalized areas. Many regions around the world continue to lack access to a stable, affordable supply of electricity; millions of people still live without reliable power. These regions can benefit from renewable energy systems, along with energy storage, as a decentralized solution and potentially be powered by off-grid or microgrid systems where energy is both generated and stored locally. This can be used to close the energy access gap and provide reliable electricity for some communities that are at risk of being stranded in the energy transition. Energy storage systems can facilitate more equitable energy systems as well as support economic development in remote regions by reducing costs and making opportunities available for off-grid applications.

The economic implications of the ongoing development and deployment of energy storage technologies are also substantial. The stage of generic energy, where storage systems continue to improve efficiency and cost-effectiveness, will subjectively lower all prices, especially for markets with higher carbon footprints. Energy storage systems can be a cost-effective alternative to meet the increased demand compared to peaking power plants, which require high operating and maintenance costs. Energy storage can reduce the cost of electricity and, therefore, benefit both consumers and utilities through demand-side management, such as shifting energy uses to off-peak energy use or storing excess generation.

Energy storage technologies are key to the transition to a low-carbon economy. They are a vital cog in the larger machine that must undergo the process of integrating renewable energy sources into the grid. That process must ensure that those sources can deliver • dispatchable power. Storage technologies are getting cheaper and providing more significant returns than conventional models, which makes them more valuable in shaping global energy trends moving forward. Through this potential to facilitate increased renewable energy penetration, low emissions, improve energy access, and drive economic growth, energy storage is critical to achieving a sustainable energy future.

As a final note, energy storage has remarkable potential to reshape the global energy landscape. This is how we can ensure that the future is greener and more sustainable. The way forward will require collaboration between industry, governments and consumers to establish an ecosystem that enables the widespread deployment of energy storage technologies to unlock renewable energy benefits for all.

CONCLUSION

This chapter has explored in-depth the key function of energy storage within the integration of renewable energy and outlined the various storage technologies and their challenges and strategic solutions to improve their performance. More energy-density low-cost components like glass fibre in a cost-efficient manner have a direct correlation with the future of storage systems. We have learned the importance of energy storage over the years. It is an integral component of any energy ecosystem. It serves as a bridge between source and sinks, balancing supply and demand and ensuring reliability and stability, supporting a grid with increasing amounts of renewable generation. On top of that, we were able to list the reasons preventing widespread adoption, such as the cost, regulatory challenges, and technological limitations, and present ways to address these obstacles through innovation, policy support, and partnerships in the industry.

Energy storage is an absolute necessity for realizing a sustainable future. As solar, wind, and other renewable energy sources become more and more woven into the fabric of the global energy mix, the ability to store and manage energy effectively will be a cornerstone of stabilizing grids, minimizing carbon footprint, and guaranteeing energy accessibility for all. Energy storage systems are, therefore, not only a crucial enabler for the integration of renewable energy but also a transformational tool for transforming into more resilient and low-carbon energy systems globally.

Finally, innovation will drive the transition in energy storage technologies. With continued investment in R&D and cross-sector collaboration, we will see breakthroughs that enhance efficiency, scalability, and affordability. As these developments progress, energy storage will become ever more integral to the worldwide transition towards a sustainable energy future, guaranteeing that renewable energy can satisfy the requirements of a growing and gradually energy-aware world.

REFERENCES

Abraham, K. M., & Jiang, Z. (1996). A polymer Electrolyte-Based rechargeable Lithium/Oxygen battery. *Journal of the Electrochemical Society, 143*(1), 1–5. DOI: 10.1149/1.1836378

Adeyinka, A. M., Esan, O. C., Ijaola, A. O., & Farayibi, P. K. (2024). Advancements in hybrid energy storage systems for enhancing renewable energy-to-grid integration. *Sustainable Energy Research, 11*(1), 26. Advance online publication. DOI: 10.1186/s40807-024-00120-4

Akinsooto, N. O., Ogundipe, N. O. B., & Ikemba, N. S. (2024). Regulatory policies for enhancing grid stability through the integration of renewable energy and battery energy storage systems (BESS). *International Journal of Frontline Research and Reviews, 2*(2), 022–044. DOI: 10.56355/ijfrr.2024.2.2.0023

Alnaqbi, S., Alasad, S., Aljaghoub, H., Alami, A., Abdelkareem, M., & Olabi, A. (2022). Applicability of hydropower generation and pumped hydro energy storage in the Middle East and North Africa. *Energies, 15*(7), 2412. DOI: 10.3390/en15072412

Arshi, O., Rai, A., Gupta, G., Pandey, J. K., & Mondal, S. (2024). IoT in energy: A comprehensive review of technologies, applications, and future directions. *Peer-to-Peer Networking and Applications, 17*(5), 2830–2869. Advance online publication. DOI: 10.1007/s12083-024-01725-8

Aurecon. (2024). *Hornsdale Power Reserve: Year 1 technical and market impact case study*. Retrieved from https://hornsdalepowerreserve.com.au/wp-content/uploads/2022/12/Aurecon-Hornsdale-Power-Reserve-Impact-Study-year-1.pdf

Bensalah, N. (2024). Challenges and Opportunities in Energy Economics: Balancing cost, sustainability, and innovation in the global energy transition. *Modern Economy, 15*(11), 1147–1180. DOI: 10.4236/me.2024.1511059

Chen, S., Wang, W., Zhang, X., & Wang, X. (2024). High-performance supercapacitors based on Graphene/Activated Carbon hybrid electrodes prepared via dry processing. *Batteries, 10*(6), 195. DOI: 10.3390/batteries10060195

Cheng, M., & Jiang, Y. (2021). 3D-printed solid-state electrolytes for electrochemical energy storage devices. *Journal of Materials Research/Pratt's Guide to Venture Capital Sources, 36*(22), 4547–4564. DOI: 10.1557/s43578-021-00355-7

Detka, K., & Górecki, K. (2023). Selected Technologies of Electrochemical Energy Storage—A Review. *Energies, 16*(13), 5034. DOI: 10.3390/en16135034

Ding, Y. (2021). Chapter 1. Thermodynamics for thermal energy Storage. In *RSC energy and environment series* (pp. 1–14). DOI: 10.1039/9781788019842-00001

Farghali, M., Osman, A. I., Chen, Z., Abdelhaleem, A., Ihara, I., Mohamed, I. M. A., Yap, P., & Rooney, D. W. (2023). Social, environmental, and economic consequences of integrating renewable energies in the electricity sector: A review. *Environmental Chemistry Letters*, *21*(3), 1381–1418. DOI: 10.1007/s10311-023-01587-1

Fekete, B. M., Bacskó, M., Zhang, J., & Chen, M. (2023). Storage requirements to mitigate intermittent renewable energy sources: Analysis for the US Northeast. *Frontiers in Environmental Science*, *11*, 1076830. Advance online publication. DOI: 10.3389/fenvs.2023.1076830

Hossain, E., Faruque, H., Sunny, M., Mohammad, N., & Nawar, N. (2020). A comprehensive review of energy storage systems: Types, comparison, current scenario, applications, barriers, potential solutions, policies, and prospects. *Energies*, *13*(14), 3651. DOI: 10.3390/en13143651

Hunt, J. D., Nascimento, A., Zakeri, B., Jurasz, J., Dąbek, P. B., Barbosa, P. S. F., Brandão, R., De Castro, N. J., Filho, W. L., & Riahi, K. (2022). Lift Energy Storage Technology: A solution for decentralized urban energy storage. *Energy*, *254*, 124102. DOI: 10.1016/j.energy.2022.124102

Jacobson, M. Z., Delucchi, M. A., Cameron, M. A., & Frew, B. A. (2015). Low-cost solution to the grid reliability problem with 100% penetration of intermittent wind, water, and solar for all purposes. *Proceedings of the National Academy of Sciences of the United States of America*, *112*(49), 15060–15065. DOI: 10.1073/pnas.1510028112 PMID: 26598655

Jiang, B., & Raza, M. Y. (2024). Renewable energy for sustainable development in China: Discourse analysis. *PLoS One*, *19*(11), e0298347. DOI: 10.1371/journal.pone.0298347 PMID: 39602411

Kalibek, M. R., Ospanova, A. D., Suleimenova, B., Soltan, R., Orazbek, T., Makhmet, A. M., Rafikova, K. S., & Nuraje, N. (2024). Solid-state hydrogen storage materials. *Discover Nano*, *19*(1), 195. Advance online publication. DOI: 10.1186/s11671-024-04137-y PMID: 39592515

Kumar, A. R., & Shrimali, G. (2024). *Role of policy in the development of business models for battery storage deployment: The California case study* (Working Paper). Sustainable Finance Initiative, Stanford University. Retrieved from https://sfi.stanford.edu/sites/sfi/files/media/file/role_of_policy_in_development_of_business_models_for_battery_storage_deployment_-_the_california_case_study_working_paper_0_0_0_0.pdf

Kumar, G. V. B., & Palanisamy, K. (2020). A review of energy storage participation for ancillary services in a microgrid environment. *Inventions (Basel, Switzerland)*, *5*(4), 63. DOI: 10.3390/inventions5040063

Lew, D., Brinkman, G., Ibanez, E., Hodge, B. M., Hummon, M., Florida, A., & Heaney, M. (2013)... *Western Wind and Solar Integration Study Phase*, *2*. Advance online publication. DOI: 10.2172/1095399

Libber, M., Gariya, N., & Kumar, M. (2024). A comprehensive analysis of supercapacitors with current limitations and emerging trends in research. *Journal of Solid State Electrochemistry*. Advance online publication. DOI: 10.1007/s10008-024-06107-x

Machín, A., Morant, C., & Márquez, F. (2024). Advancements and Challenges in Solid-State Battery Technology: An In-Depth Review of solid electrolytes and anode innovations. *Batteries*, *10*(1), 29. DOI: 10.3390/batteries10010029

Mlilo, N., Brown, J., & Ahfock, T. (2021). Impact of intermittent renewable energy generation penetration on the power system networks – A review. *Technology and Economics of Smart Grids and Sustainable Energy*, *6*(1), 25. Advance online publication. DOI: 10.1007/s40866-021-00123-w

Nassani, A. A., Imran, M., Khan, S., Zaman, K., Khan, H. U. R., & Haffar, M. (2025). Financial integration and economic growth: Impact of renewable energy investments, technology transfer, and climate change on Europe and Central Asian economies. *Financial Innovation*, *11*(1), 41. Advance online publication. DOI: 10.1186/s40854-024-00733-0

Nikolaos, P. C., Marios, F., & Dimitris, K. (2023). A review of pumped hydro storage systems. *Energies*, *16*(11), 4516. DOI: 10.3390/en16114516

Olabi, A. G., Allam, M. A., Abdelkareem, M. A., Deepa, T. D., Alami, A. H., Abbas, Q., Alkhalidi, A., & Sayed, E. T. (2023). Redox Flow Batteries: Recent development in main components, emerging technologies, diagnostic techniques, Large-Scale applications, and Challenges and barriers. *Batteries*, *9*(8), 409. DOI: 10.3390/batteries9080409

Osman, A. I., Ayati, A., Farrokhi, M., Khadempir, S., Rajabzadeh, A. R., Farghali, M., Krivoshapkin, P., Tanhaei, B., Rooney, D. W., & Yap, P. (2024). Innovations in hydrogen storage materials: Synthesis, applications, and prospects. *Journal of Energy Storage*, *95*, 112376. DOI: 10.1016/j.est.2024.112376

Oyedotun, K. O., Ighalo, J. O., Amaku, J. F., Olisah, C., Adeola, A. O., Iwuozor, K. O., Akpomie, K. G., Conradie, J., & Adegoke, K. A. (2022). Advances in supercapacitor Development: Materials, processes, and applications. *Journal of Electronic Materials*, *52*(1), 96–129. DOI: 10.1007/s11664-022-09987-9

Prakash, K., Ali, M., Siddique, M. N. I., Chand, A. A., Kumar, N. M., Dong, D., & Pota, H. R. (2022). A review of battery energy storage systems for ancillary services in distribution grids: Current status, challenges and future directions. *Frontiers in Energy Research*, p. *10*. DOI: 10.3389/fenrg.2022.971704

Qazi, S. H., Rodríguez, R. C., & Vandevelde, L. (2025). Vanadium Redox Flow Batteries: Performance Insights and Innovative Applications for a Greener Future. In *Lecture notes in civil engineering* (pp. 53–60). DOI: 10.1007/978-981-97-8348-9_7

Rabi, A., Radulovic, J., & Buick, J. (2023). Comprehensive Review of Compressed Air Energy Storage (CAES) technologies. *Thermo*, *3*(1), 104–126. DOI: 10.3390/thermo3010008

Rampai, M. M., Mtshali, C. B., Seroka, N. S., & Khotseng, L. (2024). Hydrogen production, storage, and transportation: Recent advances. *RSC Advances*, *14*(10), 6699–6718. DOI: 10.1039/D3RA08305E PMID: 38405074

Raza, M. S., Abid, M. I., Akmal, M., Munir, H. M., Haider, Z. M., Khan, M. O., Alamri, B., & Alqarni, M. (2024). A comprehensive assessment of storage elements in hybrid energy systems to optimize energy reserves. *Sustainability (Basel)*, *16*(20), 8730. DOI: 10.3390/su16208730

Rekioua, D. (2023). Energy storage systems for photovoltaic and wind systems: A review. *Energies*, *16*(9), 3893. DOI: 10.3390/en16093893

Rezaei, M., Sameti, M., & Nasiri, F. (2021). Biomass-fuelled combined heat and power: Integration in district heating and thermal energy storage. *Clean Energy*, *5*(1), 44–56. DOI: 10.1093/ce/zkaa031

Safari, A., Daneshvar, M., & Anvari-Moghaddam, A. (2024). Energy intelligence: A systematic review of artificial intelligence for energy management. *Applied Sciences (Basel, Switzerland)*, *14*(23), 11112. DOI: 10.3390/app142311112

Seh, Z. W., Jiao, K., & Castelli, I. E. (2023). Artificial intelligence and machine learning in energy storage and conversion. *Energy Advances*, *2*(9), 1237–1238. DOI: 10.1039/D3YA90022C

Sibilla, M., & Kurul, E. (2023). Towards Social Understanding of Energy Storage Systems—A Perspective. *Energies*, *16*(19), 6868. DOI: 10.3390/en16196868

Tang, C., Wang, H., Huang, J., Qian, W., Wei, F., Qiao, S., & Zhang, Q. (2019). 3D Hierarchical Porous Graphene-Based Energy Materials: Synthesis, functionalization, and application in energy storage and conversion. *Electrochemical Energy Reviews*, *2*(2), 332–371. DOI: 10.1007/s41918-019-00033-7

Vaghela, P., Pandey, V., Sircar, A., Yadav, K., Bist, N., & Kumari, R. (2023). Energy storage techniques, applications, and recent trends: A sustainable solution for power storage. *MRS Energy & Sustainability : a Review Journal*, *10*(2), 261–276. DOI: 10.1557/s43581-023-00069-9

Wang, H., Ozkan, C. S., Zhu, H., & Li, X. (2023). Advances in solid-state batteries: Materials, interfaces, characterizations, and devices. *MRS Bulletin*, *48*(12), 1221–1229. DOI: 10.1557/s43577-023-00649-7

Wei, P., Abid, M., Adun, H., Awoh, D. K., Cai, D., Zaini, J. H., & Bamisile, O. (2023). Progress in energy storage technologies and methods for renewable energy systems application. *Applied Sciences (Basel, Switzerland)*, *13*(9), 5626. DOI: 10.3390/app13095626

Worku, M. Y. (2022). Recent Advances in Energy Storage Systems for Renewable Source Grid Integration: A Comprehensive review. *Sustainability (Basel)*, *14*(10), 5985. DOI: 10.3390/su14105985

Xie, Z., Jin, Q., Su, G., & Lu, W. (2024). A review of Hydrogen Storage and Transportation: Progresses and challenges. *Energies*, *17*(16), 4070. DOI: 10.3390/en17164070

Yadav, K., Sircar, A., & Shah, M. (2024). Geothermal power production, hybridization, and storage. In *Geothermal fields of India* (pp. 45–70). Springer., DOI: 10.1007/978-3-031-53364-8_3

Yang, M., Hunger, R., Berrettoni, S., Sprecher, B., & Wang, B. (2023). A review of hydrogen storage and transport technologies. *Clean Energy*, *7*(1), 190–216. DOI: 10.1093/ce/zkad021

Yang, S., Liu, X., Li, S., & Zhang, C. (2022). Advanced battery management system for electric vehicles. In *Key Technologies on New Energy Vehicles*. DOI: 10.1007/978-981-19-3490-2

Yao, L., Yang, B., Cui, H., Zhuang, J., Ye, J., & Xue, J. (2016). Challenges and progress of energy storage technology and its application in power systems. *Journal of Modern Power Systems and Clean Energy*, *4*(4), 519–528. DOI: 10.1007/s40565-016-0248-x

Zheng, N., Qin, X., Wu, D., Murtaugh, G., & Xu, B. (2022). Energy Storage State-of-Charge Market Model. *arXiv (Cornell University)*. https://doi.org//arxiv.2207.07221DOI: 10.48550

KEY TERMS AND DEFINITIONS

Battery Technologies: Various types of batteries are used to store energy, including lithium-ion, solid-state, and flow batteries.

Energy Conversion: The process of changing energy from one form to another, such as converting solar energy into electrical energy.

Energy Storage: The capture of energy produced at one time for use at a later time, essential for balancing supply and demand in energy systems.

Grid Stability: The ability of the energy grid to maintain a consistent supply of electricity despite fluctuations in demand or supply.

Hydrogen Storage: The storage of hydrogen gas for use as an energy carrier is crucial for renewable energy systems.

Renewable Energy Integration: The process of incorporating renewable energy sources into the existing energy grid to ensure a sustainable and reliable energy supply.

Supercapacitors: Energy storage devices that store energy electrostatically and offer rapid charge and discharge capabilities.

Sustainability: The ability to meet the needs of the present without compromising the ability of future generations to meet their own needs.

Chapter 16
Revolutionizing Solar Cells:
Novel Materials in Next-Generation Energy Storage Solutions

R. Venkatesh
https://orcid.org/0000-0002-0355-8857
PSNA College of Engineering and Technology, Dindigul, India

Natarajan Sampathkumar
https://orcid.org/0000-0001-7329-7712
SSM College of Arts and Science, Dindigul, India

ABSTRACT

The quest for higher, more sustained, and affordable solar cells has opened the floodgates to an incredible period of innovation in energy storage technologies. This chapter, "Revolutionizing Solar Cells: Novel Materials in Next-Generation Energy Storage Solutions," reviews frontier materials driving this evolution of solar energy systems. It focuses on the new material science developments that achieve higher efficiency and durability in solar cells by developing materials that have huge challenges of efficiency, reliability, and scalability. The chapter reviews in detail a good number of materials that have been used, such as perovskites, organic photovoltaics, and quantum dots, and their respective evolutions of advanced nanomaterials, and discusses their impact on energy storage. It provides insights from cutting-edge research and practical applications in the quest for sustainable and futuristic solar energy solutions.

DOI: 10.4018/979-8-3693-9316-1.ch016

1. INTRODUCTION

1.1 Overview of Solar Energy and Energy Storage Challenges

Solar energy was a leading alternative source of clean, sustainable energy. The potential remains huge today. Photovoltaic (PV) cells convert sunlight directly into electricity, and solar thermal systems create heat that can be used to generate electricity or for heating purposes (figure.1). Some benefits of solar energy include reduced greenhouse gas emissions and climate change mitigation (Mawire A, 2015). It reduces the environmental impact of the source of traditional energy, which minimizes air and water pollution.

Figure 1. Solar energy collection system

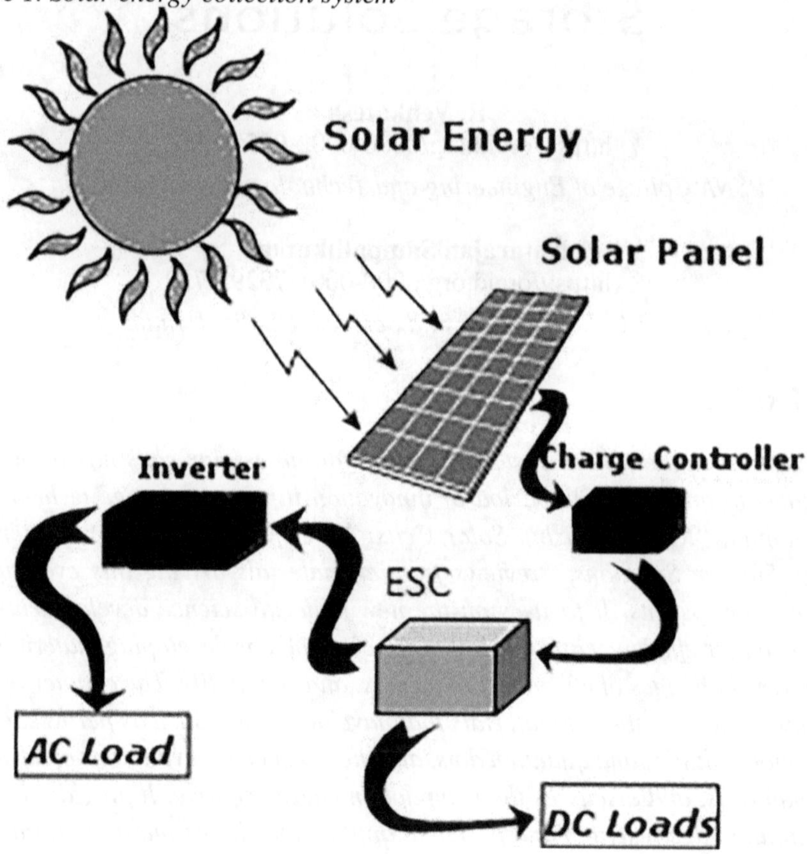

The long-term cost of the system can be considerably reduced due to the fact that sunlight is free and plentiful whereas oil drilling and coal mining carry on a recurring cost. Intermittency is the primary challenge that accompanies the integration of solar energy with existing power grids, along with the inevitable compromise (Das et al., 2014). Due to this, solar energy cannot be available at all times, particularly when it is in peak demand (Kouskso et al., 2014). To make up for this, efficient means of storing energy have to be provided, such as advanced types of batteries like lithium and solid-state, pumped hydro storage, and compressed air storage. These systems have the potential of storing extra energy generated during sunny hours and releasing it when demand is much higher or when sunlight is unavailable.

The current technologies, despite the significant advancements made in energy storage, still have some drawbacks: such as storage capacities, efficiency on energy loss during storage and retrieval, and cost put a constraint on large-scale use. Additionally, there is also the environmental impact during production of a battery, including sourcing raw materials and its end-of-life recycling or disposal. More efficient, durable, and lower-cost storage solutions are required to ensure proper unlocking of solar energy sources (Vijayakumar et al., 2018). Besides, optimization in energy delivery and consumption depends upon smart grid technology. Another vital factor about it is overcoming some barriers with a more reliable energy system based on the sun power's variation.

1.2 Importance of Material Innovation in Solar Cells

New materials are enhancing the efficiency, cost, and performance of solar cells, which makes solar energy more accessible and effective. For instance, perovskite solar cells absorb light and convert it into electricity better than traditional silicon-based cells. Organic photovoltaics and thin-film technologies are cheaper to produce and use fewer resources, making solar energy more affordable. Improved durability makes solar panels more resistant to environmental factors like moisture and UV radiation, hence reducing the costs of maintenance (Mohammad & Mahjabeen, 2023). New materials have also made flexible solar panels that can bend along curved surfaces. This means wearable and portable solutions are possible with these materials, which are sustainable, non-toxic, and recyclable. This reduces the environmental impact in manufacturing and at the end when they are discarded. With increased efficiency and scalability, the production of solar panels will increase, paving the way to a greener future and a more widespread use of solar energy.

2. HISTORICAL CONTEXT AND EVOLUTION

2.1 Development of Solar Cell Materials

The research in the field of development of next-generation solar cell materials is one of the most energetic and relentless efforts toward upping both efficiency and affordability of photovoltaic devices. The development of cell materials, although very intensive, is full of research with the objective of making it more efficient and having lower costs than those of the already existing ones, with efficiency in the energy conversion, reduction in cost, and addressing environmental and other concerns (Raj et al., 2021). Major developments in this application area of solar cell materials are summarized below.

Silicon-Based Solar Cells

a) **Monocrystalline Silicon**: The material used in this type is single-crystal silicon, with very high efficiency and a very long life a common variant applied in industry. Material purity and cell design improvements have been improved enormously, making their performance and efficiency very close to passivated emitter rear cell (PERC) technology.
b) **Polycrystalline Silicon**: These cells are made up of silicon crystals that are melted together. They are cheaper to manufacture than the monocrystalline but are usually of a lower efficiency. Improvements are done continuously to increase the quality of the crystal and, at the same time, optimize the costs of production.

Thin-Film Solar Cells

a) **Cadmium Telluride (CdTe)**: CdTe is low cost to fabricate, and efficiencies are fairly high, especially for thin films. R&D work will try to increase its efficiency, while limiting the environmental implications of using cadmium.
b) **Copper Indium Gallium Selenide (CIGS)**: CIGS cells are very efficient as well as very flexible, but the processing is intricate as well as costly. Recent new development has been worked on material composition and processing techniques to increase efficiency while at the same time trying to make a reduction in the cost.

c) **Amorphous Silicon (a-Si)**: These cells are good for thin-film applications, flexible, and are light. The cells are light but less efficient when compared to the crystalline silicon cells. They have a lot of applications areas. For instance, they can be used in solar calculators and flexible panels. Some research is going on for their improved stability and efficiency.

Perovskite Solar Cells

a) **Hybrid Organic-Inorganic Perovskites**: his class of materials has become very exciting owing to the high efficiency achieved with reasonably low production cost. Inherent low-temperature processability makes perovskites ripe for a plethora of applications. Linked to their stability and scalability issues, toxicity associated with lead-based perovskites sets the future directions.
b) **Organic Semiconductors**: Organic Photovoltaics work through the absorption of light in organic material to yield an electron/hole pair followed by charge separation and collection. They combine the flexibility and light weight of plastics with an easy, low-cost processing. Research is addressing these materials to increase both the efficiency and life of organic devices. Other research on device architectures and new organic materials is overcoming some fundamental limitations of these organic cells in relation to their inorganic counterparts.

Quantum Dot Solar Cells

a) **Nanomaterials**: Quantum Dots are a type of nanomaterial, semiconductor nanocrystals with the ability to absorb and then emit light. This could greatly improve efficiency due to their tunable optical properties. Research in this direction covers enhancement of their stability, integration into solar cells, and also problems related to material synthesis and scalability.
b) **Stacked Cells**: The multijunction cells are cells designed with more than one layer of photovoltaic material; in this case, each of these layers is purposed to capture a different slice of the solar spectrum. This is a trick that can seriously boost efficiency. In recent times, research has been specific to improving the materials applied on these layers and sourcing less expensive ways for producing them.
c) **Flexible Substrates**: Lightweight, thin solar panels being developed would make them simple to affix to a wide array of substrates, and they are already being tested with cloth and on curved surfaces. Flexible material development and

materials printable for this purpose are needed to enable the rapid integration of solar technologies into new products.

Sustainable and Eco-Friendly Materials

Recycling and Alternatives: The research on sustainable materials aims at finding alternatives to toxic or rare materials used in solar cells and developing efficient methods of recycling. For example, lead-free perovskites and recyclable parts of solar cells are being developed now to save the environment.

Development of materials in solar cells, hence, is a general, enormous field aimed at increasing efficiency and reducing the cost while at the same time addressing the created problems of this technology in regard to the environment (Paranthaman et al., 2016). Progress in the science of materials and processing for fabrication has continuously pushed development in solar technologies toward becoming more practical and plausible for clean energy.

2.2 Milestones in Energy Storage Technologies

Milestones in energy storage technologies have been major steps toward more efficient and reliable systems of power supply. The discovery of the lead acid battery by Gaston Planté in 1859 gave a fillip to the invention of viable rechargeable battery technology. The nickel-cadmium batteries, which appeared in the 1950s, offered much better life and improvement in performance. The 1960s saw the beginning of pumped hydro storage, whereby excess electricity is used to raise water to an upper level so that, at a later time, the energy is released for grid stability. Then comes the late 20th and early 21st century innovations of new technologies, including lithium-ion batteries, which provided great improvements in energy density and lifetime and would later revolutionize portable electronics and electric vehicles (Breez, 2018). Other recent developments include solid-state batteries forecasted to improve safety and energy density and flow batteries, scalable solutions for long-duration energy storage. All these put together are continuing to increase the impact of energy storage in integrating renewables with a reliable and firm power supply (Budt, 2016).

Table 1. Milestones in energy storage technologies

Milestone	Technology	Year	Description and Advancements
Early Development	Lead-Acid Batteries	1859	Invented by Gaston Planté, this was the first practical rechargeable battery. Used extensively in automotive and stationary applications.
Early 20th Century	Flywheel Energy Storage	1900s	Early systems stored energy in the rotational motion of a flywheel. Utilized for power regulation and backup energy.
Mid-20th Century	Nickel-Cadmium Batteries	1950s	Developed as a reliable rechargeable battery technology, known for its durability and performance in various applications.
1960s	Pumped Hydro Storage	1960s	Utilizes excess electricity to pump water to a higher elevation, which is then released to generate electricity when needed. A key technology for grid stabilization.
1970s	Sodium-Sulfur Batteries	1970s	High-temperature batteries offering high energy density and efficiency. Primarily used in large-scale energy storage applications.
1980s	Nickel-Metal Hydride Batteries	1980s	Introduced as an alternative to nickel-cadmium batteries, providing higher energy density and less environmental impact.
1990s	Lithium-Ion Batteries	1990s	Revolutionized portable electronics and electric vehicles with high energy density, lightweight, and long cycle life.
2000s	Supercapacitors	2000s	Provide rapid charging and discharging capabilities, ideal for applications requiring quick bursts of energy.
2010s	Solid-State Batteries	2010s	Emerging technology offering improved safety, higher energy density, and longer life by using solid electrolytes instead of liquid ones.
2010s	Flow Batteries	2010s	Feature liquid electrolytes stored in external tanks, offering scalability and long-duration storage for grid applications.
2020s	Advanced Lithium-Ion Variants	2020s	Innovations include higher energy density cells, faster charging, and improved safety features, enhancing performance for electric vehicles and renewable integration.
2020s	Solid-State Flow Batteries	2020s	Combining solid-state and flow battery technologies to offer high energy density and long-duration storage with improved safety and scalability.

Table 1 captures key developments and advancements in energy storage technologies, reflecting their evolution and impact on modern energy systems.

3. PEROVSKITE SOLAR CELLS

3.1 Structure and Properties of Perovskites

Perovskites are a class of crystalline materials with an inimitable crystal structure and tremendous breadth of applications so far in solar cells and the field of optoelectronic areas. A briefing on its structure and property is given below:

Crystal Structure:

Figure 2. Perovskite ionic structure

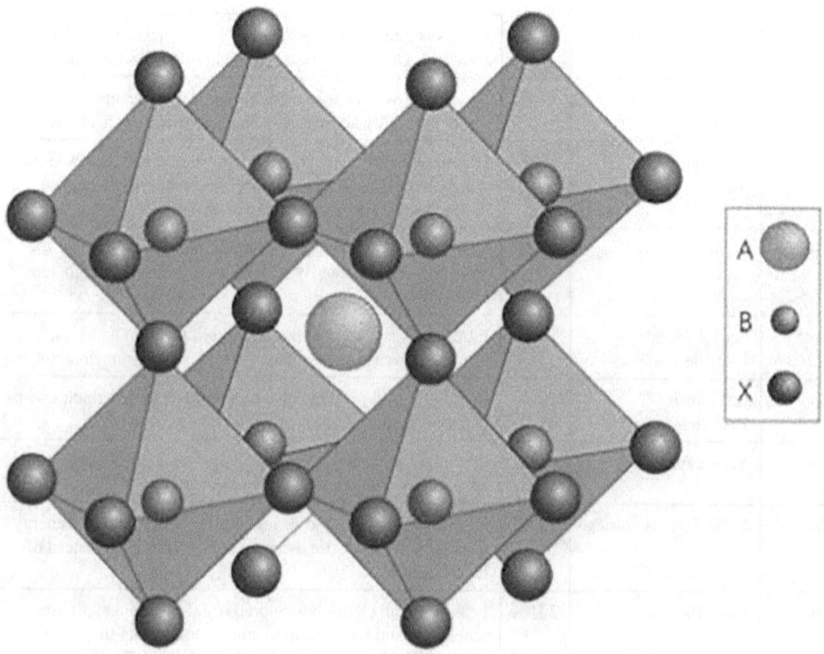

General Formula: The general formula for perovskite materials is ABX_3, where:

- **A** is a large cation (typically a metal or an organic molecule),
- **B** is a smaller cation (often a metal ion),
- **X** is an anion (usually oxygen, but can also be halides or other elements).

Structural Arrangement: In the perovskite structure, the **A** cations are located in the corners of a cube, the **B** cations are positioned in the center of the cube, and the **X** anions occupy the face centers of the cube (figure.2). This results in a three-dimensional network of corner-sharing octahedra formed by the **B** cations and **X** anions, with the **A** cations interspersed in the gaps. (Azhari et al., 2020).

Properties of Perovskites

Optical Properties:

a) **Absorption**: The light absorption coefficient of perovskites is very high. Thus, they are highly capable of absorbing a wide range of spectrums covering visible and near-infrared regions of the light. This property has also positioned this material for solar applications.
b) **Emission**: Most the perovskites have high photoluminescence and electroluminescence efficiencies and thus proved to be good for light-emitting devices and display applications.

Electrical Properties:

a) **High Carrier Mobility**: Most perovskite materials usually possess high charge carrier mobility, one of the most paramount factors that guarantee successful application in electronic devices. It allows fast movement of electrons and holes, hence high efficiency in the working of devices like solar cells and LEDs.
b) **Tunable Bandgap**: The electronic bandgap in the perovskite materials can be changed with the variation of composition of the A, B, and X ions, hence allowing tailoring of its electronic and optical characteristics toward a particular application.

Thermal Properties:

a) **Stability**: Most of the inorganic perovskites exhibit good thermal stability. However, some of the hybrids may degrade upon excessive periods of heat or humidity.
b) **Temperature Dependence**: Temperature dependence can alter the crystalline structure of perovskite compounds and hence their overall effectiveness in applications. Research is done nowadays for making them thermally stable.

Mechanical Properties:

a) **Flexibility**: Hybrid organic component-based perovskite has flexibility and therefore suitable for flexible electronic devices application as well as flexible photovoltaic system.
b) **Durability**: Inorganic perovskites are usually more stable than organic ones, but long-term stability in real conditions is under investigation nowadays.

Chemical Properties:

a) **Solubility and Processing**: The perovskites are solution processable, especially the hybrid ones, which enable thin film formation and large-area coatings. This feature is very useful in fabricating and scaling up several applications like solar cells.
b) **Reactivity**: Some perovskites, especially those with lead or other heavy metals, do raise some concerns about reactivity and stability in terms of the environment and health. Currently, research is underway to develop lead-free alternatives or less toxic ones.

Applications

a) **Solar Cells**: The application of perovskite in solar photovoltaic cells is highly recognized, and an improved light absorption and charge transport characteristics result in highly efficient solar cells.
b) **Light-Emitting Devices**: The strong luminescence brings perovskite into a suitable position for LEDs and other display technologies.
c) **Photodetectors and Lasers**: The excellent optoelectronic characteristics of the perovskite materials are the cause of their application in photodetectors and laser technology.

The key advantage for perovskites is their flexible crystalline lattice, coupled with other features promoting practical and useful current-adjustable characteristics of electronic features that can bring in flexibility with considerable optical absorption (Bisquert, 2017). With such great characteristics, they turn out to be highly promising materials for different state-of-the-art applications, however ongoing research aims at resolving the problems of stability and environmental impact.

3.2 Advances in Perovskite Materials

The research in perovskite materials has been intensive in various photovoltaic and optoelectronic domains, among many more state-of-the-art technologies. It is propelled by the incessant research towards enhancing efficiency, stability and scalability of the perovskite-based devices.

Enhanced Efficiency

a) **Tandem Solar Cells**: Perhaps one of the more significant advances in solar technology is when perovskites are combined with silicon in tandem solar cells. In such configurations, a perovskite layer is deposited over a silicon cell to increase the portion of the sun's radiation utilized and thus considerably increasing the overall efficiency. Indeed, tandem solar cells have shown efficiencies above 30%, indicating substantial enhancement for conventional silicon cells used on their own.

b) **Perovskite Composition Optimization**: The researchers have been optimizing the perovskite composition for functional efficiency. Adding mixed halides increases light absorption with stability, such as iodide-bromide mixtures. Adding new cation combinations shows better efficiencies with better thermal stability, like formamidinium and cesium.

Improved Stability and Durability

a) **Stabilized Hybrid Perovskites**: Hybrid organic-inorganic perovskites have been plagued by problems of stability for a long time. More recent developments involve encapsulation techniques and stabilizers protecting against moisture and degradation by UV light, indeed yielding much more stable perovskite solar cells with far longer lifetimes.

b) **Lead-Free Perovskites**: Considerations of the environmental and health implications of lead-based perovskites set forth research goals of making lead-free alternatives. The further research of tin- and bismuth-based materials will give high efficiencies with much lower toxicity. While it is new, a promising alternative in the developmental stage.

Advanced Processing Techniques

a) **Scalable Manufacturing**: Recently, improvements in the methodology for processing perovskite materials have been noticed. Effective mass production is under way. Some of the currently fine-tuning techniques include roll-to-roll printing and spray coating to enable large-area perovskite films that will be commercially applied.

b) **Solution Processing**: The solution-processing-based method is optimized in the case of perovskite films to realize uniformity and quality. Some of the processes involved are antisolvent treatment and controlled crystallization, and the advances in these processes are covered within the forming of the film and reduction of defects for decreased attainment of high-performance devices.

Novel Perovskite Structures

a) **2D and 2D/3D Hybrid Perovskites**: Such context has been developed through works on 2D perovskites and their 2D/3D hybrid structures. In 2D perovskites, organic and inorganic layers are arranged alternately. Such a structure provides good stability and low toxicity. Combination of both 2D with 3D will optimize the performance and mitigate the stability issues.

b) **Quantum Dot Perovskites**: Nanometer-sized perovskite single crystals with size-tunable optical features are defined as quantum dots of perovskite. The application of such quantum dots in light-emitting and display devices is realized in view of very vivid luminosity and highly precise control of emission colour.

Perovskites are bounded by an ever-increasing advance, pushing the limit of what can be realized in photovoltaics and optoelectronic areas, among others. Efficiency and stability improvements, together with processing techniques and the exploration of new novel structures and applications, will constitute the driving engines for the next generation of perovskite-based technologies (Khirade & Raut, 2022). Researchers are sure that further works will establish the very important role perovskites hold in various high-tech areas.

3.3 Challenges and Future Directions

While perovskite solar cells (PSCs) have shown great promise due to their excellent efficiency and the lowest production cost, several critical challenges impede their wide applications. First, the long-term stability; these solar cells usually rapidly degrade upon environmental exposure to moisture, heat, or UV irradiation that seriously affects their life span. Apart from that, most of the perovskite formulas include lead, which immensely threatens the environment and human health because of the potential toxicity. Research on lead-free perovskite material is under development with a view to avoiding such risks, but hitherto, finding alternatives able to preserve high efficiency is quite a task. Besides that, scalability of production processes has so far been one of the major hindrances, since production methods in their present state are not cost-effective to allow large-scale manufacturing (Mahapatra,2022).

Encapsulation techniques for perovskite solar cells in advanced ways for stability, optimized material compositions to reduce or completely avoid the use of toxic elements, and refinement of fabrication processes toward better scalability and cost reduction are some other prospective future directions of research on perovskite solar cells (Afreen et al., 2021). It is the overcoming of such challenges through innovation in material science, engineering, and manufacturing that holds the key to the realization of full potential for perovskite solar technology and its integration into the Global Energy Market.

4. ORGANIC PHOTOVOLTAICS

4.1 Fundamentals of Organic Materials

Organic materials are basically carbon-based compounds finding their applications in modern electronics, photovoltaics, and material sciences. The general properties of organic materials arise from the nature of their molecular structure and the interaction between the molecules.

Molecular Structure: Organic materials are typified by their carbon-hydrogen frames, which usually include a conjugation consisting of alternate single and double bonds that provide the channels through which delocalized electrons can migrate. This conjugation is responsible for their special electronic and optical properties, including high absorbency and emission of light in applications.

Electronic Properties: Most organic materials strongly depend on their molecular structure. In organic semiconductors, for example, the electric conduction or insulation depends on the disposition of π-electrons in conjugated systems. The electronic properties of these materials span from insulators to conductors and are used in OLEDs, organic solar cells, and field-effect transistors.

Optical Properties: Organic materials generally exhibit great light absorption and photoluminescence due to their applications deemed critical in OLEDs and OPVs. The electronic band structure interaction prescribes organic materials with light and molecular interactions, enabling them to be absorbed at particular wavelengths of light while emitting at other wavelengths.

Processing and Fabrication: Organic devices are generally processed by solution processing, spin coating, or printing techniques that are arguably cheaper and more flexible than traditional semiconductor fabrication. This freedom in processing enables lightweight, flexible, and large-area devices.

Stability and Degradation: One of the challenges with organic materials is their stability. These materials are susceptible to degradation by environmental factors such as moisture, oxygen, and UV light, which in turn may affect their performances and lifetime. Therefore, active research by various encapsulation techniques and chemical modifications is highly underway to improve the stability of these materials.

Applications: Organic material finds its applications in an amazingly broad area, starting with electronics, covering OLEDs for display and lighting, organic photovoltaic as one form of energy conversion, organic semiconductors for flexible electronics. The properties enable innovative applications such as organic sensors and organic memory devices.

All fundamentals of organic materials involve knowledge of molecular structure, electronic and optical properties, processing techniques, and stability problems. Such knowledge can further develop the applications and performances of such materials in many technologies (Andersson & Fahlman, 2016).

4.2 Recent Innovations in OPV Materials

The recent innovation in organic photovoltaics has targeted efficiency, stability, and scalability. This developing research makes the device quite practical and feasible for wide application.

Table 2. Recent innovations in organic photovoltaic materials

Year of Innovation	Focus	Material Type	Key Features	Applications
2023	High Efficiency	Non-Fullerene Acceptors	Improved light absorption and charge transport; higher PCE.	Flexible solar panels, wearable electronics
2023	Stability	Cross linkable Polymers	Enhanced thermal and chemical stability; longer device lifetime	Outdoor solar panels, building-integrated photovoltaics
2022	Low-Cost Processing	Solvent-Processed Small Molecules	Solution-processable, low-cost production; high scalability	Large-area solar panels, consumer electronics
2023	High Absorption	Low-Bandgap Polymers	Broad absorption spectrum, especially in the infrared range	Energy-efficient window coatings, architectural applications
2024	Flexible Devices	High-Performance Nanosheets	High mechanical flexibility; enhanced power density and stability	Rollable solar panels, portable chargers
2023	Environmental Impact	Biodegradable Materials	Reduced environmental footprint; compostable or recyclable	Eco-friendly solar panels, temporary installations
2024	Tandem Cells	Hybrid Organic-Inorganic Perovskites	Combined organic and inorganic materials for higher efficiency	High-efficiency solar modules, space applications

Advanced Polymer and Small Molecule Materials

a) **New Conjugated Polymers:** The research of new conjugated polymers will optimize the bandgap and improve charge transport properties. For example, conjugated polymers based on the fused-ring structure significantly improve the performance of the OPV device, such as those containing diketopyrrolopyrrole (DPP) or benzodithiophene (BDT).

b) **High-Performance Small Molecules:** Small molecules are designed to possess high absorption coefficients and superior energy levels, such as non-fullerene acceptors (NFAs). Most of these have outperformed the conventional fullerene-based acceptors in power conversion efficiency (PCE) performance.

Tandem and Multi-Junction Cells

a) **Tandem Architectures:** Combining OPV cells with any other type of solar cell-silicon or perovskite-in a tandem configuration has the effect of making more efficient use of a broader spectrum of sunlight and, therefore, improving overall efficiency. Accordingly, these tandem structures are capable of achieving higher efficiencies than single-junction OPV cells.

b) **Layer Stacking:** Innovations in layer stacking and interfacial engineering presently underpin an advance in the performance of tandem cells by improving light management and charge extraction.

Moisture and Oxygen Barriers

a) **Advanced Encapsulation:** Novel techniques and materials for encapsulation, including barrier films and laminates, are proposed to protect OPVs from moisture and oxygen, known to degrade organic materials. These barriers will be able to extend the operational life of OPVs significantly.

b) **Stable Materials:** Research is done for new stable organic materials and additives that can enhance the robustness of the active layer to environmental stress. For instance, the use of stable polymers in organic semiconductor materials, or modification in the molecular structure, can yield good results in extending lifetime.

Hybrid and Composite Materials

a) **Hybrid Materials:** Hybrid materials can be made by incorporating organic materials with inorganic nanoparticles or nanostructures. The latter would enhance stability and further improve the performance of OPVs. For instance, the addition of metal oxides or carbon nanotubes to organic material will increase charge transport and environmental stability.
b) **Composite Films:** Composite films may thus provide improved stability with such organic semiconductor-based composite materials by the addition of additives or protectants and still retain their efficiency.

Solution Processing Techniques

a) **Roll-to-Roll Printing:** Advances in Roll-to-Roll (R2R) printing technologies have enabled the large-scale production of OPVs on flexible substrates. Being a cost-effective and scalable fabrication technique, this qualifies for commercial applications.
b) **Inkjet Printing:** Refinement in inkjet printing is being done for precision deposition of OPV material. Such precision helps fabricate intricate device structures with high resolution while offering minimum material waste.

Flexible and Lightweight Substrates

a) **Flexible Substrates:** Inventions in flexible substrates, like plastic or metallic foils, further extend the area of application of OPVs in wearables up to flexible photovoltaic panels.
b) **Lightweight Materials:** The development of lightweight organic materials, while maintaining efficiency and stability, is highly relevant for weight-sensitive applications.

Nanostructured Materials

a) **Nanostructured Light Management:** Inclusion of nanostructures, such as plasmonic nanoparticles or photonic crystals, further enhances the light absorption and results in an effective increase of the light path length within the OPV active layer.

b) **Nanostructured Electrodes:** Novel concepts in nanostructured electrodes, such as nano-patterned transparent conductors, allow for better charge collection and reduced losses.

Computational and Theoretical Advances

a) **Computational Modeling:** Advanced computation models and simulations are used to predict the properties of new organic materials and optimize them. It speeds up the process of finding new materials with the most desirable characteristics.
b) **Machine Learning:** Machine learning algorithms are now being utilized to analyze large datasets comprising material properties and performance metrics that will help in identifying promising candidates for OPV applications.

Emerging Materials and Technologies

a) **New NFAs:** There has been continuous development of new non-fullerene acceptors with better absorption spectra and superior energy levels, hence significant improvement in OPV efficiency. These are also increasingly used to replace traditional fullerene-based acceptors.
b) **Hybrid Perovskites:** Organic-inorganic hybrid perovskite materials have been more and more explored to improve the efficiency and stability of OPVs further. These can yield high efficiency while providing better stability than obtained with traditional organic materials alone.

Recent advances in organic photovoltaics have resulted in significant improvements in efficiency, stability, and processing. Advances in material science, fabrication techniques, and device architecture are seen to lead toward the development of more efficient and durable OPVs. Such developments have the potential for commercial applications in general use for renewable energy (Wang, (2015). Table 2 highlighting the focus of each innovation, the types of materials involved, their key features, applications.

4.3 Performance Metrics and Commercialization Prospects

The efficiency level of sunlight converted by a solar cell to usable electrical energy is measured as power efficiency converter (PCE), an important performance indicator for OPV cells. The best laboratory OPVs have already demonstrated efficiencies higher than 18%, reaching some 20% or more (Sampaio & González, 2022). Even though these improvements are very significant, they remain significantly lower than those achieved by traditionally produced silicon-based solar cells, whose efficiencies range well over 25%. Another critical factor for OPVs is their stability and longevity, as their practical use depends on how well they maintain performance over time, especially when exposed to moisture, heat, and UV radiation. Innovations in encapsulation and material stability are ongoing to address these issues.

Manufacturing cost reduction and upscaling of production are essential for OPVs to be commercially viable. Low-cost techniques like roll-to-roll and inkjet printing have potential for large-scale production without loss of performance. OPVs are suited for niche applications, including flexible, lightweight solar panels in building-integrated photovoltaics (BIPV), wearable electronics, and portable chargers (Kuhn, 2021). However, OPVs will compete with established technologies and newer innovations like perovskite solar cells. Improving the efficiency and stability of OPVs with lower costs can make them competitive. Government incentives and clean energy policies can further improve the commercial feasibility of OPVs, leading to consumer acceptance and integration with other forms of energy technologies.

While organic photovoltaics hold promises for diverse and flexible applications, commercial successes are going to have to be complemented by the development of performance metrics, cost-effective manufacturing processes, and strategic market integration (Drury et al., 2011). Should these challenges be overcome and the unique advantages of OPVs used, they can contribute considerably toward a sustainable future.

5. QUANTUM DOTS AND NANOMATERIALS

5.1 Introduction to Quantum Dots and Their Mechanisms

Quantum dots are tiny semiconductor particles exhibiting quantum mechanical behavior, resulting in unusual electronic and optical properties. Measuring between 2 and 10 nanometers, the small scale of QDs produces discrete energy levels and size-dependent optical properties, rendering them useful for numerous applications in optoelectronics, biological imaging, and photovoltaics. The energy gap between the valence and conduction bands is greater for decreasing size of quantum dots, with a shift in the absorption and emission spectra. Smaller-sized quantum dots absorb and emit at shorter wavelengths (blue shifted), while larger-sized ones at longer wavelengths (red shifted).

Quantum dots have high efficiency in converting absorbed light into emitted light, making them useful for applications in displays, imaging, and lasers. Their charge carrier dynamics, such as electron-hole recombination, play a role in their optical properties, and the separation of charge carriers is important for improving the efficiency of solar cells (García de Arquer, 2021). Surface passivation by organic ligands or inorganic coatings stabilizes quantum dots and prevents energy losses at defect sites. Apart from photovoltaics where they contribute to enhanced light absorption, quantum dots are applied in sensors to detect environmental changes and produce optical signals for applications in chemical and biosensing.

Quantum dots are semiconductor particles at the nanometer scale with special electronic and optical properties due to quantum confinement effects. The possibility of precisely tuning their optical properties, together with high quantum yields and versatility in surface modification, makes them very attractive in a wide range of applications involving advanced displays and imaging, through renewable energy toward sensing technologies (Horoz, 2017).

5.2 Enhancements in Solar Cell Efficiency

Accordingly, one of the most significant ways to enhance the performance and economic viability of solar energy technologies is by increasing the efficiency of solar cells. There are a number of approaches and innovations for improving the efficiency of solar cells under consideration, each dealing with a different part of the process of energy conversion from the sun (Foggiato, 2009). Recent breakthroughs in solar technology emphasize improved efficiency and stability through novel materials and creative designs. In this context, materials such as perovskites and organic photovoltaics hold great promise for higher efficiencies than traditional silicon-based cells. Perovskites, in particular, have achieved high efficiencies rapidly due to their favorable electronic properties.

Quantum dot and tandem solar cells absorb light more efficiently. Quantum dots absorb tailored specific wavelengths, while tandem cells stack many layers to capture a larger portion of the solar spectrum. The techniques applied for light management-enhanced absorption contain anti-reflective coatings and nanostructures, whereas advanced charge carrier management and optimized interfaces minimize energy losses. Roll-to-roll processing and new techniques for manufacturing combined with flexible design options and various thermal management solutions have further boosted the flexibility and efficiency of solar cells (Maize et al., 2023). Hybrids that marry solar cells to energy storage or thermal technologies could provide improved efficiency and stability and hence make the source of energy in solar panels even more viable.

5.3 Integration with Existing Technologies

Combining new solar technologies with the present systems improves efficiency and widens their applications. Among these, perovskite or tandem cells might be integrated with silicon-based panels for improved energy capture or used along with solar photovoltaics, coupled with thermal collectors. The advanced solar systems must combine with an electrical grid for integration purposes, involving new panels, smart grid technology, and energy storage solutions. Retrofitting old solar panels with new, efficient technologies saves cost and disruption. New solar solutions should also be compatible with the existing mounting structures, such as flexible panels for unconventional surfaces. Building-integrated photovoltaics, such as solar windows and roofing tiles, enhance aesthetics and functionality. The cost and efficiency considerations, coupled with industry standards, make these technologies viable and compatible. Finally, the systems must be designed for easy maintenance and future upgrades as technology evolves. By paying due attention to these aspects, new solar technologies can then be integrated into existing systems, allowing better performance, increased energy efficiencies, and further dissemination of solar solutions (Akram & Hasanuzzaman, 2022).

6. ADVANCED NANOMATERIALS

6.1 Types of Nanomaterials Used in Solar Cells

Nanomaterials have a great chance of becoming a critical factor in the further development of solar cell technologies due to their specific nanoscale properties that can improve light absorption, charge transport, and efficiency in general (Ma, 2021).

a) **Quantum Dots**: Quantum dots are semiconductor particles in nanometer dimensions with size-tunable optical properties. They can be engineered to absorb specific wavelengths of light more effectively compared to their bulk counterparts, which increases light absorption and thereby enhances solar cell efficiencies. Quantum dot solar cells can probably also achieve multiple exciton generation (MEG) process whereby a single incident photon generates more than one electron-hole pair-and thus further increase energy-conversion efficiency.

b) **Nanowires**: Nanowires are one-dimensional nanostructures with high surface-to-volume ratios and distinctive electrical properties. Semiconductor nanowires could be employed in solar cells to achieve highly efficient light-harvesting structures and enhanced charge transport. They also contribute to increasing light absorption through the scattering of light inside the device, which increases the probability of photon absorption.

c) **Nanorods**: The nanorods are elongated nanoparticles that can exhibit strongly anisotropic optical and electronic properties. These have been applied to solar cells to improve the light absorption and carrier mobility. Their shape allows better light trapping, thus more control over the light-material interaction.

d) **Nanoplates**: While nanoplates are flat, two-dimensional nanostructures that could offer improved light absorption and charge separation, application can be made in a wide variety of solar cells, including but not limited to those with layered active material-based solar cells or hybrid material active material-based solar cells, for better overall efficiency.

e) **Nanofibers**: Nanofibers have elongated fibers whose diameters fall in the nanometric scale. Their application in solar cells comes by the making of high-surface-area structures that favor improved light absorption and allow better charge transport. The ability to incorporate nanofibers into flexible solar cells opens up possibilities in wearable and portable applications.

f) **Nanocomposites**: In general, nanocomposites are those materials which are fabricated by adding nanoparticles into any material to achieve better properties for specific applications. In solar cells, nanocomposites may be able to improve overall performance by combining the advantages of various different nanomaterials. As one example, embedding metal nanoparticles in semiconductors can enhance light absorption due to plasmonic effects.

g) **Graphene and Graphene Oxide**: Graphene is a single layer of carbon atoms arranged in a hexagonal lattice, while graphene oxide with extra oxygen functionalities has the potential to enhance electrical conductivity and mechanical flexibility in solar cells. Because of their very high surface area and excellent electrical properties, their applications have been explored in several different solar cell layers, including transparent conductive electrodes and charge transport layers.

h) **Carbon Nanotubes**: Carbon nanotubes (CNTs) have a cylindrical nanostructure with very high electrical and thermal conductivity. In solar cells, they enhance charge transportation, mechanical strength, and the overall efficiency of the device. CNTs can be incorporated in various layers of solar cells for enhanced performance and stability.

i) **Titanium Dioxide Nanoparticles**: It is a photoanode material that finds broad adoption in dye-sensitized solar cells because of its high surface area and mesoporous structures, enhancing the efficient dye molecules' adsorption and electron transportation capability.

j) **Zinc Oxide Nanostructures**: ZnO nanostructures include nanoparticles, nanorods, and nanotubes, and all these have been employed in different types of solar cells. Common applications include electron transport layers or parts of the photoanode in DSSCs, enhancing light absorption and charge separation.

These nanomaterials possess varied tunable properties for different aspects of solar cell performance, such as light absorption and charge transport, flexibility, and stability. Further development and optimization might enable these nanomaterials to realize higher efficiency in solar energy technologies (Batmunkh, M. 2020).

6.2 Benefits and Limitations

Advances in solar cell technology due to nanomaterials include improvement in light absorption, increased charge carrier mobility, and flexibility. Solar cells can be made to absorb a broad range of the sun's radiation by tuning the size and shape of quantum dots and nanowires. Graphene and carbon nanotubes enhance charge transport, which further reduces the recombination losses and increases efficiency in energy conversion. Nanomaterials also allow for lightweight flexible solar panels, making them particularly suitable for wearable applications as well as for building-integrated photovoltaics. Moreover, they lower material expenses because they reduce semiconductor material utilization, which is expensive, and simplified manufacturing processes, like inkjet printing or roll-to-roll processing, further reduce production costs.

However, nanomaterial synthesis techniques face challenges in cost-effectiveness for large-scale production, constant consistency, and quality, along with maintaining compatibility with established solar technologies. The cost of high-quality nanomaterials is relatively high at an initial level, and perovskites, for instance, suffer from stability problems leading to deterioration over time. Environmental risks such as toxicity or contamination due to heavy metals in quantum dots also need to be mitigated, and appropriate standards and regulations also have to be developed in conjunction. Despite these hurdles, nanomaterials' potential to revolutionize solar energy conversion remains strong, requiring continued research and development to overcome these limitations and enhance their practical and commercial applications.

6.3 Recent Research and Developments

Recent research into nanomaterials for solar cells has been characterized by a significant impetus toward the goal of improved efficiency, stability, and flexibility. Among them are optimizations of perovskite nanomaterials that have shown remarkable performance gains in solar cells due to higher light absorptivity and improved charge transport. Innovations in nanowire and nanorod technologies have led to better light trapping and enhanced charge separation hence further boosting the performance of solar cells. Graphene and carbon nanotubes have been integrated into different layers of solar cells to enhance electrical conductivity and mechanical flexibility, which enables the production of more durable and flexible solar panels. Also, in the combination of nanoparticles with conventional materials for making advanced nanocomposites, novel stability and performance improvements of solar cells have been observed (Maafa, 2022). Such development is representative of a highly dynamic and fast-growing field at the same time when research is continually conducted on overcoming the existing limitations and further developing the technology of solar energy.

7. HYBRID AND COMPOSITE MATERIALS

7.1 Combining Different Material Types

Hybrid and composite materials have been in the frontier of trying to advance solar cell technology by combining different types of materials to harness their strengths. Examples include hybrid perovskite-silicon solar cells that incorporate perovskite layers into silicon substrates to exploit a broad light absorption range of perovskites along with robust electronic properties of silicon towards high efficiency. Hybrid organic-inorganic photovoltaic cells include inorganic materials, usually titanium dioxide, to enhance the organic photovoltaics features on flexibility and lightweight properties. Such features in nanocomposites-especially those combined with graphene or carbon nanotubes and traditional semiconductors-make sure that the performance would be high due to increased light absorption and increased mobility of charge carriers. Tandem solar cells can accomplish record efficiencies by stacking multiple layers of different materials, like perovskite on silicon, for the capture of an even broader spectrum of sunlight (Swathi et al., 2012). These hybrid and composite approaches, going beyond the state-of-the-art efficiency and functionality, shall enable completely new application routes toward flexible, lightweight, and printable solar technologies-quantum leaps toward more versatile and scalable solar energy solutions.

7.2 Synergistic Effects and Performance Improvement

Most of the time, the improvements in hybrid and composite solar cells are observed simply because of the complementary properties when combined. For example, hybrid perovskite-silicon solar cells possess a wide light-harvesting capability due to perovskites, while silicon ensures better charge transport. This synergy improves the light harvesting and hence increases the conversion of electricity, leading to higher power outputs. Another group of organic-inorganic hybrid cells is made with flexible organic material combined with inorganic materials, like titanium dioxide, which has improved charge separation that can be integrated into various applications, starting from wearable devices to flexible electronics. Nanocomposites combining carbon nanotubes or graphene with conventional semiconductors significantly enhance the mobility of charge carriers and light absorption due to the minimized energy losses. Tandem solar cells achieve higher efficiencies and wider spectrums of captured sunlight by layering various materials, again often perovskite and silicon, together (Wright & Uddin, 2012). Overall, these combined materials make for a huge leap in performance for the solar cell and thus extend the ways in which solar technology can be used.

7.3 Studies of Hybrid Material Applications

Research into the application of hybrid material in solar cells has shown very good development, with a broadening of their functionalities contributing to overall performance. Hybrid materials merge different kinds of substances, often organic and inorganic, taking advantage of the complementarity of properties to realize performances superior to single-material systems (Aamir, 2023).

Perovskite-Silicon Hybrid Solar Cells: Perovskite-silicon tandem solar cells involve studies on perovskites with conventional silicon to provide higher light absorption and efficiency. Perovskites exhibit an extremely broad absorption spectrum, enabling the capture of a considerable portion of sunlight, which otherwise would have been passed over when working with pure silicon. This combination considerably enhances the light-harvesting capability, leading to higher energy conversion efficiencies.

Organic-Inorganic Hybrid Solar Cells: Organic-inorganic hybrid solar cells incorporate both organic materials, usually in the form of polymers or small molecules, and inorganic semiconductors like titanium dioxide or zinc oxide. The organic constituents provide flexibility and lightweight characteristics to these solar cells, while the inorganic constituents enhance charge separation and transport in such solar cells.

Nanocomposites in Solar Cells: Carbon nanotubes or graphene were shown to enhance the mobility of charge carriers and light absorption upon addition to conventional semiconductors. Such nanocomposites have been highlighted for their potential to reduce resistive losses toward better efficiency in solar cells by enhanced light-to-electricity conversion.

Tandem Solar Cells: In the research work, much emphasis has been put on tandem solar cells involving the stacking of multilayers made of different materials, such as perovskite and silicon. The cells take advantage of the respective light absorption properties to cover more of the solar spectrum. Studies have indeed shown that tandem configurations can realize record efficiencies by combining the strengths of each material layer effectively, hence becoming highly efficient and competitive with conventional solar technologies.

Hybrid Photovoltaic Systems: Hybrid photovoltaic systems have great potential, which will integrate solar cells with energy storage or thermal collectors, among other technologies. In general, the enhancement of energy capture and conversion shall be realized toward more stable and reliable power output. For example, the integration of photovoltaic cells with thermal systems could improve general energy yield through both electrical and thermal usage.

Hybrid material application studies continue to advance solar cell technology. Combining different materials with complementary properties, researchers develop more efficient, versatile, and cost-effective solar energy solutions that may meet a wide range of application needs and drive the future of solar technology (Ito, 2021).

8. SCALABILITY AND COMMERCIALIZATION

8.1 Challenges in Scaling Up Novel Materials

Scaling up the deployment of the new material for solar cells is not an easy matter, as several challenges lie on the way to make successful lab developments translate into large-scale industrial production. Preparing perovskite films or other new materials in a controlled environment that is impossible to repeat at any significant scale will be a challenge. Cost-effective, reproducible methods for mass production are urgently sought after (Agrawal, 2016). Furthermore, novel materials can be costly since these involve rare componentry; one must be mindful of waste generation and costs. Incorporation into the established solar technology with compatibility with these manufacturing processes is critical for large-scale deployment.

Other challenges include environmental and health issues, especially for perovskites that contain lead. More research is needed on safer alternatives or better encapsulation methods to reduce risks. New materials also need to be tested for durability and stability in real-world conditions to ensure long-term performance (Anandan, 2007). Regulatory approvals and certification can further delay large-scale adoption because new materials must meet safety, performance, and environmental standards. The last aspect is scaling up production to build the infrastructure and supply chains to enable mass manufacturing. All of these challenges would be solved with cross-field collaborations involving materials science, engineering, economics, and policy in order to make novel solar materials cheap and efficient.

8.2 Economic and Environmental Considerations

It remains of essence that the development and deployment of solar cell technologies are guided by economic and environmental considerations. There should be a balance between cost-effectiveness and sustainability, or else widespread adoption and the realization of the full potential of solar energy may not be attained. In terms of development and adoption, it has to be driven by economics and environment. This will ensure that the technologies developed are cost-efficient and sustainable. Most economists see economic viability as a key issue in making solar energy competitive. Increased efficiency in solar cells can translate to higher energy output and return on investment. Lower maintenance costs and longer lifetimes add to the economic feasibility of solar technology, while government policies, market conditions, and improvements in recycling can also affect costs and adoption rates.

The raw material extraction, energy-intensive manufacturing, and final disposal and recycling of solar panels must minimize their impact on the environment. Environmentally friendly sourcing of materials and better recycling processes can minimize harm to the environment. Solar energy can contribute to a cleaner, greener future by prioritizing sustainable materials, efficient recycling, and meeting environmental standards. The balancing of these economic and environmental factors will make solar energy more widely adopted and sustainable in the long term.

8.3 Successful Examples and Industry Trends

There have been a number of successful examples and changing trends in the solar industry that depict progress made in technological advances, market adoption, and innovation. Successful examples and evolving trends represent how technological improvements, more efficient cost structures, and increasingly innovative approaches are helping the sector inch closer towards adequate, affordable, and sustainable solar energy solutions (McGehee, 2021).

8.3.1 Perovskite-Silicon Tandem Solar Cells

a) **Example**: The collaboration between the National Renewable Energy Laboratory (NREL) with different leading academic institutions resulted in major gains achieved in perovskite-silicon tandem solar cells, with efficiencies well over 30%, marking the potential of high-performance solar technology.
b) **Impact**: These successful tandem cells very much prove the real possibility of marrying the broad absorption spectrum of perovskites to well-established efficiency in silicon-a combination that could revolutionize solar energy with more power per unit area, particularly priceless in space-constrained applications.

8.3.2 First Solar's CdTe Thin-Film Panels

a) **Example**: First Solar has been the leading manufacturer of CdTe thin-film solar panels. Their modules have exhibited high performances and were very reliable, attaining efficiencies of approximately 18%, and the company has focused on scaling up to global demand.
b) **Impact**: The success of First Solar with CdTe panels underlines the potential for thin-film technology in the solar market. The company's focus on low-cost production coupled with its sustainability has played a role in bringing down the prices and making solar energy more accessible.

8.3.3 Tesla's Solar Roof

a) **Example**: Tesla's Solar Roof embeds solar cells into roof tiles, marrying aesthetics with functionality. The tiles are made to resemble more traditional roofing materials while harnessing the power of electricity.
b) **Impact**: Tesla's Solar Roof is one of those critical developments to be made in order to embed solar into everyday life. Tesla has made it far more feasible and attractive at the residential level with the aesthetic concerns of the panels and further adaptability of the solar installation systems.

8.3.4 SunPower's Maxeon Panels

a) **Example**: SunPower's Maxeon solar panels boast of a high efficiency and resilience, reaching up to 22.8%. The back-contact cell design in these panels differs from others in how it limits energy loss and maximizes performance.
b) **Impact**: SunPower is seen to have set an industrial benchmark with its high-efficiency panels. Further developments in the design and production of panels contributed to the wider diffusion of high-efficiency solar technology.

Successful examples of various solar cell technologies and industrial trends underline a continuous development and innovation. Emphasizing efficiency improvements, integrated solar technology in everyday products, advanced energy storage, and sustainability, the solar industry is making considerable leaps forward towards a more sustainable and accessible energy future.

9. FUTURE OUTLOOK

9.1 Emerging Trends in Material Science for Solar Cells

In material science, future trends are poised to revolutionize the solar industry with new materials and improved versions of existing materials for better performance, efficiency, durability, and lower costs. One of the major developments in this area is the improvement in perovskite materials for stability and less moisture sensitivity. Perovskite-silicon tandem solar cells are showing higher efficiencies because of the best features of both materials. Quantum dot technology improved with better size, composition, and surface treatments, leading to better light absorption and electricity conversion, especially for thin-film photovoltaics (Shahzad, 2017). Organic photovoltaics were also significantly up-graded for material efficiency and stability, offering flexible, lightweight applications such as wearable electronics and building-integrated systems. These innovations make solar cells more efficient and versatile, which promotes broader adoption of solar energy toward a more sustainable future.

9.2 Potential Breakthroughs and Research Areas

Research in solar cell technology is making significant strides, focusing on materials, structure, and system integration to improve efficiency, cost, and versatility. Perovskite solar cells are being enhanced for better stability and scalability, with advancements in lead-free materials, environmental protection, and tandem cells combined with silicon for higher efficiency. Quantum dot solar cells are being optimized for improved light absorption and energy conversion, aiming for flexible, lightweight panels. Organic photovoltaics (OPVs) are seeing progress in material development and charge transport to improve efficiency and stability, making them suitable for flexible and integrated applications. Tandem solar cells, which stack multiple materials to capture a broader solar spectrum, are being refined for better performance. Research on flexible and lightweight solar cells is expanding their use in areas like wearables and portable power sources. Advanced light management techniques, such as better coatings and light-trapping structures, are increasing energy conversion. Integrating solar with energy storage devices is a key area for ensuring steady, reliable energy supply. Efforts in material recycling and sustainability are focused on reducing the environmental impact of solar technologies. Additionally, research into artificial photosynthesis aims to create solar technologies that produce not only electricity but also useful byproducts. These innovations promise to enhance solar energy's efficiency, accessibility, and adaptability, driving forward the energy systems of the future.

9.3 Long-Term Implications for Energy Storage and Solar Technology

The future of energy storage and solar technologies will revolutionize the electrical grid, ensuring stability and reliability. Advanced storage solutions will manage the intermittency of solar energy, providing a steady supply. As is done with any product developed, the production, use, and disposal of these technologies require efficient Lifecycle Assessment. Manufacturers can find ways to reduce emissions and waste by using LCA. Decentralized energy systems in which homes and businesses will generate and store their own energy significantly reduce reliance on the central grid, therefore improving energy security and resilience. As the cost of solar panels and storage technologies declines, these solutions will be more affordable, thus reducing the consumption of fossil fuels. LCA will ensure that as these technologies grow, the environmental impact will be minimal. This shift will stabilize the energy grid and reduce dependency on fossil fuels, therefore directly benefiting both users and the environment in the long run.

10. CONCLUSION

10.1 Summary of Key Innovations

The latest advances in solar cell technology address all the issues associated with higher efficiency, cost, and flexibility. Perovskite materials with excellent light absorption and ease of production are now being developed for overcoming problems related to moisture sensitivity and degradation. Researchers have made tandem cells where perovskites are combined with silicon for achieving conversion efficiencies over 30%. Quantum dot solar cells have also experienced advances in size, composition, and surface treatment to enhance efficiency, allowing for the production of thin, flexible, portable solar panels. For energy storage, next-generation solid-state and high-capacity flow batteries are improving energy density, safety, and lifespan, thus helping store intermittent solar power and provide reliability to the grid. Collectively, all of these advances are revolutionizing energy management, making solar power more sustainable, adaptable, and reliable for the future.

10.2 Implications for the Future of Solar Energy

The future of solar energy is quite promising, for it has much to do with the continued technological developments and changing economics. Emerging materials for solar cell technology, such as perovskites, quantum dots, and organic photovoltaics, will achieve higher efficiency by allowing solar panels to capture more energy from the same sunlight exposure. This may make solar power even more competitive with other forms of energy and applicable to unusual surfaces or portable devices. With a decline in the cost of solar technology and energy storage, more industries and consumers will be able to afford the systems, hence increasing the adoption of renewable energy. Improved integration of solar and storage systems will improve grid stability and energy reliability, supporting the shift to decentralized energy systems. All these will make solar energy a central part of the global energy mix, bringing long-term environmental and economic benefits.

10.3 Final Thoughts and Recommendations for Researchers

New technologies in solar energy have tremendous promise, but problems persist. Research is still necessary to increase the stability and longevity of materials. Improved integration of solar panels with storage systems will improve energy usage and the reliability of the grid. Sustainability initiatives should emphasize environmentally friendly recycling and minimizing the environment's footprint. Cost savings could be realized in manufacturing and design improvements that bring solar energy into

REFERENCES

Aamir, M., Sher, M., & Akhtar, J. (2023). Lead Free Hybrid Copper Halide Perovskite Nanosheets: A Potentially New Material for Solar Cell Applications. https://doi.org/DOI: 10.2139/ssrn.4456748

Afreen, M., Ali, J., & Bilal, M. (2021). *Challenges of Stability in Perovskite Solar Cells. Fundamentals of Solar Cell Design*. Portico., DOI: 10.1002/9781119725022.ch13

Agrawal, N., Zubair Ansari, M., Majumdar, A., Gahlot, R., & Khare, N. (2016). Efficient up-scaling of organic solar cells. *Solar Energy Materials and Solar Cells*, *157*, 960–965. DOI: 10.1016/j.solmat.2016.07.040

Akram, M. W., & Hasanuzzaman, M. (2022). Fundamentals of thermal energy and solar system integration. Technologies for Solar Thermal Energy, 1–24. https://doi.org/DOI: 10.1016/B978-0-12-823959-9.00003-9

Anandan, S. (2007). Recent improvements and arising challenges in dye-sensitized solar cells. *Solar Energy Materials and Solar Cells*, *91*(9), 843–846. DOI: 10.1016/j.solmat.2006.11.017

Andersson, M., & Fahlman, M. (2016). Organic Semiconducting Materials. Organic and Printed Electronics, 11–45. https://doi.org/DOI: 10.1201/b20043-3

Azhari, A. W., Then, F. S. X., Halin, D. S. C., Sepeai, S., & Ludin, N. A. (2020). Tin and germanium substitution in lead free perovskite solar cell: Current status and future trends. *IOP Conference Series. Materials Science and Engineering*, *957*(1), 012057. DOI: 10.1088/1757-899X/957/1/012057

Batmunkh, M. (2020). Advances in Emerging Solar Cells. *Nanomaterials (Basel, Switzerland)*, *10*(3), 534. DOI: 10.3390/nano10030534 PMID: 32192074

Bisquert, J. (2017). Fundamental Model of a Solar Cell. The Physics of Solar Cells, 71–84. https://doi.org/DOI: 10.1201/b22380-4

Breeze, P. (2018). An Introduction to Energy Storage Technologies. Power System Energy Storage Technologies, 1–11. https://doi.org/DOI: 10.1016/B978-0-12-812902-9.00001-8

Budt, M., Wolf, D., Span, R., & Yan, J. (2016). A review on compressed air energy storage: Basic principles, past milestones and recent developments. *Applied Energy*, *170*, 250–268. DOI: 10.1016/j.apenergy.2016.02.108

Das, S., Dey, A., Biswas, A., & Mohanty, A. (2014). A novel design of Super-Capacitor used to enhance solar-energy restoration of photo-voltaic cells. 2014 1st International Conference on Non-Conventional Energy (ICONCE 2014), 61–64. https://doi.org/DOI: 10.1109/ICONCE.2014.6808683

Drury, E., Denholm, P., & Margolis, R. (2011). *Impact of Different Economic Performance Metrics on the Perceived Value of Solar Photovoltaics. Office of Scientific and Technical Information.* OSTI., DOI: 10.2172/1028528

Foggiato, J. (2009). Manufacturing Equipment Components of New Materials for Solar Cell Efficiency Enhancements and Lower Cost. ECS Meeting Abstracts, MA2009-01(11), 562–562. https://doi.org/DOI: 10.1149/MA2009-01/11/562

García de Arquer, F. P. (2021). III-V Semiconductor Quantum Dots for Infrared Photodetection. *Proceedings of the NanoGe Fall Meeting 2021*. https://doi.org/DOI: 10.29363/nanoge.nfm.2021.094

Horoz, S. (2017). Investigation of Semiconductor Quantum Dots for Quantum Sensitized Solar Cells (QDSSCs). *Cumhuriyet Science Journal*, •••, 121–129. DOI: 10.17776/csj.363334

Ito, S. (2021). *Stability of Hybrid Perovskite Solar Cells. Hybrid Perovskite Solar Cells*. Portico., DOI: 10.1002/9783527825851.ch15

Khirade, P. P., & Raut, A. V. (2022). Perovskite structured materials: synthesis, structure, physical properties and applications. In *Recent advances in multifunctional perovskite materials*. IntechOpen.

Kousksou, T., Bruel, P., Jamil, A., El Rhafiki, T., & Zeraouli, Y. (2014). Energy storage: Applications and challenges. *Solar Energy Materials and Solar Cells*, *120*, 59–80. DOI: 10.1016/j.solmat.2013.08.015

Kuhn, T. E., Erban, C., Heinrich, M., Eisenlohr, J., Ensslen, F., & Neuhaus, D. H. (2021). Review of technological design options for building integrated photovoltaics (BIPV). *Energy and Building*, *231*, 110381. DOI: 10.1016/j.enbuild.2020.110381

Ma, D. (2021). Solar Energy and Solar Cells. *Nanomaterials (Basel, Switzerland)*, *11*(10), 2682. DOI: 10.3390/nano11102682 PMID: 34685123

Maafa, I. M. (2022). All-Inorganic Perovskite Solar Cells: Recent Advancements and Challenges. *Nanomaterials (Basel, Switzerland)*, *12*(10), 1651. DOI: 10.3390/nano12101651 PMID: 35630874

Mahapatra, A., Kumar, S., Kumar, P., & Pradhan, B. (2022). Recent progress in perovskite solar cells: Challenges from efficiency to stability. *Materials Today. Chemistry, 23*, 100686. DOI: 10.1016/j.mtchem.2021.100686

Maize, K., Mi, Y., Cakmak, M., & Shakouri, A. (2022). Real-Time Metrology for Roll-To-Roll and Advanced Inline Manufacturing: A Review. Advanced Materials Technologies, 8(2). Portico. https://doi.org/DOI: 10.1002/admt.202200173

Mawire, A. (2015). Solar Thermal Energy Storage for Solar Cookers. Solar Energy Storage, 327–358. https://doi.org/DOI: 10.1016/B978-0-12-409540-3.00014-1

McGehee, M. (2021). *Perovskite on Silicon Tandem Solar Cells. Office of Scientific and Technical Information.* OSTI., DOI: 10.2172/1830219

Mohammad, A., & Mahjabeen, F. (2023). From silicon to sunlight: exploring the evolution of solar cell materials. JURIHUM: Jurnal Inovasi dan Humaniora, 1(2), 316-330. http://jurnalmahasiswa.com/index.php/Jurihum/article/view/409

Paranthaman, M. P., Wong-Ng, W., & Bhattacharya, R. N. (Eds.). Semiconductor Materials for Solar Photovoltaic Cells. (2016). Springer Series in Materials Science. Springer International Publishing. https://doi.org/DOI: 10.1007/978-3-319-20331-7

Raj, M. K. A., Rathanasamy, R., & Chinnasamy, M. (2021). *Development of Solar Cell. Materials for Solar Energy Conversion.* Portico., DOI: 10.1002/9781119752202.ch2

Sampaio, P. G. V., & González, M. O. A. (2022). A review on organic photovoltaic cell. International Journal of Energy Research, 46(13), 17813–17828. Portico. https://doi.org/DOI: 10.1002/er.8456

Shahzad, N. (2017). Nanostructured Electrodes for Emerging Solar Cells. Nanomaterials in Energy Devices, 150–181. https://doi.org/DOI: 10.1201/9781315153445-5

Swathi, S. K., Ranjith, K., Kumar, P., & Ramamurthy, P. C. (2012). Novel thiophene derivative hybrid composite solar cells. *Solar Energy Materials and Solar Cells, 96*, 101–107. DOI: 10.1016/j.solmat.2011.09.021

Vijayakumar, M., Adduru, J., Rao, T. N., & Karthik, M. (2018). Solar Cells: Conversion of Solar Energy into Electrical Energy Storage: Supercapacitor as an Ultrafast Energy-Storage Device Made from Biodegradable Agar-Agar as a Novel and Low-Cost Carbon Precursor (Global Challenges 10/2018). Global Challenges, 2(10). Portico. https://doi.org/DOI: 10.1002/gch2.201870204

Wang, J. (2015). Recent innovations in materials science. Materials Research Innovations, 19(sup9), S9-1-S9-1. https://doi.org/DOI: 10.1179/1432891715Z.0000000001932

Wright, M., & Uddin, A. (2012). Organic - inorganic hybrid solar cells: A comparative review. *Solar Energy Materials and Solar Cells*, *107*, 87–111. DOI: 10.1016/j.solmat.2012.07.006

Chapter 17
Roll of Carbon-Based Supercapacitors in Regenerative Breaking for Electrical Vehicles

Kutubuddin Sayyad Liyakat Kazi
https://orcid.org/0000-0001-5623-9211
BMIT, Solapur, India

ABSTRACT

While the study does a good job of highlighting the most important advantages of carbon-based supercapacitors, it would be beneficial to conduct a more in-depth analysis of their limits, such as the energy density of carbon-based supercapacitors in comparison to typical lithium-ion batteries. To have a more nuanced knowledge of the trade-offs that are involved in the process of incorporating these systems into electric vehicle architectures, a balanced perspective would be beneficial. A solid foundation for understanding the potential of this technology is provided by the carbon-based supercapacitors that are used in electric vehicles for regenerative braking systems. It opens how to prospective research and innovation in this area of energy storage by concentrating on their benefits while also admitting the difficulties that they face along the process.

INTRODUCTION

When it comes to the field of energy storage technologies, carbon-based supercapacitors are emerging as an essential component. These supercapacitors strike a balance between conventional capacitors and their counterparts, batteries. They

DOI: 10.4018/979-8-3693-9316-1.ch017

have been positioned as a possible alternative for applications that require rapid energy delivery and high cycle stability due to their unique characteristics, which include a high surface area, superior conductivity, and environmental friendliness.

- Composition of the material and form of the material: Activated carbon, carbon nanotubes, graphene, and a variety of carbon composites are the principal materials that are utilised in the production of carbon-based supercapacitors. These materials are favored due to their remarkable electrochemical performance and huge specific surface area, both of which enable them to store charge effectively. Activated carbon, in particular, has been utilised extensively due to its abundance and cost-effectiveness. However, developments in graphene and carbon nanotubes have the potential to greatly improve the energy density and power performance of supercapacitors. Activated carbon has been widely utilised.
- Characteristics of performance to consider: The capacity of carbon-based supercapacitors to produce high power density while simultaneously delivering quick charge and discharge rates is one of the most significant features of these devices. A supercapacitor may be charged in a matter of seconds, in contrast to batteries, that may require several hours to charge. This makes supercapacitors an excellent choice for applications that demand immediate energy bursts. In addition, they have remarkable cycle stability, which allows them to withstand hundreds of thousands of cycles without seeing a substantial decline in performance. This allows them to have a longer lifespan in terms of their operational capabilities. When compared to ordinary lithium-ion batteries, carbon-based supercapacitors have typically been found to be inferior in terms of energy density. However, they are superior in terms of power density and lifetime. As a result of this limitation, a significant amount of researches has been conducted to enhance the capabilities of energy storage through a variety of strategies. These strategies include hybrid designs that combine supercapacitors using batteries or the incorporation of pseudocapacitive materials that can enhance energy density.
- Environmental impact and sustainability: The remarkable environmental friendliness of carbon-based supercapacitors is one of its most notable characteristics. The raw ingredients are frequently obtained from renewable resources or are naturally less toxic than those that are utilised in conventional battery technologies. This results in a reduction in the potential for environmental damage. A further contribution to sustainability is made by the high cycle life of supercapacitors, which helps to reduce waste throughout their lifetime.

Carbon-based supercapacitors can find applications in a wide range of fields due to their versatility. According to Prashant (2023) and Kazi (2025), some of these fields include grid energy storage, consumer electronics, renewable energy systems, and electric cars. As they serve to complement intermittent renewable energy sources such as solar and wind, they are becoming increasingly important in supporting energy transition activities. This is because they facilitate smoother energy delivery and stabilise power networks, which are both important for achieving energy transition.

The market for supercapacitors is experiencing substantial expansion, and the innovation that is driving this growth is being driven by breakthroughs in nanomaterials and device architecture. A rising confluence of battery and supercapacitor technologies is being created as a result of researchers' particular focus on increasing energy density while preserving the quick charge-discharge capabilities that are unique to supercapacitors.

All things considered, carbon-based supercapacitors are a revolutionary method of energy storage because they combine the advantages of high power density, rapid charge/discharge rates, and endurance with a framework that is environmentally sustainable. The current research and technical improvements promise to widen their use and effectiveness, even though there are still obstacles to be overcome, particularly in terms of improving energy density. Carbon-based supercapacitors are positioned to play a major role in influencing the future of energy management and storage systems. This is because the demand for energy storage solutions that are both efficient and sustainable is continuing to rise.

In this era, there has been a growing interest in carbon-based supercapacitors (CBSCs), notably in the context of regenerative braking systems. This interest has been prompted by the demand for more efficient energy storage options in electric and hybrid vehicles.

During the braking process, regenerative braking systems work to absorb kinetic energy and convert it into electrical energy. This energy can then be stored for later use, which ultimately increases the overall efficiency of the vehicle. Historically, this energy has been stored in lithium-ion batteries; however, the utilisation of supercapacitors has the potential to considerably optimise the process due to their capacity to supply rapid bursts of energy and their low internal resistance (Odnala et al., 2025).

CBSCs make use of carbonaceous materials as electrodes, including activated carbons, carbon nanotubes, and charcoal. Because of their high conductivity and vast surface area, they can facilitate high energy and power densities, which makes them suitable for applications in which the transmission of energy at a quick rate is essential. When it comes to regenerative braking, the capacity to quickly take and release energy is of the utmost importance, and CBSCs really shine in this regard.

CARBON MATERIALS IN SUPERCAPACITOR TECHNOLOGIES

Because of their remarkable capacity to store and release electrical energy in a short amount of time, supercapacitors, which are sometimes referred to as ultracapacitors or electrochemical capacitors, earned a lot of interest in recent years. Electrochemical double-layer capacitors (EDLC) and pseudocapacitors are utilised by supercapacitors, in contrast to traditional capacitors, which are dependent on dielectric materials. The selection of materials that are utilised in the construction of these capacitors has a significant impact on the performance of the supercapacitors. Carbon-based materials have become the material selection over electrodes in supercapacitors because of the favorable qualities that they possess. These properties include a high surface area, great electrical conductivity, and porosity that can be adjusted depending on the application. The purpose of this review is to examine the many types of carbon materials that are used in supercapacitors, including their structures, the benefits they offer, and the difficulties they present.

1. The Most Important Component of Supercapacitors Is Activated Carbon

Activated carbon may be the carbon element that is utilized in supercapacitors the most frequently. Because of its large surface area, which can be greater than 2000 m^2/g, it offers a sufficient amount of space for ion adsorption. Activated carbon can be created using either physical or chemical activation procedures. It is derived from carbon-rich sources such as coconut shells, coal, and biomass. Listed below are a few favorable aspects:

As a result of having a large effective surface area, capacitance values are quite high.

- Availability and cost-effectiveness of the product.
- Straightforward Production methods.
- Also, it presents a few difficulties:
- Electrical conductivity that is relatively poor in comparison to that of other carbon materials.
- An increase in energy density that is limited beyond a certain point.

2. Graphene

In the future, graphene will be the material of choice. Graphene, which is composed of a single layer of carbon atoms organized in a honeycomb lattice in two dimensions, has been praised for its remarkable electrical conductivity, mechanical strength, and huge theoretical surface area (about 2630 m2/g). The inclusion of this material into the electrodes of supercapacitors has produced encouraging results. Take a look at the following advantages:

- Greater energy and power density compared to activated carbon.
- Improved charge/discharge speeds as a result of its superior conductivity.

Also, it has several challenges, such as high production costs and difficulties in manufacturing on a big scale. Additionally, it is difficult to achieve appropriate porosity and surface functionalization for optimal ion access because of the challenges it presents.

3. The Potential of Carbon Nanotubes to Improve Performance

Carbon nanotubes, also known as CNTs, are structures made of graphene that are cylindrical in shape and are renowned for their mechanical and electrical capabilities. You can use them as separate electrodes or as composites with other materials. Both applications are possible. Take a look at the following advantages:

- High electrical conductivity and mechanical strength (Mishra, 2024).
- Large surface area and characteristic tubular structure that facilitates ion transport by facilitating the movement of ions.

Additionally, it presents a few difficulties:

- The difficulty of the synthesis process and the inclination to agglomerate, can reduce the effective surface area.
- Concerns over the cost and scalability of the product.

In addition to being lightweight and versatile, carbon aerogels, the polymerization and carbonization of organic precursors result in the formation of carbon aerogels, which are made up of extremely lightweight materials that have an extremely porous structure. The combination of a high surface area and a low density is something that they display. Take a look at the following advantages:

Because of their low weight, they are suited for use in tiny devices. Additionally, they have an exceptional porosity, which enables great ionic accessibility. Additionally, it presents a few difficulties:

- There is a high cost of production, and it is difficult to keep the structural integrity intact while it is being used.

- The energy density is low in comparison to that of certain more sophisticated materials.

4. Biochar and Other Sources of Carbon That are Renewable

The pyrolysis of biomass results in the production of biochar, which provides a sustainable option for use in supercapacitor applications. Because of its porous nature and the fact that its qualities may be altered based on the feedstock, it is an attractive option. Take a look at the following advantages:

The potential for functionalization to improve performance, in addition to the fact that it is a renewable source with a lower impact on the environment.

Additionally, it presents a few difficulties:

- The quality and performance of the material can vary depending on the feedstock and the production method.
- Ongoing research is required to optimize its properties for use in supercapacitor applications.

The constant search for greater performance in energy storage technology is reflected in the wide variety of carbon materials that are used in supercapacitors. Advanced materials such as graphene, carbon nanotubes, and carbon aerogels have the potential to significantly improve performance, even though activated carbon continues to be the industry standard due to its availability and cost-effectiveness. There are, however, still difficulties associated with the cost, scalability, and practical application of the solution. In the future, research that is directed toward hybrid approaches that combine the strengths of various carbon materials could pave the way for the next generation of supercapacitors that have improved energy and power density. This would make them ideal candidates for a wide variety of applications, including renewable energy systems and electric vehicles (Kazi ,2024b; Kutubuddin, 2024c).

ROLE OF SUPERCAPACITORS IN ELECTRIC VEHICLES

As the development of electric vehicles(EVs) continues at a breakneck speed, the search for more effective methods of energy storage has never been more important than it is right now. The use of supercapacitors has arisen as a key complement to conventional lithium-ion batteries, which are among the many technologies that are currently being investigated. In the context of electric vehicles, this paper digs into the benefits, difficulties, and potential future applications of supercapacitors. Figure 1 shows the supercapacitor connection in EV (Sultanabanu, 2023; Sultanabanu, 2023a).

Figure 1. Supercapacitor-battery energy storage system in EV

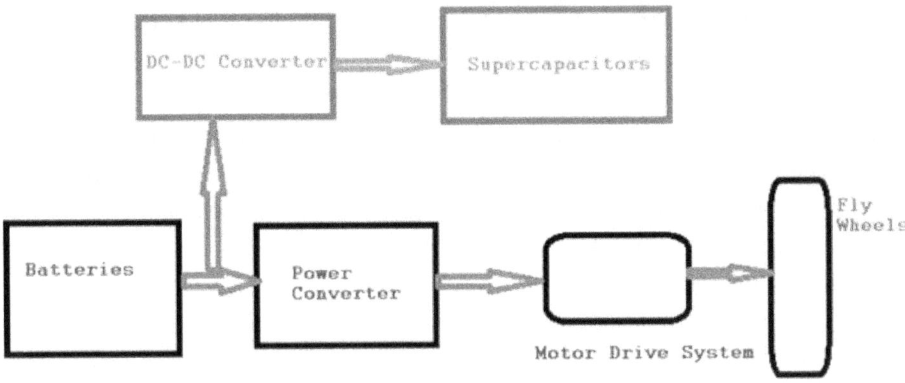

Supercapacitors, which are often referred to as ultracapacitors, are devices that are used to store energy and bridge the gap between traditional capacitors and batteries. Instead of storing energy through chemical reactions, they store energy through the separàtion of electrostatic charges, which enables them to undergo rapid charge and discharge cycles. Because of this property, they are particularly appealing for applications that call for immediate bursts of energy.

In electric vehicles, some of the benefits of using supercapacitors include the following:

1. Rapid Charging and Discharging: Because supercapacitors are capable of charging and discharging in a matter of seconds, they are ideally suited for applications that demand the transfer of energy in a short amount of time. As per Sultanabanu(2023e), these quick responses have the potential to improve the overall performance of electric vehicles, particularly in situations involving acceleration and regenerative braking. Supercapacitors have a lifespan of over a million charge-discharge cycles, which is significantly longer than the few thousand cycles that are generally seen in regular batteries. This is one of the reasons why they are better than traditional batteries in terms of longevity. Because of its durability, the vehicle will have a lower impact on the environment and cheaper replacement costs throughout its operational lifetime.
2. High Power Density: Supercapacitors have a power density that is far higher than that of batteries, which enables them to provide power instantaneously. Particularly advantageous for high-performance electric vehicles that need to

accelerate quickly without putting a strain on the primary battery systems is the fact that this is possible.

3. Temperature Resilience: In comparison to conventional batteries, supercapacitors often exhibit superior performance in terms of their ability to withstand high temperatures. The reliability of EVs in a wide range of settings is improved as a result of this property, which makes them suitable for a wide range of environmental circumstances.
4. Efficiency in terms of both weight and space: The small design of supercapacitors can help to reduce the overall weight of the vehicle, becoming the most important factor in determining the efficiency of electric vehicles. It is also possible for them to free up room for additional elements or improve the user experience inside of buildings thanks to their capacity to store energy densely in smaller volumes.

The implementation of supercapacitors in popular electric vehicles may be hindered by several problems, even though they have potential benefits.

1. A lower energy density: When compared to lithium-ion batteries, supercapacitors have a lower energy density, which is one of the most significant constraints of these electrochemical devices. They are superior in terms of power density; however, the quantity of energy that is stored per unit mass is substantially lower, which makes them less suited for applications that require long-range communication when employed on their own.
2. Cost: The production costs of supercapacitors are often greater than those of conventional batteries, which can be a barrier to the broad deployment of this technology. It is possible that these costs will reduce as technology progresses and production increases, which will have the effect of making supercapacitors more competitive.
3. Integration Challenges: Incorporating supercapacitors into existing electric vehicle architectures necessitates the development of intricate engineering solutions to guarantee that they complement rather than compete with battery systems. To maximize performance gains, it is essential to hybridize the two technologies effectively.

Electric buses and sports cars are two examples of applications that could benefit greatly from the incorporation of supercapacitors into electric vehicles. This is especially true for applications that require large power bursts. In addition, ongoing developments in materials science, like as the creation of supercapacitors based on graphene, have the potential to improve energy density while maintaining the ability to charge and discharge at a rapid rate.

Dual-energy storage systems, which incorporate the benefits of both supercapacitors and batteries, are now being investigated by manufacturers in response to the ongoing expansion of the electric vehicle market. Hybrid technologies like these have the potential to optimise energy management, thereby increasing the range and improving performance while simultaneously reducing costs and weight.

Increasing the performance and efficiency of electric vehicles through the use of supercapacitors is an exciting and potentially fruitful avenue. The fact that they can deliver electricity at a rapid rate, in addition to having an amazing lifespan and being resistant to temperature, makes them an extremely significant tool in the journey towards environmentally friendly transportation. Even if there are still obstacles to overcome, the current research and development activities are anticipated to result in the creation of creative solutions that will make it easier to incorporate supercapacitors into electric vehicle technology. Although the landscape of the automotive industry is always shifting, the synergy between supercapacitors and batteries will likely play a significant role in determining the future of electric mobility.

CARBON-BASED SUPERCAPACITORS IN ELECTRIC VEHICLES

Because the market for EVs is rapidly moving in the direction of a more environmentally friendly future, the search for effective energy storage solutions is now even more important. Carbon-based supercapacitors are emerging as a crucial actor in the hunt for improved performance, sustainability, and cost-effectiveness in EVs. These supercapacitors are among the promising technologies in this domain.

Supercapacitors are characterized by their incredibly high power density, their capacity for quick charging and discharging, and their exceptional cycle stability. Additionally, the incorporation of carbon-based materials, particularly activated carbon, Graphene, and carbon nanotubes, results in these supercapacitors exhibiting increased surface area and conductivity, which in turn results in a major improvement in their performance. Activated carbon, carbon nanotubes, and graphene are some of the materials that are utilized by carbon-based supercapacitors to attain improved energy storage performance. Because of their capacity to supply energy in short bursts, they are particularly well-suited for applications that need rapid acceleration and deceleration, like those seen in electric vehicles.

The operation of carbon-based supercapacitors is based on the principles of electrostatic storage, and they provide charge storage through the utilization of high-surface-area carbon materials. The following are some of the benefits that they offer over conventional batteries:

1. High Power Density: Because CBSCs can provide or absorb power in a nearly instantaneous manner, they are ideally suited for applications that require the storage and release of energy in a short amount of time.
2. Extended lifespan: CBSCs have a substantially longer lifespan than conventional batteries, with tens of thousands of charge and discharge cycles. This significantly reduces the frequency with which they need to be replaced.
3. Environmental Benefits: The carbon materials that are employed typically originate from sustainable sources, which contributes to a diminished influence on the environment in comparison to batteries that are based on metal.

Benefits in Electric Vehicle Applications

1. Rapid Energy Delivery: The capacity of carbon-based supercapacitors to deliver a significant amount of power in a period that is quite short is one of the most notable advantages of these devices. The ability to accelerate quickly and improve overall driving dynamics are both made possible by this property, which is especially advantageous for EVs during acceleration.
2. Extended Cycle Life: In contrast to conventional lithium-ion batteries, which experience deterioration over time, carbon-based supercapacitors can withstand hundreds of thousands to millions of charge-discharge cycles without experiencing substantial damage. This longevity has the effect of lowering the costs associated with maintenance and reducing the frequency with which batteries need to be replaced.
3. Rapid Charging: Because of their capacity to charge at a rapid rate, supercapacitors are well-suited for situations in which downtime is a problem. It is possible to measure charge durations in seconds to minutes, which has the potential to dramatically improve both the user experience and the accessibility of your vehicle.
4. High Thermal Stability: Carbon-based supercapacitors can function well across a wide temperature range, which makes them well-suited for the many different climates and situations in which electric vehicles run.
5. Environmental Sustainability: The utilization of carbon-based materials that are both abundant and kind to the environment in the production of supercapacitors is a factor that contributes to the sustainability profile of electric vehicles. Relying on less rare minerals is a step towards decreasing the environmental impact of car manufacture. This is in comparison to conventional battery technologies, which rely substantially more on rare materials.

Despite the various benefits they offer, carbon-based supercapacitors are confronted with several obstacles when applied to electric vehicles (EVs):

1. Energy Density Limitations: Although supercapacitors are superior to lithium-ion batteries in terms of power density, their energy density is still lower than that of lithium-containing batteries. Because of this, they are unable to store the same amount of energy in the same amount of space, which is an essential component for electric vehicles that have a significantly longer range.
2. Cost: Even though costs have been reduced, the manufacturing methods for carbon-based materials of high quality can still be quite pricey. As the technology continues to advance, it will be essential to successfully strike a balance between performance and price to achieve widespread adoption.
3. Integration with Preexisting Systems: The incorporation of supercapacitors into electric vehicle (EV) systems necessitates careful design to provide optimal performance in conjunction with conventional batteries. The creation of hybrid systems that capitalize on the advantages offered by both technologies presents several engineering hurdles.

In the realm of energy storage for electric vehicles, carbon-based supercapacitors are becoming an increasingly viable future. As a result of their capacity to supply quick power, long cycle life, and environmental benefits, they come across as an appealing alternative for improving the performance of electric vehicles and ensuring their sustainability by Kutubuddin(2024a(Kutubuddin,,(2024b). However, before they can become widely used, there are several obstacles that need to be overcome, including those concerning energy density and cost. As research is carried out and technology advances, it is quite probable that carbon-based supercapacitors are going to play a significant part in the coming generations of electric vehicles. This will pave the way for a future in which automobiles are both cleaner and more efficient.

According to Kazi S(2024b), as a result of the ongoing developments in materials science and engineering, it is anticipated that their full potential will be unlocked, which will make them an interesting sector for both manufacturers and consumers in the quickly expanding electric vehicle market.

CARBON-BASED SUPERCAPACITORS IN REGENERATIVE BRAKING FOR ELECTRIC VEHICLES

As electric vehicles(EVs) continue growing in popularity as a more environmentally friendly alternative to conventional automobiles powered by internal combustion engines, the demand for energy storage systems that are both efficient and effective has become of the utmost importance. substantial progress has been achieved in the integration of carbon-based supercapacitors, especially within the realm of regenerative braking systems. This is one of the areas in which substantial breakthroughs have been made. The purpose of this analysis is to investigate the function that carbon-based supercapacitors play in regenerative braking, as well as their benefits, problems, and prospective advances in the foreseeable future.

The term "regenerative braking" refers to a procedure that recovers energy that would otherwise be wasted during the braking process and then reroutes that energy into the power system of the vehicle. Traditional methods of braking in an electric vehicle (EV) convert kinetic energy into heat, which then dissipates into the surrounding environment. Regenerative braking systems, on the other hand, can transform this kinetic energy into new electrical energy. This electrical energy can either be used to recharge the battery or stored in auxiliary storage devices such as supercapacitors.

Below is a list of the most important advantages:

1. Rapid Charge and Discharge Rates: Carbon-based supercapacitors can swiftly absorb energy during the braking process and allow it to be released instantly. This makes them suitable for applications that require rapid energy transfer, such as regenerative braking.
2. High Cycle Life: In contrast to conventional lithium-ion batteries, carbon-based supercapacitors can withstand hundreds of thousands of cycles without experiencing any substantial degradation. Because of its longevity, the costs associated with maintenance are reduced, and the entire lifespan of the energy storage system is increased.
3. Considerations Regarding the Environment Supercapacitors, particularly those that are generated from carbon materials, are typically more ecologically benign than other types of batteries. They can be manufactured using renewable resources and, when compared to traditional batteries, they have a smaller impact on the environment after their life cycle.
4. Enhancement of Energy Efficiency Carbon-based supercapacitors improve the overall energy efficiency of electric vehicles (EVs) by catching and recycling energy that would otherwise be lost as heat. This results in EVs having longer ranges and consuming less energy.

The incorporation of carbon-based supercapacitors into regenerative braking systems presents several problems, even though these elements offer several amazing advantages:

1. Energy Density: Although supercapacitors are superior to lithium-ion batteries in terms of power density, their energy density is often lower than that of lithium-containing batteries. It is possible that they may not be able to replace batteries due to this constraint; rather, they will serve as a complementing technology.

2. Cost: The manufacturing method and materials requirements for high-performance carbon-based supercapacitors can be expensive, which can affect the overall cost of electric vehicle systems. Ongoing research is being conducted to investigate alternate materials and fabrication methods that are more cost-effective.
3. Performance Under Varying Circumstances The performance of supercapacitors might deteriorate when subjected to circumstances and temperatures that are extremely high. When it comes to electric vehicles, additional study is required to improve their robustness for use in all-weather situations.

There is reason to be optimistic about the future of carbon-based supercapacitors in the field of regenerative braking. The research is concentrating on a few crucial topics, including:

1. Hybrid Systems: If you combine supercapacitors and batteries in a hybrid system, you can make use of the benefits that both technologies offer. This allows you to maximize the amount of energy that can be stored for a variety of driving circumstances and increase the range of your vehicle.
2. The development of advanced materials: The ongoing innovation in nanomaterials and composites is projected to improve the electrochemical properties of carbon-based supercapacitors, which will ultimately result in improved performance metrics and a wider range of applications.
3. Improved Manufacturing Techniques: The development of production processes that are both scalable and economically viable would improve the possibility of incorporating supercapacitors into electric vehicles that are being sold to the general public.
4. Intelligent Energy Management Systems: Recent developments in energy management algorithms and systems have the potential to enhance the utilization of regenerative braking systems, hence guaranteeing the highest possible level of efficiency when carbon-based supercapacitors are utilized.

A significant amount of potential exists for the utilization of carbon-based supercapacitors in the improvement of regenerative braking systems in electric vehicles. Their capacity to store and release energy in a short amount of time, in conjunction with their high cycle durability, places them in a position to be an essential component in the quest for transportation solutions that are both more efficient and more environmentally friendly. By overcoming the hurdles that currently exist, technological advancements will open the road for wider adoption, which will ultimately contribute to a future that is more ecologically friendly in terms of mobility. The combination of these supercapacitors and other techniques of energy storage has the potential to result in the development of novel designs and systems that will further revolutionize the landscape of electric vehicles (Kazi K S,2025; Kazi K,2025;).

ENERGY DELIVERY IN REGENERATIVE BRAKING

Regenerative braking is one of the most inventive advances in modern automotive and transportation technology. It plays a significant part in improving the efficiency and sustainability of electric and hybrid vehicles, making it one of the most important technological advancements in recent history. In its most fundamental form, regenerative braking includes a technique that uses the kinetic energy of the vehicle to transform it into usable electrical energy during the process of slowing. This electrical energy can then be stored for use at a later time. The purpose of this review is to investigate the mechanics of energy transfer in regenerative braking systems, as well as the benefits and difficulties associated with this technology, as well as its promise for the future (Bhawana et al,2025; Tamboli et al,2024; Prasad et al,2024;) .

To do its job, regenerative braking involves changing the function of the electric motor in the car. Through the process of braking, the motor acts in reverse, thereby transforming the kinetic energy that is generated by the motion of the vehicle back into electrical energy. This prevents the motor from expending energy to drive the car farther. This procedure requires the use of several essential components, such as an inverter, a battery or energy storage system, and the actual braking system and its components.

The device that generates regenerative braking is activated by the control system of the car whenever the driver applies the gas pedal. The inverter is responsible for converting the alternating current (AC) that is produced by the motor into direct current (DC). This DC electrical current can either be stored in the battery or used to power the electrical systems that are located within the vehicle. This mechanism of energy transfer not only helps to increase the driving range of electric vehicles but also lessens the amount of wear and tear that is placed on conventional brake components, which eventually results in cheaper expenses for maintenance (K K S,Liyakat,2024; Kasat et al,2023; Kazi K,2024a;).

A few of the benefits of regenerative braking in electric vehicles are as follows:

1. Energy Efficiency: Regenerative braking can improve total energy efficiency, which is the fundamental advantage of this method of braking. Regenerative braking systems allow vehicles to achieve a large reduction in energy consumption by catching and recycling energy that would otherwise be wasted as heat

during traditional braking. This allows the vehicles to reach a lower overall energy consumption.

2. Extended Driving Range: The capacity of electric vehicles to reclaim energy can bring an increase in the driving range of these vehicles. By making it possible to use energy more efficiently, this function helps alleviate range anxiety, which is one of the most prominent concerns among electric vehicle owners.
3. Decreased Brake Wear: The utilization of regenerative braking has the potential to greatly reduce the amount of wear that is placed on conventional braking components. Because of this, not only does the lifespan of brake pads and rotors increase, but it also adds to the reduction of the frequency of maintenance that is required, which further improves the overall efficiency of the vehicle.

Increased energy efficiency results in a smaller carbon footprint, which is one of the environmental benefits in this context. The use of regenerative braking helps to create more environmentally friendly modes of transportation, which is in line with the worldwide aims to combat climate change.

Although regenerative braking shows a great deal of potential, there are still several obstacles to overcome:

1. Energy Capture Limitations: The amount of energy that can be caught during braking is restricted, and it can be affected by several factors including the speed of the vehicle, the weight of the vehicle, and the intensity of the braking. Conventional brakes frequently have to take over when forceful braking is being applied, which reduces the efficiency of regeneration.
2. Battery Limitations: The performance of the onboard battery is intimately related to the efficiency of energy delivery in regenerative braking systems. This is because the battery is the source of power. It is possible for there to be bottlenecks in the process of energy recovery because not all batteries can take energy at the same rate. When it comes to the amount of energy that can be captured, the health of the battery and its charge status are crucial factors.
3. Driver Acceptance: If the sensation and responsiveness of regenerative braking are different from those of traditional braking systems, then it may be necessary for some drivers to go through a learning curve. The adoption of regenerative braking by users is contingent upon the smooth transition between conventional braking and regenerative braking (Kazi K S,2025;).

The incorporation of regenerative braking systems into vehicle design leads to an increase in the complexity of the design process, which in turn can lead to an increase in the costs of production. These expenses may reduce as technology continues to improve and become more popular; nevertheless, early investments should still be taken into consideration.

The development of battery technology, improvements in vehicle design, and control algorithms are all entwined with the future of energy transfer in regenerative braking. The incorporation of regenerative systems that are more intelligently designed, ultra-fast charging batteries, and advances in energy recovery efficiency are all examples of innovations that are expected to lead to even larger benefits. Additionally, spreading the use of regenerative braking beyond the realm of personal automobiles and into the realm of public transport and freight networks might lead to significant reductions in emissions and energy consumption on a broader scale(Kazi K S,2025a; Kazi S,2024;).

As a conclusion, the supply of energy through regenerative braking is an essential component of the move towards automobiles that are more environmentally friendly. The benefits of higher efficiency, expanded range, and lower environmental effect give a compelling justification for further investment and research in this technology, even though there are still obstacles to be overcome. The use of regenerative braking is likely to play an even more major role in the development of the transportation industry in the future as technological improvements continue to be made.

EFFECT OF ENERGY DELIVERY IN REGENERATIVE BRAKING

Regenerative braking is a cutting-edge technology that is mostly utilized in hybrid and electric automobiles. It is implemented to recapture the kinetic energy that is lost during the process of deceleration. This energy, which might have been lost as heat in conventional braking systems, is converted into electrical energy and stored for future use, which increases the overall efficiency of the vehicle. The efficiency with which energy is delivered in regenerative braking not only affects performance, but also constitutes a key factor in the range of the vehicle, the longevity of the battery, and the dependability of the system.

To function, regenerative braking makes use of the electric motor of the car as a generator whenever the vehicle is decelerating. The energy that is generated by the motor's spinning is transformed into electrical energy, which is subsequently sent to the battery of the car. The effectiveness of this technology is contingent upon a number of elements, such as the speed of the vehicle, the level of charge maintained by the battery, and the conditions of the road (Kazi S,2024a; Kazi S,2025c☺.

During regenerative braking, the control method that is utilized is an essential component of the energy delivery process. Sophisticated algorithms can optimise the quantity of energy recovered based on real-time conditions, which has a substantial impact on the overall efficiency of the energy distribution. As an illustration, automobiles that are fitted with advanced control systems can modify the amount of force applied by the regenerative braking system by the required stopping distance, weight distribution, and even the behavior of the driver.

- Influence on the Performance of the Vehicle: When it comes to vehicle performance, the energy delivery that occurs as a result of regenerative braking can have a significant impact. One of the factors that contribute to a discernible increase in the driving range of electric cars (EVs) is improved energy recovery efficiency. It has been demonstrated through research that automobiles equipped with powerful regenerative braking systems have the potential to enhance their range by around 10-30%, depending on the methods of driving and the conditions. Moreover, efficient regenerative braking makes it possible to decelerate in a more seamless manner, which improves both the driver's comfort and control. Creating systems that seem more natural to drivers can be accomplished by manufacturers through the process of fine-tuning the energy distribution. These systems can emulate the familiar feel of traditional braking while simultaneously minimizing energy loss.
- The Longevity of Batteries and Their Health: Additionally, the transfer of energy via regenerative braking affects the longevity of the battery. In the long run, the battery may experience wear and degeneration if it is subjected to frequent cycling. On the other hand, a regenerative braking system that is well-designed can assist reduce this issue by balancing the flow of energy and lowering the depth of battery discharge. Furthermore, regenerative braking systems can be configured to charge the battery in little increments rather than in massive bursts, which helps to create a more consistent charging environment. Not only does this consideration in energy supply lengthen the life of the battery, but it also improves the safety and reliability of the vehicle, which is particularly important to achieve widespread customer acceptance (Kazi S,2024d; Kazi S,2025;).

The delivery of energy in regenerative braking systems is not without its difficulties, even though it has several advantages. Several factors can influence the performance of regenerative braking systems, including variations in road surfaces, weather conditions(Kazi S,2025a;), and driver habits(Neejara,2024). The presence of ice or damp conditions might result in decreased traction, which in turn can lead to a reduction in efficacy. Furthermore, aggressive driving can nullify the advantages of energy recovery, highlighting the importance of driver education and adaptive control systems in the automotive industry.

There are several facets to consider while discussing the influence of energy delivery in regenerative braking, which is a topic that has significant ramifications for the future of hybrid and electric vehicles. According to Kazi K S(2024), as technology continues to advance, it will become increasingly important to place a strong emphasis on optimizing energy recovery systems to improve vehicle performance, extend range, and ensure that batteries last for a longer period of time. Further refinement of regenerative braking systems will be achieved by ongoing research and development in control strategies, materials, and vehicle architecture. This will ultimately play a crucial role in the shift towards environmentally friendly transportation options. The effectiveness of regenerative braking could be a crucial element in influencing purchasing decisions, which is why manufacturers should prioritize its optimization. As the demand for electric vehicles among consumers continues to grow, this is a factor that becomes increasingly important.

EFFECT OF CARBON-BASED SUPERCAPACITOR ENERGY DELIVERY IN REGENERATIVE BRAKING

In the pursuit of environmentally friendly energy solutions, the importance of energy storage and distribution systems that are both efficient and effective has significantly increased. In the context of these approaches, carbon-based supercapacitors emerged as a potentially useful technology, particularly in the context of the implementation of regenerative braking systems in hybrid and electric vehicles. The purpose of this review is to investigate the performance characteristics, advantages, and implications of utilizing carbon-based supercapacitors for the supply of energy during regenerative braking.

During the process of a vehicle's deceleration, regenerative braking systems are characterized by their ability to transform kinetic energy into stored energy. Batteries have traditionally been how this process has been made easier; however, the research of alternate storage devices has been pushed by obstacles such as limited charge/discharge rates, energy losses, and temperature control. Because of the exceptional electrochemical properties that they possess, supercapacitors, and more specifically those that are based on carbon materials, offer a method that can be implemented (Keerthana et al,2025;).

A high surface area, electrical conductivity, and cycle stability are some of the characteristics that are associated with carbon-based supercapacitors. These supercapacitors are often manufactured from materials such as activated carbon, graphene, and carbon nanotubes. Because of these characteristics, they can store significant amounts of energy in a short amount of time and release it effectively.

As a result, they are especially well-suited for applications that require fast charge and discharge cycles, like regenerative braking.

Performance in the Delivery of Energy

1. Capabilities on a Rapid Charge and Discharge Spectrum: The capacity of carbon-based supercapacitors to charge and discharge at a rate that is far faster than that of traditional batteries is one of the most significant advantages of these batteries. In the process of regenerative braking, where the vehicle is required to rapidly convert and store the energy that is generated during deceleration, this trait is especially advantageous. Because of their high power density, supercapacitors can supply energy to the electrical systems of the car or to the primary battery that is located within the vehicle.
2. Improved Efficiency: It has been demonstrated that the incorporation of carbon-based supercapacitors into regenerative braking systems results in an improvement in the overall efficiency of energy recovery. According to studies, these supercapacitors are capable of recovering up to 95% of the kinetic energy that is lost during braking, which is far higher than the 70-80% efficiency that is seen in conventional battery systems. Consequently, this results in increased driving ranges and decreased energy consumption, both of which are essential for hybrid and electric vehicles.
3. Extended Cycle Life: In comparison to lithium-ion batteries, carbon-based supercapacitors can withstand a substantially greater number of charge and discharge cycles. This makes them more robust and reliable over longer periods. This decreases the amount of maintenance that is required and increases the longevity of elements of the vehicle.

The utilization of carbon-based supercapacitors in regenerative braking systems presents several obstacles, even though they offer a multitude of additional advantages:

- Energy Density: Although supercapacitors are significantly superior to batteries in terms of power density, their energy density is still lower than that of batteries. It is also possible that there are restrictions on the amount of energy that can be stored, which calls for hybrid techniques that mix supercapacitors and batteries to achieve the highest possible level of performance.
- The Availability of Materials and the Cost: There is a possibility that the creation of high-quality carbon materials will be both costly and resource-abundant. This may affect the economic viability of widespread adoption in commercial vehicles.

- Sensitivity to Temperature: Just like the performance of all other electronic components, the performance of supercapacitors can be affected by extremely high temperatures. This could be a cause for concern in certain working settings.

The development of carbon-based supercapacitors is a revolutionary step forward in the field of energy storage technology. This is especially true for improving the effectiveness and functionality of regenerative braking strategies. In electric and hybrid vehicles, their rapid charge/discharge capabilities, high efficiency, and extended cycle life provide them with the potential to serve as a viable supplement, if not a replacement, to conventional battery systems.

The incorporation of carbon-based supercapacitors into regenerative braking is projected to play a significant part in the development of environmentally friendly transportation solutions. This is because continuing research and development efforts are addressing difficulties related to energy density and cost. Future technological breakthroughs may make it possible to create automobiles that not only improve overall performance, driving efficiency, and environmental sustainability but also harness energy more efficiently when braking.

EFFECT OF CARBON-BASED SUPERCAPACITORS ON RANGE ENHANCEMENT IN REGENERATIVE BRAKING

A large amount of attention has been drawn to the convergence of energy storage solutions and regenerative braking technology as the automotive industry continues to move towards more environmentally friendly practices. Carbon-based supercapacitors(CBSCs) have emerged as a promising alternative among the many energy storage systems due to their excellent energy density, rapid charge/discharge capabilities, and long cycle life. This is because of the various advantages that these supercapacitors possess. Within the context of regenerative braking systems, which are an essential component of electric and hybrid vehicles, this review investigates the impact that carbon-based supercapacitors have on the efficiency of range improvement.

Incorporating CBSCs into the Regenerative Braking System, Several benefits can be gained from the incorporation of carbon-based supercapacitors into regenerative braking systems, including the following:

1. CBSCs can absorb energy while braking considerably more quickly than lithium-ion batteries, which is one of the advantages of using them. This rapid energy recovery helps to reduce the amount of energy that is lost and maximizes the overall efficiency of the system.

2. Instantaneous Power Supply: CBSCs can rapidly release stored energy, which improves vehicle performance without putting strain on the primary battery pack. This occurs when the vehicle is accelerating or when additional power is required. The load conditions on batteries are alleviated as a result of this dual functioning, which causes the batteries' operating lifespan to be extended.
3. The third benefit is an increase in range. CBSCs can effectively improve the operational range of electric and hybrid vehicles by increasing the efficiency of energy capture and delivering extra power during acceleration. It is possible to lessen the number of times the battery needs to be charged by having the capability to collect more energy during braking, which allows for longer rides between breaks.
4. Effectiveness of Regenerative Braking Systems The utilization of CBSCs has the potential to improve the efficiency of regenerative braking systems by enhancing the responsiveness of these systems. Because of this, the deceleration process is more seamless, which results in a more comfortable experience for both the driver and the passengers.

There are still several obstacles to overcome to successfully incorporate carbon-based supercapacitors into regenerative braking systems, even though the benefits are quite promising.

1. Energy Density Limitations: Although the power density of CBSCs is high, their energy density is often lower than that of lithium-ion batteries. To obtain optimal performance, it is necessary to take a hybrid strategy, in which both technologies coexist.
2. Reasons Related to Costs: The production procedures for high-performance carbon materials may be more expensive than those for conventional battery components, which may result in an increase in the overall cost of the vehicle.
3. System Integration: To successfully integrate CBSCs with pre-existing braking and energy management systems, it is necessary to implement sophisticated control techniques, which may take a large amount of engineering work.

To summarise, carbon-based supercapacitors have the potential to significantly improve the efficiency of regenerative braking systems in electric and hybrid vehicles. This is a potentially game-changing development. Because of their capacity to rapidly store and release energy, they not only increase the effectiveness of the braking system, but they also help the overall energy management and the extension of the vehicle's range. Despite the fact that there are obstacles concerning cost and energy density, it is expected that these issues will be overcome by continual breakthroughs in material science and engineering technology. In the future, research and development will be especially important for the widespread use of CBSCs, which will position them as an essential component in the search for transportation solutions that are both more environmentally friendly and more efficient. All things considered, the advantages of adding carbon-based supercapacitors into regenerative braking systems highlight the vital role that these components play in promoting the future of environmentally responsible transportation (Kutubuddin,2024a; Liyakat,2023a; Liyakat,2024;).

EFFECT OF CARBON-BASED SUPERCAPACITOR ON ENERGY DELIVERY IN REGENERATIVE BRAKING

There has been a significant uptick in interest in energy storage technologies, particularly in electric vehicles (EVs) and hybrid electric vehicles (HEVs), as a result of the global shift towards more sustainable energy systems. These vehicles are equipped with a crucial component known as regenerative braking, which enables the recuperation of kinetic energy that is generally lost during the braking process. In this scenario, carbon-based supercapacitors emerged as a potentially useful alternative for improving the supply of energy. This article examines the impact that carbon-based supercapacitors have on the supply of energy in regenerative braking systems, analyzing their performance as well as the benefits and difficulties associated with using them.

The Function of the CBSC in the Process of Regenerative Braking

1. Energy Recovery Efficiency: Supercapacitors that are based on carbon can improve the efficiency of energy recovery during braking cycles. When the vehicle slows down, the kinetic energy is transformed into electrical energy, which can then be stored in supercapacitors (electrical energy storage devices). Carbon-based supercapacitors can swiftly absorb the energy that is generated because of their rapid charge and discharge rates. This helps to reduce losses and results in an overall improvement in the efficiency of the system.
2. Power Management: The instantaneous power generated during regenerative braking might result in a large amount of variation for the driver. It is possible for carbon-based supercapacitors to properly handle these variations, which guarantees that energy is stored rapidly while simultaneously preserving the stability of the system. Because of this feature, traditional battery systems are subjected to less stress, which in turn extends their lives and improves their performance.

3. Smooth Energy Delivery: The capacity of supercapacitors to deliver high currents immediately can be of assistance in giving the energy that is required for short bursts of acceleration following braking. In addition to improving the overall performance of the car, this rapid response also makes for a more comfortable driving experience.

There are several obstacles that are linked with carbon-based supercapacitors, despite the fact that they offer several useful benefits. Even though their energy density is higher than that of conventional capacitors, it is still smaller than that of batteries. Because of this constraint, it may be necessary to make use of hybrid systems, which combine the properties of both supercapacitors and batteries to take advantage of the advantages offered by each technology.

An additional element that may operate as a barrier to wider adoption in mainstream automotive applications is the expensive cost of high-performance carbon materials. Research into alternate materials and methods for lowering production costs is now being conducted, and findings from this research will be essential for widespread application(Liyakat,2025; Liyakat KK,2025; Liyakat,2025d;).

A significant contribution to the enhancement of the energy delivery capabilities of regenerative braking systems in electric and hybrid vehicles is made by supercapacitors that are based on carbon. As a result of their capacity to rapidly store and release energy, the overall efficiency, performance, and durability of the vehicle are drastically improved. Even though there are still obstacles to overcome, such as energy density and cost, ongoing research, and technological breakthroughs continue to pave the way for a larger integration of these devices in automotive applications. A viable path towards achieving more environmentally friendly and energy-efficient transportation in the future is represented by the synergy between carbon-based supercapacitors and regenerative braking technologies. There is little doubt that the full potential of supercapacitors in regenerative systems will be realized when the automotive industry transitions towards electrification. This will surely contribute to the development of vehicles that are more efficient in terms of energy consumption.

DESIGN STEPS IN CARBON-BASED SUPERCAPACITORS FOR ENERGY DELIVERY IN REGENERATIVE BRAKING

Regenerative braking systems, also known as RBS, have developed as a crucial technology in the transportation industry as a result of the increased demand for energy efficiency and sustainability in this sector. Not only do they improve energy recovery, but they also make a major contribution to the overall efficiency of hybrid and electric vehicles. The carbon-based supercapacitors (CBS) that serve as energy storage solutions are at the core of this breakthrough. These supercapacitors are distinguished by their greater power density and exceptional cyclability. However, to optimize the design of these supercapacitors for maximum performance in applications involving regenerative braking, it is necessary to have a comprehensive grasp of several essential issues (Mahant,2025; Khadake et al,2024; Upadhyaya et al,2025;)

Increasing energy economy and sustainability by recapturing energy that would otherwise be lost during braking, regenerative braking systems have emerged as a crucial technology in modern electric and hybrid vehicles. These systems have become increasingly important in recent years. The most important factors that determine the effectiveness of these systems is the energy storage device that is utilized. Carbon-based supercapacitors are a popular option due to the remarkable performance features that they possess. The purpose of this review was to investigate the essential design phases involved in optimizing carbon-based supercapacitors for energy delivery in applications involving regenerative braking shown in Figure 2.

Figure 2. Design steps

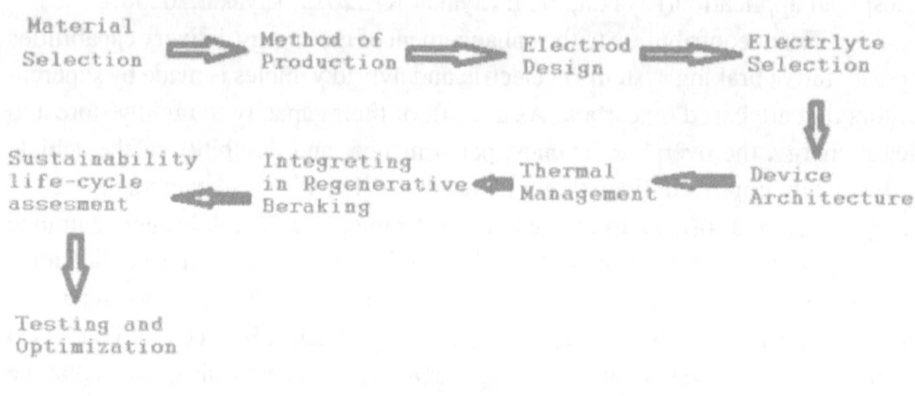

1. The selection of the material: Beginning with the selection of suitable carbon materials is the first step in the design process. When it comes to the performance of supercapacitors, the selection of carbon materials is of the utmost importance. Each of the three materials—Graphene, carbon nanotubes, and activated carbon—possesses distinctive structural features that have an impact on the surface area and the distribution of pore sizes. As a result of their high surface area, conductivity, and mechanical stability, activated carbon, carbon nanotubes, graphene, and carbon aerogels are among the materials that are utilized in supercapacitors the most frequently. Several factors, including energy density, power density, and cycle life, are greatly impacted by the features of

each material. Materials must be able to resist rapid charge and discharge cycles while still maintaining maximum performance to be used for regenerative braking. A greater surface area makes it possible for a greater amount of ions to be absorbed, which is essential for increasing the specific capacitance and power density by a significant amount. Further enhancement of conductivity and energy storage capacity can be achieved through the utilization of hybrid materials or composites that incorporate conductive polymers. This results in the supercapacitors becoming more effective in the rapid charge and discharge cycles that are typical of regenerative braking.

2. Methods of Production: The manufacturing method has a direct influence on the microstructure and morphology of carbon-based materials, which in turn affects the electrochemical properties of these materials. Techniques like chemical vapor deposition (CVD), templating methods, and sol-gel processes can be modified to build carbon structures that have surface properties and porosity that have been optimized. Controlled manufacturing procedures are necessary for getting high performance in practical applications because they provide uniformity and consistency, which are essential for achieving high performance (Shinde et al,2025;).
3. To improve the performance of supercapacitors, it is essential to incorporate an electrode design that is both efficient and effective. The use of nanostructured materials results in a large increase in surface area, which in turn makes it easier to access ions. Increasing the capacitance and energy density of supercapacitors can be accomplished by including conductive polymers or hybridizing carbon materials with pseudocapacitive materials. This will result in the supercapacitors being better adapted to meet the fluctuations in energy that are required by braking systems. Additionally, it is essential to implement strategies such as optimizing the pore size and shape to improve ion transport.
4. Electrolyte Selection: The selection of the electrolyte is an important factor in determining the performance of the semiconductor supercapacitor. Even though they have a high ionic conductivity, aqueous electrolytes frequently restrict the voltage window. On the other hand, organic electrolytes enable larger working voltages, which in turn results in a higher energy density. Additionally, ionic liquid electrolytes are attracting interest due to their thermal stability and wider electrochemical stability window. These characteristics can be especially advantageous in high-performance braking systems which are becoming increasingly popular.
5. Device Architecture: The architecture of the supercapacitor, which includes the arrangement and configuration of electrodes and electrolytes, is fundamental to the process of optimizing the supply of energy and power. It is possible that this step will entail the creation of symmetric or asymmetric supercapacitors,

both of which have benefits that are distinct from one another in terms of energy density and power density. Enhanced performance can also be achieved by the utilization of multi-layer designs and the combination of supercapacitors with other technologies for the storage of energy, such as batteries. To achieve a higher power density, it is necessary to take into consideration the physical design of carbon-based supercapacitors. This involves optimizing the thickness of the electrodes, the materials used for the separators, and the overall structure of the device. The surface area that is exposed to the electrolyte can be increased by the use of multi-layer structures or three-dimensional architectures. This provides the opportunity for quicker ion transport and enhances the efficiency of energy storage during times of high power demands. In addition, the incorporation of features such as integrated thermal management systems can help alleviate problems that are associated with the development of heat and increase the longevity of the devices (Sayyad 2025c; Sayyad,2025e; Sayyad,2025d;).

6. Thermal Management: - Thermal management becomes extremely important during regenerative braking because of the quick energy uptake and delivery that occurs during this process. Efficient methods of heat dissipation must be incorporated into the design to minimize overheating, which can have a detrimental impact on both performance and sustainability. Utilizing materials that have high thermal conductivity, in conjunction with making strategic design decisions that either improve airflow or incorporate phase change materials, can be of assistance in effectively managing thermal loads.
7. The integration of regenerative braking systems includes the following: The degree to which carbon-based supercapacitors can combine with regenerative braking systems is the determining factor in the final effectiveness of these devices. In this context, the evaluation of the supercapacitors' capacity to deal with transient loads and quick charge-discharge cycles is included. It is vital to have a proper system-level design, which includes power management techniques and control algorithms, to guarantee that the supercapacitors will operate without any interruptions and will engage in the most effective manner possible alongside ordinary batteries.
8. Sustainability and Lifecycle Assessment: Because the world is moving more and more towards developing more environmentally friendly manufacturing techniques, it is imperative that the environmental impact of carbon-based products not be ignored. Considerations that are of the utmost importance include eco-friendly sourcing, recycling techniques, and disposal of supercapacitors at the end of their useful lives. To guarantee that the implementation of carbon-based supercapacitors is in accordance with broader environmental objectives, designers need to make the use of sustainable materials and methods a top priority.

9. Testing and optimization: - In final, it is necessary to conduct comprehensive testing to evaluate the performance of the proposed supercapacitors under situations that are present in the actual world. It is necessary to conduct a thorough analysis of numerous parameters, including energy density, cycle life, efficiency, and operational stability. To refine designs prior to the development of actual prototypes, simulation tools, and modeling can be utilized. In addition, the optimization of the production processes to improve scalability and decrease costs is of the utmost importance for the realization of commercial viability. One of the most significant factors that determines the power density of CBS is the electrolyte that is used. When opposed to organic or ionic liquid electrolytes, traditional aqueous electrolytes have the potential to limit voltage windows, but they also offer advantages from a cost and safety standpoint. By carefully selecting an electrolyte that is suitable for the carbon matrix, designers can improve ionic conductivity while simultaneously maintaining compatibility with the matrix. Furthermore, the ionic size and mobility of the electrolyte ions have a considerable influence on the rate capabilities of the supercapacitors, which in turn affects their performance during the phases of rapid energy recovery that occur during regenerative braking (see, Odnala et al, 2025; Keerthana et al,2025; for more).

To achieve the goal of increasing power density in regenerative braking systems, the design considerations for carbon-based supercapacitors involve a multidisciplinary approach that includes materials science, chemical engineering, and system design. Researchers and engineers can significantly improve the performance of supercapacitors by concentrating on the selection of materials, fabrication techniques, electrolyte optimization, and architectural innovations. This makes supercapacitors an invaluable tool in the search for transportation solutions that are more efficient and less harmful to the environment. Several interrelated phases are required in the design of carbon-based supercapacitors for energy delivery in regenerative braking systems. These steps range from the selection of materials to the management of thermal energy. The effective execution of each phase brings about major gains in the performance and dependability of regenerative braking technologies. Each step presents its own unique set of problems and opportunities. For the future of electric and hybrid vehicles, it will be essential to conduct additional research and innovation in this field. This is because the demand for energy storage systems that are both efficient and effective is expected to continue to develop. Exploration of novel materials and design approaches is ongoing, and it holds the potential to release even higher efficiency and sustainability in the transportation sector. As the automobile industry continues to push the boundaries of innovation, the influence of cutting-edge energy storage systems like CBS will become increasingly important in determining the trajectory of sustainable mobility in the future.

CARBON-BASED SUPERCAPACITOR ON POWER DENSITY OF ENERGY DELIVERY IN REGENERATIVE BRAKING

When it comes to meeting the growing demand for environmentally friendly energy solutions, particularly in the realm of hybrid systems and electric vehicles(EVs), the development of advancements in energy storage technologies is necessary. Carbon-based supercapacitors (CBSCs) have emerged as a potential alternative among these technologies due to their high power density, extended cycle life, and quick charge-discharge capabilities as suggested by K Kazi(2024). All of these characteristics set them apart from other technologies. This article focuses on the function that carbon-based supercapacitors play in increasing the power density of the energy supply during regenerative braking. This is a crucial feature in many modern automobiles that are designed to recover and reuse energy that would otherwise be lost.

In the context of regenerative braking applications, the power density of an energy storage system, which refers to the rate at which energy may be delivered, is an essential critical factor in determining the effectiveness of the system. Traditional lead-acid or lithium-ion batteries have power densities that are much lower than those of CBSCs, which typically display power densities in the range of 10 kW/kg to over 30 kW/kg of power density. Through the utilization of this capability for rapid energy delivery, the vehicle can swiftly transform the kinetic energy that is generated during the braking process into electrical energy without slowing the braking process.

It has been established through research that including CBSCs in a regenerative braking system can improve overall efficiency by efficiently controlling the energy capture and discharge phases of the system. For instance, a supercapacitor can immediately absorb huge amounts of energy during heavy braking, whereas a battery can process energy storage and supply for propulsion over a longer period of time. In this hybrid method, the strengths of both technologies are optimized, which results in an improvement in the performance of the vehicle as well as an extension of the lifecycle of the energy storage devices that are involved.

There are several difficulties involved in incorporating carbon-based supercapacitors into regenerative braking systems, even though these supercapacitors have many benefits. For example, while CBSCs are excellent at delivering power, they typically have a lower energy density in comparison to batteries. As a result, hybrid systems that integrate both technologies are required. In addition, the cost of high-performance carbon materials has the potential to have an effect on the comprehensive economics of the system. Research is now being conducted with the goal of developing materials and manufacturing processes that are more cost-effective, while simultaneously improving energy density, in order to make CBSCs a more viable standalone alternative for utilization in future applications.

In terms of the environment, the use of carbon-based materials, particularly those that are derived from environmentally friendly techniques, is in accordance with the objectives of lowering the carbon footprints that are connected with conventional energy storage. Through the enhancement of power density and efficiency in energy recovery applications such as regenerative braking, CBSCs help to the overall reduction of emissions and energy consumption in the automobile industry.

The use of carbon-based supercapacitors is a game-changing technology that has the potential to improve the power density of energy delivery in regenerative braking systems. Because of their capacity to rapidly store and release energy, they are a good choice for integration with other energy storage solutions. This would pave the way for electric car systems that are more efficient and sustainable. Even though there are still obstacles to overcome in terms of cost and the requirement for hybrid configurations, continued breakthroughs in material science are anticipated to move CBSCs toward more widespread deployment. Carbon-based supercapacitors will play a vital role in influencing the future of energy storage and delivery inside regenerative braking systems. This shift towards electrification in the automobile industry is expected to continue in the coming years.

DESIGN STEPS IN CARBON-BASED SUPERCAPACITORS FOR ENHANCED POWER DENSITY IN REGENERATIVE BRAKING

It is becoming increasingly important for electric and hybrid vehicles to incorporate regenerative braking technology as the globe moves towards energy solutions that are more environmentally friendly and efficient. One of the most important aspects of optimizing this system is the utilization of carbon-based supercapacitors, which have become well-known for their capacity to transfer energy in a short amount of time and with a high power density. Within the context of regenerative braking systems, this review focuses on the design processes that are involved in carbon-based supercapacitors and the way these design steps influence the power density of energy delivery shown in Figure 3.

Figure 3. Design steps

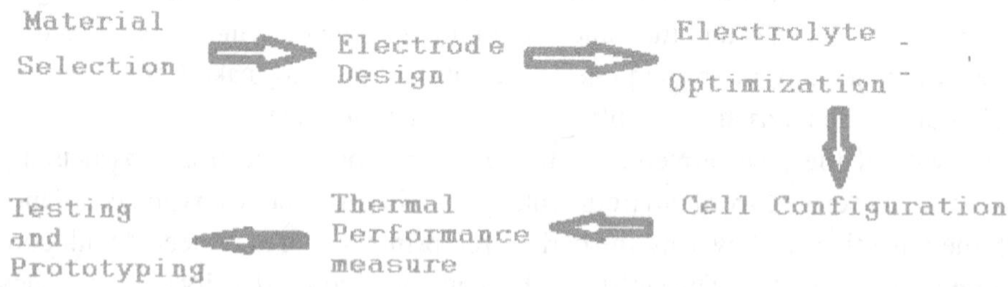

1. The selection of the material: In carbon-based supercapacitors, the selection of electrode materials has a significant impact on the performance of the device. Activated carbon, graphene, and carbon nanotubes are examples of porous carbon materials that have high surface areas, which contribute to an increase in their capacity to store electronic charges. These materials need to have their porosity and surface functionality optimized to enhance the amount of surface area that is accessible for ion adsorption, which will ultimately increase the overall power density of the supercapacitor. The selection of the proper carbon materials, which may include activated carbon, carbon nanotubes, and graphene, is the initial step in the process of constructing supercapacitors that require carbon as their primary ingredient. The capacity of the supercapacitor to store energy is directly influenced by the surface area, conductivity, and mechanical strength of each material. These characteristics determine the supercapacitor's overall performance. Capacitance can be greatly increased by materials with a higher surface area, which enables larger energy storage per unit volume. This is essential for efficient power transmission during regenerative braking events.
2. The Design of the Electrode: Immediately following the selection of the material, the subsequent stage is the design of the electrodes. When it comes to evaluating the overall performance of the supercapacitor, the shape, thickness, and porosity of the electrodes are extremely important features to consider. There is a correlation between thinner electrodes with high porosity and faster ion movement, which in turn leads to an increase in power density. In addition, the utilization of a sandwich structure or three-dimensional topologies can maximize the active surface area, which ultimately results in increased energy delivery rates during the braking process. When carbon-based materials are fabricated, the microstructure and morphology of these materials are directly influenced by the fabrication process, which in turn influences the electrochemical properties of these materials. Techniques like chemical vapor deposition (CVD), templat-

ing methods, and sol-gel processes can be modified to build carbon structures that have surface properties and porosity that have been optimized. Controlled manufacturing procedures are necessary for getting high performance in practical applications because they provide uniformity and consistency, which are essential for achieving high performance.

3. Optimisation of the Electrolyte: -The selection of the electrolyte is an important factor in determining the overall efficiency of the supercapacitor. Additionally, the electrolyte plays a significant role in determining the ion mobility and conductivity of the supercapacitor. When it comes to electrolytes, the option of In comparison to organic electrolytes, aqueous electrolytes typically have voltage windows that are lower, even though they are safe and efficient. Ion migration within the device is influenced by many factors, including the electrolyte's conductivity, ionic strength, and stability. An electrolyte that has been optimized and has a high ionic conductivity will minimize the internal resistance, which will allow for faster charge and discharge cycles. This is vital during the rapid energy recovery that occurs in regenerative braking systems. But the latter can provide potential dangers to one's safety. In recent years, there have been significant developments in ionic liquids and mixed electrolytes, which have shown encouraging results in terms of balancing the voltage window and conductivity, hence improving the distribution of energy during braking.
4. Regarding the configuration of the cell, the design of the cell architecture affects the effectiveness of the energy delivery. For instance, symmetric or asymmetric designs of supercapacitor cells can be optimized for regenerative braking dependent on the individual application requirements that are being satisfied. It is possible that asymmetric cells, which combine different materials for the anode and cathode, have the potential to increase energy density without compromising power density. The configuration of the supercapacitor cells is another factor that affects the power density levels. The design must strike a balance between the voltage needs and the capacity to deliver high current, regardless of whether the configurations are in series or parallel. When it comes to applications that involve regenerative braking, a parallel layout can be especially advantageous since it enables a higher current capacity, which in turn results in more effective recovery of energy during situations that involve a high amount of braking demand.
5. The thermal performance of supercapacitors is a factor that is sometimes disregarded, even though it has a significant impact on the power density of these devices. During the high charge and discharge rates that are associated with regenerative braking, it is possible to minimize overheating by using efficient heat dissipation procedures. These tactics may include the incorporation of thermally conductive materials or cooling techniques. Supercapacitors are guaranteed to

have dependable performance and a long lifespan if they are kept at the proper operating temperatures consistently.

6. Testing and Prototyping: - Once the design has been finalized, it is necessary to conduct exhaustive testing and prototyping to evaluate the performance of the carbon-based supercapacitors under conditions that imitate regenerative braking. When determining whether or not the supercapacitor is capable of meeting the appropriate performance criteria, it is necessary to analyze parameters such as energy density, power density, cycle life, and efficiency. Furthermore, iterative testing has the capability of revealing alternative design improvements that could further improve performance.

When it comes to architecture and design, it is essential to take into consideration the physical design of carbon-based supercapacitors to enhance power density. This involves optimizing the thickness of the electrodes, the materials used for the separators, and the overall structure of the device. The surface area that is exposed to the electrolyte can be increased by the use of multi-layer structures or three-dimensional architectures. This provides the opportunity for quicker ion transport and enhances the efficiency of energy storage during times of high power demands. In addition, the incorporation of features such as integrated thermal management systems can help alleviate problems that are associated with the development of heat and increase the longevity of the devices.

Integrating with Regenerative Braking Systems: The degree to which carbon-based supercapacitors can successfully integrate with regenerative braking systems is the single most important factor in determining their overall effectiveness. In this context, the evaluation of the supercapacitors' capacity to deal with transient loads and quick charge-discharge cycles is included. It is vital to have a proper system-level design, which includes power management techniques and control algorithms, to guarantee that the supercapacitors will operate without any interruptions and will engage in the most effective manner possible alongside ordinary batteries.

Assessing the lifecycle and sustainability of an organization: In light of the fact that the world is moving more and more towards environmentally responsible industrial techniques, it is imperative that the environmental impact of carbon-based products not be passed over. Considerations that are of the utmost importance include eco-friendly sourcing, recycling techniques, and disposal of supercapacitors at the end of their useful lives. To guarantee that the implementation of carbon-based supercapacitors is in accordance with broader environmental objectives, designers need to make the use of sustainable materials and methods a top priority.

To summarize, the design stages that are involved in the development of carbon-based supercapacitors have a significant influence on the power density of energy delivery in regenerative braking systems. To achieve the goal of increasing power density in regenerative braking systems, the design considerations for carbon-based supercapacitors involve a multidisciplinary approach that includes materials science, chemical engineering, and system design. Researchers and engineers can significantly improve the performance of supercapacitors by concentrating on the selection of materials, the techniques of fabrication, the optimization of electrolytes, the optimization of electrode and electrolyte configurations, the effective configuration of cells, the management of thermal properties, and the implementation of architectural innovations. This makes supercapacitors an invaluable tool in the search for transportation solutions that are more

As the automobile industry continues to push the boundaries of innovation, the influence of cutting-edge energy storage systems like CBS will become increasingly important in determining the trajectory of sustainable mobility in the future. Engineers have the ability to dramatically increase the responsiveness and efficiency of supercapacitors in electric vehicles by selecting materials with great care. The incorporation of advanced carbon-based supercapacitor designs will play a pivotal role in enhancing the sustainability and performance of regenerative braking systems, which will contribute to the overarching goal of reducing carbon emissions in the transportation sector. This will be accomplished as the technology of energy storage continues to advance.

It is ultimately the thorough evaluation of each design stage that gives a pathway toward high-performance energy solutions. These solutions are essential for meeting the expanding demands of transportation technologies of the next generation.

CARBON-BASED SUPERCAPACITORS FOR TEMPERATURE SENSITIVITY IN REGENERATIVE BRAKING

Regenerative braking systems have risen to the forefront of automotive technology as a result of the development of electric and hybrid vehicles. It is not only that these systems improve the energy economy, but they also contribute to the longevity of the components that make up the brakes. One of the most important components of energy storage, supercapacitors, may be found at the core of many of these kinds of devices. The purpose of this review is to examine the role that carbon-based supercapacitors play in overcoming temperature sensitivity issues in applications that involve regenerative braking processes.

When it comes to regenerative braking, temperature sensitivity is one of the key obstacles that must be overcome. Temperature variations have the potential to have a substantial impact on the efficiency and performance of energy storage systems. At lower temperatures, performance may be hindered due to increased viscosity in electrolytes and slower reaction times. On the other hand, higher temperatures can lead to a rise in internal resistance, which can result in decreased power output and capacity fading. For this reason, it is essential for efficient regenerative braking systems to have a solid understanding of the temperature sensitivity of energy storage devices and to take measures to mitigate this sensitivity.

When compared to their equivalents made of metal oxide, carbon-based supercapacitors have been shown to have superior thermal stability, according to research investigations. Through individualized material synthesis and electrolyte selection, it is possible to broaden the temperature range in which they provide their services. An example of this would be the utilization of ionic liquids as electrolytes, which has demonstrated potential in terms of enhancing performance at both high and low temperatures. In addition, sophisticated carbon structures such as graphene have the potential to improve thermal management within the supercapacitor, which in turn enables greater performance despite swings in temperature.

Several benefits can be gained from the incorporation of carbon-based supercapacitors into regenerative braking systems, including the following:

1. Efficiency: These supercapacitors can swiftly absorb and release energy, which is critical for catching the rapid bursts of energy that occur during braking.
2. Longevity: High cycle stability means that the performance of the supercapacitors is maintained during a large number of charging and discharging cycles, which is a common occurrence in urban driving situations with frequent stops.
3. Temperature Management: Improved thermal stability reduces the amount of performance deterioration, which enables consistent efficiency regardless of the conditions of the environment. When it comes to locations that see high-temperature swings, this is very advantageous.
4. Weight Considerations: As a result of continued breakthroughs in carbon nanomaterials, manufacturers are now able to create lightweight designs without sacrificing energy storage capacity, which is an essential component in the field of automobile engineering.

Despite the benefits, there are still some difficulties to overcome. There is a need for additional research into the scalability of improved carbon materials, the cost of production, and the optimization of hybrid systems that mix supercapacitors with conventional batteries. Additionally, there is a need for additional studies to be conducted on the effects that temperature cycling has on the structural integrity of these materials over the long term.

When it comes to solving the issue of temperature sensitivity in regenerative braking systems, carbon-based supercapacitors are a promising option that might be implemented. The one-of-a-kind characteristics that they possess make it possible for them to effectively absorb and store energy, which guarantees that automobiles can maximize energy recovery while retaining performance in a wide range of temperature conditions. With the ongoing advancement of research and technology in this area, we may anticipate even more significant advancements in regenerative braking systems, which will ultimately contribute to the development of transportation solutions that are more environmentally friendly. The combination of material science and automotive engineering will likely play a significant part in determining the direction that electric mobility will go in the future.

DESIGN STEPS IN CARBON-BASED SUPERCAPACITORS FOR TEMPERATURE SENSITIVITY IN REGENERATIVE BRAKING

Innovative technologies such as regenerative braking systems are becoming increasingly popular as the automobile industry continues to move in the direction of electrification and increased energy economy. When compared to the many different types of energy storage devices that are utilized in these systems, carbon-based supercapacitors (CSCs) offer several distinct advantages due to their capacity for rapid charge and discharge as well as their high power density. The temperature sensitivity of these supercapacitors, on the other hand, is an important factor that requires careful consideration. This paper takes a look at the design stages that are required in optimizing carbon-based supercapacitors for temperature sensitivity, specifically about applications that entail regenerative braking.

As the demand for transportation that is both energy-efficient and environmentally friendly continues to grow, regenerative braking systems have emerged as an important technology. These systems can harness energy that is generally lost during the braking process. Carbon-based supercapacitors (CBSCs), which are at the heart of this technology, have gained popularity because they have a high power density, the ability to charge and discharge quickly, and a long cycle life. The fact that CBSCs are sensitive to changes in temperature is, however, one of the most significant obstacles that must be overcome to successfully incorporate them into regenerative braking systems. This review focuses on the most important design considerations that should be taken into account to improve the performance of carbon-based supercapacitors regardless of the temperature.

- An Explanation of Regenerative Braking: Regenerative braking systems are designed to recover kinetic energy that is lost during the slowdown of a vehicle and convert it into electrical energy that may be stored for later use. One of the most important factors that determines how effective this process is is the technology that is used to store energy. Supercapacitors are particularly well-suited for this application because of their capacity to rapidly absorb and release energy. As a result, they are an effective supplement to conventional batteries.

- The Relationship Between Temperature and Supercapacitors: Variations in temperature have the potential to have a substantial impact on the efficiency and durability of energy storage devices. As a result of the rapid cycling between charge and discharge that occurs during regenerative braking, supercapacitors go through a period of rapid temperature fluctuation. It is possible for high temperatures to result in increased impedance and decreased capacitance, whilst low temperatures can make it more difficult for charge carriers to move around more freely. Therefore, it is essential to gain an understanding of these temperature impacts and find ways to mitigate them in order to improve the performance and dependability of CSCs.

Steps in the Design Process (Figure 4) are as Follows-

1. The Selection of Materials: When it comes to the thermal behaviour of supercapacitors, the selection of carbon materials is also a very important factor. The thermal stability and electrochemical performance of supercapacitors are both greatly impacted by the selection of materials that are based on carbon. The thermal and electrical properties of activated carbons, carbon nanotubes, and graphene are all distinct from one another. For example, graphene, despite the fact that it has superior conductivity and thermal management, can be quite expensive and difficult to manufacture on a large scale. Carbon nanotubes, activated carbon, and graphene are examples of advanced carbons that not only have high surface areas but also display thermal qualities that are not only favourable but also advantageous. Increasing thermal conductivity and mitigating the effects of temperature changes can be accomplished through research that focuses on hybrid materials or composite structures. For example, integrating conductive polymers or metal oxides is one way to incorporate these materials. The ability of these materials to sustain their performance across a wide temperature range is significantly improved. It is also possible to boost thermal conductivity by using conductive additives, which will help with the process of heat dispersion.

Figure 4. Design steps

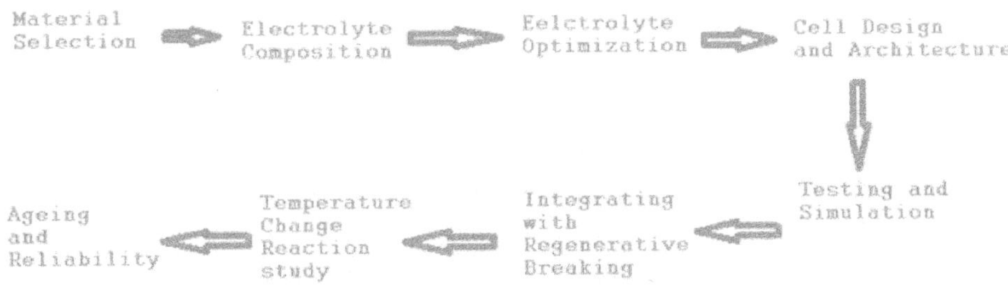

2. The Composition of the Electrolyte: The electrolyte is an essential component in the performance of the supercapacitor, and the thermal sensitivity of the electrolyte can have a significant impact on the overall efficiency. In spite of the fact that organic electrolytes have a high energy density, they frequently experience a decline in performance when exposed to situations of extremely high temperatures. On the other hand, aqueous electrolytes have the ability to function effectively across a wider temperature range, but they may put a limit on the energy density. The investigation of ionic liquids and gel electrolytes, which provide thermal stability as well as a wider operational window, might prove to be beneficial for applications involving regenerative braking.
3. Optimisation of the Electrolyte: The selection of the electrolyte has a considerable impact on the temperature sensitivity of supercapacitors. When compared to aqueous electrolytes, ionic liquids, and organic electrolytes are known to have a larger temperature range during which they can function appropriately. In order to optimise the overall performance of CSCs, it is possible to pick electrolytes that have a low viscosity at higher temperatures and a higher conductivity at lower temperatures.
4. The architecture and design of the cell: effective thermal management in supercapacitor cells is essential for ensuring that the cells continue to operate at appropriate temperatures. It is necessary to take into consideration heat dissipation techniques while designing the layout of the cell. These strategies may include the incorporation of thermal interface materials or the utilisation of cooling systems that are able to transfer temperature uniformly across the modules. When designing the architecture of the cell, it is vital to take heat management into consideration. Maintaining ideal operating conditions can be facilitated through the utilisation of layered structures that encourage the spread of temperatures in a consistent manner and the incorporation of cooling devices. It is also possible for improved structural design to reduce the amount of thermal stress that occurs

during rapid charge and discharge cycles. Additionally, the materials used for packaging should not only be able to provide mechanical stability but should also be able to promote thermal conductivity. This is done in order to prevent hotspots during high-cycle activities, which are typical of regenerative braking.

5. Testing and Simulation: It is exceptionally important to conduct exhaustive testing under a wide range of temperature conditions. It is possible to make an accurate prediction of how temperature influences the performance of supercapacitors by making use of thermal imaging and modeling technologies. With the use of these data, designs can be improved and appropriate materials can be chosen to reduce the impact of any negative impacts.
6. Integration with Regenerative Braking Systems: Finally, supercapacitors need to be incorporated with the regenerative braking systems of vehicles in a way that is completely seamless. To do this, it is necessary to take into account the thermal dynamics of the entire system, which includes the engine, the brakes, and the energy management systems, in order to guarantee that variations in temperature do not jeopardise performance.
7. The reaction to changes in temperature: The performance of a supercapacitor can deteriorate when the temperature fluctuates, which can have an effect on the charge and discharge rates. It is possible to gain insights into how to optimise the design for minimal voltage loss and maximise energy recovery by doing dynamic simulations of temperature response under situations of fast braking. The development of control algorithms that take into account variations in temperature can also be of assistance in properly controlling energy flows, assuring protection, and maintaining performance stability.
8. Ageing and Reliability: Thermal fluctuations can hasten the aging process of supercapacitors and shorten their lifespan. It is of the utmost importance to investigate methods that can improve the lifetime under varying heat conditions. Some examples of such methods include optimising cycling regimens or using protective coatings. Performing long-term testing under a variety of heat conditions can produce vital data for reliability evaluations, which can then be used to guide future design decisions.

For the record, with the incorporation of carbon-based supercapacitors into regenerative braking systems, there is a viable route for improving energy efficiency and sustainability in the transportation sector. Nevertheless, it is essential to manage temperature sensitivity through careful design considerations in order to maximise their performance and durability. The development of carbon-based supercapacitors with enhanced temperature sensitivity for use in applications involving regenerative braking brings both intriguing challenges and opportunities. Researchers and engineers have the ability to pave the way for the next generation of robust supercapacitors that can integrate seamlessly into advanced braking systems. This can be accomplished by strategically selecting materials, innovating electrolytes, optimising electrolytes, designing cells with careful consideration, conducting rigorous tests, integrating systems in a seamless manner, responding to thermal dynamics, and ensuring reliability. Both the effectiveness and dependability of regenerative braking systems can be improved through the development of energy storage devices that are both sturdy and reliable. As the automotive industry continues to undergo transformations, the optimisation of energy storage technology will become of the utmost importance. This will ensure that advancements will result in vehicles that are more environmentally friendly and efficient. The process of mastering temperature sensitivity is a difficult one, but it is an essential step in the direction of realising the full potential of carbon-based supercapacitors in practical applications. A holistic strategy that integrates performance, reliability, and sustainability should be prioritised in future innovations and studies in this subject in order to meet the ever-changing requirements of energy-efficient vehicles by combining all three of these factors.

DISCUSSION

A revolution has been triggered in the automotive sector as a result of the introduction of electric cars(EVs), which has promoted a considerable shift towards more environmentally friendly alternative modes of transportation. The use of supercapacitors has emerged as an essential component among the many technologies that are being investigated to improve the performance of electric vehicles. The purpose of the article is to examine the function of supercapacitors in electric vehicles, with a particular emphasis on their benefits, difficulties, and possible applications in the future.

Supercapacitors, which are often referred to as ultracapacitors, are devices that are used to store energy and bridge the gap between conventional capacitors and those used in batteries. The capacity of these devices to store and transfer energy in a short amount of time makes them particularly appealing for applications that require significant power output for a relatively short period of time. In the context of electric vehicles, this quality is complementary to lithium-ion batteries, which are superior in terms of energy density but may be slower in terms of the rate at which they provide energy.

One of the most significant benefits of using supercapacitors in electric vehicles is the remarkable charge and discharge speeds that they offer. Because of this, rapid acceleration is possible, which is essential for the requirements of modern driving. It is possible for supercapacitors to store the energy that is produced during regenerative braking when they are linked with batteries. This allows the energy to be used immediately. This not only increases the performance of the car but also makes the vehicle more energy efficient and extends the life of the battery by lowering the number of charging cycles it undergoes. Furthermore, in comparison to batteries, supercapacitors have a longer cycle life, which makes them a potential alternative for increasing the lifespan of the whole energy storage system in electric vehicles (EVs).

The ability of supercapacitors to work well in a larger variety of environmental circumstances is possible due to their temperature resistance, which is another significant advantage of these storage devices. Because electric vehicles are used in a wide range of climates, this feature is very useful for ensuring that the vehicle's performance and safety remain consistent.

However, in order for supercapacitors to reach their full potential in electric vehicles, there are obstacles that need to be overcome. If we compare them to lithium-ion batteries, their energy density is far lower. This is the most significant restriction. Although they are capable of delivering power quickly, they are not able to store as much energy, which results in a shorter range when they are utilised on their own. The current focus of research is on hybrid energy storage devices that combine the benefits of both supercapacitors and batteries in order to maximise performance while minimising the drawbacks of each.

Despite the fact that the price of supercapacitors is reducing, it is still possible for them to be more expensive than conventional methods of energy storage. It will be necessary to make additional advances in material science and manufacturing methods in order to achieve widespread adoption and commercial-scale production. These advancements should aim to reduce prices while simultaneously improving performance.

When considering the future, the utility of supercapacitors in electric vehicle applications is very promising. With continuous research into innovative materials like graphene and carbon nanotubes, it is possible that supercapacitors will soon be able to boast higher energy densities, which could pose a threat to the position of lithium-ion technology as the dominant technology. Furthermore, the incorporation of supercapacitors into future electric vehicle designs may result in the development of ingenious configurations that are tailored specifically to urban driving patterns. This is because urban driving patterns are characterised by a greater demand for regenerative braking and rapid acceleration.

As a conclusion, supercapacitors are an essential component in the development of electric vehicles. It is possible for them to considerably improve electric vehicle performance, efficiency, and lifetime when they are successfully integrated with battery systems. They offer a number of distinct advantages. There is a high probability that supercapacitors will become an integral component of the ecosystem of electric mobility as technology continues to advance and the automotive landscape continues to undergo transformation. This will pave the way for a more environmentally friendly future in the transportation sector.

When it comes to the development of electric vehicles(EVs), the incorporation of innovative energy storage systems proves to be an essential component. This is because the automotive industry is continuing its transition towards more environmentally friendly energy solutions. Among these systems, carbon-based supercapacitors are growing in popularity as a potentially useful technology. These supercapacitors combine a high power density with the ability to charge and discharge quickly, as well as an environmentally favorable profile. The purpose of this review is to investigate the current state of affairs regarding carbon-based supercapacitors in electric vehicles, focussing on their benefits, limitations, and potential implications for the future.

The storage of energy is accomplished using carbon-based supercapacitors, which are often referred to as electric double-layer capacitors (EDLCs). These capacitors make use of carbon materials, such as activated carbon, carbon nanotubes, or graphene. With their ability to deliver rapid bursts of energy, which is vital for applications that require high power output in short intervals, they operate on electrostatic principles, which enables them to deliver energy. Supercapacitors are superior to ordinary batteries in terms of power density, which makes them a great choice for improving the performance of electric vehicles during acceleration and regenerative braking. Ordinary batteries are primarily concerned with energy density.

The great cycle stability of carbon-based supercapacitors is one of the most significant features of these devices. They can typically withstand more than one million charge/discharge cycles without experiencing substantial degradation. Because of its endurance, there is a decreased requirement for replacements, which results in a reduction in the long-term expenses for both users and manufacturers of vehicles.

Additionally, these supercapacitors have a broad temperature range that they can operate in, which guarantees dependable performance in a variety of different temperatures and environments. Their capabilities of rapid charging and discharging increase driving experiences by offering immediate power boosts. This is especially advantageous in situations where driving in the city involves continuous stop-and-go traffic, which is a common occurrence.

Another key advantage is that they are good for the environment. This is because many carbon materials can be acquired in a sustainable manner, and they do not contain the poisonous heavy metals that are commonly found in conventional batteries. The goals of the automobile industry as a whole, which include lowering environmental effects and promoting sustainability, are successfully aligned with this.

Although carbon-based supercapacitors have a number of attractive characteristics, they also have a number of problems. One of the most significant problems is that, in comparison to lithium-ion batteries, they have a relatively poor energy density. Because of this constraint, their capacity to act as independent energy storage solutions for longer journeys is restricted. As a result, a hybrid strategy is required, in which supercapacitors assist batteries rather than completely replacing them.

Furthermore, despite the fact that the cost of manufacture has gradually decreased as a result of developments in manufacturing techniques, it continues to be a barrier to general use. To ensure the successful incorporation of supercapacitors into mainstream electric vehicle applications in the future, it is vital to develop technologies that are more efficient, scalable, and economically viable for the production of high-quality carbon materials.

The use of carbon-based supercapacitors in electric vehicles appears to have a bright future, particularly in light of the development of scientific advances that continue to improve their qualities. Strategies such as hybrid systems, which mix batteries with supercapacitors, are gaining steam. These systems allow for effective energy management that takes advantage of the characteristics of both technologies.

Furthermore, it is anticipated that developments in nanotechnology and material science will provide an increase in the energy density of carbon-based supercapacitors, so rendering them more suited for a wider range of applications within the electric vehicle market.

The automotive industry is currently grappling with the dual issues of energy economy and sustainability. In this context, regenerative braking has emerged as a crucial technology, particularly in the context of electric cars(EVs). One of the most encouraging developments in this field is the incorporation of supercapacitors that are based on carbon. The purpose of this review is to investigate the role that carbon-based supercapacitors play in improving regenerative braking systems, as well as their possibilities for the future, as well as their advantages and limitations.

Electric vehicles are able to recover kinetic energy during deceleration by utilising regenerative braking systems. This allows the vehicle to convert the kinetic energy into electrical energy that can be stored for later use. Not only does this procedure increase the driving range of the car, but it also lessens the amount of wear that is placed on the conventional braking components. Nevertheless, the efficiency of regenerative braking is greatly dependent on the energy storage technology that is currently being utilised.

Their one-of-a-kind characteristics, which include high power density, rapid charge/discharge rates, and long cycle life, have the potential to result in more effective energy recovery, which will ultimately contribute to an improvement in the overall performance of electric vehicles. Although there are still obstacles to overcome, the progress that is being made in this field indicates that carbon-based technology could have a promising future in the field of sustainable transportation. There is a good chance that greater electric mobility will be made possible as a result of the synergy that exists between novel materials and modern automotive technology as the industry continues to develop.

Within the field of electric and hybrid vehicles, regenerative braking has emerged as a crucial technology that has made a substantial contribution to the efficiency of energy use and the sustainability of the environment. In this debate on energy transfer systems in regenerative braking, several important aspects that demand a comprehensive analysis are addressed.

The advantages of using regenerative braking are numerous and varied. The increased energy efficiency of hybrid and electric vehicles is one of the most persuasive factors that has been mentioned by the researchers. The total range and performance of these cars are improved by the use of regenerative braking, which is accomplished by capturing energy that would otherwise be lost as heat. Throughout the course of the conversation, it is clearly outlined how this technology may lessen the need for external charging stations, which will ultimately result in a reduction in operational expenses for customers.

Regenerative braking also has substantial ramifications for the environment, which is another important consideration. Not only does the capacity to recover and reuse energy contribute to a reduction in energy consumption, but it also significantly reduces the emissions of greenhouse gases that are related to the creation of power. The debate provides insightful information regarding how the implementation of regenerative braking across bigger fleets can result in significant reductions in emissions for the environment.

The presentation of both the benefits and the difficulties associated with regenerative braking technology enables a comprehensive understanding of this essential invention in the automobile industry. The investigation of potential future trends is not only a source of encouragement for the ongoing progress that is being made in energy recovery, but it also highlights the critical role that partnership plays in the process of building a sustainable transportation future. This dialogue is a commendable contribution to the area, as it lays a solid platform for ongoing conversations and research on the subject of regenerative braking technology.

One of the technologies that is causing a revolution in the automobile industry is the one that is being discussed in relation to energy delivery in regenerative braking. The potential of this technology to recover energy and use it in an effective manner not only enhances the performance of the vehicle but also helps the larger objective of achieving sustainability in the transportation sector. Even if there are still obstacles to overcome, the future of regenerative braking offers promise as a solution that could play a pivotal role in the journey toward vehicles that are more environmentally friendly and efficient. There is no doubt that the ongoing development of this technology will unquestionably play a big part in the formation of the subsequent generation of transportation mechanisms.

The power density of energy transfer in regenerative braking is an important topic that has to be discussed in order to make considerable progress in the efficiency of current automobiles. Not only does a high power density help to greater energy recovery and vehicle performance, but it also works to promote the larger aims of sustainability and reducing carbon emissions. In order to realize the full potential of regenerative braking systems, it will be essential to address the challenges that are associated with high power density as technology continues to advance.

It is quite likely that continued research and innovation in power density will play a crucial role in creating the future of energy-efficient transportation. This is because the automotive industry is progressing towards technologies that are more refined and efficient. This investigation promises to provide a driving experience that is more environmentally friendly and efficient, which is in line with the current global trend towards greener mobility solutions.

The distinctive qualities of CBSCs, which include their high power density, quick charge/discharge capabilities, and long cycle life, make them an appealing alternative to the conventional battery technologies that are now in use. The current state of CBSCs is investigated in this paper, with a particular emphasis placed on the temperature sensitivity of CBSCs and their performance in applications including regenerative braking.

CBSCs are very sensitive to temperature, which can have a negative impact on their performance, efficiency, and longevity. This is one of the most significant issues they face. There is a considerable relationship between temperature changes and performance parameters such as capacitance, energy density, and operational stability. It is possible that CBSCs will exhibit increased ionic mobility when subjected to higher temperatures, which will result in enhanced conductivity and a better capacity for charge storage. On the other hand, temperatures that are extremely high can contribute to instability and deterioration of the materials that make up the electrodes, which can result in a decrease in performance over time.

On the other hand, low temperatures might impede the passage of ions inside the electrolyte, which can lead to an increase in the internal resistance and a decrease in the charge/discharge rates during the process. It is of the utmost importance to develop CBSC designs and electrolyte formulations that are capable of retaining optimal performance across a broad temperature spectrum. This is especially important for applications that are carried out in regions that experience a variety of climates.

However, their temperature sensitivity is an important factor that must be taken into consideration, since it is one of the most important factors that determines whether or not they are viable for use in practical applications. In order to solve these problems and ensure that CBSCs are able to perform dependably throughout a wide range of environmental conditions, it is vital to continue research and development in the fields of material science and engineering. In conclusion, carbon-based supercapacitors have the potential to play a crucial role in the development of environmentally friendly transportation systems in the future, provided that additional breakthroughs are made.

In recent years, researchers have been concentrating their efforts on developing novel material composites and hybrid designs in order to improve the thermal stability and performance of CBSCs. For example, the utilisation of nanostructured carbon composites or the combination of carbon materials with polymer electrolytes has demonstrated potential for enhancing both the energy density and the thermal stability of the material. Furthermore, developments in $scCO_2$ (supercritical carbon dioxide) methods for synthesis are opening the way for the production of high-performance electrodes that are not only efficient but also durable throughout a wide range of environments and temperatures.

CONCLUSION

In a nutshell, carbon-based supercapacitors provide a compelling alternative for the electric vehicle industry. These supercapacitors offer advantages in terms of sustainability, short recharge times, and high power output. Despite the fact that there are still obstacles to overcome in terms of their energy density and production costs, continued research and development are paving the way towards their more widespread incorporation into electric vehicle platforms. As efforts to find energy storage solutions that are both effective and environmentally friendly continue, carbon-based supercapacitors have emerged as a crucial component for the development of environmentally friendly transportation in the future. In the near future, the trajectory of electric mobility will be determined by the equilibrium of utilising different energy storage technologies, with carbon-based supercapacitors playing an essential part in this evolution. The development of high-performance electrodes that are merely efficient yet additionally resilient across a wide range of temperatures is being made possible by recent breakthroughs in the technologies for the synthesis of $scCO_2$, which stands for supercritical carbon dioxide.

ACKNOWLEDGEMENT

I would like to dedicate this work to my Mother, father and sisters, who encourages me to work on this field, which is outside from my boundaries of my research field. They gave me support throughout the work and motivate me to review the capacitor usages in Supercapacitors.

Funding: Nil

Conflict: No

REFERENCES

Kasat, K., Shaikh, N., Rayabharapu, V. K., & Nayak, M. (2023). *Implementation and Recognition of Waste Management System with Mobility Solution in Smart Cities using Internet of Things, 2023 Second International Conference on Augmented Intelligence and Sustainable Systems.* ICAISS., Available at https://ieeexplore.ieee.org/document/10250690/, DOI: 10.1109/ICAISS58487.2023.10250690

Kazi, K. (2024a). AI-Driven IoT (AIIoT) in Healthcare Monitoring. In Nguyen, T., & Vo, N. (Eds.), *Using Traditional Design Methods to Enhance AI-Driven Decision Making* (pp. 77–101). IGI Global., available at https://www.igi-global.com/chapter/ai-driven-iot-aiiot-in-healthcare-monitoring/336693, DOI: 10.4018/979-8-3693-0639-0.ch003

Kazi, K. (2024b). Machine Learning (ML)-Based Braille Lippi Characters and Numbers Detection and Announcement System for Blind Children in Learning. In Sart, G. (Ed.), *Social Reflections of Human-Computer Interaction in Education, Management, and Economics*. IGI Global., DOI: 10.4018/979-8-3693-3033-3.ch002

Kazi, K. (2024c). Modelling and Simulation of Electric Vehicle for Performance Analysis: BEV and HEV Electrical Vehicle Implementation Using Simulink for E-Mobility Ecosystems. *In L. D., N. Nagpal, N. Kassarwani, V. Varthanan G., & P. Siano (Eds.), E-Mobility in Electrical Energy Systems for Sustainability (pp. 295-320). IGI Global.* Available at: https://www.igi-global.com/gateway/chapter/full-text-pdf/341172DOI: 10.4018/979-8-3693-2611-4.ch014

Kazi, K. (2025). Machine Learning-Powered IoT (MLIoT) for Retail Apparel Industry. In Tarnanidis, T., Papachristou, E., Karypidis, M., & Manda, V. (Eds.), *Sustainable Practices in the Fashion and Retail Industry* (pp. 345–372). IGI Global Scientific Publishing., DOI: 10.4018/979-8-3693-9959-0.ch015

Kazi, K. S. (2025). Braille-Lippi Numbers and Characters Detection and Announcement System for Blind Children Using KSK Approach: AI-Driven Decision-Making Approach. In T. Murugan, K. P., & A. Abirami (Eds.), Driving Quality Education Through AI and Data Science (pp. 531-556). IGI Global Scientific Publishing. DOI: 10.4018/979-8-3693-8292-9.ch023

Kazi, K. S. (2025). AI-Driven IoT (AIIoT)-Based Decision-Making System for High BP Patient Healthcare Monitoring: KSK1 Approach for BP Patient Healthcare Monitoring. In Mzili, T., Arya, A., Pamucar, D., & Shaheen, M. (Eds.), *Optimization, Machine Learning, and Fuzzy Logic: Theory, Algorithms, and Applications* (pp. 71–102). IGI Global Scientific Publishing., DOI: 10.4018/979-8-3693-7352-1.ch003

Kazi, K. S. (2025a). Advancing Towards Sustainable Energy With Hydrogen Solutions: Adaptation and Challenges. In Özsungur, F., Chaychi Semsari, M., & Küçük Bayraktar, H. (Eds.), *Geopolitical Landscapes of Renewable Energy and Urban Growth* (pp. 357–394). IGI Global Scientific Publishing., DOI: 10.4018/979-8-3693-8814-3.ch013

Kazi, S. (2024). Machine Learning-Based Pomegranate Disease Detection and Treatment. In Zia Ul Haq, M., & Ali, I. (Eds.), *Revolutionizing Pest Management for Sustainable Agriculture* (pp. 469–498). IGI Global., DOI: 10.4018/979-8-3693-3061-6.ch019

Kazi, S. (2024a). Computer-Aided Diagnosis in Ophthalmology: A Technical Review of Deep Learning Applications. In Garcia, M., & de Almeida, R. (Eds.), *Transformative Approaches to Patient Literacy and Healthcare Innovation* (pp. 112–135). IGI Global., Available at https://www.igi-global.com/chapter/computer-aided-diagnosis-in-ophthalmology/342823, DOI: 10.4018/979-8-3693-3661-8.ch006

Kazi, S. (2024b). IoT Driven by Machine Learning (MLIoT) for the Retail Apparel Sector. In Tarnanidis, T., Papachristou, E., Karypidis, M., & Ismyrlis, V. (Eds.), *Driving Green Marketing in Fashion and Retail* (pp. 63–81). IGI Global., DOI: 10.4018/979-8-3693-3049-4.ch004

Kazi, S. (2024d). Artificial Intelligence (AI)-Driven IoT (AIIoT)-Based Agriculture Automation. In Satapathy, S., & Muduli, K. (Eds.), *Advanced Computational Methods for Agri-Business Sustainability* (pp. 72–94). IGI Global., DOI: 10.4018/979-8-3693-3583-3.ch005

Kazi, S. (2025). Machine Learning-Driven Internet of Medical Things (ML-IoMT)-Based Healthcare Monitoring System. In Soufiene, B., & Chakraborty, C. (Eds.), *Responsible AI for Digital Health and Medical Analytics* (pp. 49–86). IGI Global Scientific Publishing., DOI: 10.4018/979-8-3693-6294-5.ch003

Kazi, S. (2025a). Transformation of Agriculture Effectuated by Artificial Intelligence-Driven Internet of Things (AIIoT). In Garwi, J., Dzingirai, M., & Masengu, R. (Eds.), *Integrating Agriculture, Green Marketing Strategies, and Artificial Intelligence* (pp. 449–484). IGI Global Scientific Publishing., DOI: 10.4018/979-8-3693-6468-0.ch015

Kazi, S. (2025c). AI-Driven-IoT (AIIoT)-Based Decision Making in Drones for Climate Change: KSK Approach. In Aouadni, S., & Aouadni, I. (Eds.), *Recent Theories and Applications for Multi-Criteria Decision-Making* (pp. 311–340). IGI Global., DOI: 10.4018/979-8-3693-6502-1.ch011

Keerthana, R. K. V., Bhagyalakshmi, K., Papinaidu, M. V. V., & Liyakat, K. K. S. (2025). Machine learning based risk assessment for financial management in big data IoT credit. SSRN *Electronic Journal.* DOI: 10.2139/ssrn.5086671

Khadake, S. B., Chounde, A. B., & Suryagan, A. A., M. H. M. and M. R. Khadatare, (2024). AI-Driven-IoT(AIIoT) Based Decision Making System for High-Blood Pressure Patient Healthcare Monitoring, *2024 International Conference on Sustainable Communication Networks and Application (ICSCNA)*, Theni, India, 2024, pp. 96-102, DOI: 10.1109/ICSCNA63714.2024.10863954

Kutubuddin, K. (2024c). Vehicle Health Monitoring System (VHMS) by Employing IoT and Sensors, *Grenze International Journal of Engineering and Technology,* Vol 10, Issue 2, pp- 5367-5374.Grenze ID: 01.GIJET.10.2.429. Available at: https://thegrenze.com/index.php?display=page&view=journalabstract&absid=3371&id=8

Liyakat, K. K. S. (2023a). Machine Learning Approach Using Artificial Neural Networks to Detect Malicious Nodes in IoT Networks. In Shukla, P. K., Mittal, H., & Engelbrecht, A. (Eds.), *Computer Vision and Robotics. CVR 2023. Algorithms for Intelligent Systems*. Springer., DOI: 10.1007/978-981-99-4577-1_3

Mahant, M. A. (2025). Machine Learning-Driven Internet of Things (MLIoT)-Based Healthcare Monitoring System. In Wickramasinghe, N. (Ed.), *Digitalization and the Transformation of the Healthcare Sector* (pp. 205–236). IGI Global Scientific Publishing., DOI: 10.4018/979-8-3693-9641-4.ch007

Odnala, S., Shanthy, R., Bharathi, B., Pandey, C., Rachapalli, A., & Liyakat, K. K. S. (2025). Artificial Intelligence and Cloud-Enabled E-Vehicle Design with Wireless Sensor Integration. SSRN *Electronic Journal*. DOI: 10.2139/ssrn.5107242

Parihar, B., Kiran, A., Valaboju, S., Rashid, S. Z., Liyakat, K. K. S., & DR, A. S. L. (2025). Enhancing Data Security in Distributed Systems Using Homomorphic Encryption and Secure Computation Techniques. In ITM Web of Conferences (Vol. 76, p. 02010). EDP Sciences.

Prasad, K. R., Karanam, S. R., Ganesh, D., Liyakat, K. K. S., Talasila, V., & Purushotham, P. (2024, May). AI in public-private partnership for IT infrastructure development. *The Journal of High Technology Management Research*, *35*(1), 100496. DOI: 10.1016/j.hitech.2024.100496

Shinde, S. S., Nerkar, P. M., Kazi, S. S., & Kazi, V. S. (2025). Machine Learning for Brand Protection: A Review of a Proactive Defense Mechanism. In Khan, M., & Amin Ul Haq, M. (Eds.), *Avoiding Ad Fraud and Supporting Brand Safety: Programmatic Advertising Solutions* (pp. 175–220). IGI Global Scientific Publishing., DOI: 10.4018/979-8-3693-7041-4.ch007

Tamboli, D. A., & Sawant, V. A. M. H. M. and S. Sathe, (2024). AI-Driven-IoT(AIIoT) Based Decision-Making- KSK Approach in Drones for Climate Change Study, *2024 4th International Conference on Ubiquitous Computing and Intelligent Information Systems (ICUIS)*, Gobichettipalayam, India, 2024, pp. 1735-1744, DOI: 10.1109/ICUIS64676.2024.10866450

Upadhyaya, A. N., Surekha, C., Malathi, P., Suresh, G., Suriyan, K., & Liyakat, K. K. S. (2025). Pioneering cognitive computing for transformative healthcare innovations. SSRN *Electronic Journal.* DOI: 10.2139/ssrn.5086894

Chapter 18
Supercapacitors

G. Shashank
https://orcid.org/0009-0006-8815-980X
Chandigarh Group of Colleges, Mohali, India

B. V. Sai Thrinath
https://orcid.org/0000-0003-2451-0534
Mohan Babu University, India

S. Seenivasan
https://orcid.org/0000-0002-2155-9781
Rathinam Technical Campus, India

P. Selvakumar
https://orcid.org/0000-0002-3650-4548
Department of Science and Humanities, Nehru Institute of Technology, Coimbatore, India

Umesh Sankpal
R.P. Gogate College of Arts and Science, India & R.V. Jogalekar College of Commerce, India

T. C. Manjunath
https://orcid.org/0000-0003-2545-9160
Rajarajeswari College of Engineering, India

ABSTRACT

The fundamental principle behind a supercapacitor's operation involves the formation of an electric double layer at the electrode-electrolyte interface, where charge is stored electrostatically. These materials are chosen for their high surface area and electrical conductivity. The electrolyte, which can be either aqueous or organic, allows for ion transport between the electrodes. In an aqueous electrolyte, supercapacitors exhibit high capacitance and power density, while organic electrolytes can provide a wider voltage window, enhancing energy density. The separation of the charges in a supercapacitor is achieved through a process known as electrostatic charge accumulation. Cycle life compared to batteries, often exceeding one million charge-discharge cycles, which contributes to their reliability and longevity in various applications. Despite their many benefits, supercapacitors do face limitations, particularly in terms of energy density.

DOI: 10.4018/979-8-3693-9316-1.ch018

INTRODUCTION TO SUPERCAPACITORS: PRINCIPLES AND APPLICATIONS

The fundamental principle behind a supercapacitor's operation involves the formation of an electric double layer at the electrode-electrolyte interface, where charge is stored electrostatically. These materials are chosen for their high surface area and electrical conductivity. The electrolyte, which can be either aqueous or organic, allows for ion transport between the electrodes. In an aqueous electrolyte, supercapacitors exhibit high capacitance and power density, while organic electrolytes can provide a wider voltage window, enhancing energy density. The separation of the charges in a supercapacitor is achieved through a process known as electrostatic charge accumulation. Cycle life compared to batteries, often exceeding one million charge-discharge cycles, which contributes to their reliability and longevity in various applications. Despite their many benefits, supercapacitors do face limitations, particularly in terms of energy density. For instance, in a hybrid vehicle, a supercapacitor can handle the high power demands during acceleration and braking, while the battery can provide the longer-term energy storage needed for continuous driving. In addition to automotive applications, supercapacitors are increasingly being utilized in various other fields, including renewable energy systems, consumer electronics, and power backup systems. Furthermore, in power backup systems, supercapacitors provide an instantaneous power source during brief outages, ensuring uninterrupted operation of critical systems. The ongoing advancements in supercapacitor technology, including the development of new electrode materials and electrolytes, continue to enhance their performance and broaden their range of applications. Research into improving the energy density of supercapacitors, as well as exploring new materials and designs, holds promise for making these devices even more versatile and effective in the future.

HISTORY AND EVOLUTION OF SUPERCAPACITORS: FROM CONCEPT TO REALITY

The journey of supercapacitors from theoretical concept to practical application is a story of scientific innovation and technological evolution (Adetayo et al., 2019). The origins of supercapacitors can be traced back to the early 20th century when the fundamental principles of electrochemical energy storage were first being explored. The earliest capacitors relied on dielectric materials to store energy electrostatically, but these devices had limited energy storage capabilities and were primarily used for filtering and signal processing. It wasn't until the 1950s and 1960s that the groundwork for modern supercapacitors began to take shape with advancements in materials science and electrochemistry. The term "supercapacitor" gained prominence in the 1970s, a period marked by significant progress in the understanding and development of high-capacitance electrochemical devices. Researchers like Howard Becker and Conway et al. began to investigate materials that could offer higher capacitance than traditional capacitors. The concept of EDLCs was based on earlier discoveries related to the double-layer capacitor effect, where high capacitance was achieved through the use of porous electrode materials with large surface areas. The 1980s saw substantial advancements with the development of activated carbon as the primary electrode material for supercapacitors. This era also saw the introduction of the first commercial supercapacitors, which found niche applications in electronics and power management systems. In the 1990s, the field of supercapacitors experienced rapid growth and diversification. Similarly, graphene's outstanding electrical conductivity and large surface area made it a promising material for supercapacitor electrodes. During this time, supercapacitors began to find applications beyond consumer electronics, including use in automotive systems, where their ability to deliver quick bursts of power proved beneficial for regenerative braking and start-stop systems (Al-Refai et al., 2021). The early 2000s marked a period of commercialization and increased industrial interest in supercapacitors. Companies and research institutions focused on scaling up production and improving the cost-effectiveness of supercapacitors. This era also saw advancements in electrolyte technology, with the introduction of organic electrolytes that allowed for higher voltage operation and improved energy density.

Hybrid systems that combined supercapacitors with traditional batteries became increasingly popular, leveraging the strengths of both technologies to optimize performance. These hybrid systems were particularly useful in applications requiring both high power and high energy density, such as in electric and hybrid vehicles. The 2010s and beyond have witnessed continued innovation in supercapacitor technology (Sugumaran et al., 2024). Researchers have explored a wide range of new materials, including metal-organic frameworks (MOFs) and conducting polymers, to further enhance the performance of supercapacitors. MOFs offer the potential for exceptionally high surface areas and tunable properties, while conducting polymers provide the ability to engineer capacitance through chemical modifications. Advances in nanotechnology and fabrication techniques have also contributed to improvements in supercapacitor performance, making them more efficient and cost-effective. Today, supercapacitors are a vital component in various fields, including renewable energy systems, consumer electronics, and power backup solutions. They are used to smooth out fluctuations in power from renewable sources, provide quick bursts of power in electronic devices, and offer backup power in critical systems. In summary, the evolution of supercapacitors from theoretical concepts to practical devices is a testament to the advancements in materials science, electrochemistry, and engineering. From the early days of exploring high-capacitance materials to the modern era of sophisticated hybrid systems and cutting-edge materials, supercapacitors have undergone significant transformation. As technology continues to advance, the future of supercapacitors holds great promise, with potential innovations that could further enhance their performance and broaden their applications in various sectors.

FUNDAMENTALS OF SUPERCAPACITOR OPERATION: CHARGE STORAGE AND TRANSFER

The operation of supercapacitors is fundamentally rooted in their ability to store and transfer electrical energy through electrochemical processes that differ significantly from those in traditional capacitors and batteries. At the core of a supercapacitor's function is the electric (Vijayakumar et al., 2024) double layer (EDL) that forms at the interface between a high-surface-area electrode material and an electrolyte. In aqueous electrolytes, which generally have lower voltage windows but higher ionic conductivity, supercapacitors exhibit high capacitance. Organic electrolytes, on the other hand, enable higher voltage operation and improve the overall energy density of the device. This is why materials with high surface areas, such as activated carbon with its porous structure or graphene with its two-dimensional lattice, are highly effective in enhancing the performance of supercapacitors. This process is relatively rapid, allowing supercapacitors to charge and discharge quickly compared to batteries. When the supercapacitor discharges, the stored energy is released as the ions migrate back into the electrolyte, and the electric double layer is neutralized (Bhadra et al., 2020).. This rapid charge-discharge capability is one of the defining features of supercapacitors, making them ideal for applications requiring quick bursts of power. Another key aspect of supercapacitor operation is the role of the electrolyte. The electrolyte not only facilitates ion movement but also affects the overall performance of the supercapacitor (Chen et al., 2019). The choice between aqueous and organic electrolytes impacts the voltage window and capacitance of the supercapacitor. Aqueous electrolytes typically have lower voltage limits due to water's electrolysis at high voltages but offer high ionic conductivity and stability. Organic electrolytes, while offering a broader voltage range, may have lower ionic conductivity and can be more expensive. Supercapacitors also benefit from advanced materials and nanotechnologies that improve their performance. For example, the use of carbon nanotubes and graphene can further enhance the surface area and conductivity of the electrodes, resulting in higher capacitance and power density. This process is driven by electrostatic interactions and is characterized by rapid charge and discharge capabilities. The performance of supercapacitors is significantly influenced by the choice of electrode materials and electrolytes, as well as advancements in material science and nanotechnology.

MATERIALS SCIENCE OF SUPERCAPACITORS: ELECTRODES, ELECTROLYTES, AND SEPARATORS

The materials science behind supercapacitors is integral to their performance, with key components including electrodes, electrolytes, and separators. Each material plays a crucial role in determining the efficiency, capacity, and overall functionality of the supercapacitor. Advances in materials science have been pivotal in enhancing the capabilities of supercapacitors, making them more efficient and versatile. Electrodes are the most critical component of supercapacitors, responsible for the majority of their energy storage capabilities. The electrodes must possess high surface area, electrical conductivity, and chemical stability to facilitate efficient charge storage and transfer (Cai et al., 2019). Traditional supercapacitors use activated carbon as the electrode material due to its high surface area and porosity, which provides ample space for charge accumulation. Electrolytes in supercapacitors are crucial for facilitating ion movement between the electrodes and determining the voltage window and overall performance of the device. Electrolytes can be broadly classified into aqueous, organic, and ionic liquid types. However, they are limited by their narrow voltage window, typically around 1.2 to 1.6 volts, due to the risk of water electrolysis at higher voltages (Annamalai et al., 2018).Organic electrolytes, which include compounds like acetonitrile or ethylene carbonate, offer a wider voltage window, typically ranging from 2.5 to 3.5 volts, and can enhance the energy density of supercapacitors, represent an emerging electrolyte option.

They offer a wide voltage range, high thermal stability, and low volatility, although their high cost and relatively lower ionic conductivity can be limiting factors. Separators are an essential component that prevents electrical short-circuiting between the electrodes while allowing ion transport (Paramasivam Selvakumar et al., 2024). Separators must be electrically insulating, chemically inert, and mechanically stable. They are typically made from materials such as cellulose, polypropylene, or polyethylene. The separator material must also possess good ionic conductivity to ensure efficient ion movement between the electrodes. Recent advancements have focused on developing separators with enhanced porosity and improved mechanical strength to better support the high-power density requirements of modern supercapacitors. Nanotechnology and advanced fabrication techniques have played a significant role in the evolution of supercapacitor materials. Nanostructured materials, including nanowires, nanorods, and nanosheets, offer high surface areas and enhanced electrochemical properties, leading to improvements in capacitance and energy density. In summary, the materials science of supercapacitors encompasses the development and optimization of electrodes, electrolytes, and separators. Each material type contributes uniquely to the supercapacitor's performance, influencing factors such as capacitance, energy density, power density, and operational voltage range. Advances in material science, including the use of high-surface-area carbon materials, conductive polymers, and novel electrolytes, continue to drive improvements in supercapacitor technology.

DESIGN AND ARCHITECTURE OF SUPERCAPACITORS: CELL CONFIGURATION AND SCALING

The design and architecture of supercapacitors play a critical role in determining their performance, efficiency, and applicability across different use cases. The configuration of supercapacitor cells and the methods used for scaling up these devices are fundamental aspects that influence their overall capabilities (Ramakrishnan et al., 2022). Supercapacitors can be designed in various configurations, each tailored to specific requirements in terms of energy density, power density, and form factor. Cell Configuration refers to the arrangement and design of the basic unit within a supercapacitor. The most common cell configurations are the symmetrical and asymmetrical designs. In a symmetrical supercapacitor, both electrodes are made from the same material and have the same capacitance. This design is straightforward and is often used in standard applications where uniform performance is required. Conversely, an asymmetrical supercapacitor utilizes different materials for the two electrodes, typically combining a high-surface-area material like activated carbon with a pseudocapacitive material such as manganese dioxide or conductive polymers (Cossutta et al., 2020). This asymmetry can enhance the overall energy density and voltage window of the supercapacitor, making it suitable for applications where higher energy storage is needed. Cell architecture also involves the arrangement of the electrodes and electrolyte within the device. Planar supercapacitor have a flat structure with electrodes and separators stacked in parallel, while wound supercapacitors feature electrodes and separators rolled into a cylindrical shape. Planar configurations are commonly used in devices requiring thin, flexible supercapacitors, such as in wearable electronics or flexible displays.

Wound configurations are typically employed in applications where a compact, high-capacity device is needed, such as in automotive systems or large-scale energy storage. Scaling supercapacitors involves increasing the size of individual cells or combining multiple cells to meet specific power and energy requirements. In scaling up, the design must account for factors like maintaining uniformity in electrode material distribution, ensuring consistent electrolyte saturation, and managing thermal dissipation (Darkwa et al., 2019). The process often involves creating supercapacitor modules or packs where multiple cells are connected in series or parallel to achieve desired voltage and capacitance levels. Connecting cells in series increases the overall voltage capability, while connecting them in parallel enhances the capacitance and energy storage capacity. This modular approach allows for customization and flexibility in designing supercapacitor systems for various applications. Electrolyte distribution is also a crucial aspect of scaling supercapacitors. In large-scale devices, ensuring that the electrolyte is evenly distributed and effectively reaches all parts of the electrodes is essential for maintaining performance and reliability. Advanced manufacturing techniques, such as roll-to-roll processing for thin-film supercapacitors or precision coating methods for electrode materials, help ensure consistent electrolyte application and optimal performance in scaled-up devices. The thermal management of supercapacitors is another important consideration, especially in large-scale or high-power applications. Efficient heat dissipation mechanisms must be incorporated into the design to prevent overheating and ensure reliable operation. Techniques such as incorporating heat sinks, using thermally conductive materials, and designing for optimal airflow are commonly employed to manage thermal issues. Advanced fabrication techniques and materials continue to push the boundaries of supercapacitor design. Innovations such as 3D printing and nanostructured material offer new possibilities for creating high-performance supercapacitors with tailored properties.

ELECTROCHEMICAL CHARACTERIZATION OF SUPERCAPACITORS: PERFORMANCE METRICS AND TESTING

Electrochemical characterization of supercapacitors is crucial for understanding and optimizing their performance. This process involves assessing various metrics and conducting tests to evaluate the device's efficiency, stability, and overall suitability for specific applications. The key performance metrics for supercapacitors include capacitance, energy density, power density, cycle life, and rate capability. Each of these parameters provides insight into different aspects of the supercapacitor's functionality and effectiveness. EIS provides a detailed analysis of the impedance of the supercapacitor across a range of frequencies, helping to identify the contributions of different resistive and capacitive elements. CV involves applying a varying voltage to the supercapacitor and measuring the resulting current, allowing for the determination of capacitance from the area under the current-voltage curve (Das et al., 2018). It indicates how quickly the supercapacitor can release stored energy, which is essential for applications requiring rapid bursts of power. Power density is often evaluated through discharge tests where the supercapacitor is subjected to high current loads and the power output is measured. The power density can be calculated from the discharge curve, which provides information on how quickly the supercapacitor can deliver energy relative to its size or weight (Selvakumar, 2023). EIS provides a frequency-dependent analysis of the supercapacitor's impedance, revealing information about its internal resistance, capacitance, and the dynamics of charge transfer. CV involves sweeping the voltage and measuring the current response, allowing for the evaluation of electrochemical behavior and capacitance characteristics. These techniques help identify the factors influencing performance and guide the optimization of supercapacitor design. This method provides practical insights into the supercapacitor's performance under real-world conditions (Down et al., 2018). Thermal and environmental testing are also essential for assessing the stability and reliability of supercapacitors. Supercapacitors are subjected to extreme temperatures, humidity, and other environmental factors to ensure their performance remains consistent across different conditions. in summary, electrochemical characterization involves a comprehensive analysis of supercapacitor performance through various metrics and testing methods. Capacitance, energy density, power density, cycle life, and rate capability are key parameters that define the performance of supercapacitors.

APPLICATIONS OF SUPERCAPACITORS IN ENERGY STORAGE AND POWER SYSTEMS

Their applications span various fields, including automotive, renewable energy, consumer electronics, and power management systems, where their ability to deliver quick bursts of power and support frequent cycling is highly valued. This capability helps improve overall energy efficiency and reduces the demand on the vehicle's battery, extending its lifespan. Supercapacitors are also employed in start-stop systems, which shut down the engine when the vehicle is stationary and restart it quickly when needed (Eskandaria et al., 2020). This functionality not only reduces fuel consumption but also minimizes emissions. In some high-performance vehicles, supercapacitors are used to provide short bursts of power for acceleration or to stabilize power supply during high-load conditions. Similarly, in wind energy systems, they help manage the variability in wind speeds, providing a buffer against sudden changes and improving the overall stability of the energy grid. In the realm of consumer electronics, supercapacitors offer significant advantages provide rapid charging and discharging capabilities, which can lead to longer battery life and faster charging times. Supercapacitors are also utilized in energy harvesting applications, where they store energy from small, ambient sources like solar cells or kinetic energy converters. This stored energy can then be used to power low-energy devices or sensors, contributing to more efficient and sustainable technology. This rapid response capability helps protect sensitive electronic equipment and data from damage or loss. In grid energy storage, supercapacitors contribute to stabilizing and enhancing the efficiency of the electrical grid. They are employed to address rapid fluctuations in power demand and supply, acting as a buffer to smooth out these variations and support grid stability. Their high power density makes them particularly effective in applications requiring quick energy release, such as frequency regulation and voltage support (Fan et al., 2018). By integrating supercapacitors with traditional battery storage systems, it is possible to create hybrid systems that leverage the strengths of both technologies, optimizing performance across a range of conditions. Smart grid technologies also benefit from the inclusion of supercapacitors, which help manage and stabilize the distribution of electricity. They can be used in voltage regulation and frequency regulation, providing a rapid response to changes in the grid and ensuring consistent power quality. In summary, the diverse applications of supercapacitors in energy storage and power systems underscore their versatility and effectiveness.

SUPERCAPACITORS IN TRANSPORTATION: ELECTRIC VEHICLES, AEROSPACE, AND MORE

Supercapacitors are making significant strides in various transportation sectors, including electric vehicles (EVs), aerospace, and other transportation systems, due to their unique life (Farjadian et al., 2020). Their ability to deliver quick bursts of energy and efficiently handle frequent cycling is transforming the landscape of transportation technology, providing benefits in performance, efficiency, and reliability. This energy is then quickly released to assist in acceleration or to power auxiliary systems, thereby improving overall energy efficiency and reducing the load on the main battery. This capability not only enhances the driving range but also contributes to better vehicle performance. Additionally, supercapacitors are used in start-stop systems, which shut down the engine during periods of idling and restart it quickly when needed. This reduces fuel consumption and emissions, contributing to more sustainable vehicle operation. Supercapacitors are also being explored for peak power support, where they provide short bursts of high power to assist with rapid acceleration or to handle sudden power demands, complementing the energy storage provided by traditional batteries. In the aerospace sector, supercapacitors offer significant advantages for both commercial and space applications. For commercial aviation, they are used in auxiliary power units (APUs), which provide electrical power for onboard systems when the main engines are not running.

Supercapacitors can handle the rapid charge and discharge cycles associated with APUs, improving reliability and reducing maintenance needs (Gu et al., 2018). In space applications, supercapacitors are utilized for power conditioning and energy storage in spacecraft and satellites. They help manage power surges and provide stable energy output, which is crucial for the operation of sensitive equipment and communication systems. Supercapacitors' ability to function in extreme temperatures and their long cycle life make them well-suited for the demanding conditions of space missions. In railway transportation, supercapacitors are used to improve the efficiency and performance of trains and trams. They are employed in regenerative braking systems, similar to those in EVs, where they capture energy during braking and store it for reuse during acceleration.

This reduces energy consumption and operational costs, while also decreasing wear and tear on braking systems. Supercapacitors are also used in stationary energy storage systems at railway stations, where they help manage energy loads and support grid stability. Their high power density enables rapid response to power fluctuations, contributing to a more reliable and efficient rail network. In maritime transportation, supercapacitors are being explored for use in hybrid propulsion systems and energy storage for auxiliary power. By storing energy generated during periods of low power demand or from regenerative braking systems, supercapacitors can assist with power management and improve overall fuel efficiency. They also contribute to smoother and more efficient operation of onboard systems, helping to reduce emissions and operational costs. In urban transportation, supercapacitors are utilized in electric buses and trams to enhance performance and operational efficiency. They support regenerative braking systems, improving energy efficiency and extending the operational range of these vehicles (Hamsan et al., 2020). Additionally, supercapacitors can provide rapid charging capabilities, enabling quick turnaround times for urban transit systems and reducing downtime. In emerging transportation technologies, such as autonomous vehicles and flying cars supercapacitors offer potential benefits due to their ability to handle rapid charge and discharge cycles. They are used to smooth out fluctuations in power from renewable sources, provide quick bursts of power in electronic devices, and offer backup power in critical systems. In summary, the evolution of supercapacitors from theoretical concepts to practical devices is a testament to the advancements in materials science, electrochemistry, and engineering. From the early days of exploring high-capacitance materials to the modern era of sophisticated hybrid systems and cutting-edge materials, supercapacitors have undergone significant transformation. As technology continues to advance, the future of supercapacitors holds great promise, with potential innovations that could further enhance their performance and broaden their applications in various sectors.

RENEWABLE ENERGY SYSTEMS AND SUPERCAPACITORS: SOLAR, WIND, AND GRID INTEGRATION

Supercapacitors are increasingly being integrated into renewable energy systems, enhancing their efficiency, stability, and performance across various applications such as solar and wind energy, and in grid integration. In solar energy systems, supercapacitors address several key challenges related to energy storage and management. Additionally, supercapacitors can be employed to provide Short-term power support for grid stabilization and to manage sudden changes in power demand. They can also enhance the performance of solar photovoltaic (PV) systems by improving the efficiency of energy harvesting and reducing the impact of voltage fluctuations. In solar-powered devices, such as remote sensors or off-grid installations, supercapacitors offer a practical solution for managing energy supply and demand. In wind energy systems, supercapacitors play a crucial role in managing the variability and intermittency associated with wind power generation. Wind speeds can fluctuate significantly, leading to variable power output from wind turbines. Supercapacitors can capture and store energy during periods of high wind speed and release it when wind speeds drop, thus stabilizing the power output and ensuring a more consistent energy supply (Han et al., 2020). This capability is particularly beneficial for wind farms integrated into the electrical grid, where supercapacitors help mitigate the impact of sudden changes in wind power and support grid stability. By smoothing out power fluctuations, supercapacitors contribute to a more reliable and efficient integration of wind energy into the broader energy infrastructure. The integration of intermittent renewable energy sources, such as solar and wind, can cause fluctuations in grid voltage and frequency, potentially leading to instability and reliability issues. Supercapacitors can provide frequency regulation and voltage support by rapidly responding to changes in grid conditions. Energy storage systems that incorporate supercapacitors can also enhance the performance of hybrid storage solutions. Supercapacitors handle rapid charge and discharge cycles, providing quick bursts of power and improving overall efficiency, while batteries offer higher energy density for long-term storage. EV charging stations are another application where supercapacitors contribute to renewable energy integration. This capability helps balance the supply of renewable energy with the demand for fast charging, ensuring efficient and reliable operation of EV charging infrastructure (Ikram et al., 2020). Advanced grid management systems are increasingly incorporating supercapacitors to enhance the resilience and reliability of power grids. By providing rapid response to power disturbances and supporting grid stabilization efforts, supercapacitors play a crucial role in modernizing the grid and accommodating the growing share of renewable energy sources.

SUPERCAPACITORS IN CONSUMER ELECTRONICS: MOBILE DEVICES, WEARABLES, AND IOT

Supercapacitors are becoming increasingly integral to consumer electronics, revolutionizing the functionality and performance of mobile devices, wearables, and internet of things (IoT) applications. offer distinct advantages that address several challenges faced by modern electronic devices. By enhancing energy management and improving device longevity, supercapacitors are reshaping the landscape of consumer electronics (Jagadale et al., 2019). This capability helps reduce the burden on traditional batteries, allowing them to operate more efficiently and extend their overall lifespan. For instance, supercapacitors can handle power surges during intensive operations, ensuring that the battery is not strained and thus contributing to its longevity. They are also used in quick charge feature, where they provide a rapid boost of power to the device, shortening charging times and enhancing convenience for users. Additionally, supercapacitors can support backup power functions, such as preserving settings and data during sudden power interruptions or device restarts. Wearable technology benefits significantly from the integration of supercapacitors. Devices like smartwatches, fitness trackers, and health monitors require compact and efficient energy storage solutions due to their small form factors and the need for frequent power cycling. Supercapacitors provide a practical solution by offering rapid charging capabilities and high power density in a small volume. They enhance the battery life of wearables by handling quick energy bursts for sensors, displays, and wireless communication, reducing the need for frequent recharging.

Their ability to operate effectively over many charge-discharge cycles is particularly valuable for wearables, which often experience regular power cycling due to their continuous use throughout the day. Furthermore, supercapacitors contribute to low-power operation in wearables, allowing devices to maintain performance while optimizing energy consumption. Internet of things (IoT) devices are another area where supercapacitors are making a significant impact (Jha et al., 2018). IoT devices, which include sensors, smart home appliances, and connected infrastructure, often require efficient energy management to ensure reliable operation over long periods. Supercapacitors are used to store energy from intermittent sources, such as energy harvested from environmental sources (e.g., solar cells or kinetic energy converters), and provide power to IoT devices when needed. This capability enables IoT devices to function effectively in remote or off-grid locations, where access to continuous power sources may be limited. Supercapacitors also enhance the performance of wireless communication modules in IoT devices, supporting the rapid transmission of data and improving overall system responsiveness. Their durability and long cycle life ensure that IoT devices remain operational and reliable, even in harsh or challenging environments. Power management in consumer electronics is greatly improved with the use of supercapacitors (Zhang et al., 2018). In addition to their role in enhancing battery life and performance, supercapacitors contribute to more efficient energy management by storing and releasing energy during peak demand periods. This helps stabilize power supply and reduces energy waste. For instance, in smart home systems, supercapacitors can manage power for various smart devices, ensuring that they operate seamlessly and efficiently. This aligns with the growing emphasis on environmental sustainability and reducing the ecological impact of electronic devices (Jin et al., 2021). In summary, supercapacitors offer substantial benefits to consumer electronics, including mobile devices, wearables, and IoT applications. By optimizing energy management and contributing to sustainability, supercapacitors are playing a pivotal role in advancing consumer electronics technology and meeting the evolving needs of modern users. As technological advancements continue, the integration of supercapacitors in these devices is expected to further enhance their functionality and reliability, driving innovation in the consumer electronics sector.

MEDICAL APPLICATIONS OF SUPERCAPACITORS: IMPLANTABLE DEVICES AND WEARABLE TECH

Supercapacitors are increasingly finding applications in the medical field, particularly in implantable devices and wearable technology, transforming medical technology, improving patient outcomes, and enabling advancements in health monitoring and treatment. In implantable medical devices supercapacitors offer significant advantages over traditional batteries. Devices such as pacemakers, neurostimulators, and implantable cardioverter defibrillators (ICDs) require reliable and efficient power sources to function continuously within the human body. Supercapacitors can support these devices by providing high power density and rapid energy delivery, which is essential for devices that need to operate under varying load conditions and require frequent bursts of power (Karade et al., 2018). The long cycle life of supercapacitors ensures that they can endure the repetitive charge-discharge cycles of these devices without significant degradation, reducing the need for frequent replacements or invasive procedures to replace power sources. Furthermore, supercapacitors can enhance the reliability of implantable devices by ensuring consistent power output and minimizing the risk of power-related failures. Energy harvesting and management in implantable devices is another area where supercapacitors excel. Many implantable devices are designed to be powered by energy harvested from the body, such as kinetic energy from movement or thermal energy from body heat. Supercapacitors are well-suited to store and manage this harvested energy efficiently (Zare et al., 2020). For instance, they can capture and store energy generated from body movements or vibrations, which can then be used to power the device or recharge an internal battery. This capability reduces the reliance on external power sources and enhances the self-sufficiency of implantable devices. Wearable medical technology also benefits from the integration of supercapacitors. Wearables such as fitness trackers, smartwatches, and health monitoring devices require compact and efficient energy storage solutions due to their small form factors and frequent use. Supercapacitors provide a solution by offering rapid charging and high power density in a small volume, which is crucial for maintaining continuous operation and supporting advanced features such as real-time health monitoring, GPS tracking, and communication functions.

The rapid charge-discharge capabilities of supercapacitors ensure that wearables can quickly handle peak power demands, such as those required for GPS tracking or high-resolution displays, without compromising performance or battery life. Health monitoring and diagnostics in wearables are significantly enhanced by supercapacitors. These devices often incorporate sensors that continuously monitor various health metrics such as heart rate, blood oxygen levels, and physical activity (Lee et al., 2019). Supercapacitors can efficiently manage the power needs of these sensors and support data transmission to connected devices. They help ensure that wearables remain functional and responsive over long periods, providing valuable real-time health insights and contributing to proactive health management. Implantable sensors and drug delivery systems are additional areas where supercapacitors offer substantial benefits. Implantable sensors used for continuous monitoring of physiological parameters, such as glucose levels or blood pressure, require reliable and long-lasting power sources (Karaoglan et al., 2018). Supercapacitors can provide the necessary energy for these sensors, ensuring accurate and consistent readings. In drug delivery systems, supercapacitors enable the precise control of drug release by providing rapid power for the operation of micro-pumps or other delivery mechanisms. This capability is crucial for managing chronic conditions and improving patient compliance with treatment regimens. Biocompatibility is a critical consideration for implantable devices, and supercapacitors are increasingly being designed with biocompatible materials to ensure safe integration within the human body. Advances in material science and engineering are leading to the development of supercapacitors that meet stringent safety and compatibility standards, reducing the risk of adverse reactions and improving patient outcomes. Energy efficiency and sustainability are also important factors in the medical applications of supercapacitors.

ADVANCES IN SUPERCAPACITOR TECHNOLOGY: NANOMATERIALS, GRAPHENE, AND 2D MATERIALS

Advances in supercapacitor technology are being driven by innovations in materials science, particularly through the development of nanomaterials, graphene, and 2D materials. These advancements are enhancing the performance, efficiency, and application potential of supercapacitors, making them increasingly viable for a range of energy storage solutions. Nanomaterials have revolutionized supercapacitor technology by significantly improving the specific surface area and electrochemical performance of electrodes. Nanostructured materials, including nanoparticles, nanotubes, and nanowires, offer a high surface area-to-volume ratio, which enhances the charge storage capacity of supercapacitors.. Similarly, metal oxides such as manganese dioxide (MnO_2) and nickel oxide (NiO) in nanowire forms exhibit high pseudocapacitance, contributing to enhanced energy storage and power density. D materials beyond graphene, such as transition metal dichalcogenides (TMDs) and layered double hydroxides (LDHs), are making significant contributions to the advancement of supercapacitor technology(Leonard. D. P., et al.,2018). These materials offer additional pseudocapacitive behavior, which complements the electrical double-layer capacitance typically seen in supercapacitors. The layered structure of TMDs provides a large interlayer spacing that facilitates ion transport and enhances the overall capacitance of the supercapacitor. This results in increased capacitance and improved performance in supercapacitors. LDHs can be tailored through various synthesis methods to optimize their electrochemical properties, making them suitable for high-performance energy storage applications. Hybrid materials that combine nanomaterials, graphene, and 2D materials are pushing the boundaries of supercapacitor technology even further. Graphene-metal oxide composites have demonstrated improved electrochemical performance due to the synergistic effects between the graphene's high conductivity and the metal oxide's high pseudocapacitance. Electrolyte advancements are also crucial in optimizing the performance of supercapacitors with advanced materials. The development of onic liquids and superionic electrolytes with high conductivity and wide electrochemical stability windows complements the improved electrode materials (Maksoud et al., 2021). These advanced electrolytes enhance the overall performance of supercapacitors by enabling higher operating voltages and improving energy density. Manufacturing and scalability of these advanced materials are key factors in the commercial viability of next-generation supercapacitors. Techniques such as chemical vapor deposition (CVD) for graphene synthesis and solvothermal methods for nanomaterial fabrication are being refined to produce high-quality materials at scale (Yadav, 2020). The integration of these materials into supercapacitor devices involves developing effective scaling-up processes and ensuring that the performance improvements achieved in laboratory settings are translated into commercially viable products. In summary, advances in supercapacitor technology are being driven by innovations in nanomaterials, graphene, and 2D materials. These advancements enhance the performance, efficiency, and application potential of supercapacitors, leading to improvements in specific capacitance, power density, and energy storage capabilities.

CHALLENGES AND LIMITATIONS OF SUPERCAPACITORS: COST, ENERGY DENSITY, AND DURABILITY

Despite their promising advantages, supercapacitors face several challenges and limitations that impact their broader adoption and application. Key issues include high cost, relatively low energy density compared to batteries, and durability concerns. Addressing these challenges is crucial for advancing supercapacitor technology and expanding its use across various applications. Cost remains a significant barrier to the widespread adoption of supercapacitors. The production of advanced materials used in supercapacitors, such as high-purity graphene, carbon nanotubes, and transition metal oxides, can be expensive. The synthesis processes for these materials often require specialized equipment and conditions, contributing to high manufacturing costs (Mohanty et al., 2021). For instance, methods such as chemical vapor deposition (CVD) used for producing high-quality graphene are costly and complex, which translates into higher prices for supercapacitor components. Furthermore, the fabrication of supercapacitor cells and the integration of these advanced materials into commercially viable products involve additional costs. As a result, supercapacitors are often more expensive compared to conventional capacitors and batteries, which limits their affordability for certain applications. Energy density is another significant limitation of supercapacitors. Supercapacitors store energy through electrostatic charge separation in the electric double layer or through pseudocapacitance, which generally results in lower energy storage compared to the electrochemical reactions in batteries.

This limitation means that supercapacitors are not well-suited for applications requiring long-term, high-energy storage, such as in electric vehicles or portable electronics where extended operation between charges is necessary. These hybrid systems leverage the high power density of supercapacitors and the high energy density of batteries, optimizing overall performance for specific applications. However, the integration and management of such hybrid systems introduce additional complexity and cost. Durability and lifespan are also concerns for supercapacitors, though they generally exhibit superior cycle life compared to batteries (Namsheer et al., 2021). The durability of supercapacitors is influenced by factors such as the degradation of electrode materials, electrolyte stability, and environmental conditions (Xu et al., 2020). For example, in certain applications, high-voltage operation can lead to the degradation of the electrolyte or electrode materials, impacting the overall lifespan and performance of the supercapacitor. Additionally, supercapacitors can suffer from performance degradation in extreme temperatures or harsh environmental conditions, which may limit their suitability for specific applications. The performance stability of supercapacitors over time is another aspect that requires attention. While supercapacitors can endure many charge-discharge cycles with minimal degradation, factors such as electrode material deterioration, electrolyte breakdown, and mechanical stress can affect their long-term reliability. Scalability and manufacturing challenges are additional issues that impact the widespread use of supercapacitors. In summary, while supercapacitors offer significant advantages in terms of power density and rapid charge-discharge capabilities, they face challenges related to cost, energy density, and durability. Addressing these limitations requires ongoing research and development to improve material performance (Oca et al., 2019), reduce manufacturing costs, and enhance overall reliability. By overcoming these obstacles, supercapacitors can fulfill their potential as a versatile and effective energy storage solution, expanding their applications and contributing to advancements in various fields.

FUTURE DIRECTIONS AND EMERGING TRENDS IN SUPERCAPACITOR RESEARCH AND DEVELOPMENT

The future of supercapacitor technology is set to be shaped by ongoing research and emerging trends aimed at overcoming current limitations and expanding their applications. Key directions in this field include advancements in materials science, novel device architectures, integration with other energy storage technologies, and innovative manufacturing techniques. Nanomaterials are another area of intense research, with significant efforts focused on carbon nanotubes (CNTs), nanowires, and nanofibers (Pang et al., 2020). These materials can be engineered to achieve high surface areas and efficient charge transfer, addressing the energy density limitations of traditional supercapacitors. Additionally, the development of composite materials that integrate these nanomaterials with other advanced substances, such as conductive polymers or metal oxides, aims to leverage the unique properties of each component to enhance overall performance. Innovative device architectures are also a critical area of exploration. Researchers are developing new supercapacitor designs that improve energy density and efficiency. Flexible and wearable supercapacitors are gaining attention due to their potential applications in emerging technologies such as flexible electronics and smart textiles. These devices often utilize stretchable or bendable materials that maintain performance while conforming to different shapes and surfaces (Wu et al., 2019). Asymmetric supercapacitors, which combine different types of electrode materials to balance energy and power density, are being investigated to achieve higher overall performance. For instance, supercapacitors can handle rapid charge-discharge cycles and provide high power output, while batteries offer higher energy density for longer-term storage. These hybrid systems are being explored for applications in electric vehicles, renewable energy integration, and grid stabilization (Reece et al., 2018)A. dvanced manufacturing techniques are essential for scaling up the production of next-generation supercapacitors and making them commercially viable. Innovations in printing and coating technologies, such as screen printing or roll-to-roll processing, are being developed to produce supercapacitor components more efficiently and cost-effectively. Batch synthesis and automated fabrication methods are also being optimized to ensure consistency and quality in large-scale production. Additionally, researchers are exploring sustainable and cost-effective materials and processes to reduce the environmental impact and cost of supercapacitor production. Integration with emerging technologies is a key focus for future supercapacitor research. In the realm of Internet of Things (IoT) and smart devices, supercapacitors are being designed to work seamlessly with low-power electronics and sensors, offering rapid energy storage and efficient power management (Pazhamalai et al., 2018). For wearable technology, supercapacitors are being developed to provide reliable and lightweight energy solutions for devices such as health monitors and smartwatches. Furthermore, the role of supercapacitors in electric vehicles (EVs) is expanding, with ongoing research aimed at improving their performance in hybrid energy systems and fast-charging applications. In summary, the future of supercapacitor technology is poised to benefit from advancements in materials science, novel device architectures, integration with other energy storage technologies, and innovative manufacturing techniques (Xia et al., 2018). These directions aim to address current limitations, such as cost and energy density, and expand the applicability of supercapacitors across various fields. As research and development continue to progress, supercapacitors are expected to play a crucial role in enhancing energy storage solutions and enabling new technological advancements.

THE FUTURE OF ENERGY STORAGE AND POWER DELIVERY WITH SUPERCAPACITORS

Energy storage is a central area where supercapacitors are expected to make significant contributions. Traditional energy storage solutions, such as batteries, while effective in providing high energy density, often face limitations in terms of power density and cycle life. Supercapacitors, with their ability to deliver rapid bursts of power and withstand numerous charge-discharge cycles, complement these traditional systems by addressing their shortcomings.

In electric vehicles, supercapacitors are poised to enhance performance by supporting fast acceleration and regenerative braking systems. Their ability to rapidly absorb and release energy makes them ideal for capturing and utilizing energy during braking, which can then be used to support acceleration (Wang et al., 2018). Devices such as smartphones, tablets, and wearables demand efficient and reliable energy storage solutions that support both high performance and long battery life. Supercapacitors can enhance these devices by providing rapid charge and discharge capabilities, improving the performance of high-power features, and extending battery lifespan. As technology advances, we can expect to see more compact and flexible supercapacitors integrated into consumer electronics, supporting innovations such as wearable tech and flexible displays Medical devices represent another critical application area for supercapacitors. Implantable devices, such as pacemakers and neurostimulators, require reliable and long-lasting power sources. Supercapacitors' high power density and long cycle life make them well-suited for these applications, where they can provide consistent and efficient energy delivery. Furthermore, wearable medical devices that monitor health metrics can benefit from supercapacitors' rapid charging capabilities, supporting continuous health monitoring without frequent battery replacements. Energy harvesting technologies are poised to benefit significantly from the integration of supercapacitors. Devices that capture and store energy from sources such as solar, kinetic, or thermal energy can use supercapacitors to efficiently store and manage this energy. By providing a means to quickly capture and utilize harvested energy, supercapacitors can enhance the efficiency and effectiveness of these technologies, contributing to the development of self-sustaining energy systems (Saha et al., 2018). Emerging technologies such as Internet of Things (IoT) devices and smart grids also stand to gain from advancements in supercapacitor technology. IoT devices, which often operate on low power and require frequent energy bursts, can benefit from supercapacitors' ability to deliver rapid power and extend operational life. Similarly, supercapacitors can support the development of smart grids by providing energy storage solutions that enable more efficient energy distribution and load balancing (Vul et al., 2018). Research and development will continue to drive the future of supercapacitors, focusing on enhancing energy density, reducing costs, and improving durability. Innovations in materials science, such as the development of advanced nanomaterials, 2D materials, and hybrid composites, are expected to further enhance the performance of supercapacitors. Additionally, advancements in manufacturing techniques and scalable production methods will be crucial in making supercapacitors more accessible and cost-effective for widespread use. Position them as a key technology in addressing the diverse demands of modern energy systems. By complementing traditional energy storage solutions and supporting advancements in various applications, supercapacitors are set to play a transformative role in enhancing energy efficiency, reliability, and sustainability across multiple sectors.

CONCLUSION

The unique characteristics of supercapacitors make them ideal for a variety of applications. In Transportation, they enhance the performance of electric vehicles by supporting regenerative braking systems and enabling quick acceleration. The integration of supercapacitors in hybrid systems alongside batteries optimizes energy storage and power delivery, improving efficiency and extending the lifespan of battery systems. Renewable energy systems benefit from supercapacitors' ability to stabilize the grid and manage energy fluctuations, thus facilitating the integration of intermittent renewable sources like wind and solar. In consumer electronics, supercapacitors enable rapid charging and contribute to longer-lasting, more reliable devices, while in medical technology, their reliability and long cycle life support the functionality of implantable devices and wearable health monitors. Advancements in materials science** are driving the future of supercapacitors, with innovations in nanomaterials, graphene, and 2D materials offering the potential to significantly enhance performance. These materials contribute to improved energy density, power density, and overall efficiency. The development of hybrid and flexible supercapacitor designs, along with novel manufacturing techniques, and energy harvesting technologies further extends the utility of supercapacitors, providing versatile solutions that can meet diverse energy needs.The future of supercapacitors will likely be characterized by their increasingly sophisticated applications and integration into various technologies. The combination of advanced materials, innovative designs, and improved manufacturing processes is expected to enhance their performance and reduce costs, making supercapacitors more accessible and commercially viable. As research progresses, we can anticipate the emergence of new applications and enhancements that will solidify the role of supercapacitors in the future of energy storage and power delivery. In summary, supercapacitors represent a pivotal technology in the evolution of energy storage solutions. Their distinct advantages in power density and cycle life, coupled with ongoing advancements in materials and design, position them as essential components in the future of energy systems. By addressing existing challenges and embracing emerging trends, supercapacitors will continue to play a critical role in enhancing energy efficiency, reliability, and sustainability across a wide array of applications.

REFERENCES

Adetayo, A., & Runsewe, D. (2019). Synthesis and Fabrication of Graphene and Graphene Oxide: A Review. *Open Journal of Composite Materials*, *9*(2), 207–229. DOI: 10.4236/ojcm.2019.92012

Al-Refai, H. H., Ganash, A. A., & Hussein, M. A. (2021). Polythiophene and its derivatives – Based nanocomposites in electrochemical sensing: A mini review. Materials Today Communications, 26, 101935(1-11).

Annamalai, K. P., Zheng, X., Gao, J., Chen, T., & Tao, Y. (2018). Nanoporous Ru-Mn oxide nanoparticles/reduced graphene oxide for high-energy symmetric supercapacitors. *Carbon*, *144*, 185–192. DOI: 10.1016/j.carbon.2018.11.073

Bhadra, J., Alkareem, A., & Al-Thani, N. (2020). A review of advances in the preparation and application of polyaniline based thermoset blends and composites. *Journal of Polymer Research*, *27*(122), 1–20. DOI: 10.1007/s10965-020-02052-1

Cai, W., Yao, Y. X., Zhu, G. L., Yan, C., Jiang, L. L., He, C., Huang, J. Q., & Zhang, Q. (2019). Areview on energy chemistry of fast-charging anodes. *Chemical Society Reviews*, *49*(12), 3806–3833. DOI: 10.1039/C9CS00728H PMID: 32478786

Chen, C. R., Qin, H., Cong, H. P., & Yu, S. H. (2019). A highly stretchable and real-time healable supercapacitor. *Advanced Materials*, *31*(19), 1900573–1900583. DOI: 10.1002/adma.201900573 PMID: 30920707

Cossutta, M., Vretenar, V., Centeno, T. A., Kotrusz, P., McKechnie, J., & Pickering, S. J. (2020). A comparative life cycle assessment of graphene and activated carbon in a supercapacitor application. Journal of Cleaner Production, 242, 118468(1-10).

Darkwa, K. M., Zequine, C., Kahol, P. K., & Gupta, R. K. (2019). Supercapacitor energy storage device using biowastes: A sustainable approach to green energy. *Sustainability*, *11*(414), 1–22.

Das, M., & Mandal, R. (2018). A comparative performance analytics of direct, with battery, supercapacitor, and baterry-supercapacitor enabled photovoltaic water pumping systems using centrifugal pump. *Solar Energy*, *171*, 302–309. DOI: 10.1016/j.solener.2018.06.069

Down, M. P., Rowley-Neale, S. J., Smith, G. C., & Banks, C. E. (2018). Fasbrication of graphene oxide supercapacitor devices. *ACS Applied Energy Materials*, *1*(2), 707–714. DOI: 10.1021/acsaem.7b00164

Eskandaria, E., Kosarib, M., Farahania, M. H. D. A., Khiavic, N. D., Saeedikhanid, M., Katale, R., & Zarinejadf, M. (2020). A review on polyaniline-based materials applications in heavy metals removal and catalytic processes. *Separation and Purification Technology*, *231*(115901), 1–27. DOI: 10.1016/j.seppur.2019.115901

Fan, L., Lin, K., Wang, J., Ma, R., & Lu, B. (2018). A nonaqueous potassium-based battery-supercapacitor hybrid device. *Advanced Materials*, *30*(1800804), 1–7. PMID: 29603424

Farjadian, F., Abbaspour, S., Sadatlu, M. A. A., Mirkiani, S., Ghasemi, A., Hoiseini-Ghahfarokhi, M., Mozaffari, N., Karimi, M., & Hamblin, M. R. (2020). Recent Developments in Graphene Oxide: Properties, Synthesis, and Modifications: A review. *ChemistrySelect*, *5*(33), 10200–10219. DOI: 10.1002/slct.202002501

Gu, H., Zhang, H., Gao, C., Liang, C., Gu, J., & Guo, Z. (2018). New Functions of Polyaniline. *ES Materials & Manufacturing*, *1*, 3–12.

Hamsan, M. H., Aziz, S. B., Kadir, M. F. Z., Brza, M. A., & Karim, W. O. (2020). The study of EDLC device fabricated from plasticized magnesium ion conducting chitosan based polymer electrolyte. *Polymer Testing*, *90*(106714), 1–10. DOI: 10.1016/j.polymertesting.2020.106714

Han, R., Liu, F., Wang, X., Huang, M., Li, W., Yamauchi, Y., Sun, X., & Huang, Z. (2020). Functionalised Hexagonal Boron Nitride for Energy Conversion and Storage. *Journal of Materials Chemistry. A, Materials for Energy and Sustainability*, *8*(29), 14384–14399. DOI: 10.1039/D0TA05008C

Ikram, R., Jan, B. M., & Ahmad, W. (2020). An overview of industrial scalable production of graphene oxide and analytical approaches for synthesis and characterization. *Journal of Materials Research and Technology*, *9*(5), 11587–11610. DOI: 10.1016/j.jmrt.2020.08.050

Jagadale, A., Zhou, X., Xiong, R., Dubal, D. P., Xu, J., & Yang, S. (2019). Lithium ion capacitors (LICs): Development of materials. *Energy Storage Materials*, *19*, 314–329. DOI: 10.1016/j.ensm.2019.02.031

Jha, R. K., Wan, M., Jacob, C., & Guha, P. K. (2018). Ammonia vapour sensing properties of in situ polymerized conducting PANI-nanofiber/WS2 nanosheet composites. *New Journal of Chemistry*, *42*(1), 735–745. DOI: 10.1039/C7NJ03343E

Jin, X., Song, L., Yang, H., Dai, C., Xiao, Y., Zhang, X., Han, Y., Bai, C., Lu, B., Liu, Q., Zhao, Y., Zhang, J., Zhang, Z., & Qu, L. (2021). Stretchable supercapacitor at -30°C. *Energy & Environmental Science*, *14*(5), 3075–3085. DOI: 10.1039/D0EE04066E

Kanipriya, M., Sugumaran, D., Pandiyarajan, V., Yogananda, B. S., Selvakumar, P., & Mohan, S. (2024, January). Analysis of Electromyography Signals for Control of Mechanical Prosthesis using Machine Learning Techniques. In 2024 2nd International Conference on Intelligent Data Communication Technologies and Internet of Things (IDCIoT) (pp. 1059-1067). IEEE.

Karade, S. S., & Sankapal, B. R. (2018). Materials Mutualism through EDLC-Behaved MWCNTs with pseudocapacitive MoTe2 nanopebbles: Enhanced supercapacitive performance. *ACS Sustainable Chemistry & Engineering*, *6*(11), 15072–15082. DOI: 10.1021/acssuschemeng.8b03366

Karaoglan, N., & Bindal, C. (2018). Synthesis and optical characterization of benzene sulfonic acid doped. *Engineering Science and Technology, an International Journal*, *21*(6), 1152–1158. DOI: 10.1016/j.jestch.2018.09.010

Lee, S. L., & Chang, C. J. (2019). Recent Developments about Conductive Polymer Based Composite Photocatalysts. *Polymers*, *11*(206), 1–21. DOI: 10.3390/polym11020206 PMID: 30960189

Leonard, D. P., Wei, Z., Chen, G., Du, F., & Ji, X. (2018). Water-in-Salt electrolyte for potassium-ion batteries. *ACS Energy Letters*, *3*(2), 373–374. DOI: 10.1021/acsenergylett.8b00009

Maksoud, M. I. A., Fahim, R. A., Shalan, A. E., Elkodous, M. A., Olojede, S. O., Osman, A. I., Farrell, C., Al-Muhtaseb, A. H., Awed, A. S., Ashour, A. H., & Rooney, D. W. (2021). *Advanced materials and technologies for supercapacitors used in energy conversion and storage: a review*. Environmental.

Mohanty, A., Jaihindh, D. P., Fu, Y. P., Senanayak, S. P., Mende, L. S., & Ramadoss, A. (2021). An extensive review on three dimension architectural Metal-Organic Frameworks towards supercapacitor application. *Journal of Power Sources*, *488*, 229444. DOI: 10.1016/j.jpowsour.2020.229444

Namsheer, K., & Rout, C. S. (2021). Conducting polymers: A comprehensive review on recent advances in synthesis, properties and applications. *Royal Society of Chemistry Advances*, *11*, 5659–5697. PMID: 35686160

Oca, L., Guillet, N., Tessard, R., & Iraola, U. (2019). Lithium-ion capacitor safety assessment under electrical abuse tests based on ultrasound characterization and cell opening. *Journal of Energy Storage*, *23*, 29–36. DOI: 10.1016/j.est.2019.02.033

Pang, A. L., Arsad, A., & Ahmadipour, M. (2020). Synthesis and factor affecting on the conductivity of polypyrrole: A short review. *Polymers for Advanced Technologies*, *32*(4), 1428–1454. DOI: 10.1002/pat.5201

Pazhamalai, P., Krishnamoorthy, K., Manoharan, S., & Kim, S. J. (2018). High energy symmetric supercapacitor based on mechanically delaminated few-layeres MoS2 sheets in organic electrolyte. *Journal of Alloys and Compounds*, *771*, 803–809. DOI: 10.1016/j.jallcom.2018.08.203

Ramakrishnan, T., Mohan Gift, M. D., Chitradevi, S., Jegan, R., Hency Jose, P. S., Nagaraja, H. N., & Hailegiorgis, S. M. (2022). Study of numerous resins used in polymer matrix composite materials. *Advances in Materials Science and Engineering*, *2022*(1), 1088926.

Reece, R., Lekakou, C., & Smith, P. A. (2018). A structural supercapacitor based on activated carbon fabric and solid electrolyte. *Materials Science and Technology*, *35*(3), 368–375. DOI: 10.1080/02670836.2018.1560536

Reece, R., Lekakou, C., & Smith, P. A. (2020). A high-performance structural supercapacitor. *ACS Applied Materials & Interfaces*, *12*(23), 25683–25692. DOI: 10.1021/acsami.9b23427 PMID: 32407618

Saha, S., Samanta, P., Murmu, N. C., & Kuila, T. (2018). Areview on the heterostructure nanomaterials for supercapacitor application. *Journal of Energy Storage*, *17*, 181–202. DOI: 10.1016/j.est.2018.03.006

Selvakumar, P. (2023). Biomedical applications of polymers in biosensors, cancer vaccines and drug delivery systems. *Biopolimery i Kletka*, *39*(4), 245–256. DOI: 10.7124/bc.000AA0

Selvakumar, P., Mariappan, P., Shunmugam, S., Paul, R., & Jabez, S. (2024). Removal of chemicals from effluent using photobioreactor technology to improve environmental and health impacts. *Zeitschrift für Physikalische Chemie*, *2024*(7), 1295–1302. Advance online publication. DOI: 10.1515/zpch-2023-0450

Vijayakumar, G., Rajkumar, M., Rajiv Chandar, N., Selvakumar, P., & Duraisamy, R. (2024). Muniyandi Rajkumar, Rajiv Chandar N, Selvakumar P and Ramesh Duraisamy, (2024). Environmentally friendly TDS removal from waste water by electrochemical ion exchange batch-type recirculation (EIR) technique. *Environmental Science. Water Research & Technology*, *10*(4), 826–835. DOI: 10.1039/D3EW00793F

Vul, A., & Dideikin, A. T. (2018). Graphene oxide and derivatives: the place in graphene family. Fronteirs in Physics, 103389/fphy.2018.00149.

Wang, X., Ma, Y., Sheng, X., Wang, Y., & Xu, H. (2018). Ultrathin polypyrrole nanosheets via Space-confined synthesis for efficient photothermal therapy in the second near-infrared window. *Nano Letters*, *18*(4), 2217–2225. DOI: 10.1021/acs.nanolett.7b04675 PMID: 29528661

Wu, S., Chen, Y., Jiao, T., Zhou, J., Cheng, J., Liu, B., Yang, S., Zhang, K., & Zhang, W. (2019). An aqueous Zn-Ion hybrid supercapacitor with high energy density and ultrastability up to 80.000 cycles. *Advanced Energy Materials*, *9*(47), 1902915. DOI: 10.1002/aenm.201902915

Xia, D. D., Gong, F., Pei, X., Wang, W., Li, H., Zeng, W., Wu, M., & Papavassiliou, D. V. (2018). Molybdenum and tungsten disulfides-based nanocomposite films for energy storage and conversion: A review. *Chemical Engineering Journal*, *18*, 1–75. DOI: 10.1016/j.cej.2018.04.207

Xu, B., Wanga, H., Zhua, Q., Suna, N., Anasoric, B., Hua, L., Wanga, F., Guanb, Y., & Gogotsic, Y. (2018). Reduced graphene oxide as a multi-functional conductive binder for supercapacitor electrodes. *Energy Storage Materials*, *12*, 128–136. DOI: 10.1016/j.ensm.2017.12.006

Xu, B., Zhang, H., Mei, H., & Sun, D. (2020). Recent progress in metal-organic framework-based supercapacitor electrode materials. *Coordination Chemistry Reviews*, *420*, 213438. DOI: 10.1016/j.ccr.2020.213438

Yadav, S., & Devi, A. (2020). Recent advancements of metal oxides/Nitrogen-doped graphene nanocomposites for supercapacitor electrode materials. *Journal of Energy Storage*, *30*, 101486. DOI: 10.1016/j.est.2020.101486

Ye, W., Wang, H., Ning, J., Zhong, Y., & Hu, Y. (2020). New TYPES Of hybrid electrolytes for supercapacitors. *Journal of Energy Chemistry*, *57*, 219–232. DOI: 10.1016/j.jechem.2020.09.016

Zare, E. N., Makvandi, P., Ashtari, B., Rossi, F., Motahari, A., & Perale, G. (2020). Progress in conductive polyaniline-based nanocomposites for biomedical applications: A review. *Journal of Medicinal Chemistry*, *63*(1), 1–22. DOI: 10.1021/acs.jmedchem.9b00803 PMID: 31502840

Zhang, L., Hu, X., Wang, Z., Sun, F., & Dorrell, D. G. (2018a). A review of supercapacitor modeling, estimation, and applications: A control/management perspective. *Renewable & Sustainable Energy Reviews*, *81*, 1868–1878. DOI: 10.1016/j.rser.2017.05.283

Chapter 19
The Role of Supercapacitors in Future Energy Solutions

J. Mangaiyarkkarasi
https://orcid.org/0000-0003-1431-9584
Nadar Mahajana Sangam S. Vellaichamy Nadar College, India

J. Shanthalakshmi Revathy
https://orcid.org/0000-0003-1724-7117
Velammal College of Engineering and Technology, India

ABSTRACT

Supercapacitors are transforming energy storage by bridging the gap between batteries and advanced solutions. This chapter, The Role of Supercapacitors in Future Energy Solutions, explores their unique properties, advantages, and applications. Known for rapid energy delivery, long lifecycles, and efficiency across temperatures, supercapacitors are ideal for high-power-density, low-maintenance applications. They rely on electrostatic charge rather than chemical reactions, making them distinct from batteries like lithium-ion. While their energy density is lower, advancements in nanomaterials like graphene are improving performance. Environmentally, supercapacitors use abundant, non-toxic materials, offering recyclability and sustainability advantages over batteries. Hybrid supercapacitors, blending the strengths of batteries and capacitors, promise better energy and power densities. This chapter highlights their role in a renewable, electrified future, emphasizing innovation to unlock their potential in sustainable energy systems.

DOI: 10.4018/979-8-3693-9316-1.ch019

1. INTRODUCTION

The energy storage industry has experienced a revolution through supercapacitors or electrochemical capacitors or ultracapacitors because these devices contain distinct attributes that could assist next-generation energy systems meet their requirements. The power generation and cycling efficiency of supercapacitors creates interest in alternative battery usage in specific applications (Y. A. Kumar et al., 2024). The complete operation of renewable power systems which depend on solar and wind generation demands advanced energy storage systems because they need flexible operational capabilities for reliability and efficiency. Extensive operations and fast replenishment ability position supercapacitors as optimal solutions for meeting power storage requirements (Yan et al., 2024). The study demonstrates how supercapacitors can create next-generation energy systems through a fundamental assessment of their operating capacity together with their advantages and limitations in contrast with regular batteries alongside their functions for grid stabilization and renewable power integration. Key to their functionality is the ability to handle sudden power demands and balance fluctuations in renewable energy output, making them particularly valuable in hybrid systems where rapid energy shifts are common (Gopi & Ramesh, 2024). However, supercapacitors have issues with energy density, which restricts their capacity to store significant amounts of energy for prolonged periods, even while they excel in power density and endurance. This gap could be filled by recent developments in materials science and design, which would increase storage capacity and make supercapacitors more practical for a larger range of uses. Supercapacitors function as a sustainable energy storage alternative with capabilities to transform power storage processes in wearable technology and electric vehicles among other industrial uses (Lee et al., 2024).

In delving into the current and future landscape of supercapacitors, this chapter aims to highlight how these devices could be pivotal in advancing energy systems that are not only resilient and reliable but also aligned with global sustainability goals.

1.1 Literature Review

The role of supercapacitors in modern energy solutions has been the focus of extensive research, highlighting their potential as a battery substitute or supplementary technology in energy storage applications. According to studies, supercapacitors are perfect for applications needing short-term, high-power outputs because of their remarkable durability, quick charge and discharge cycles, and high power densities. According to Ramachandran et al, (Ramachandran et al., 2024) the CuO/SnO2/MXene material could be used in advanced supercapacitor and energy storage applications. Supercapacitors demonstrate impressive potential as grid solutions due to their effective regulation of renewable energy source variations such as solar power and wind energy. To overcome the primary drawback of supercapacitors—their lower energy density in comparison to conventional batteries—researchers have investigated developments in electrode materials, including graphene, carbon nanotubes, and metal oxides (Otgonbayar et al., 2023)These advancements are essential for expanding their uses beyond high-power requirements to other energy-intensive industries, including as large-scale renewable energy storage and electric vehicles (EVs) (N. Kumar et al., 2022).

Current research investigates merging supercapacitors with hybrid energy storage systems (HESS) because they share advantages between batteries and supercapacitors. These hybrid systems could enable faster charging, longer life spans, and enhanced performance under varying load conditions, as discussed in recent reviews(Saadati pour et al., 2023). Thus, supercapacitors are increasingly recognized as a pivotal technology in achieving sustainable, efficient, and resilient energy solutions for the future.

2. UNDERSTANDING SUPERCAPACITORS

2.1 Key Features

Ultracapacitors or supercapacitors have different characteristics in comparison with traditional energy storage systems which include batteries. The device suits applications that require quick energy supplies and powerful output as it functions across wide temperatures provides prolonged service life and maintains high energy volume at quick charging and discharging cycles. Supercapacitors display their model design in Figure 1.

Figure 1. Supercapacitor model

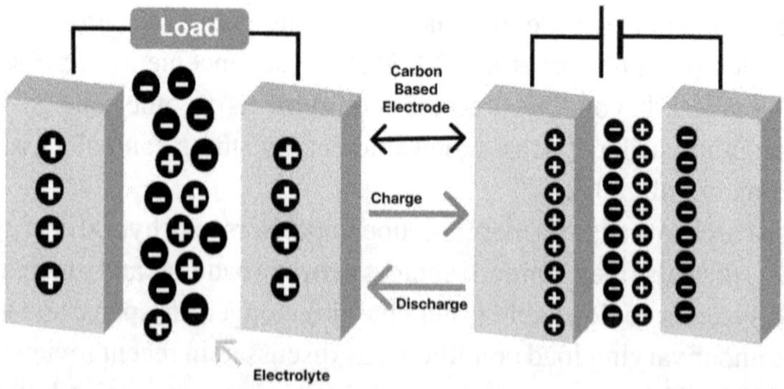

High Power Density: Supercapacitors deliver power levels surpassing conventional batteries so they work well to generate fast bursts of energy. Supercapacitors are excellent at transferring energy at a rate known as power density. This is particularly useful in power-intensive applications, such regenerative braking or acceleration in EVs.

Rapid Charging and Discharging: Supercapacitors store energy using electrostatic charge separation, which enables them to charge and discharge nearly instantly, in contrast to batteries, which depend on sluggish electrochemical reactions (Shaheen et al., 2023). This characteristic is beneficial in systems that require frequent power cycling, such as in load leveling in renewable energy grids, where they help stabilize power output from intermittent sources like solar and wind.

Long Cycle Life: Supercapacitors offer their most favorable trait through their exceptional long-term durability. The lifespan of supercapacitors reaches beyond one million cycles even though they function without showing any signs of degradation. The frequent energy cycling in applications makes supercapacitors financially beneficial because they maintain a long cycle life (Du et al., 2019).

Wide Operating Temperature Range: The operating temperature range of supercapacitors extends from -40 to +65°C. This resilience makes them suitable for harsh environments and applications where temperature control is challenging, such as in automotive or industrial settings.

Environmentally Friendly and Low Maintenance: Supercapacitors have fewer toxic components compared to batteries, and their extended lifespan reduces waste and replacement needs. The ability of compressed air to operate using renewable energy sources and to be reversibly recycled makes its use environmentally responsible for energy storage applications (Saadati pour et al., 2023).

2.2 Advantages of Supercapacitors

These are the main advantages of supercapacitors which are listed below

High Power Density: The ability of supercapacitors to generate short bursts of energy enables them to power rapid power production systems found in electric car regenerative brakes.

Fast Charging and Discharging: They can charge and discharge almost instantaneously due to electrostatic energy storage, ideal for applications needing frequent and rapid cycling.

Long Cycle Life: Supercapacitors showcase extended operation cycles exceeding hundreds of thousands or millions of cycles till they reach failure points which bears no visible signs of damage.

Wide Operating Temperature Range: Batteries work well under hard or fluctuating conditions because their operational temperatures span from -40 to +65°C.

Low Maintenance Requirements: Supercapacitors are generally low-maintenance, as they require fewer replacements and have fewer components susceptible to wear and tear.

Environmentally Friendly: Compared to batteries, supercapacitors contain fewer toxic materials and have a longer lifespan, contributing to reduced waste and lower environmental impact (Figueroa-González et al., 2022).

Enhanced Safety: Supercapacitors are safer for a variety of applications because they are less likely to experience thermal runaway or overheating, which lowers the risk of fire or explosion.

Energy Efficiency: They offer high charge/discharge efficiency (often above 95%), making them more efficient for applications where energy conservation is critical.

Lightweight and Compact: Supercapacitors are appropriate for portable electronics and electric cars since they are frequently smaller and lighter than batteries with comparable power ratings.

Versatile Applications: They are beneficial in a variety of applications due to their distinct qualities, ranging from consumer electronics and industrial machinery to automotive and renewable energy systems (Hsieh et al., 2022).

3. COMPARISON OF SUPERCAPACITORS AND BATTERIES IN ENERGY STORAGE

A comparison of supercapacitors and batteries in energy storage is necessary to comprehend the necessity of supercapacitors, and this is shown in Table 1.

The table lists the advantages and disadvantages of batteries and supercapacitors. It demonstrates that batteries are better suited for long-term, energy-dense storage, whereas supercapacitors perform best in high-power, short-term applications.

Table 1. Comparison between supercapacitors and batteries.

Feature	Supercapacitors	Batteries
Energy Density	Low; typically 1-10 Wh/kg	High; ranges from 100-265 Wh/kg
Power Density	Very high; can deliver rapid bursts of power	Moderate; suitable for steady, sustained energy flow
Charge/Discharge Time	Very fast; seconds to minutes	Slow; usually minutes to hours
Cycle Life	Long; up to 1,000,000 cycles without significant degradation	Shorter; typically 500-2,000 cycles
Efficiency	High; often >95%	Moderate; typically around 80-90%
Energy Loss	Low; low self-discharge over time	Moderate; higher self-discharge than supercapacitors
Weight	Lightweight	Heavier for the same energy storage capacity
Temperature Range	Wide; operates well from -40°C to +65°C	Narrower; performance drops in extreme temperatures
Maintenance	Low maintenance requirements	Moderate to high maintenance (especially lithium-ion)
Environmental Impact	Fewer toxic materials and longer lifespan	Contains more hazardous materials; shorter lifespan
Applications	Quick energy delivery (e.g., EVs, load leveling)	Long-term energy storage (e.g., grid storage, EVs)
Safety	Less prone to overheating or thermal runaway	Risk of overheating and thermal runaway
Cost	Higher initial cost per unit of energy	Generally lower cost per unit of energy

4. APPLICATIONS OF SUPERCAPACITORS IN RENEWABLE ENERGY SYSTEMS

The following lists some of the main uses for supercapacitors in renewable energy systems.

Grid Stabilization and Frequency Regulation: Fast energy charging of supercapacitors helps stabilize grid frequency which results from intermittent solar and wind renewable energy generation.

Energy Storage for Solar Power Systems: They can help even out the power production by storing extra energy from solar panels during the hours of greatest sunlight and swiftly releasing it when demand increases.

Wind Turbine Pitch Control Systems: In order to maintain maximum efficiency and shield the turbine from damage during strong winds, supercapacitors are employed to modify the pitch of the turbine blades during variations in wind speed (Gan et al., 2024).

Renewable Energy Microgrids: Supercapacitors help manage load balancing and demand response in microgrids by providing quick energy storage and release, especially in isolated or off-grid renewable setups.

Hybrid Energy Storage Systems (HESS): In hybrid renewable energy storage systems, supercapacitors work in tandem with batteries to manage brief, high-power demands, prolonging battery life and enhancing system efficiency.

Wave and Tidal Energy Systems: For wave and tidal power installations, supercapacitors stabilize energy output to match grid requirements, compensating for the natural fluctuations in energy generation (Fatih Guven et al., 2024).

Power Smoothing in Wind Farms: By rapidly releasing or storing power, supercapacitors smooth out variations in power generation from wind farms, ensuring a more consistent energy flow to the grid(Rashid Khan & Latif Ahmad, 2024).

Backup Power for Critical Loads: Supercapacitors provide instant backup power to renewable energy facilities in the event of a power interruption, allowing safe shutdown or transition to auxiliary power systems.

4.1 Role of Supercapacitors In Electrical Vehicles

Supercapacitors function as essential power storage units in electric vehicles because of their fast charging capabilities alongside high energy retention (Xu & Shen, 2021). Here are 10 key applications of supercapacitors in EVs:

Regenerative Braking Energy Storage: Utilizing supercapacitors extends battery lifespan by storing brake-regenerated power which the system discharges for acceleration.

Boosting Acceleration: The voltage spikes provided by auxiliary batteries reduce the strain on the main power source during acceleration and while tackling hills.

Start-Stop Systems: Used in hybrid and electric vehicles to provide instant power for restarting the engine or motor, improving fuel efficiency and reducing emissions.

Battery Load Balancing: They help manage sudden fluctuations in power demand, reducing stress on lithium-ion batteries and enhancing their longevity.

Cold Start Assistance: In extremely cold temperatures, where batteries struggle, supercapacitors provide the necessary energy for vehicle startup.

Wireless Charging Support: Supercapacitors enable fast and efficient energy transfer in wireless charging systems by smoothing out fluctuations in power delivery.

Energy Buffering in Electric Buses and Trams: Public transport systems use supercapacitors to store energy at stops and quickly discharge it for movement between stations.

Voltage Stabilization in Power Electronics: They act as buffers to stabilize voltage variations in powertrain electronics, ensuring consistent performance and protecting sensitive components.

Hybrid Energy Storage Systems (HESS): Combined with batteries, supercapacitors create an optimized energy storage system, balancing energy density (batteries) with power density (supercapacitors).

Quick-Charge Infrastructure Support: Supercapacitors support ultra-fast charging stations by buffering power surges and reducing grid load during peak demand.

5. RECENT ADVANCEMENTS IN SUPERCAPACITOR MATERIALS AND TECHNOLOGY

The development of supercapacitors progressed into practical uses because of better energy storage capacity, powerful performance, and longer device lifespans through recent materials technology breakthroughs (Alam et al., 2023). Innovations in electrode materials, electrolyte formulations, and device architecture largely drive these improvements.

Advanced Electrode Materials: The feasibility of supercapacitors increases for various applications through material development which advances parallel to technological improvements that increase power storage while extending operational lifetime. Graphene, in particular, has garnered attention for its remarkable properties, enabling faster electron movement and larger energy storage potential(Balamurugan et al., 2019). Researchers focus on transition metal oxides together with manganese oxide and ruthenium oxide and metal-organic frameworks (MOFs) since these materials offer better stability and capacitance characteristics but remain expensive to produce and difficult to scale. The figure in Figure 2 displays multiple materials that serve as electrodes for supercapacitors.

Figure 2. Electrode materials for supercapacitors

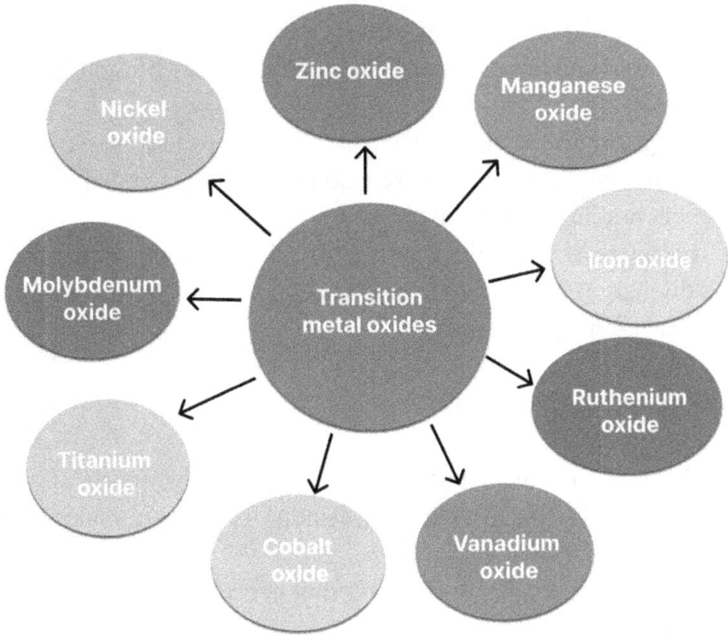

Hybrid Supercapacitors: Manufacturers develop hybrid supercapacitors that unite features of batteries and supercapacitors to provide high energy density from batteries alongside high power density from supercapacitors. Hybrid devices with increased energy density and sustained power along with lifetime appear feasible through the incorporation of battery-designed materials like lithium compounds in the supercapacitor structure.

5.1 Hybrid Energy Storage Using Supercapacitors

The combination of battery cells and supercapacitors in hybrid energy storage systems creates advanced applications for electric cars and other systems as per (Şahin et al., 2022). Combined operations of these technologies enable efficient long-lasting power management through stable charging and discharging mechanisms.

5.1.1 Key Benefits of HESS

Enhanced Power and Energy Balance: Batteries provide high energy density, while supercapacitors deliver quick bursts of power during acceleration or regenerative braking. This combination improves overall energy efficiency.

Extended Battery Life: Supercapacitors absorb sudden power fluctuations, reducing battery stress and degradation. This extends battery lifespan and minimizes replacement costs.

Improved Regenerative Braking Efficiency: Supercapacitors serve electric buses, trams and electric vehicles through their capability to instantaneously provide energy extracted during braking operations.

Faster Charging and Discharging: Unlike batteries, supercapacitors charge within seconds and discharge quickly. In a hybrid system, this feature helps reduce waiting times and enhances energy availability.

5.1.2 Real-World Applications

Electric Buses: Cities like Shanghai and Paris use hybrid electric buses that integrate supercapacitors for fast charging at stops, reducing reliance on large battery packs.

Formula 1 Cars: Supercapacitors are used in hybrid racing cars to store kinetic energy and deliver instant power boosts.

Hybrid Passenger Vehicles: Mazda's i-ELOOP system combines a supercapacitor with a conventional battery to improve fuel efficiency by storing braking energy.

HESS provides transport applications together with other systems with performance-based sustainable energy storage through its integration of superior supercapacitor and battery attributes.

Figure 3. Hybrid supercapacitors

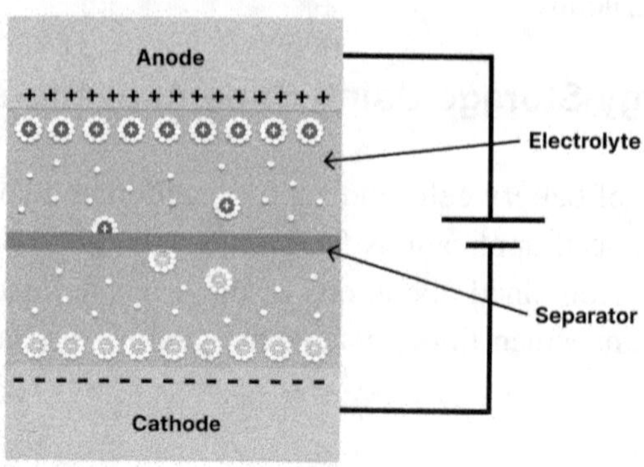

Solid-State Electrolytes: Traditional supercapacitors use liquid electrolytes, which can limit stability and safety, particularly at high temperatures or in harsh environments. Recent research focuses on solid-state electrolytes, which improve thermal stability, reduce leakage risks, and potentially increase energy density. Solid-state supercapacitors also open up possibilities for flexible and wearable devices, as they can be incorporated into compact and bendable designs(Wang et al., 2023) .

Nanostructured Designs: Nanotechnology has enabled the design of nanostructured electrodes that maximize surface area and optimize ion movement, enhancing both capacitance and charge/discharge rates. Examples include nanowires, nanofibers, and porous structures that allow for faster ion diffusion and increased energy storage in smaller, lighter packages.

These advancements in materials and technology are making supercapacitors more competitive in energy storage, with improved energy density, efficiency, and versatility. Supercapacitors may become more and more significant as research advances in a variety of applications, from smart grids and portable electronics to electric vehicles and renewable energy systems (Peng et al., 2023).

6. CHALLENGES IN SUPERCAPACITOR INTEGRATION AND SCALABILITY

While supercapacitors offer promising advantages for energy storage, their integration and scalability in real-world applications face several challenges (Rudra et al., 2024). These limitations include issues with energy density, material costs, device architecture, and compatibility with existing energy infrastructure.

Low Energy Density: Supercapacitors hold particularly low energy storage capacity compared to conventional batteries which represents a major disadvantage in their application. Because supercapacitors deliver power fast yet maintain minimal power storage they do not match well with energy needs that require sustained power distribution. This limitation restricts their scalability in areas like EVs and grid storage, where higher energy capacity is essential. Bridging this gap requires advancements in materials and hybrid systems to enhance energy storage without compromising other performance aspects(Chen et al., 2021).

High Material Costs: Although they are expensive to produce, advanced electrode materials including graphene, carbon nanotubes, and transition metal oxides have showed promise in enhancing supercapacitor performance. These materials are often challenging to synthesize and scale up, particularly for large-scale or cost-sensitive applications. Additionally, manufacturing processes that involve nanostructuring and material purity requirements can add significant expense, making the production of high-performance supercapacitors cost-prohibitive for widespread use.

Complex Device Architecture: The design and integration of hybrid energy storage systems are made more difficult by the frequent hybridization or combination of supercapacitors and batteries. Ensuring compatibility between supercapacitors and batteries, optimizing control systems, and managing different charge/discharge rates and thermal characteristics require sophisticated design and engineering. This complexity can increase production costs and necessitate more extensive testing and regulation.

Limited Existing Infrastructure Compatibility: Traditional energy storage systems and power management infrastructure are typically designed around battery-based systems, making it difficult to integrate supercapacitors seamlessly. Supercapacitors require specialized power management systems and often cannot simply replace batteries in existing applications, especially those involving prolonged, steady energy delivery. Adapting infrastructure to incorporate supercapacitors often requires new hardware and software, which can be costly and time-consuming (Li et al., 2022).

Environmental and Safety Regulations: While supercapacitors generally have fewer toxic materials than some battery chemistries, their production still involves environmentally sensitive processes. Ensuring that materials and manufacturing processes meet safety and environmental standards is essential, particularly as regulatory requirements for sustainable energy storage tighten.

Despite these challenges, ongoing research in materials science, device engineering, and cost reduction is advancing the potential of supercapacitors. By overcoming these challenges, scalable, sustainable energy solutions may become possible, ranging from electric cars and portable devices to grid stabilization and renewable energy storage.

7. THE FUTURE OF SUPERCAPACITORS IN SUSTAINABLE ENERGY GRIDS

Supercapacitors position themselves for sustainable energy success because they improve traditional energy storage limitations. Supercapacitors can be extremely helpful in tackling the issues of sporadic energy supply, rapid power needs, and system stability in grids that are becoming more and more dependent on renewable energy sources like solar and wind. These batteries excel for sustainable network integration because they hold long cycles, provide high power density and achieve fast charging and discharging rates.

Grid Stability and Frequency Regulation: Super capacitors act as a crucial benefit for energy networks by regulating changes in voltage and frequency which result from variable renewable energy sources. Wind and solar power systems produce unpredictable output which causes supply to differ from demand levels. Supercapacitors can rapidly inject or absorb energy, helping to smooth out these fluctuations and maintain grid stability. Their fast response times make them especially valuable in frequency regulation, where even minor delays can disrupt grid performance (Kim et al., 2023).

Enhancing Renewable Energy Utilization: Supercapacitors can maximize the grid's incorporation of renewable energy by collaborating with batteries. While batteries provide sustained energy over longer periods, supercapacitors handle short-term, high-power bursts, reducing the strain on batteries and extending their lifespan. This hybrid energy storage approach enables more efficient use of renewable energy, allowing grids to rely less on fossil-fuel-based backup power.

Supporting Microgrids and Off-Grid Systems: In remote or decentralized regions, supercapacitors are valuable components in microgrids powered by renewable sources. They offer instant power for critical loads, such as healthcare facilities or communication systems, and can be paired with solar or wind systems to provide stable power without relying on a central grid. Their durability and low maintenance needs make them ideal for remote locations with limited access to replacement parts or frequent servicing.

Reducing Environmental Impact: Due to their longer operating life and lower hazardous content, supercapacitors are typically more ecologically benign than many battery chemistries. They help reduce electrical waste and promote a more sustainable approach to energy storage by lowering the need for frequent replacements.

Enabling Ultra-Fast EV Charging Infrastructure: As EVs become more common, sustainable grids will require infrastructure that can support ultra-fast charging demands (Shi et al., 2021). Supercapacitors can supply the high-power bursts needed for rapid EV charging, improving the efficiency of charging stations connected to renewable energy grids. By handling peak loads during charging, they help prevent grid overloads and reduce the dependency on traditional, fossil-fuel-based power plants.

Supercapacitors hold a vital role for the implementation of sustainable energy networks because their material development and technological advancements will continue. By offering reliable, high-speed energy support, supercapacitors can help create a more resilient, efficient, and environmentally friendly energy infrastructure, contributing significantly to the global transition toward sustainable energy.

8. RECENT STUDIES AND REAL WORLD CASE STUDIES

The electric vehicle industry shows significant interest in supercapacitors as both ultracapacitors because they provide extended economic lives alongside quick charging and powerful energy storage benefits. Recent studies and real-world implementations underscore their potential to enhance EV performance and efficiency (Velasco et al., 2021).

Recent Studies

The Energy Storage Materials publication offers a thorough review of battery-supercapacitor hybrid energy storage system (HESS) development. The combination of batteries with supercapacitors produces beneficial power-storage relationships for EVs which improves system operational time and capability.

The scientific research examined the development of fresh supercapacitor electrode materials. The research appeared in Materials Today Energy, while showcasing modern advances in supercapacitor devices specifically the creation of solid-state flexible designs and innovative materials to boost device energy density and operational efficiency.

The research team studied how supercapacitor capacitive performance relates to carbon electrode physicochemical properties through the application of machine learning techniques. The analysis detected that potential window and nitrogen doping and specific surface area act as essential variables in determining capacitance values for designing electrode materials.

Real-World Case Studies

In the realm of public transportation, Hong Kong has implemented electric trams equipped with supercapacitors to capture and release energy during braking. This system has led to significant reductions in energy consumption and decreased wear on the main batteries, showcasing the practical benefits of supercapacitor integration.

Similarly, China has been experimenting with electric buses, known as Capabus, which utilize onboard electric double-layer capacitors. These buses quickly recharge at designated stops equipped with charging infrastructure, allowing them to operate efficiently without continuous overhead lines. This implementation demonstrates the feasibility of supercapacitor-powered public transit systems.

In the automotive industry, Mazda's i-ELOOP system employs supercapacitors to recover and store braking energy in the Mazda 6 model. This regenerative braking system reportedly reduces fuel consumption by approximately 10%, illustrating the potential of supercapacitors to enhance fuel efficiency in passenger vehicles.

Studies demonstrate how supercapacitors function as fundamental units to improve electric vehicles by managing their energy storage needs and creating green transport systems.

CONCLUSION

Supercapacitors play a crucial role in advancing sustainable energy solutions because they address major challenges related to storing renewable energy and energy distribution. Due to their exclusive characteristics including high power density and swift charging and discharging abilities with extended cycle life supercapacitors are fundamental elements for various applications which include renewable energy system improvements as well as electric vehicles and power stabilization and frequency management within grids. The need for effective, dependable, and flexible energy storage solutions will only increase as the globe moves toward a low-carbon economy. Supercapacitors, with their ability to quickly respond to fluctuations in power supply and demand, are well-suited to complement existing battery technologies and address the intermittency of renewable energy sources. Moreover, their environmentally friendly attributes and low maintenance needs align with global sustainability goals, making them an attractive option for various applications. Supercapacitors' energy density and overall performance should be improved by ongoing developments in materials science, device architecture, and hybrid energy storage systems, which will make it easier for them to be used more widely in sustainable energy infrastructures. By integrating supercapacitors into energy systems, we can create a more resilient, efficient, and environmentally conscious energy landscape that meets the demands of the 21st Century. Ultimately, the role of supercapacitors in our sustainable energy future is not only significant but essential, as they pave the way for innovative solutions that support the global transition to renewable energy and sustainable development.

REFERENCES

Alam, A., Saeed, G., Kim, K., Kim, J. K., Park, H. S., & Lim, S. (2023). Direct Ink Writing (DIW) printed high-performance asymmetric supercapacitor based on 0D@2D silver-nanoparticles@MXene as anode and 0D@2D MnO2-nanoparticles@ MXene as cathode materials. *Journal of Energy Storage*, *72*, 108227. DOI: 10.1016/j.est.2023.108227

Balamurugan, J., Nguyen, T. T., Aravindan, V., Kim, N. H., Lee, S. H., & Lee, J. H. (2019). All ternary metal selenide nanostructures for high energy flexible charge storage devices. *Nano Energy*, *65*, 103999. DOI: 10.1016/j.nanoen.2019.103999

Chen, S., Xin, S., Yang, L., Guo, Y., Zhang, W., & Sun, K. (2021). Multi-sized planar capacitive pressure sensor with ultra-high sensitivity. *Nano Energy*, *87*, 106178. DOI: 10.1016/j.nanoen.2021.106178

Du, X., Chen, Q., Xiang, C., Zou, Y., Xu, F., Xie, J., & Sun, L. (2019). Formation of Highly Dispersed Ultrafine MnO2 Nanoparticles on Nitrogen-doped Porous Carbon for Supercapacitor Applications. *International Journal of Electrochemical Science*, *14*(5), 4195–4205. DOI: 10.20964/2019.05.07

Fatih Guven, A., Abdelaziz, A. Y., Mahmoud Samy, M., & Barakat, S. (2024). Optimizing energy Dynamics: A comprehensive analysis of hybrid energy storage systems integrating battery banks and supercapacitors. *Energy Conversion and Management*, *312*, 118560. DOI: 10.1016/j.enconman.2024.118560

Figueroa-González, E., Oliva, A. I., Rodriguez-Gonzalez, V., Gomez-Solis, C., Garcia, C. R., & Oliva, J. (2022). Using recycled tetrapak and Ag/BaMoO4 nanoparticles to make efficient and flexible solid state supercapacitors. *Journal of Energy Storage*, *47*, 103544. DOI: 10.1016/j.est.2021.103544

Gan, Z., Ren, X., Liu, M., Sun, M., Shen, W., & Li, Z. (2024). Advances in design of copper-based nanostructures and nanocomposites for high-performance supercapacitors. *Journal of Energy Storage*, *99*, 113383. DOI: 10.1016/j.est.2024.113383

Gopi, C. V. M., & Ramesh, R. (2024). Review of battery-supercapacitor hybrid energy storage systems for electric vehicles. *Results in Engineering*, *24*, 103598. DOI: 10.1016/j.rineng.2024.103598

Hsieh, C.-E., Chang, C., Gupta, S., Hsiao, C.-H., Lee, C.-Y., & Tai, N.-H. (2022). Binder-free CoMn2O4/carbon nanotubes composite electrodes for high-performance asymmetric supercapacitor. *Journal of Alloys and Compounds*, *897*, 163231. DOI: 10.1016/j.jallcom.2021.163231

Kim, Y., Parale, V. G., Kim, T., Kim, S.-H., Choi, H., Patil, U. M., Kanamori, K., & Park, H.-H. (2023). Synthesis of Zn-Co-S nanowire bundle type aerogel electrodes for asymmetric supercapacitors. *Chemical Engineering Journal, 474*, 145875. DOI: 10.1016/j.cej.2023.145875

Kumar, N., Kim, S.-B., Lee, S.-Y., & Park, S.-J. (2022). Recent Advanced Supercapacitor: A Review of Storage Mechanisms, Electrode Materials, Modification, and Perspectives. *Nanomaterials (Basel, Switzerland), 12*(20), 3708. DOI: 10.3390/nano12203708 PMID: 36296898

Kumar, Y. A., Roy, N., Ramachandran, T., Hussien, M., Moniruzzaman, M., & Joo, S. W. (2024). Shaping the future of energy: The rise of supercapacitors progress in the last five years. *Journal of Energy Storage, 98*, 113040. DOI: 10.1016/j.est.2024.113040

Lee, Y.-S., Alagarasan, J. K., Dasha Kumar, K., Ramachandran, T., Kim, H.-J., Khan, M. S., Anil Kumar, Y., Lee, M., & Kim, S. (2024). Redox behavior in hydrothermally synthesized thistle flower-like Co3O4–NiO-GO composite for advanced supercapacitor electrodes. *Journal of Physics and Chemistry of Solids, 192*, 112114. DOI: 10.1016/j.jpcs.2024.112114

Li, L., Jin, Y., Zeng, C., Ren, D., Zhou, K., Zhang, Q., Liu, J., Zhou, Q., & Wang, H. (2022). Porous hierarchical Ni2P nanosheet arrays combined with Cu3P layers on copper foam as a binder-free battery-type cathode for high-energy-density asymmetrical all-solid-state electrochemical capacitors. *Journal of Energy Storage, 52*, 104783. DOI: 10.1016/j.est.2022.104783

Otgonbayar, Z., Yang, S., Kim, I.-J., & Oh, W.-C. (2023). Recent advances in 2D MXene and solid state electrolyte for energy storage applications: Comprehensive review. *Chemical Engineering Journal, 472*, 144801. DOI: 10.1016/j.cej.2023.144801

Peng, P., Yang, F., Wang, Z., & Wei, D. (2023). Integratable Paper-Based Iontronic Power Source for All-In-One Disposable Electronics. *Advanced Energy Materials, 13*(42), 2302360. Advance online publication. DOI: 10.1002/aenm.202302360

Ramachandran, T., Pachamuthu, M. P., Karthikeyan, G., Hamed, F., & Rezeq, M. (2024). Synergistic effects in CuO/SnO2/Ti3C2Tx nanohybrids: Unveiling their potential as supercapacitor cathode material. *Materials Science in Semiconductor Processing, 179*, 108486. DOI: 10.1016/j.mssp.2024.108486

Rashid Khan, H., & Latif Ahmad, A. (2024). Supercapacitors: Overcoming current limitations and charting the course for next-generation energy storage. *Journal of Industrial and Engineering Chemistry*. Advance online publication. DOI: 10.1016/j.jiec.2024.07.014

Rudra, S., Seo, H. W., Sarker, S., & Kim, D. M. (2024). Supercapatteries as Hybrid Electrochemical Energy Storage Devices: Current Status and Future Prospects. *Molecules (Basel, Switzerland)*, *29*(1), 243. DOI: 10.3390/molecules29010243 PMID: 38202828

Saadati pour, M., Gilak, M. R., Pedram, M. Z., & Naikoo, G.Saadati pour. (2023). Enhancement electrochemical properties of supercapacitors using hybrid CuO/Ag/rGO based nanoporous composite as electrode materials. *Journal of Energy Storage*, *74*, 109330. DOI: 10.1016/j.est.2023.109330

Şahin, M., Blaabjerg, F., & Sangwongwanich, A. (2022). A Comprehensive Review on Supercapacitor Applications and Developments. *Energies*, *15*(3), 674. DOI: 10.3390/en15030674

Shaheen, I., Hussain, I., Zahra, T., Memon, R., Alothman, A. A., Ouladsmane, M., Qureshi, A., & Niazi, J. H. (2023). Electrophoretic Fabrication of ZnO/CuO and ZnO/CuO/rGO Heterostructures-based Thin Films as Environmental Benign Flexible Electrode for Supercapacitor. *Chemosphere*, *322*, 138149. DOI: 10.1016/j.chemosphere.2023.138149 PMID: 36804630

Shi, M., Zhao, M., Jiao, L., Su, Z., Li, M., & Song, X. (2021). Novel Mo-doped nickel sulfide thin sheets decorated with Ni–Co layered double hydroxide sheets as an advanced electrode for aqueous asymmetric super-capacitor battery. *Journal of Power Sources*, *509*, 230333. DOI: 10.1016/j.jpowsour.2021.230333

Velasco, A., Ryu, Y. K., Boscá, A., Ladrón-de-Guevara, A., Hunt, E., Zuo, J., Pedrós, J., Calle, F., & Martinez, J. (2021). Recent trends in graphene supercapacitors: From large area to microsupercapacitors. *Sustainable Energy & Fuels*, *5*(5), 1235–1254. DOI: 10.1039/D0SE01849J

Wang, X., Zhu, B., Huang, Y., Shen, L., Chai, Y., Han, J., Yu, J., Wang, Z., & Chen, X. (2023). High-performance self-powered integrated system of pressure sensor and supercapacitor based on Cu@Cu2O/graphitic carbon layered porous structure. *Journal of Colloid and Interface Science*, *632*, 140–150. DOI: 10.1016/j.jcis.2022.11.064 PMID: 36413940

Xu, H., & Shen, M. (2021). The control of lithium-ion batteries and supercapacitors in hybrid energy storage systems for electric vehicles: A review. *International Journal of Energy Research*, *45*(15), 20524–20544. DOI: 10.1002/er.7150

Yan, Z., Luo, S., Li, Q., Wu, Z., & Liu, S. (2024). Recent Advances in Flexible Wearable Supercapacitors: Properties, Fabrication, and Applications. *Advancement of Science*, *11*(8), 2302172. Advance online publication. DOI: 10.1002/advs.202302172 PMID: 37537662

About the Contributors

Mohsen Mhadhbi obtained his Ph.D. degree from the University of Sfax, Faculty of Sciences of Sfax, Tunisia. He is currently Assistant Professor in National Institute of Research and Physicochemical Analysis, Tunisia. His research interests include material engineering, ceramics, composites, energy, powder technology, simulation. Her published works in national and international impacted journals and Books. He is a teacher in material sciences. Hence, he supervised several researchers. He is a member of various scientific journals and associations and has been serving as an editorial board member of repute

***

B.V. Sai Thrinath received B.Tech degree in EEE from Sri Sarathi Institute of Engineering & Technology, Nuzvid affiliated to JNTUK, KAKINADA, Andhrapradesh in 2012. M.Tech Degree in EMD from Newton's Institute of Engineering, Guntur affiliated to JNTUK, KAKINADA Andhrapradesh in 2014. He is currently working as Assistant Professor in EEE department in Mohan Babu University,(Formerly Sree Vidyanikethan Engineering College), Tirupati, Andhrapradesh and pursuing his Ph.D. degree at the Department of Electrical Engineering, Andhra University, Vishakapatnam, Andhrapradesh, India. His research interests include, FACTS, Power quality, Electrical Machines & Drives, Distributed Generation, Power Systems, Power Electronics & IOT.

P. Balakrishnan, an accomplished academic and researcher, is an Assistant Professor in the Department of Mechanical Engineering at SRM Institute of Science and Technology, Chennai. He specializes in flow dynamics, noise reduction, solar energy systems, and alternative fuels. Dr. Balakrishnan earned his Ph.D. from IIT Madras, focusing on jet noise reduction using coaxial swirlers. His work includes a postdoctoral tenure at SUSTech, China, where he achieved significant noise reduction in aircraft wings. With numerous publications in high-impact journals, including works with an impact factor of 9.6, his research has garnered international recognition, including the Young Professional Grant Award from INCE in 2016. He holds a gold medal for securing the university's first rank in his M.E., and his patented innovations contribute to noise mitigation and renewable energy technologies. Dr. Balakrishnan has also been instrumental in developing research facilities, conducting motivational lectures, and contributing to academic course development, reflecting his dedication to education and innovation.

Hari Shankar Biswas is Assistant Professor in Chemistry Surendranath College, Kolkata. Dr. Biswas is an esteemed Assistant Professor in Chemistry at Surendranath College, Kolkata, with a multifaceted academic background and a strong dedication to research and education. He embarked on his academic journey by obtaining his B.Sc. degree in Chemistry Honours from R.K.M.V.C. College in 2005, followed by his M.Sc. degree from Presidency College, Kolkata, in 2008. Dr. Biswas further pursued his scholarly pursuits by earning a Ph.D. in Material Science from the Saha Institute of Nuclear Physics, University of Calcutta, in 2015. His professional experience spans various roles, including a stint as a School teacher at Keshub Academy, Kolkata, from 2008 to 2010, before transitioning into academia as an Assistant Professor at Surendranath College. Dr. Biswas specializes in the research domain of Carbon thin Film, Graphene, and Graphene Oxide-based nano Composites, focusing on their synthesis, characterization, and application across Environmental, Biological, and Technological fields. Dr. Biswas's scholarly contributions are significant, with authorship of 15 book chapters and over 31 papers published in esteemed International Journals. He is actively involved in academic societies, holding life membership in the Indian Science Congress Association and the Indian Chemical Society, along with an affiliated membership with the Royal Society of Chemistry. In addition to his research and academic endeavors, Dr. Biswas plays a vital role in the dissemination of scientific knowledge as a series book Editor and reviewer for reputable international publishers and journals. He serves as an Editorial board member for the American Journal of Chemical Engineering, Reseapro Journal, Lincoln University College, etc., and is affiliated with Rotary International. Dr. Biswas's editorial contributions extend to editing various books, including titles such as "Emerging Concepts in Chemical and Biological Science" and "Modern Approaches in Chemical and Biological Sciences," among others. Through his comprehensive involvement in research, education, and editorial responsibilities, Dr. Hari Shankar Biswas continues to make valuable contributions to the field of Chemistry, fostering advancements and innovations for the betterment of society.

Shib Shankar Biswas is an Assistant Professor in the Department of Physics at Surendranath College, Kolkata, India. With expertise in condensed matter physics and nanomaterials, he has made significant contributions to research in energy storage materials and advanced functional materials. His academic journey includes a strong background in theoretical and experimental physics, focusing on the development and characterization of novel materials. Apart from his research, Dr. Biswas is actively involved in teaching and mentoring students in physics and interdisciplinary scientific studies. His work has been published in reputed journals, and he continues to explore innovative approaches to material science for sustainable technological advancements.

Vineeta S. Chauhan received her PhD degree in the field of Power System Engineering, B.E. degree in Electrical Engineering from Rajasthan University and M. Tech degree with specialization in Power System from Rajasthan Technical University. She is currently working as faculty in Electrical Engineering Department. She is the author/co-author of more than 25 publications in National and International conferences and journals. She had total Academic experience of 17 years. She is a life time member of Indian Society of Technical Education. Her area of Research interest includes Electrical Machines, Power system Protection, Distributed Generation, Interconnected Power System, Smart grids etc.

Siddharthsingh K. Chauhan received his B.E., M.E. & Ph.D. in 2003, 2005 and 2014, respectively, in the field of Electrical Engineering. He is presently working as Associate Professor & Head at Department of Electrical Engineering, Institute of Technology, Nirma University. His research areas include power quality improvement devices, advanced current controllers, digital signal processor-based power electronic systems, power electronics applications to power systems, and applications of artificial intelligence to power electronic systems.

Susanta Das is a seasoned educator with a proven record of teaching and mentoring diverse students, working at multiple institutions, and providing service to the various initiatives of organizations. He received his Ph.D. and M.A. degrees from Western Michigan University (WMU), USA, and M.Sc. degree from Banaras Hindu University (BHU), India all in Physics. He continued his research as a Marie-Curie post-doctoral fellow at Stockholm University, Stockholm, Sweden on beam diagnostics for the DESIREE (Double ElectroStatic Ion Ring ExpEriment) in the project DITANET (Diagnostic Techniques for particle Accelerators – a Marie-Curie initial training NETwork), at the Indian Institute of Science Education and Research-Kolkata, India on high-pressure physics as a Project Scientist-B, and at the University of Electro-Communications, Tokyo, Japan on ion-surface interactions as a

Post-doctoral fellow. Throughout his career, Dr. Das worked and collaborated with many researchers, Ph.D., master, and visiting students from different countries. He further visited many countries to discuss research and an international conference participant (UK, Italy, Germany, Greece, Belgium, Bulgaria, Romania, Brazil etc.). He co-authored several research articles in WoS/Scopus indexed international journals, conference proceedings, and scholarly book chapters published by renowned international publishing houses. Beyond his research endeavors, Dr. Das has a storied history of service to academic and administrative committees, exhibiting his commitment to the growth and development of educational institutions. His experience includes tenure at Central University South Bihar, Sri Sri University, and P.K. University before assuming his current role at Ajeenkya DY Patil University. He received the Marie-Curie post-doctoral fellowship in Sweden, Science Academies' Summer Research Fellowship in India, Gwen Frostic Doctoral Fellowship, Department Graduate Research and Creative Scholar Award by WMU, and Leo R. Parpart Doctoral Fellowship by Dept. of Physics, WMU, among many others, throughout his academic journey. At present, Dr. Das continues to delve into cutting-edge fields, with a keen interest in nanotechnology, quantum computing, and data science. His multifaceted contributions, spanning teaching, research, and administrative leadership, showcase a dedicated professional who is instrumental in advancing the vision and mission of the institutions he serves.

Pooja Dehankar is working in department of Artificial Intelligence and Data Science, Ajeenkya DY Patil School of Engineering, affiliated to Savitribai Phule Pune University, Pune. She has completed BE in Computer Engineering from Priyadarshini College of Engineering, Nagpur, and ME in Information Technology from Sinhgad College of Engineering, Pune. She has done Post Graduate Diploma in Advance Computing. She is currently pursuing Ph.D. from Symbiosis International University, Pune. She has 16 years of experience in academics. Her areas of interest are Artificial Intelligence, Cyber Security, and Data Mining. She has published papers in Scopus Journals. She actively participates in workshops, faculty development programs, short-term training programs, conferences, and webinars.

E. Nirupama is a distinguished academician with over 22 years of experience in Commerce and Management Education. She is currently serving as a Professor in the MBA Program at Saveetha School of Management, bringing extensive expertise in Finance and Accounting. A multi-award-winning educator and researcher, she has published widely in reputed national and international journals. Dr. Nirupama has actively contributed to Academic Leadership, Curriculum Development, and Faculty Development initiatives.

Boopathy Govindarajan earned his undergraduate degree in mechanical engineering from University of Madras, his master's degree in aeronautical engineering from Madras Institute of Technology, and his doctorate from the Department of Aeronautical Engineering, St. Peter's Institute of Higher Education and Research, Chennai. At Vel Tech University, a professorship is currently held in the department of aeronautical engineering. In addition to guiding 5 Ph.D Scholars, guided 8 PG & 90 UG scholars on their significant projects, he has written in 30 international publications. He is a Life Member of the Society of Automotive Engineers INDIA and the Aeronautical Society of India.

Shashank G is an experienced Assistant Professor in the Department of Management Studies at ISBR Business School, Bengaluru, with 5 years of academic expertise. He specializes in teaching MBA and MCom subjects, including Marketing Management, Consumer Behavior, and Financial Markets. Additionally, he serves as a Placement Coordinator and guides student projects. Prof. Shashank has published research on green marketing and has undertaken studies on digital marketing trends in FMCG. With a strong educational background in MBA and BCom, he is skilled in MS Office and Tally ERP9. His personal attributes include confidence, communication skills, and leadership.

Nesrine Gaaliche, PhD Dr Nesrine Gaaliche is an Assistant professor at Bahrain Polytechnic. She is a Mechanical Engineer with PhD in Mechanics and Systems Engineering from National Engineering School of Sousse. Previously she was a lecturer at the University of Bahrain, Kingdom of Bahrain. She delivered lectures and supervised several industry projects and practical labs. She was a project engineer at the aluminum profiles company. She received her Master of Science in Mechanics and Engineering System from the National Engineering School of Sousse, university of Sousse, Tunisia, in 2015, and her PhD in Mechanical Engineering "Mechanics and Systems Engineering" from the National Engineering School of Sousse, university of Sousse, Tunisia, in 2020. Her research interests include Renewable Energy, FEM Analysis, Thermoplastic Bearing, Dynamic Performance, Fatigue and vibration condition monitoring. Nesrine is a regular conference speaker and her researches have been published in several highly indexed international journals (e.g. Journal of Thermal Analysis and Calorimetry, Journal of Testing and Evaluation, Journal of Theoretical and Applied Mechanics, international Journal of Acoustics and vibration, etc.).

Balaji Ganesan is a prominent aerodynamicist and professor at the Hindustan Institute of Technology and Science (HITS) in Chennai, India. With a specialization in aerodynamics, fluid dynamics, and propulsion systems, Dr. G. Balaji earned his doctoral degree in Aerospace Engineering, focusing on high-speed aerodynamics, and has extensive experience in both academic and applied research within the field. At HITS, Dr. G. Balaji is involved in teaching, mentoring, and guiding students in aerospace and mechanical engineering programs. His research encompasses a wide range of topics within aerodynamics, including computational fluid dynamics (CFD), experimental fluid mechanics, and aircraft design. Dr. G. Balaji has published numerous research articles in reputed international journals and has presented at various prestigious conferences worldwide. His work often addresses critical issues in aerodynamics and propulsion, aiming to improve performance, efficiency, and safety in aerospace applications

Alphonsa Diana Haokip works as an Assistant Professor at Don Bosco College (Autonomous), Maram, Manipur. Before this, Dr (Sr) Alphonsa was the Headmistress of Christ King High School, Chingjaroi Khullen, Manipur. She has completed her doctorate, "Attitude, Problem Solving Ability, and Achievement in Mathematics Learning among the Secondary School Students of Thadou Kuki Tribe in Manipur". Her areas of interest are mathematics learning for secondary school students, problem-solving ability, academic achievement, mathematics teachers' concerns, paradigm shift in higher education, and gender issues.

Mangaiyarkkarasi J is currently working as an Associate Professor of Physics in Nadar Mahajana Sangam S. Vellaichamy Nadar College, Nagamalai, Madurai- 625 019, Tamil Nadu, India. I have 24 years of teaching experience. My major research interest includes synthesis of ceramic materials, powder X-ray analysis and electron density studies. I have published 13 research papers in the refereed International journals and 14 scopus indexed book chapters. I have contributed study materials (UG & PG Physics) for Madurai Kamaraj University Distance Education Course.

Shanthalakshmi Revathy J has a Bachelor of Engineering (B.E.) in Information Technology from Mohamed Sathak Engineering College, Kilakkarai, and a Master of Engineering (M.E.) in Computer Science and Engineering from PSG College of Technology, Coimbatore. My current research focus lies in the dynamic realm of data analytics. I take pride in my academic contributions, having authored more than 10 papers in prestigious international journals. My expertise extends to 15 book chapters, highlighting my comprehensive understanding of the subject matter. Furthermore, I have secured 2 patents, affirming my innovative prowess in the field. My academic journey reflects my unwavering dedication to advancing knowledge in data analytics.

Shruti Jose is a Research Enthusiast and an Assistant Professor in the Department of Professional Studies under the School of Commerce, Finance and Accountancy at CHRIST (Deemed to be) University, Bangalore (India). Her research interests lie in banking, finance, technology adoption, consumer behaviour, payments, eCommerce, artificial intelligence, and innovation research. ORCID iD:

Joy Josephine is a distinguished educationist with extensive qualifications, including a B.R.Sc., M.A., M.Ed., and M.Phil. She has exemplified her expertise through her role as District Coordinator for school training programs in Karaikal and Nagapattinam and as the Principal of Good Shepherd English School, Karaikal, where she served with dedication for over two decades. Dr. Josephine is an Executive Member of Pondy Sahodaya and Vice President of Rockcity Sahodaya. Her leadership has inspired and guided 70 schools to engage in various CBSE training programs. Her thoughtful and adaptable approach to leadership has garnered widespread admiration and respect. Her exceptional contributions to education have been recognized with numerous awards. These include the prestigious Abdul Kalam Award, presented by former Tamil Nadu Governor Mr Konijeti Rosaiah for her social work, the Vidhya Shironmani Award in 2018, the Best Principal Award by International Chinthana Prakashna, the Best Teacher Award from IIHM Bangalore in 2020, and the Smt. Sushma Swaraj Women Icon Award in 2022. Beyond her educational achievements, Dr. Josephine is actively involved in social work. She has contributed to the C.M. Relief Fund, raised awareness in villages to combat dengue, supported senior citizens through Help Age India, and provided counselling and companionship to elderly and physically challenged individuals. Her commitment to making a positive impact extends throughout her community involvement.

Jayabal K is a Professor in Mechanical Engineering at Indian Institute of Information Technology Design & Manufacturing Kancheepuram, Chennai, India. He obtained his PhD from IIT Madras and carried out Post-doctoral research at TU Dortmund, Germany. He has more than 20 years of Research and Teaching experience. His research interests include Advanced finite element methods, Smart materials, Material modeling, Energy Harvesting, Cold Spray Coating. He has several publications to his credit and executed funded projects in the filed of smart material modeling and experimentation, Stent design and fabrication, wearable thermoelectric devices.

Kazi Kutubuddin Sayyad Liyakat has completed his B.E., M.E., and Ph.D. in E&TC Engineering and is nowadays working as a Professor & Head of Department, E&TC Engineering Department and was Dean R&D. He is Post-Doctoral Fellow worked on "IoT in Healthcare Applications". His area of Interest is IoT, AI and ML. He has published more than 210+ papers in various Journals. Also published 26 books in the field of Engineering. He has published 15 Indian Patents, 2 South

African Grant Patent, 2 Indian Copyright Patents and 8 UK Grant Patent. He worked as a Reviewer for Scopus Conferences and Journal. Also work as Editorial Board Member for various Journals. He got 2 Best Researcher Award, Best Faculty Award and Appreciation Letter from MoE's Innovation Cell, Govt. of India.

Amit Kumar Kundu is Assistant Professor & HOD Department of Chemistry Sripat Singh College, Jiaganj, Murshidabad. Dr. Amit Kumar Kundu is an Assistant professor and HOD at the Department of Chemistry, Sripat Singh College, Jiaganj, Murshidabad. Dr. Kundu did B.Sc. from Krishnagar Government College, University of Kalyani in 2007 and completed MSc. Degree from Department of Chemistry, University of Kalyani in 2009. After that he received his PhD in Analytical Chemistry in 2017 from University of Kalyani. His research mainly focuses on Environmental Analytical Chemistry, carbon nano materials, Graphene, Graphene Oxide-basednano composites Synthesis, Characterization and its application. He has huge experience in Groundwater and surface water onsite sampling, sediment sampling from vadose zone and aquifer. He is also proficient in operating and managing various advanced analytical instruments. To date, he has published near about eighteen peer-reviewed journal papers and he is the author of five book chapters. He received World Environment and Livelihood Award form 'International Foundation for Environment and Ecology'.

Arvind M is studying B Tech I T in Hindustan University.

Ahmed M. Abdelrahman is an experienced Postdoctoral Researcher with a demonstrated history of working in the mechanical or industrial engineering industry. Skilled in AutoCAD, Computer-Aided Design (CAD), LabVIEW, Condition Monitoring, and Management. Strong research professional with a Doctorate (PhD) focused in Mechanical Engineering from Universiti Teknologi Malaysia (UTM).

Nisha Malik a proficient individual with strong focus on achieving outcomes, possessing notable expertise in Mentoring, Counselling, Teaching, Training, Administration, Business Management etc. I have demonstrated exceptional leadership capabilities while overseeing and directing Students, Start-ups & Business Management in National; International Universities; Organizations. I have a proven track record of enhancing students' personality, start-ups to bootstrap, team performance, and cultivating collaborative alliances with internal and external stakeholders

T.C. Manjunath was born on 6th February 1967. He did his BE from RV College of Engg. in 1989, ME in Robotics & Automation from LD College of Engg. in 1992 with First Rank and PhD in Systems & Control Engg. from IIT Bombay in the year 2007. He is currently working as the Dean Research R D and as Professor in the Dept. of Computer Science & Engineering with specialization in IoT, Cyber Security & Blockchain Technology in Rajarajeswari College of Engineering, Bengaluru, Karnataka. He has completed B.E. Degree from R.V. College of Engg., M.E. Govt.'s LD College of Engg., Gujarat Univ., Ahmedabad in First Rank and Ph.D. in Systems and Control Engineering from the Indian Institute of Technology Bombay IIT Bombay He has got a teaching academic, research and administrative experience of more than 35+ years 10 years as Principal, 8 years as HoD rest as faculty and has published more than 1200+ research papers in various national & international journals and conferences, with 175+ patents, 25 book-chapters, 25 awards and also published 25 text books in various field of science, engineering and technology, completed projects worth Rs. 9 Crores till date apart from guiding 10 PhD scholars. He is a member of 30 Professional Society Chapters and is a fellow of IE, IETE, AMIEE and a chartered engineer. He got the best research scholar award in 2006 in IIT Bombay amongst all the research scholars of the institute. He has got 10 awards in academics and research activities.

Babitha Nadampalli is a law academician and researcher with Constitutional and Administrative Law expertise. She holds LLM degree and is pursuing her Ph.D. in Law from Christ University. Babitha has taught law as a seasoned educator at University Law College, Bangalore University, and Christ University, School of Law. Her research focuses on Public Health Education and the Right to Health, aiming to contribute to healthcare and human rights discourse.

Santosh Kumar Nathsharma is a distinguished academician in the field of Chemistry. He is currently serving as a faculty member at Stewart Science College, located in Cuttack, Odisha. With his expertise and dedication to teaching, Dr. Nathsharma has made significant contributions to the academic community. He is committed to fostering a deep understanding of chemistry among his students, and his work reflects a strong passion for education and scientific exploration. The college is located in Cuttack, Odisha, with the postal code 753001.

P. Selvakumar is currently an Associate Professor in the Department of Science and Humanities at Nehru Institute of Technology, Coimbatore, Tamil Nadu, India. Since 2023, he has served as the Department IQAC Coordinator at the institute. Prior to joining Nehru Institute of Technology, he held positions as an Assistant Professor in the Department of Science and Humanities various engineering colleges in Coimbatore, Tamil Nadu, India, spanning from 2011 to 2023. Dr.P. Selvakumar's academic and research interests lie primarily in the field of phytochemistry. He has made significant contributions to his field, with an impressive record that includes

30 journal publications, 12 patents, and 20 conference papers. Additionally, he has organized 5 seminars and conferences, further demonstrating his commitment to academic and scientific exchange. Throughout his career, Dr. P.Selvakumar has been recognized for his dedication and excellence in teaching, having been honored with several awards, including the Best Teacher Awards. He holds a prominent position in the academic community, contributing actively to research.

P. VidhyaPriya is a dedicated academician with a strong foundation in both computer science and business management. She holds an MBA from Avinashilingam Deemed University, Coimbatore, and earned her Ph.D. from Bharathiar University, Coimbatore, in 2010. Before transitioning to academia, she gained valuable industry experience working with a textile firm for over three years. In 2000, she joined Kongu Engineering College, where she has since built an illustrious career in research and teaching. With over 24 years of academic experience, Dr. VidhyaPriya's research interests lie in corporate finance, corporate governance, and behavioral finance. She has published 57 research papers in reputed national and international journals, including Scopus-indexed publications. Throughout her career, she has actively engaged with the academic community by attending numerous Faculty Development Programs (FDPs) at leading B-Schools, including IIM Bangalore, IIT Madras, Pondicherry University, and Mangalore University. Dr.VidhyaPriya has also presented papers at 30 national and international conferences and has organized multiple conferences, seminars, FDPs, workshops, and training programs. She has successfully completed two ICSSR-sponsored projects focused on start-up entrepreneurship and the Stand-Up India Scheme for entrepreneurs.

Sivaprakasam Palani received the B.E. M.E., and Ph.D. degree in Faculty of Mechanical Engineering from the College of Engineering, Guindy, Anna University, Chennai, India. He is currently working as an Associate Professor with Addis Ababa Science and Technology University, Ethiopia. He has been supervising Master and Ph.D. students in the area of Machining, Composite materials and Surface Engineering. He has 15 years of teaching and research experience. He has published and presented over 100 technical papers in international journals, conferences, and book chapters. His research interests include Micro/nano manufacturing, Machining, Composite materials, Surface Engineering, and Optimization techniques.

R. Venkatesh has established a distinguished academic career, culminating in significant achievements. He holds an M.Phil. in Physics, an M.Sc. in Applied Physics, and a B.Sc. in Applied Physics from esteemed institutions including Nehru Memorial College (Autonomous), Puthanampatti, Tiruchirappalli, and Adaikala Matha College, Vallam, Thanjavur, all affiliated with Bharathidasan University,

Tiruchirappalli, Tamil Nadu, earned in 2006, 2005, and 2002 respectively. Currently, he is actively pursuing a part-time Ph.D. program in Physics at Nehru Memorial College (Autonomous), Puthanampatti, Tiruchirappalli, under Bharathidasan University, demonstrating his ongoing dedication to scholarly pursuits. As an Assistant Professor of Physics at PSNA College of Engineering and Technology (Autonomous) in Dindigul, Tamil Nadu, India, R. Venkatesh leverages over 19 years of teaching experience to enrich the academic environment. His commitment to the field is evident through numerous publications in respected journals and conferences, as well as authored chapters for reputable publishers. He is widely recognized in the academic community for his profound dedication and expertise.

Sreethi Rebeka R is pursuing her Ph.D in Commerce from CHRIST (Deemed to be University) & Faculty of Commerce at Mount Carmel College, Autonomous. Completed her graduation from Stella Maris College, Autonomous, M.Com from Loyola College, Autonomous, M.Phil from University of Madras, and SLET in Commerce. She has participated and presented research papers at national and international conferences and has published research papers and book chapters in reputed journals.

V. Muthu Ruben is a distinguished Associate Professor at the School of Law, Christ University, situated in Bengaluru, Karnataka, India. Driven by an extensive scholarly foundation and a steadfast dedication to legal education, Dr. Ruben makes substantial contributions to both the academic community and the domain of law. His proficiency transcends numerous spheres within the realm of law, and he significantly influences the educational journeys of students enrolled at Christ University. Motivated by an ardent interest in legal research, he actively participates in scholarly endeavors, thereby making significant contributions to the ongoing legal dialogue. In addition to his scholarly responsibilities, Dr. V. Muthu Ruben is renowned for his commitment to cultivating an atmosphere that encourages intellectual engagement. His mentorship and counsel motivate pupils to explore into the complexities of the law, fostering the development of critical thinking skills and a profound comprehension of legal principles. In his role as an Associate Professor, Dr. Ruben provides the School of Law with an abundance of expertise, experience, and a forward-thinking perspective, thereby enhancing the institution's dedication to providing superior legal education.

S. Seenivasan is a dedicated academic professional in the Department of Mechanical Engineering at Rathinam Technical Campus, located in Coimbatore, Tamil Nadu (Postal Code: 641021). With a strong commitment to education and research, Dr. Seenivasan contributes significantly to the development of future engineers through innovative teaching and mentorship.

N. Sampathkumar earned his graduation, post-graduation and Doctoral Degrees from Bharathidasan University Trichy in the field of Chemistry. He is currently serving as Professor in Department of Chemistry at SSM College of Arts and Science, Dindigul since 2022. He is a Life member of ISTE. He has published more than 18 research papers in reputed international journals including ACS and Luminescence. His fields of research include Phytochemistry, Crystallography, Nanomaterial, Material Science and Biological Science.

Umesh Sankpal is currently worked as Associate Professor in Chemistry in the R. E. Society's Gogate – Jogalekar College (Autonomous), Ratnagiri, Maharashtra affiliated to University of Mumbai, Mumbai. Author obtained his M.Sc. in Organic Chemistry in 2005 while Ph.D. in Material Sciences in the year of 2011 from Shivaji University, Kolhapur, Maharashtra. He has published 21 International and 10 National research papers in the field of Semiconducting Materials which has potential applications in electromagnetic devices, chemical gas sensors and heterogeneous catalysts. His Scopus based h – index is 13 while citation is 783. Two book chapters on his credit from National and International publishers.

SC Vetrivel is a faculty member in the Department of Management Studies, Kongu Engineering College (Autonomous), Perundurai, Erode Dt. Having experience in Industry 20 years and Teaching 16 years. Awarded with Doctoral Degree in Management Sciences in Anna University, Chennai. He has organized various workshops and Faculty Development Programmes. He is actively involved in research and consultancy works. He acted as a resource person to FDPs & MDPs to various industries like, SPB ltd, Tamilnadu Police, DIET, Rotary school and many. His areas of interest include Entrepreneurship, Business Law, Marketing and Case writing. Articles published more than 100 International and National Journals. Presented papers in more than 30 National and International conferences including IIM Bangalore, IIM Kozhikode, IIM Kashipur and IIM Indore. He was a Chief Co-ordinator of Entrepreneurship and Management Development Centre (EMDC) of Kongu Engineering College, he was instrumental in organizing various Awareness Camps, FDP, and TEDPs to aspiring entrepreneurs which was funded by NSTEDB – DST/GoI

MohitSharma is a research scholar at the Institute of Management Studies and Research, Maharshi Dayanand University, Rohtak, Haryana, India. His research focuses on the dynamic relationship between brand image and purchase intentions, particularly in the context of brand extensions within the FMCG sector.

Mohit actively engages in bibliometric research, analyzing publication trends and identifying research gaps in his area of interest. He has co-authored several scholarly works and continues to contribute to the academic community. Mohit can be reached at mohitattri.rs.imsar@mdurohtak.ac.in

Ranjit Singha is a dedicated Ph.D. Scholar in the Department of Psychology at Christ (Deemed to be University), specializing in addiction research. As the developer of the Multidimensional Addiction Behaviour Scale (MABS), his research unravels the complexities of addiction to inform effective prevention, intervention, and education strategies. With over 10 years of experience as a facilitator, Editor and Reviewer, he has conducted workshops and talks that empower individuals with knowledge and tools for healthier decision-making and addiction awareness programs. Additionally, he has led Art of Writing Skills workshops, helping participants enhance their academic and professional communication. With a solid academic record, he has published 195+ scholarly works, including 61 Scopus-indexed publications, reflecting his commitment to impactful research and dissemination of knowledge.

Surjit Singha is an academician with a broad spectrum of interests, including U.N. Sustainable Development Goals, Organizational Climate, Workforce Diversity, Organizational Culture, H.R.M., Marketing, Finance, I.B., Global Business, Business, A.I., K12 & Higher Education, Gender and Cultural Studies. Currently a faculty member at Kristu Jayanti College, Dr. Surjit also serves as an editor, reviewer, and author for prominent global publications and journals, including being an editorial review board member of the Information Resources Management Journal and contributing to various publications. With over 15 years of experience in Administration, Teaching, and Research, Dr. Surjit is dedicated to imparting knowledge and guiding students in their research pursuits. As a research mentor, Dr. Surjit has nurtured young minds and fostered academic growth. Dr. Surjit has an impressive track record of over 100+ publications, including articles, book chapters, and textbooks, holds two U.S. Copyrights, and has completed and published two fully funded minor research projects from Kristu Jayanti College.

C.V. Suresh Babu is a pioneer in content development. A true entrepreneur, he founded Anniyappa Publications, a company that is highly active in publishing books related to Computer Science and Management. Dr. C.V. Suresh Babu has also ventured into SB Institute, a center for knowledge transfer. He holds a Ph.D. in Engineering Education from the National Institute of Technical Teachers Training & Research in Chennai, along with seven master's degrees in various disciplines such as Engineering, Computer Applications, Management, Commerce, Economics,

Psychology, Law, and Education. Additionally, he has UGC-NET/SET qualifications in the fields of Computer Science, Management, Commerce, and Education. Currently, Dr. C.V. Suresh Babu is a Professor in the Department of Information Technology at the School of Computing Science, Hindustan Institute of Technology and Science (Hindustan University) in Padur, Chennai, Tamil Nadu, India.

V. Sabareeshwari currently serves as Assistant Professor in Department of Soil Science, Amrita School of Agricultural Sciences, Coimbatore. Having more than 5 years of research experience and more than 2 years of teaching experience. She got 7 awards in the field of agriculture. Her field of expertise are soil genesis, soil pedological studies as well as soil fertility mapping using advanced software like Arc GIS. She had published 22 research papers and more than 10 book chapters and books in high- impact reputed journals. She has actively participated and presented her papers in more than 20 conferences and seminars. She not only restrict her contribution only in the academic and research part, she had extension experience at farm level (lab to land) with varied crop research.

V. Srinivasan earned his undergraduate degree in mechanical engineering from S.K.P Institute of Technology affiliated to Anna University, his master's degree in CAD/CAM from Vel Tech Rangarajan Dr. Sagunthala R & D Institute of Science and Technology and currently pursuing Ph. D in Aeronautical Engineering from Vel Tech Rangarajan Dr. Sagunthala R & D Institute of Science and Technology. Currently doing research in LiDAR (Light detection and Ranging) and other remote sensing and imaging sensors, its integration with UAVs. Obstacle avoidance and autonomous navigation systems in UAV using sensor fusion are his key areas for research.

ArunV.P. is a driven and accomplished professional with a diverse educational background and extensive hands-on experience across various industries. Graduating with honors, Arun earned his Master of Business Administration (M.B.A) with a specialization in Human Resources and Marketing from the renowned Sona School of Management in Salem in 2018, where he excelled academically with an impressive 8.3 Cumulative Grade Point Average (CGPA).Throughout his academic journey, Arun displayed an unwavering commitment to learning and personal growth, actively seeking opportunities to expand his knowledge and skills beyond the confines of traditional education. He sought practical experiences to complement his theoretical understanding, such as a 45-day summer internship focused on conducting a feasibility study for R-Doc Sustainability in the market. Additionally, Arun broadened his horizons through a 7-day industrial visit to Malaysia and Singapore, immersing himself in diverse cultural and professional environments.

Arun's academic pursuits were further enriched by his involvement in hands-on projects, including a comprehensive study on Employee Job Satisfaction at Roots Cast Private Limited. His professional trajectory includes serving as a Growth Officer at Parle Agro Private Limited, where he played a crucial role in driving sales growth and market expansion. Arun is a dedicated professional, currently employed at JKKN Engineering College in the Management Studies Department. He holds expertise in his field and has contributed significantly to academia through the publication of two journal articles. Arun's work focuses on advancing knowledge and understanding in management studies, making valuable contributions to the academic community.

Index

Symbols

A

B

C

D

E

F

G

H

L

M

S

T

W

www.ingramcontent.com/pod-product-compliance
Lightning Source LLC
LaVergne TN
LVHW080356011026
849161LV00009B/2317
* 9 7 9 8 3 6 9 3 9 3 1 6 1 *